SEVENTH EDITION

Anatomy and Physiology Laboratory Manual

Robert J. Amitrano
Bergen Community College

Gerard J. Tortora
Bergen Community College

BROOKS/COLE
CENGAGE Learning

Australia • Brazil • Japan • Korea • Mexico • Singapore • Spain • United Kingdom • United States

BROOKS/COLE
CENGAGE Learning™

Anatomy and Physiology Laboratory Manual, Seventh Edition
Robert J. Amitrano and Gerard J. Tortora

Publisher: Peter Adams

Assistant Editor: Kari Hopperstead

Editorial Assistant: Kristin Marrs

Technology Project Manager: Earl Perry

Marketing Manager: Stacy Best

Marketing Assistant: Brian Smith

Marketing Communications Manager:
  Jessica Perry

Signing Representative: Melissa Lerner

Project Manager, Editorial Production:
  Jennifer Risden

Creative Director: Rob Hugel

Art Director: Lee Friedman

Print Buyer: Barbara Britton

Permissions Editor: Joohee Lee

Production Service: Graphic World Inc.

Text Designer: Carolyn Deacy

Photo Researcher: Terri Wright

Copy Editor: Graphic World Inc.

Illustrator: Graphic World Illustration Studio

Cover Designer: Gia Giasullo

Cover Image: Mike Powell/Stone+/Getty

Compositor: Graphic World Inc.

Library of Congress Control Number: 2006925079

ISBN-13: 978-0-495-11217-4

ISBN-10: 0-495-11217-8

Brooks/Cole Cengage Learning
20 Davis Drive
Belmont, CA 94002-3098
USA

Cengage Learning is a leading provider of customized learning solutions with office locations around the globe, including Singapore, the United Kingdom, Australia, Mexico, Brazil, and Japan. Locate your local office at **www.cengage.com/global**

Cengage Learning products are represented in Canada by Nelson Education, Ltd.

To learn more about Brooks/Cole, visit **www.cengage.com/brookscole**

Purchase any of our products at your local college store or at our preferred online store **www.cengagebrain.com**

Printed in the United States of America
2 3 45 13 12 11 10

Preface

Anatomy and Physiology Laboratory Manual, Seventh Edition, has been written to guide students in the laboratory study of introductory anatomy and physiology. The manual was written to accompany the leading anatomy and physiology textbooks.

Comprehensiveness

This manual examines virtually every structure and function of the human body that is typically studied in an introductory anatomy and physiology course. Because of its detail, the need for supplemental handouts is minimized; the manual is a strong teaching device in itself.

Use of the Scientific Method

Anatomy (the science of body structure) and physiology (the science of body function) cannot be understood without the practical experience of laboratory work. The exercises in this manual challenge students to understand the way scientists work by asking them to make microscopic examinations and evaluations of cells and tissues, to observe and interpret chemical reactions, to record data, to make gross examinations of organs and systems, to dissect, and to conduct physiological laboratory work and interpret and apply the results of this work.

Illustrations

The manual contains a large number and variety of illustrations. The illustrations of the body systems of the human have been carefully drawn to depict structures that are essential to students' understanding of anatomy and physiology. Numerous photographs, photomicrographs, and scanning electron micrographs are presented to show students how the structures of the body actually look. We feel that this laboratory manual has better and more complete illustrations than any other anatomy and physiology manual.

Important Features

Among the key features of this manual are (1) dissection of the white rat, and selected mammalian organs; (2) numerous physiological experiments; (3) emphasis on the study of anatomy through histology; (4) lists of appropriate terms accompany drawings and photographs to be labeled; (5) inclusion of numerous scanning electron micrographs and specimen photos; (6) phonetic pronunciations and derivations for the vast majority of anatomical and physiological terms; (7) diagrams of commonly used laboratory equipment; (8) laboratory report questions and reports at the end of each exercise that can be filled in, removed, and turned in for grading if the instructor so desires; (9) three appendixes dealing with units of measurement, a periodic table of elements, and eponyms used in the laboratory manual; and (10) emphasis on laboratory safety throughout the manual.

New to the Seventh Edition

Numerous changes have been made in the seventh edition of this manual in response to suggestions from instructors and students. A significant change is the incorporation of the most up-to-date anatomical terminology with new phonetic pronunciations and derivations. Many laboratory exercises contain new computer-based tests. The principal additions to various exercises are as follows:

Exercise 3, "Cells," has the discussion of plasma membrane transport reorganized into kinetic energy transport and transport by transporter proteins.

Exercise 9, "Muscular Tissue," has revised sections on the contraction of skeletal muscle tissue and smooth muscle tissue.

Exercise 10, "Muscular System," contains updated anatomical nomenclature in all muscle tables.

Exercise 13, "Nervous System," has revised sections on the transverse section of the spinal cord and spinal nerve attachments.

In Exercise 14, "Sensory Receptors and Sensory and Motor Pathways," the discussion of the characteristics of sensation has been updated.

In Exercise 15, "Endocrine System," the section dealing with the pituitary gland has been revised and expanded.

In Exercise 17, "Heart," the section on blood supply of the heart has been expanded.

In Exercise 18, "Blood Vessels," tables have been expanded with updated anatomical nomenclature.

In Exercise 19, "Cardiovascular Physiology," computerized laboratory exercise references have been updated and expanded.

In Exercise 24, "pH and Acid-Base Balance," additional acid-base balance exercises have been included.

Exercise 25, "Reproductive Systems," contains revised sections dealing with sperm cells and phases of the female reproductive cycle.

Changes in Terminology

In recent years, the use of eponyms for anatomical terms has been minimized or eliminated. Anatomical eponyms are terms named after various individuals. Examples include fallopian tube (after Gabriello Fallopio) and eustachian tube (after Bartolommeo Eustachio).

Anatomical eponyms are often vague and nondescriptive and do not necessarily mean that the person whose name is applied contributed anything very original. For these reasons, we have also decided to minimize their use. However, because some still prevail, we have provided eponyms, in parentheses, after the first reference in each chapter to the more acceptable synonym. Thus, you will ex-

pect to see terms such as *uterine (fallopian) tube* or *auditory (pharyngotympanic* or *eustachian) tube*. See Appendix C.

Instructor's Guide

A complementary instructor's guide by the authors to accompany the manual is available from the publisher. This comprehensive guide contains: (1) a listing of materials needed to complete each exercise, (2) suggested audio-visual materials, (3) answers to illustrations and questions within the exercises, and (4) answers to laboratory report questions.

Robert J. Amitrano
Science and Health, S229
Bergen Community College
400 Paramus Road
Paramus, NJ 07652

Gerard J. Tortora
Science and Health, S229
Bergen Community College
400 Paramus Road
Paramus, NJ 07652

Reviewer List

The authors would like to thank the following reviewers for their time and input:

Stephanie Brown	Norwalk Community College
Robert Broyles	Butler County Community College
Jessie Chappell	University of Arkansas, Monticello
Jett Chinn	College of San Mateo
Susan Forrest	Butler County Community College
Margery Herrington	Adams State College
Caron Inouye	California State University, East Bay
Jonathan McMenamin-Balano	Norwalk Community College
Peggie Orrell	University of Arkansas, Monticello
Thoniot Prabhakaran	Southwest Texas State University
Jeff Thompson	Hudson Valley Community College
Frank Voorhees	Central Missouri State University

Contents

Laboratory Safety*

In 1989, The Centers for Disease Control and Prevention (CDC) published "Guidelines for Prevention of Transmission of Human Immunodeficiency Virus and Hepatitis B Virus to Health-Care and Public-Safety Workers" (*MMWR*, vol. 36, No. 6S). The CDC guidelines recommended precautions to protect health care and public safety workers from exposure to human immunodeficiency virus (HIV), the causative agent of acquired immunodeficiency syndrome (AIDS), and hepatitis B virus (HBV), the causative agent of hepatitis B. These guidelines are presented to reaffirm the basic principles involved in the transmission of not only the AIDS and hepatitis B viruses but also any disease-producing organism.

Based on the CDC guidelines for health care workers, as well as on other standard additional laboratory precautions and procedures, the following list has been developed for your safety in the laboratory. Although specific cautions and warnings concerning laboratory safety are indicated throughout the manual, read the following *before* performing any experiments.

A. General Safety Precautions and Procedures

1. Arrive on time. Laboratory directions and procedures are given at the beginning of the laboratory period.

2. Read all experiments before you come to class to be sure that you understand all the procedures and safety precautions. Ask the instructor about any procedure you do not

*The authors and publisher urge consultation with each instructor's institutional policies concerning laboratory safety and first-aid procedures.

understand exactly. Do not improvise any procedure.

3. Protective eyewear and laboratory coats or aprons must be worn by all students performing or observing experiments.

4. Do not perform any unauthorized experiments.

5. Do not bring any unnecessary items to the laboratory and do not place any personal items (pocketbooks, bookbags, coats, umbrellas, etc.) on the laboratory table or at your feet.

6. Make sure each apparatus is supported and squarely on the table.

7. Tie back long hair to prevent it from becoming a laboratory fire hazard.

8. Never remove equipment, chemicals, biological materials, or any other materials from the laboratory.

9. Do not operate any equipment until you are instructed in its proper use. If you are unsure of the procedures, ask the instructor.

10. Dispose of chemicals, biological materials, used apparatus, and waste materials according to your instructor's directions. Not all liquids are to be disposed of in the sink.

11. Some exercises in the laboratory manual are designed to induce some degree of cardiovascular stress. Students should not participate in these exercises if they are pregnant or have hypertension or any other known or suspected condition that might compromise health. Before you perform any of these exercises, check with your physician.

12. Do not put anything in your mouth while in the laboratory. Never eat, drink, or taste chemicals; lick labels; smoke; or store food in the laboratory.

13. Your instructor will show you the location of emergency equipment such as fire extinguishers, fire blankets, and first-aid kits as well as eyewash stations. Memorize their locations and know how to use them.

14. Wash your hands before leaving the laboratory. Because bar soaps can become contaminated, liquid or powdered soaps should be used. Before leaving the laboratory, remove any protective clothing, such as laboratory coats or aprons, gloves, and eyewear.

B. Precautions for Working with Blood, Blood Products, or Other Body Fluids

1. Work only with *your own* body fluids, such as blood, saliva, urine, tears, and other secretions and excretions; blood from a clinical laboratory that has been tested and certified as noninfectious; or blood from a mammal (other than a human).

2. Wear gloves when touching another person's blood or other body fluids.

3. Wear safety goggles when working with another person's blood.

4. Wear a mask and protective eyewear or a face shield during procedures that are likely to generate droplets of blood or other body fluids.

5. Wear a gown or an apron during procedures that are likely to generate splashes of blood or other body fluids.

6. Wash your hands immediately and thoroughly if contaminated with blood or other body fluids. Hands can be rapidly disinfected by using (1) a phenol disinfectant-detergent for 20 to 30 seconds (sec) and then rinsing with water, or (2) alcohol (50% to 70%) for 20 to 30 sec, followed by a soap scrub of 10 to 15 sec and rinsing with water.

7. Spills of blood, urine, or other body fluids onto bench tops can be disinfected by flooding them with a disinfectant-detergent. The spill should be covered with disinfectant for 20 minutes (min) before being cleaned up.

8. Potentially infectious wastes, including human body secretions and fluids, and objects such as slides, syringes, bandages, gloves, and cotton balls contaminated with those substances, should be placed in an autoclave container. Sharp objects (including broken glass) should be placed in a puncture-proof sharps container. Contaminated glassware should be placed in a container of disinfectant and autoclaved before it is washed.

9. Use only single-use, disposable lancets and needles. Never recap, bend, or break the lancet once it has been used. Place used lancets, needles, and other sharp instruments in a *fresh* 1:10 dilution of household bleach (sodium hypochlorite) or other disinfectant such as phenols (Amphyl), aldehydes (glutaraldehyde, 1%), and 70% ethyl alcohol and then dispose of the instruments in a puncture-proof container. These disinfectants disrupt the envelope of HIV and HBV. The fresh household bleach solution or other disinfectant should be prepared for *each* laboratory session.

10. All reusable instruments, such as hemocytometers, well slides, and reusable pipettes, should be disinfected with a *fresh* 1:10 solution of household bleach or other disinfectant and thoroughly washed with soap and hot water. The fresh household bleach solution or other disinfectant should be prepared for *each* laboratory session.

11. A laboratory disinfectant should be used to clean laboratory surfaces *before* and after procedures and should be available for quick cleanup of any blood spills.

12. Mouth pipetting should never be done. Use mechanical pipetting devices for manipulating all liquids in the laboratory.

13. All procedures and manipulations that have a high potential for creating aerosols or infectious droplets (such as centrifuging, sonicating, and blending) should be performed carefully. In such instances, a biological safety cabinet or other primary containment device is required.

C. Precautions Related to Working with Reagents

1. Use extreme care when working with reagents. Should any reagents make contact with your eyes, flush with water for 15 min; or, if they make contact with your skin, flush with water for 5 min. Notify your instructor immediately should a reagent make contact

with your eyes or skin, and seek immediate medical attention.

2. Report all accidents to your instructor, no matter how minor they may appear.

3. When you are working with chemicals or preserved specimens, the room should be well ventilated. Avoid breathing fumes for any extended period of time.

4. Never point the opening of a test tube containing a reacting mixture (especially when heating it) toward yourself or another person.

5. Exercise care in noting the odor of fumes. Use "wafting" if you are directed to note an odor. Your instructor will demonstrate this procedure.

6. Do not force glass tubing or a thermometer into rubber stoppers. Lubricate the tubing and introduce it gradually and gently into the stopper. Protect your hands with toweling when inserting the tubing or thermometer into the stopper.

7. Never heat a flammable liquid over or near an open flame.

8. Use only glassware marked Pyrex or Kimax. Other glassware may shatter when heated. Handle hot glassware with test tube holders.

9. If you have to dilute an acid, always add acid (AAA) to water.

10. When shaking a test tube or bottle to mix its contents, do not use your fingers as a stopper.

11. Read the label on a chemical twice before using it.

12. Replace caps or stoppers on bottles immediately after using them. Return spatulas to their correct place immediately after using them and do not mix them up.

13. Mouth pipetting should never be done. Use mechanical pipetting devices for manipulating all liquids in the laboratory.

D. Precautions Related to Dissection

1. When you are working with chemicals or preserved specimens, the room should be well ventilated. Avoid breathing fumes for any extended period of time.

2. Wear rubber gloves when dissecting.

3. To reduce the irritating effects of chemical preservatives to your skin, eyes, and nose, soak or wrap your specimen in a substance such as "Biostat." If this is not available, hold your specimen under running water for several minutes to wash away excess preservative and dilute what remains.

4. When dissecting, there is always the possibility of skin cuts or punctures from dissecting equipment or the specimens themselves, such as the teeth or claws of an animal. Should you sustain a cut or puncture in this manner, wash your hands with disinfectant soap, notify your instructor, and seek immediate medical attention to decrease the possibility of infection. A first-aid kit should be readily available for your use.

5. When cleaning dissecting instruments, always hold the sharp edges away from you.

6. Dispose of any damaged or worn-out dissecting equipment in an appropriate container supplied by your instructor.

Selected Laboratory Safety Signs/Labels

⚠ CAUTION

Protect eyes.
Wear goggles at
all times.

⚠ CAUTION

Hot surface.
Do not touch.

⚠ CAUTION

Cancer suspect
agent. Trained
personnel only.

⚠ CAUTION

Radiation area.
Authorized
personnel only.

⚠ CAUTION

Biological hazard.
Authorized
personnel only.

⚠ DANGER

Highly toxic.
Handle with
care.

⚠ DANGER

Do not smoke
in this area.

⚠ DANGER

Do not smoke,
eat or drink
in this area.

⚠ DANGER

Do not pipet
liquids by
mouth.

⚠ DANGER

Corrosive. Avoid
contact with eyes
and skin.

⚠ DANGER

Flammable
material. Keep
fire away.

EMERGENCY

Eye Wash
Station.
Keep area clear.

EMERGENCY

Safety Shower.
Keep area clear.

EMERGENCY

First Aid Station.

Fire extinguisher.
Remove pin and
squeeze trigger.

Commonly Used Laboratory Equipment

Beaker

Erlenmeyer flask

Florence flask

Funnel

Graduated cylinder

Pipet

Mortar and pestle

Watch glass

Stirring rod

Test tube

Test tube brush

Test tube holder

Test tube rack

Ring stand and ring

Pinch clamp

Utility clamp

Tripod

Clay triangle

Wire gauze

Crucible tongs

Beaker tongs

Forceps

Medicine dropper

Nichrome wire

Spatula

Pronunciation Key

A unique feature of this revised manual is the phonetic pronunciations given for many anatomical and physiological terms. The pronunciations are given in parentheses immediately after the particular term is introduced. The following key explains the essential features of the pronunciations.

1. The syllable with the stongest accent appears in capital letters; for example, bilateral (bī-LAT-er-al) and diagnosis (dī-ag-NŌ-sis).

2. A secondary accent is denoted by a single quote mark ('); for example, constitution (kon'-sti-TOO-shun) and physiology (fiz'-ē-OL-ō-jē). Additional secondary accents are also noted by a single quotation mark; for example, decarboxylation (dē-kar-bok'-si-LĀ-shun).

3. Vowels marked with a line above the letter are pronounced with the long sound, as in the following common words:

   ā as in *mā*ke   ī as in *ī*vy
   ē as in *bē*   ō as in *pō*le

4. Unmarked vowels are pronounced with the short sound, as in the following words:

   e as in *bet*   o as in *not*
   i as in *sip*   u as in *bud*

5. Other phonetic symbols are used to indicate the following sounds:

   a as in *above*   yoo as in *cute*
   oo as in *soon*   oy as in *oil*

Microscopy

Objectives

At the completion of this exercise you should understand

A The parts and proper use and care of a light microscope.

B The interpretation of images viewed through a light microscope, including the concept of magnification.

NOTE *Before you begin any laboratory exercises in this manual, please read the section on LABORATORY SAFETY on page xi.*

One of the most important instruments that you will use in your anatomy and physiology course is a compound light microscope. In this instrument, the lenses are arranged so that images of objects too small to be seen with the naked eye can become highly magnified; that is, apparent size can be increased, and their minute details can be revealed. Before you actually learn the parts of a compound light microscope and how to use it properly, discussion of some of the principles employed in light microscopy (mī-KROS-ko-pē) will be helpful.

A. Compound Light Microscope

A **compound light microscope** uses two sets of lenses, ocular and objective, and employs light as its source of illumination. Magnification is achieved as follows. Light rays from an illuminator are passed through a condenser, which directs the light rays through the specimen under observation; from here, light rays pass into the objective lens, the magnifying lens that is closest to the specimen; the image of the specimen then forms on a prism and is magnified again by the ocular lens.

A general principle of microscopy is that the shorter the wavelength of light used in the instrument, the greater the resolution. **Resolution (re-** **solving power)** is the ability of the lenses to distinguish fine detail and structure, that is, to distinguish between two points as separate objects. As an example, a microscope with a resolving power of 0.3 micrometers (mī-KROM-e-ters), symbolized μm, is capable of distinguishing two points as separate objects if they are at least 0.3 μm apart. 1 μm = 0.000001 or 10^{-6} m. (See Appendix A.) The light used in a compound light microscope has a relatively long wavelength and cannot resolve structures smaller than 0.3 μm. This fact, as well as practical considerations, means that even the best compound light microscopes can magnify images only about 2000 times.

A **photomicrograph** (fō-tō-MĪ-krō'-graf), a photograph of a specimen taken through a compound light microscope, is shown in Figure 4.1. In later exercises you will be asked to examine photomicrographs of various specimens of the body before you actually view them yourself through the microscope.

1. Parts of the Microscope

Carefully carry the microscope from the cabinet to your desk by placing one hand around the arm and the other hand firmly under the base. Gently place it on your desk, directly in front of you, with the arm facing you. Locate the following parts of the microscope and, as you read about each part, label Figure 1.1 by placing the correct numbers in the spaces next to the list of terms that accompanies the figure.

Olympus CH-2 microscope.

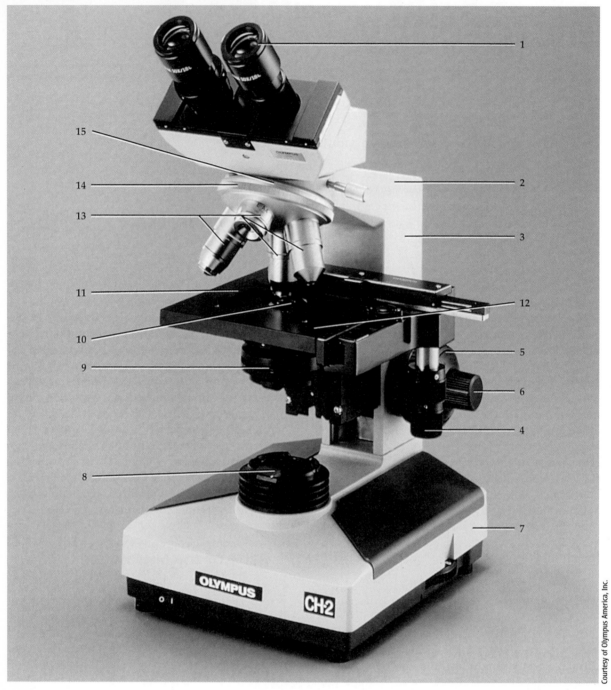

___ Arm   ___ Diaphragm   ___ Ocular
___ Base   ___ Fine adjustment knob   ___ Revolving nosepiece
___ Body tube   ___ Mechanical stage knob   ___ Stage
___ Coarse adjustment knob   ___ Nosepiece   ___ Stage clip of mechanical stage
___ Condenser   ___ Objectives   ___ Substage lamp

1. *Base.* The bottom portion on which the microscope rests.

2. *Body tube.* The portion that receives the ocular.

3. *Arm.* The angular or curved part of the frame.

4. *Inclination joint.* A movable hinge in some microscopes that allows the instrument to be tilted to a comfortable viewing position.

5. *Stage.* A platform on which microscope slides or other objects to be studied are placed. The opening in the center, called the **stage opening,** allows light to pass from below through the specimen being examined. Some microscopes have a **mechanical stage.** An adjustor knob below the stage moves the stage forward and backward and from side to side. With a mechanical stage, the slide and the stage move simultaneously. A mechanical stage permits a smooth, precise movement of a slide. Sometimes a mechanical stage is fitted with calibrations that permit the numerical "mapping" of a specimen on a slide.

6. *Stage (spring) clips.* Two clips mounted on the stage that hold the microscope slide securely in place.

7. *Substage lamp.* The source of illumination for some light microscopes with a built-in lamp.

8. *Mirror.* A feature found in some microscopes below the stage. The mirror directs light from its source through the stage opening and through the lenses. If the light source is built-in, a mirror is not necessary.

9. *Condenser.* A lens located beneath the stage opening that concentrates the light beam on the specimen.

10. *Condenser adjustment knob.* A knob that functions to raise and lower the condenser. In its highest position, it allows full illumination and thus can be used to adjust illumination.

11. *Diaphragm* (DĪ-a-fram). A device located below the condenser that regulates light intensity passing through the condenser and lenses to the observer's eyes. Such regulation is needed because transparent or very thin specimens cannot be seen in bright light. One of two types of diaphragms is usually used. An **iris diaphragm,** as found in cameras, is a series of sliding leaves that vary the size of the opening and thus the amount of light entering the lenses. The leaves are moved by a **diaphragm lever** to regulate the diameter of a central opening. A **disc diaphragm** consists of a plate with a graded series of holes, any of which can be rotated into position.

12. *Coarse adjustment knob.* A usually larger knob that raises and lowers the body tube (or stage) to bring a specimen into general view.

13. *Fine adjustment knob.* A usually smaller knob found below or external to the coarse adjustment knob and used for fine or final focusing. Some microscopes have both coarse and fine adjustment knobs combined into one.

14. *Nosepiece.* A plate, usually circular, at the bottom of the body tube.

15. *Revolving nosepiece.* The lower, movable part of the nosepiece that contains the various objective lenses.

16. *Scanning objective.* A lens, marked 5× on most microscopes (× means the same as "times"); it is the shortest objective and is not present on all microscopes.

17. *Low-power objective.* A lens, marked 10× on most microscopes; it is the next longer objective.

18. *High-power objective.* A lens, marked 40×, 43×, or 45× on most microscopes; also called a **high-dry objective;** it is an even longer objective.

19. *Oil-immersion objective.* A lens, marked 100× on most microscopes and distinguished by an etched colored circle (special instructions for this objective are discussed later); it is the longest objective.

20. *Ocular (eyepiece).* A removable lens at the top of the body tube, marked 10× on most microscopes. An ocular is sometimes fitted with a pointer or measuring scale.

2. Rules of Microscopy

You must observe certain basic rules at all times to obtain maximum efficiency and provide proper care for your microscope.

1. Keep all parts of the microscope clean, especially the lenses of the ocular, objectives, condenser, and also the mirror. *You should use the*

special lens paper that is provided and never use paper towels or cloths, because these tend to scratch the delicate glass surfaces. When using lens paper, use the same area on the paper only once. As you wipe the lens, change the position of the paper as you go.

2. Do not permit the objectives to get wet, especially when observing a **wet mount.** You must use a **cover slip** when you examine a wet mount or the image becomes distorted.

3. Consult your instructor if any mechanical or optical difficulties rise. *Do not try to solve these problems yourself.*

4. Keep *both* eyes open at all times while observing objects through the microscope. This is difficult at first, but with practice becomes natural. This important technique will help you to draw and observe microscopic specimens without moving your head. Only your eyes will move.

5. Always use either the scanning or low-power objective first to locate an object; then, if necessary, switch to a higher power.

6. If you are using the high-power or oil-immersion objectives, *never focus using the coarse adjustment knob.* The distance between these objectives and the slide, called **working distance,** is very small and you may break the cover slip and the slide and scratch the lens.

7. Some microscopes have a stage that moves while focusing, others have a body tube that moves while focusing. Be sure you are familiar with which type you are using. Look at your microscope from the side and using the scanning or low-power objective, gently turn the coarse adjustment knob. Which moves? The stage or the body tube? *Never focus downward* if the microscope's body tube moves when focusing. *Never focus upward* if the microscope's stage moves when focusing. By observing from one side you can see that the objectives do not make contact with the cover slip or slide.

8. Make sure that you raise the body tube before placing a slide on the stage or before removing a slide.

3. Setting Up the Microscope

PROCEDURE

1. Place the microscope on the table with the ocular toward you and with the back of the base at least 1 inch (in.) from the edge of the table.

2. Position yourself and the microscope so that you can look into the ocular comfortably.

3. Wipe the objectives, the top lens of the ocular, the condenser, and the mirror with lens paper. Clean the most delicate and the least dirty lens first. Apply xylol or alcohol to the lens paper only to remove grease and oil from the lenses and microscope slides.

4. Position the low-power objective in line with the body tube. When it is in its proper position, it will click. Lower the body tube using the coarse adjustment knob until the bottom of the lens is approximately 1/4 in. from the stage.

5. Admit the maximum amount of light by opening the diaphragm, if it is an iris diaphragm, or turning the disc to its largest opening, if it is a disc diaphragm.

6. Place your eye to the ocular, and adjust the light. When a uniform circle (the **microscopic field**) appears without any shadows, the microscope is ready for use.

4. Using the Microscope

PROCEDURE

1. Using the coarse adjustment knob, raise the body tube to its highest fixed position.

2. Make a temporary mount using a single letter of newsprint, or use a slide that has been specially prepared with a letter, usually the letter "e." If you prepare such a slide, cut a single letter—"a," "b," or "e"—from the smallest print available and place this letter in the correct position to be read with the naked eye. Your instructor will provide directions for preparing the slide.

3. Place the slide on the stage, making sure that the letter is centered over the stage opening, directly over the condenser. Secure the slide in place with the stage clips.

4. Align the low-power objective with the body tube.

5. Lower the body tube or raise the stage as far as it will go *while you watch it from the side,* taking care not to touch the slide. The tube should reach an automatic stop that prevents the low-power objective from hitting the slide.

6. While looking through the ocular, turn the coarse adjustment knob counterclockwise, raising the body tube. Or, turn the coarse ad-

justment knob clockwise, lowering the stage. When focusing, always *raise* the body tube or *lower* the stage. Watch for the object to suddenly appear in the microscopic field. If it is in proper focus, the low-power objective is about 1/2 in. above the slide. When focusing, always *raise* the body tube.

7. Use the fine adjustment knob to complete the focusing; you will usually use a counterclockwise motion once again.

8. Compare the position of the letter as originally seen with the naked eye to its appearance under the microscope.

   Has the position of the letter been changed?

   _____

9. While looking at the slide through the ocular, move the slide by using your thumbs, or, if the microscope is equipped with them, the mechanical stage knobs. This exercise teaches you to move your specimen in various directions quickly and efficiently.

   In which direction does the letter move when you move the slide to the left?

   _____

   This procedure, called "scanning" a slide, will be useful for examining living objects and for centering specimens so you can observe them easily.

   Make a drawing of the letter as it appears under low power in the microscopic field in the following space.

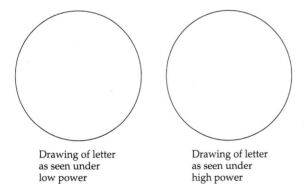

Drawing of letter
as seen under
low power

Drawing of letter
as seen under
high power

10. Change your magnification from low to high power by carrying out the following steps:

    a. Place the letter in the center of the field under low power. Centering is important because you are now focusing on a smaller area of the microscopic field. As you will see, *microscopic field size decreases with higher magnifications.*

    b. Make sure the illumination is at its maximum. Illumination must be increased at higher magnifications because the amount of light entering the lens decreases as the size of the objective lens increases.

    c. The letter should be in focus, and if the microscope is parfocal, the high-power objective can be switched into line with the body tube without changing focus. **Parfocal** means that when clear focus has been attained using any objective at random, revolving the nosepiece results in a change in magnification but leaves the specimen still in focus. If it is not completely in focus after switching the lens, a slight turn of the fine adjustment knob will focus it.

    d. If your microscope is not parfocal, observe the stage from one side and carefully switch the high-power objective in line with the body tube.

    e. While still observing from the side and using the coarse adjustment knob, *carefully* lower the objective or raise the stage until the objective almost touches the slide.

    f. Look through the ocular and focus up slowly. Finish focusing by turning the fine adjustment knob.

    g. If your microscope has an oil-immersion objective, you must follow special procedures. Place a drop of special **immersion oil** directly over the letter on the microscope slide, and lower the oil-immersion objective until it just contacts the oil. If your microscope is parfocal, you do not have to raise or lower the objectives. For example, if you are using the high-power objective and the specimen is in focus, just switch the high-power objective out of line with the body tube. Then add the oil and switch the oil-immersion objective into position; the specimen should be in focus. The same holds true when you switch from low power to high power. The special light-transmitting properties of the oil are such that light is refracted (bent) toward the specimen, permitting the use of powerful objectives in a relatively narrow field of vision. This objective is extremely close to the slide being examined, so when it is in position, take precautions *never to focus downward* while you are looking through the ocular. Whenever you finish using immersion oil, be sure to saturate a piece of lens paper with xylol or alcohol and clean the oil-immersion objective and the slide if it is to be used again.

Is as much of the letter visible under high power as under low power? Explain.

_____

Make a drawing of the letter as it appears under high power in the microscopic field in the space next to the drawing you made for low power following step 9.

11. Now select a prepared slide of three different-colored threads. Examination will show that a specimen mounted on a slide has depth as well as length and width. At lower magnification the amount of depth of the specimen that is clearly in focus, the depth of field, is greater than that at higher magnification. You must focus at different depths to determine the position (depth) of each thread.

    After you make your observation under low power and high power, answer the following questions about the location of the different threads:

    What color is at the bottom, closest to the

    slide? _____

    On top, closest to the cover slip?

    _____

    In the middle?

    _____

12. Your instructor might want you to prepare a wet mount as part of your introduction to microscopy. If so, the directions are given in Exercise 4, A.2.f, on page 56.

13. When you are finished using the microscope
    a. Remove the slide from the stage.
    b. Clean all lenses with lens paper.

    c. Align the mechanical stage so that it does not protrude.
    d. Leave the scanning or low-power objective in place.
    e. Lower the body tube or raise the stage as far as it will go.
    f. Wrap the cord according to your instructor's directions.
    g. Replace the dust cover or place the microscope in a cabinet.

5. Magnification

The total magnification of your microscope is calculated by multiplying the magnification of the ocular by the magnification of the objective used. Example: An ocular of $10\times$ used with an objective of $5\times$ gives a total magnification of $50\times$. Calculate the total magnification of each of the objectives on your microscope:

1. Ocular _____ $\times$ _____ Objective = _____

2. Ocular _____ $\times$ _____ Objective = _____

3. Ocular _____ $\times$ _____ Objective = _____

4. Ocular _____ $\times$ _____ Objective = _____

NOTE *A section of* LABORATORY REPORT QUESTIONS *is located at the end of each exercise. These questions can be answered by the student and handed in for grading at the discretion of the instructor. Even if the instructor does not require you to answer these questions, we recommend that you do so anyway to check your understanding.*

Some exercises also have a section of LABORATORY REPORT RESULTS, *in which students can record results of laboratory exercises, in addition to laboratory report questions. As with the laboratory report questions, the laboratory report results are located at the end of selected exercises and can be handed in as the instructor directs. Instructions in the manual tell students when and where to record laboratory results.*

ANSWER THE LABORATORY REPORT QUESTIONS AT THE END OF THE EXERCISE.

Microscopy

Name _____   Date _____

Laboratory Section _____   Score/Grade _____

PART 1 ▪ Multiple Choice

_____ 1. The amount of light entering a microscope may be adjusted by regulating the (a) ocular (b) diaphragm (c) fine adjustment knob (d) nosepiece

_____ 2. If the ocular on a microscope is marked 10× and the low-power objective is marked 15×, the total magnification is (a) 50× (b) 25× (c) 150× (d) 1500×

_____ 3. The size of the light beam that passes through a microscope is regulated by the (a) revolving nosepiece (b) coarse adjustment knob (c) ocular (d) condenser

_____ 4. Parfocal means that (a) the microscope employs only one lens (b) final focusing can be done only with the fine adjustment knob (c) changing objectives by revolving the nosepiece will still keep the specimen in focus (d) the highest magnification attainable is 1000×

_____ 5. Which of these is *not* true when changing magnification from low power to high power? (a) the specimen should be centered (b) illumination should be decreased (c) the specimen should be in clear focus (d) the high-power objective should be in line with the body tube

_____ 6. The ability of a microscope to distinguish between two points as separate objects is called (a) parfocal focusing (b) working distance (c) diffraction (d) resolution

PART 2 ▪ Completion

7. The advantage of using immersion oil is that it has special _____ properties that permit the use of a powerful objective in a narrow field of vision.

8. The uniform circle of light that appears when one looks into the ocular is called the

_____.

9. In determining the position (depth) of the colored threads, the _____ (red, green, yellow) colored thread was in the middle.

10. If you move your slide to the right, the specimen moves to the _____ as you are viewing it microscopically.

11. After switching from low power to high power, _____ (more or less) of the specimen will be visible.

12. Microscopic field size _____ (increases or decreases) with higher magnifications.

13. The distance between the objectives and the slide is called the _____.

14. A photograph of a specimen taken through a compound light microscope is called a(n) _____.

15. An ocular of 10× used with an objective of 40× gives a total magnification of _____ ×.

PART 3 ■ Matching

_____ 16. Ocular

_____ 17. Stage

_____ 18. Arm

_____ 19. Condenser

_____ 20. Revolving nosepiece

_____ 21. Low-power objective

_____ 22. Fine adjustment knob

_____ 23. Diaphragm

_____ 24. Coarse adjustment knob

_____ 25. High-power objective

A. Platform on which slide is placed

B. Mounting for objectives

C. Lens below stage opening

D. Brings specimen into sharp focus

E. Eyepiece

F. An objective usually marked 40×, 43×, or 45×

G. An objective usually marked 10×

H. Angular or curved part of frame

I. Brings specimen into general focus

J. Regulates light intensity

Introduction to the Human Body

Objectives

At the completion of this exercise you should understand

A The subdivisions of anatomy and physiology as scientific disciplines.

B The various levels of body organization.

C The major systems of the human body in terms of component organs and functions.

D Essential anatomical and physiological terminology related to life processes, homeostasis, the anatomical position, terms of direction, planes, body cavities, and abdominopelvic regions and quadrants.

In this exercise, you will be introduced to the organization of the human body through a study of the principal subdivisions of anatomy and physiology, levels of structural organization, principal body systems, the anatomical position, regional names, directional terms, planes of the body, body cavities, abdominopelvic regions, and abdominopelvic quadrants.

A. Anatomy and Physiology

Whereas **anatomy** (a-NAT-o-mē; *ana* = up; *tomy* = process of cutting) is the science of body *structures* and the relationships among structures, **physiology** (fiz-ē-OL-ō-jē; *physio* = nature; *logy* = study of) is the science of body functions, that is, how they work. Each structure of the body is designed to carry out a particular function.

Following are selected subdivisions of anatomy and physiology. In the spaces provided, define each term.

Subdivisions of Anatomy

Surface anatomy _____

_____

_____

_____

Gross anatomy _____

_____

_____

_____

Systemic anatomy _____

_____

_____

_____

Regional anatomy _____

_____

_____

_____

Radiographic (rā'-dē-ō-GRAF-ik) *anatomy* _____

_____

_____

_____

Developmental biology _____

9

Embryology (em'-brē-OL-ō-jē) _____

_____

_____

_____

Histology (his'-TOL-ō-jē) _____

_____

_____

_____

Pathological (path'-ō-LOJ-i-kal) *anatomy* _____

_____

_____

_____

Subdivisions of Physiology

Pathophysiology (PATH-ō-fiz-ē-ol'-ō-jē) _____

_____

_____

_____

Exercise physiology _____

_____

_____

_____

Neurophysiology (NOOR-ō-fiz-ē-ol'-ō-jē) _____

_____

_____

_____

Endocrinology (en'-dō-kri-NOL-ō-jē) _____

_____

_____

_____

Cardiovascular (kar-dē-ō-VAS-kū-lar)

physiology _____

_____

_____

Immunology (im'-ū-NOL-ō-jē) _____

_____

_____

_____

Respiratory (RES-pir-a-to'rē) *physiology* _____

_____

_____

_____

Renal (RĒ-nal) *physiology* _____

_____

_____

_____

B. Levels of Body Organization

The human body is composed of six levels of structural organization associated with one another in various ways:

1. *Chemical level.* Composed of all **atoms** and **molecules** essential for maintaining life.
2. *Cellular level.* Consists of **cells,** the basic structural and functional units of an organism and the smallest living units in the human body.
3. *Tissue level.* Formed by **tissues,** groups of cells and the materials surrounding them that work together to perform a particular function.
4. *Organ level.* Consists of **organs,** structures composed of two or more different types of tissues, having specific functions and usually having recognizable shapes.
5. *System level.* Formed by **systems;** related organs that have a common function.
6. *Organismal level.* The systems together constitute an **organism,** a total living individual.

C. Systems of the Body

Using an anatomy and physiology textbook, torso, wall chart, and any other materials that might be available to you, identify the principal organs that compose the following body systems.[1]

---

[1]You will probably need other sources, plus any aids the instructor might provide, to label many of the figures and answer some questions in this manual. You are encouraged to use other sources as you find necessary.

In the spaces that follow, indicate the organs and functions of the systems.

Integumentary system

Organs _____

_____

_____

_____

Functions _____

_____

_____

_____

Skeletal system

Organs _____

_____

_____

_____

Functions _____

_____

_____

_____

Muscular system

Organs _____

_____

_____

_____

Functions _____

_____

_____

_____

Nervous system

Organs _____

_____

_____

_____

Functions _____

_____

_____

_____

Endocrine system

Organs _____

_____

_____

_____

Functions _____

_____

_____

_____

Cardiovascular system

Organs _____

_____

_____

_____

Functions _____

_____

_____

_____

Lymphatic system and Immunity

Organs _____

_____

_____

_____

Functions _____

_____

_____

_____

Respiratory system

Organs _____

_____

_____

_____

Functions _____

_____

_____

_____

Digestive system

Organs _____

_____

_____

_____

Functions _____

_____

_____

_____

Urinary system

Organs _____

_____

_____

_____

Functions _____

_____

_____

_____

Reproductive systems

Organs _____

_____

_____

_____

Functions _____

_____

_____

_____

D. Life Processes

All living forms carry on certain processes that distinguish them from nonliving things. Using an anatomy and physiology textbook, define the six most important life processes of the human body.

Metabolism _____

_____

_____

_____

Responsiveness _____

_____

_____

_____

Movement _____

_____

_____

_____

Growth _____

_____

_____

_____

Differentiation _____

_____

_____

_____

Reproduction _____

_____

_____

_____

Using an anatomy and physiology textbook, define the following components of a feedback system.

Stimulus _____

_____

_____

_____

Controlled condition _____

_____

_____

_____

Receptor _____

_____

_____

_____

Input _____

_____

_____

_____

Control center _____

_____

_____

_____

Output _____

_____

_____

_____

E. Homeostasis

Homeostasis (hō-mē-ō-STĀ-sis; *homeo-* = sameness; *stasis* = standing still) is a condition of equilibrium (balance) in the body's internal environment. The body's regulatory processes function to maintain homeostasis in response to changing external and internal conditions. Homeostasis is achieved through the operation of feedback systems. A **feedback system** or **feedback loop** is a cycle of events in which the status of a body condition is continually monitored, changed, remonitored, reevaluated, and so on.

Effector _____

_____

_____

_____

Response _____

_____

_____

_____

If the response of the body *reverses* the original stimulus, the system is operating by **negative feedback.** If the response *strengthens* or *reinforces* the original stimulus, the system is operating by **positive feedback.**

Negative feedback systems tend to maintain conditions that require frequent monitoring and adjustments within physiological limits, such as body temperature or blood pressure. (See Figure 2.1.) Positive feedback systems, on the other hand, are important for conditions that do not require continual fine-tuning. Since positive feedback systems tend to intensify or amplify a controlled condition, they usually are shut off by some mechanism outside the system if they are part of a normal physiological response. Most feedback systems in the body are negative. Positive feedback systems can be destructive and result in various disorders, yet some are normal and beneficial. For example, during blood clotting, which helps stop loss of blood from a cut, the initial signal is amplified until the blood clot forms and bleeding is under control. Then, other substances help turn off the clotting response. Positive feedback mechanisms also contribute during birth of a baby to strengthen labor contractions and during immune responses to provide defense against pathogens.

Label Figure 2.1, the control of blood pressure by a negative feedback system.

F. Anatomical Position and Regional Names

Figure 2.2 shows anterior and posterior views of a subject in the **anatomical position.** The subject is standing erect and facing the observer with the head and eyes facing forward when in anterior view, the arms are at the sides with the palms facing forward, and the feet are flat on the floor and directed forward. The figure also shows the common names for various regions of the body. When you as the observer make reference to the left and right sides of the subject you are studying, this refers to the *subject's* left and right sides. In the spaces next to the list of terms in Figure 2.2, write the number of each common term next to each corresponding anatomical term. For example, the skull (29) is cranial, so write the number *29* next to the term *Cranial.*

G. External Features of the Body

Referring to your textbook and human models, identify the following external features of the body:

1. *Head (cephalic region* or *caput).* This is divided into the **skull** and **face.** The skull (cranium) encloses and protects the brain; the face is the anterior portion of the head that includes the eyes, nose, mouth, forehead, cheeks, and chin.

2. *Neck (collum).* This region supports the head and attaches it to the trunk. It is called the **cervical region.**

3. *Trunk.* This region is also called the torso and is divided into the **chest** (thorax), **abdomen** (venter), and **pelvis.**

4. *Upper limb (extremity).* This consists of the **armpit** (axilla), **shoulder** (acromial region or omos), **arm** (brachium), **elbow** (cubitus), **forearm** (antebrachium), and **hand** (manus). The hand, in turn, consists of the **wrist** (carpus), **palm** (metacarpus), and **fingers** (digits). Individual bones of a digit (finger or toe) are called **phalanges. Phalanx** is singular.

5. *Lower limb (extremity).* This consists of the **buttocks** (gluteal region), **thigh** (femoral region), **knee** (genu), **leg** (crus), and **foot** (pes). The foot includes the **ankle** (tarsus), **sole** (metatarsus), and **toes** (digits). The **groin** is the area on the front surface of the body marked by a crease on each side, where the trunk attaches to the thighs.

H. Directional Terms

To explain exactly where a structure of the body is located, it is a standard procedure to use **directional terms.** Such terms are very precise and

FIGURE 2.1   The control of blood pressure by a negative feedback system.

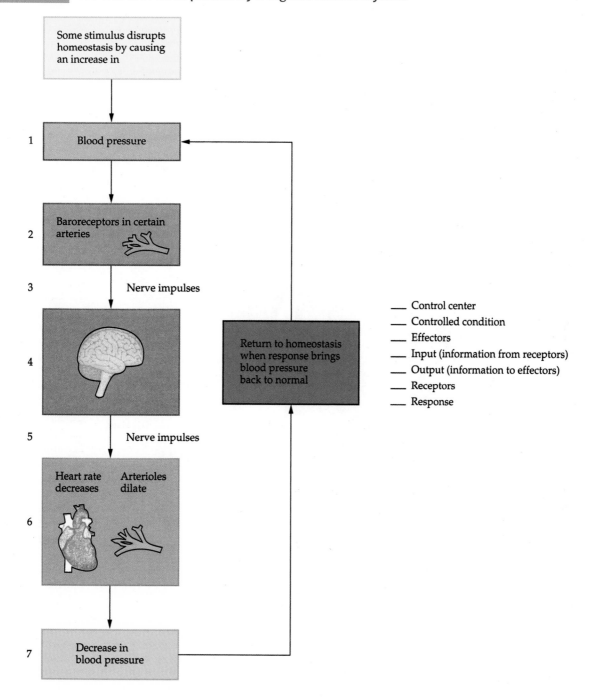

avoid the use of unnecessary words. Commonly used directional terms for humans are as follows:

1. *Superior* (soo'-PĒR-ē-or) *(cephalic* or *cranial).* Toward the head or the upper part of a structure; generally refers to structures in the trunk.

2. *Inferior* (in'-FĒR-ē-or) *(caudal).* Away from the head or the lower part of a structure; generally refers to structures in the trunk.

3. *Anterior* (an-TĒR-ē-or) *(ventral).* Nearer to or at the front surface of the body. In the **prone position,** the body lies anterior side

FIGURE 2.2   The anatomical position.

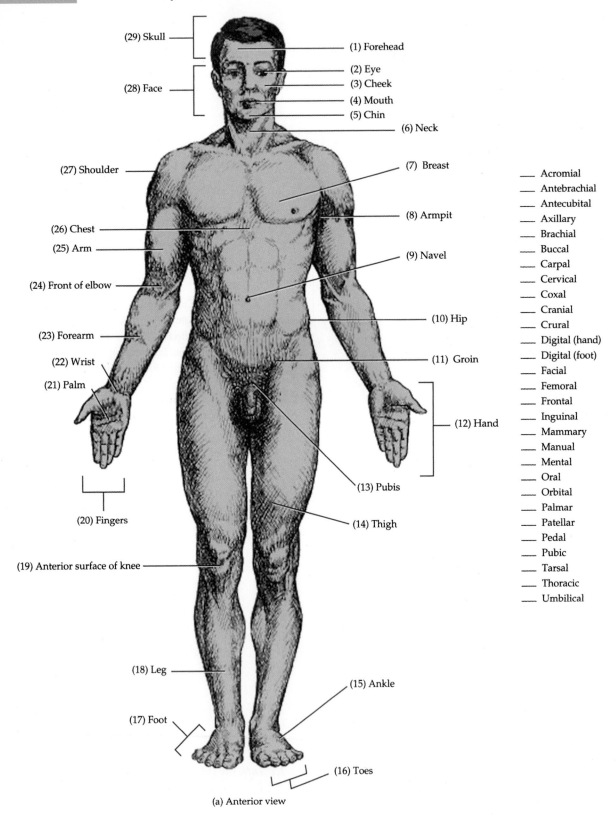

(29) Skull
(28) Face
(27) Shoulder
(26) Chest
(25) Arm
(24) Front of elbow
(23) Forearm
(22) Wrist
(21) Palm
(20) Fingers
(19) Anterior surface of knee
(18) Leg
(17) Foot

(1) Forehead
(2) Eye
(3) Cheek
(4) Mouth
(5) Chin
(6) Neck
(7) Breast
(8) Armpit
(9) Navel
(10) Hip
(11) Groin
(12) Hand
(13) Pubis
(14) Thigh
(15) Ankle
(16) Toes

(a) Anterior view

___ Acromial
___ Antebrachial
___ Antecubital
___ Axillary
___ Brachial
___ Buccal
___ Carpal
___ Cervical
___ Coxal
___ Cranial
___ Crural
___ Digital (hand)
___ Digital (foot)
___ Facial
___ Femoral
___ Frontal
___ Inguinal
___ Mammary
___ Manual
___ Mental
___ Oral
___ Orbital
___ Palmar
___ Patellar
___ Pedal
___ Pubic
___ Tarsal
___ Thoracic
___ Umbilical

FIGURE 2.2 The anatomical position. (Continued)

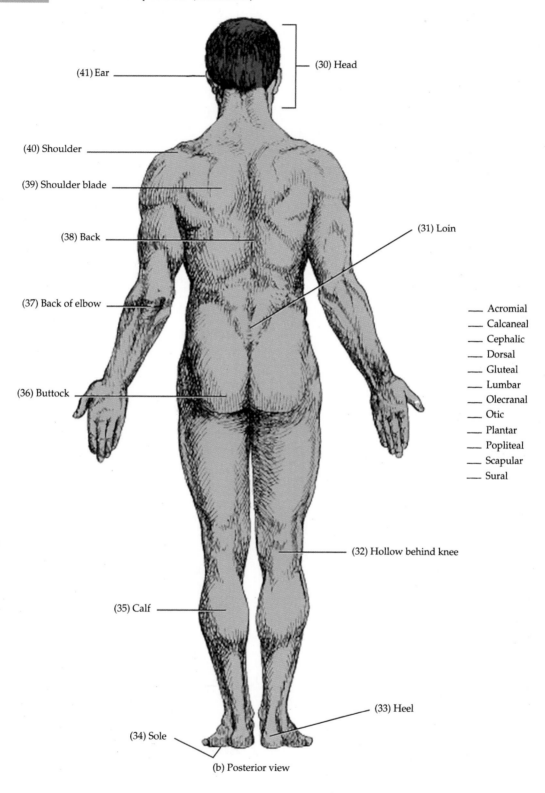

(41) Ear

(30) Head

(40) Shoulder

(39) Shoulder blade

(38) Back

(31) Loin

(37) Back of elbow

(36) Buttock

__ Acromial
__ Calcaneal
__ Cephalic
__ Dorsal
__ Gluteal
__ Lumbar
__ Olecranal
__ Otic
__ Plantar
__ Popliteal
__ Scapular
__ Sural

(32) Hollow behind knee

(35) Calf

(33) Heel

(34) Sole

(b) Posterior view

down; in the **supine position,** the body lies anterior side up.

4. *Posterior* (pos-TĒR-ē-or) *(dorsal).* Nearer to or at the back of the body.

5. *Medial* (MĒ-dē-al). Nearer the midline of the body. The **midline** is an imaginary vertical line that divides the body into equal left and right sides.

6. *Lateral* (LAT-er-al). Farther from the midline of the body.

7. *Intermediate* (in'-ter-MĒ-dē-at). Between two structures.

8. *Ipsilateral* (ip-si-LAT-er-al). On the same side of the body as another structure.

9. *Contralateral* (CON-tra-lat-er-al). On the opposite side of the body from another structure.

10. *Proximal* (PROK-si-mal). Nearer to the attachment of a limb to the trunk; nearer to the origination of a structure.

11. *Distal* (DIS-tal). Farther from the attachment of a limb to the trunk; farther from the origination of a structure.

12. *Superficial* (soo'-per-FISH-al). Toward or on the surface of the body.

13. *Deep* (DĒP). Away from the surface of the body.

Using a torso and an articulated skeleton, and consulting with your instructor as necessary, describe the location of the following by inserting the proper directional term. Use each term once only.

1. The ulna is on the _____ side of the forearm.

2. The lungs are _____ to the heart.

3. The heart is _____ to the liver.

4. The muscles of the arm are _____ to the skin of the arm.

5. The sternum is _____ to the heart.

6. The humerus is _____ to the radius.

7. The stomach is _____ to the lungs.

8. The muscles of the thoracic wall are _____ to the viscera in the thoracic cavity.

9. The esophagus is _____ to the trachea.

10. The phalanges are _____ to the carpals.

11. The ring finger is _____ between the little (medial) and middle (lateral) fingers.

12. The ascending colon of the large intestine and the gallbladder are _____.

13. The ascending and descending colons of the large intestine are _____.

I. Planes of the Body

The structural plan of the human body may be described with respect to **planes** (imaginary flat surfaces) passing through it. Planes are frequently used to show the anatomical relationship of several structures in a region to one another.

Commonly used planes are as follows:

1. *Midsagittal* (mid-SAJ-i-tal; *sagitt* = arrow). A vertical plane that passes through the midline of the body and divides the body or an organ into *equal* right and left sides.

2. *Parasagittal* (par-a-SAJ-i-tal; *para* = near). A vertical plane that does not pass through the midline of the body and divides the body or an organ into *unequal* right and left sides.

3. *Frontal (coronal;* kō-RŌ-nal; *corona* = crown). A vertical plane that divides the body or an organ into anterior (front) and posterior (back) portions.

4. *Transverse (cross-sectional* or *horizontal).* A plane that divides the body or an organ into superior (upper) and inferior (lower) portions.

5. *Oblique* (ō-BLĒK). A plane that passes through the body or an organ at an angle between the transverse plane and either the sagittal or frontal plane.

Refer to Figure 2.3 and label the planes shown.

J. Body Cavities

Spaces within the body that help protect, separate, and support internal organs are called **body cavities.** The principal body cavities are as follows:

Cranial (KRĀ-nē-al) *cavity*

Vertebral (VER-te-bral) or *spinal cavity*

FIGURE 2.3   Planes of the body.

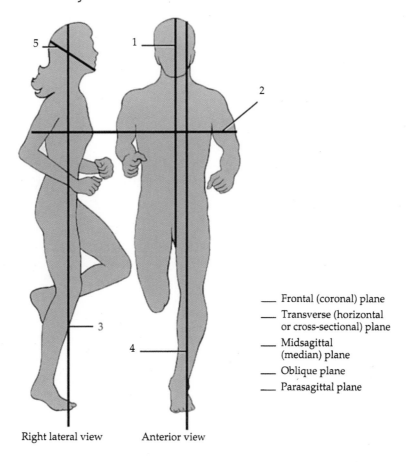

___ Frontal (coronal) plane

___ Transverse (horizontal or cross-sectional) plane

___ Midsagittal (median) plane

___ Oblique plane

___ Parasagittal plane

Right lateral view   Anterior view

Thoracic (thor-AS-ik) *cavity*
    Right pleural (PLOOR-al) cavity
    Left pleural cavity
    Pericardial (per′-i-KAR-dē-al) cavity

Abdominopelvic cavity
    Abdominal cavity
    Pelvic cavity

The mass of tissue between pleurae of the lungs and extending from the sternum (breast bone) to the vertebral column and from the neck to the diaphragm is called the **mediastinum** (mē′-dē-as-TĪ-num; *media* = middle; *stare* = stand in). It contains all structures in the thoracic cavity, except the lungs themselves. Included are the heart, thymus gland, esophagus, trachea, and many large blood and lymphatic vessels.

Label the body cavities shown in Figure 2.4. Then examine a torso or wall chart, or both, and determine which organs lie within each cavity.

Using *T* (for thoracic), *A* (for abdominal), and *P* (for pelvic), indicate which organs are found in their respective cavities.

1. ____ Urinary bladder
2. ____ Stomach
3. ____ Spleen
4. ____ Lungs
5. ____ Liver
6. ____ Internal reproductive organs
7. ____ Small intestine
8. ____ Heart
9. ____ Gallbladder
10. ____ Small portion of large intestine

K. Abdominopelvic Regions

To describe the location of viscera more easily, the abdominopelvic cavity may be divided into **nine regions** by using four imaginary lines: (1) an upper horizontal **subcostal** (sub-KOS-tal) **line** that passes just below the bottom of the rib cage through the lower portion of the stomach, (2) a lower horizontal line, the **transtubercular** (trans-too-BER-kyoo′-lar) **line,** just below the top surfaces of the hip bones, (3) a **right midclavicular**

FIGURE 2.4 Body cavities.

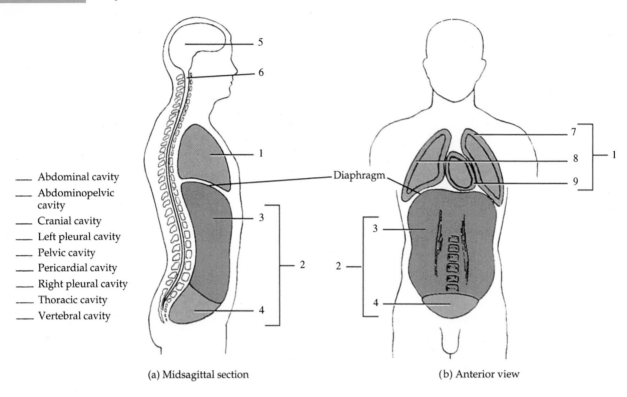

Abdominal cavity
Abdominopelvic cavity
Cranial cavity
Left pleural cavity
Pelvic cavity
Pericardial cavity
Right pleural cavity
Thoracic cavity
Vertebral cavity

(a) Midsagittal section

(b) Anterior view

(mid-kla-VIK-yoo′-lar) **line** drawn through the midpoint of the right clavicle slightly medial to the right nipple, and (4) a **left midclavicular line** drawn through the midpoint of the left clavicle slightly medial to the left nipple.

The four lines divide the abdominopelvic cavity into the following nine regions: (1) **umbilical** (um-BIL-i-kul) **region,** which is centrally located; (2) **left lumbar** (*lumbus* = loin) **region,** to the left of the umbilical region; (3) **right lumbar,** to the right of the umbilical region; (4) **epigastric** (ep-i-GAS-trik; *epi* = above; *gaster* = stomach) **region,** directly above the umbilical region; (5) **left hypochondriac** (hī′-pō-KON-drē-ak; *hypo* = under; *chondro* = cartilage) **region,** to the left of the epigastric region; (6) **right hypochondriac region,** to the right of the epigastric region; (7) **hypogastric (pubic) region,** directly below the umbilical region; (8) **left iliac** (IL-ē-ak; *iliacus* = superior part of hip bone) or **inguinal region,** to the left of the hypogastric (pubic) region; and (9) **right iliac (inguinal) region,** to the right of the hypogastric (pubic) region.

Label Figure 2.5 by indicating the names of the four imaginary lines and the nine abdominopelvic regions.

Examine a torso and determine which organs or parts of organs lie within each of the nine abdominopelvic regions.

In the space provided, list several organs or parts of organs found in the following abdominopelvic regions:

Right hypochondriac _____

_____

_____

Epigastric _____

_____

_____

Left hypochondriac _____

_____

_____

FIGURE 2.5 Abdominopelvic regions.

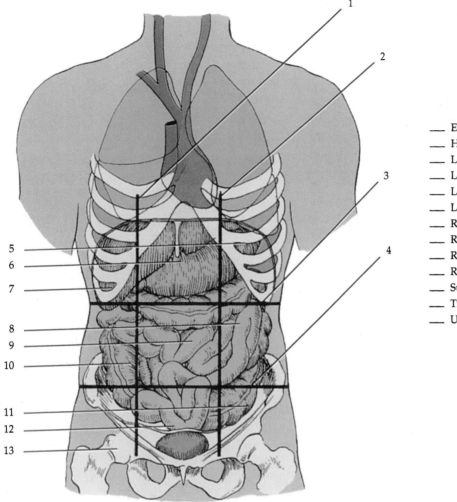

Anterior view

___ Epigastric region
___ Hypogastric region
___ Left hypochondriac region
___ Left iliac region
___ Left midclavicular line
___ Left lumbar region
___ Right hypochondriac region
___ Right iliac region
___ Right midclavicular line
___ Right lumbar region
___ Subcostal line
___ Transtubercular line
___ Umbilical region

Right lumbar _____

_____

_____

Umbilical _____

_____

_____

Left lumbar _____

_____

_____

Right iliac _____

_____

_____

Hypogastric _____

_____

_____

Left iliac _____

_____

_____

L. Abdominopelvic Quadrants

A second method to divide the abdominopelvic cavity is into **quadrants** by passing one horizontal line and one vertical line through the umbilicus (*umbilic* = navel) or belly button. The two lines thus divide the abdominopelvic cavity into a **right upper quadrant (RUQ), left upper quadrant (LUQ), right lower quadrant (RLQ),** and **left lower quadrant (LLQ).** Quadrant names are frequently used by health care professionals for locating the site of an abdominopelvic pain, tumor, or other abnormality.

Examine a torso or wall chart, or both, and determine which organs or parts of organs lie within each of the abdominopelvic quadrants.

M. Dissection of White Rat

Now that you have some idea of the names of the various body systems and the principal organs that comprise each, you can actually observe some of these organs by dissecting a white rat. **Dissect** means "to separate." This dissection gives you an excellent opportunity to see the different sizes, shapes, locations, and relationships of organs and to compare the different textures and external features of organs. In addition, this exercise will introduce you to the general procedure for dissection before you dissect in later exercises.

▲ **CAUTION!** *Please reread Section D, "Precautions Related to Dissection" at the beginning of the laboratory manual on page xiii before you begin your dissection.*

PROCEDURE

1. Place the rat on its backbone on a wax dissecting pan (tray). Using dissecting pins, anchor each of the four limbs to the wax (Figure 2.6a).

2. To expose the contents of the thoracic, abdominal, and pelvic cavities, you will have to first make a midline incision. This is done by lifting the abdominal skin with a forceps to separate the skin from the underlying connective tissue and muscles. While lifting the abdominal skin, cut through it with scissors and make an incision that extends from the lower jaw to the anus (Figure 2.6a).

3. Now make four lateral incisions that extend from the midline incision into the four limbs (Figure 2.6a).

FIGURE 2.6   Dissection procedure for exposing thoracic and abdominopelvic viscera of the white rat for examination.

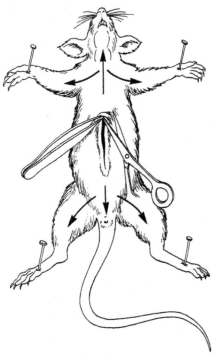

(a) Lines of incision in skin

(b) Peeling back skin and lines of incision in muscles

4. Peel the skin back and pin the flaps to the wax to expose the superficial muscles (Figure 2.6b).

5. Next, lift the abdominal muscles with a forceps and cut through the muscle layer, being careful not to damage any underlying organs. Keep the scissors parallel to the rat's backbone. Extend this incision from the anus to a point just below the bottom of the rib cage (Figure 2.6b). Make two lateral incisions just below the rib cage and fold back the muscle flaps to expose the abdominal and pelvic viscera (Figure 2.6b).

6. To expose the thoracic viscera, cut through the ribs on either side of the sternum. This incision should extend from the diaphragm to the neck (Figure 2.6b). The **diaphragm** is the thin muscular partition that separates the thoracic from the abdominal cavity. Again make lateral incisions in the chest wall so that you can lift the ribs to view the thoracic contents.

1. Examination of Thoracic Viscera

You will first examine the thoracic viscera (large internal organs). As you dissect and observe the various structures, palpate (feel with the hand) them so that you can compare their texture. Use Figure 2.7 as a guide.

a. *Thymus gland.* An irregular mass of glandular tissue superior to the heart and superficial to the trachea. Push the thymus gland aside or remove it.

b. *Heart.* A structure located in the midline, deep to the thymus gland and between the lungs. The sac covering the heart is the **pericardium,** which may be removed. The large vein that returns blood from the lower regions of the body is the **inferior vena cava;** the large vein that returns blood to the heart from the upper regions of the body is the **superior vena cava.** The large artery that carries blood from the heart to most parts of the body is the **aorta.**

c. *Lungs.* Reddish, spongy structures on either side of the heart. Note that the lungs are divided into regions called **lobes.**

d. *Trachea.* A tubelike passageway superior to the heart and deep to the thymus gland. Note that the wall of the trachea consists of rings of cartilage. Identify the **larynx** (voice box) at the superior end of the trachea and the **thyroid gland,** a bilobed structure on either side of the larynx. The lobes of the thyroid gland are connected by a band of thyroid tissue, the isthmus.

e. *Bronchial tubes.* Trace the trachea inferiorly and note that it divides into bronchial tubes that enter the lungs and continue to divide within them.

f. *Esophagus.* A muscular tube posterior to the trachea that transports food from the throat into the stomach. Trace the esophagus inferiorly to see where it passes through the diaphragm to join the stomach.

2. Examination of Abdominopelvic Viscera

You will now examine the principal viscera of the abdomen and pelvis. As you do so, again refer to Figure 2.7.

a. *Stomach.* An organ located on the left side of the abdomen and in contact with the liver. The digestive organs are attached to the posterior abdominal wall by a membrane called the **mesentery.** Note the blood vessels in the mesentery.

b. *Small intestine.* An extensively coiled tube that extends from the stomach to the first portion of the large intestine called the **cecum.**

c. *Large intestine.* A wider tube than the small intestine that begins at the cecum and ends at the rectum. The cecum is a large, saclike structure. In humans, the appendix arises from the cecum.

d. *Rectum.* A muscular passageway, located on the midline in the pelvic cavity, that terminates in the anus.

e. *Anus.* Terminal opening of the digestive system to the exterior.

f. *Pancreas.* A pale gray, glandular organ posterior and inferior to the stomach.

g. *Spleen.* A small, dark red organ lateral to the stomach.

h. *Liver.* A large, brownish-red organ directly inferior to the diaphragm. The rat does not have a gallbladder, a structure associated with the liver. To locate the remaining viscera, either move the superifical viscera aside or remove them. Use Figure 2.8 as a guide.

i. *Kidneys.* Bean-shaped organs embedded in fat and attached to the posterior abdominal wall on either side of the backbone. As will be explained later, the kidneys and a few other structures are behind the membrane that lines the abdomen **(peritoneum).** Such structures are referred to as **retroperitoneal** and are not

FIGURE 2.7 Superficial structures of the thoracic and abdominopelvic cavities of the white rat.

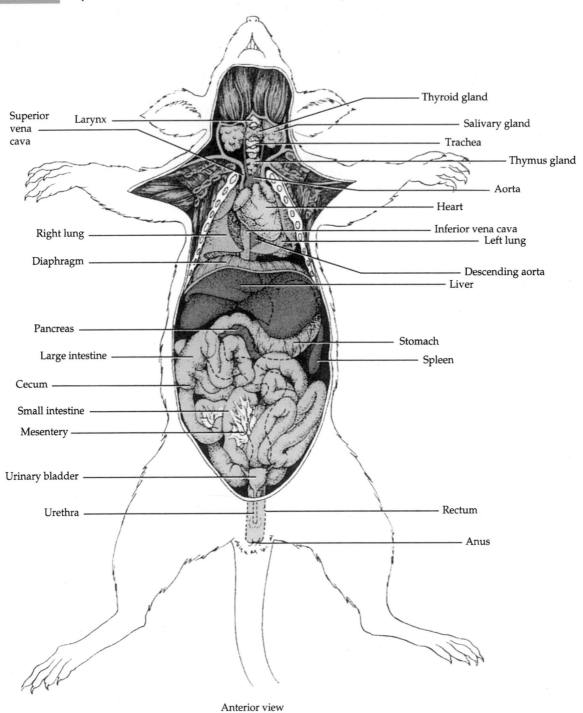

Thyroid gland

Superior vena cava

Larynx

Salivary gland

Trachea

Thymus gland

Aorta

Heart

Right lung

Inferior vena cava

Left lung

Diaphragm

Descending aorta

Liver

Pancreas

Stomach

Large intestine

Spleen

Cecum

Small intestine

Mesentery

Urinary bladder

Urethra

Rectum

Anus

Anterior view

FIGURE 2.8 Deep structures of the abdominopelvic cavity of the white rat.

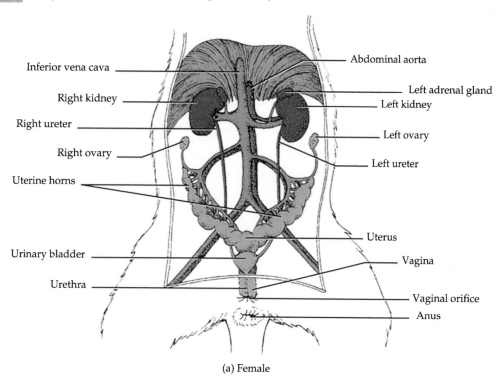

(a) Female

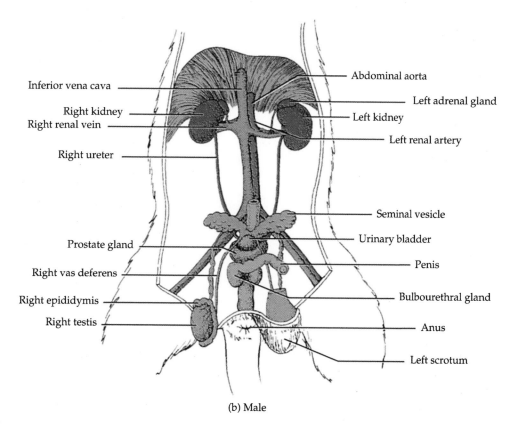

(b) Male

actually within the abdominal cavity. See if you can find the **abdominal aorta,** the large artery located along the midline behind the inferior vena cava. Also, locate the **renal arteries** branching off the abdominal aorta to enter the kidneys.

j. *Adrenal (suprarenal) glands.* Glandular structures. One is located on top of each kidney.

k. *Ureters.* Tubes that extend from the medial surface of the kidneys inferiorly to the urinary bladder.

l. *Urinary bladder.* A saclike structure in the pelvic cavity that stores urine.

m. *Urethra.* A tube that extends from the urinary bladder to the exterior. Its opening to the exterior is called the **urethral orifice.** In male rats, the urethra extends through the penis; in female rats, the tube is separate from the reproductive tract.

If your specimen is female (no visible scrotum anterior to the anus), identify the following:

n. *Ovaries.* Small, dark structures inferior to the kidneys.

o. *Uterus.* An organ located near the urinary bladder consisting of two sides (horns) that join separately into the vagina.

p. *Vagina.* A tube that leads from the uterus to the external vaginal opening, the **vaginal orifice.** This orifice is in front of the anus and behind the urethral orifice.

If your specimen is male, identify the following:

q. *Scrotum.* Large sac anterior to the anus that contains the testes.

r. *Testes.* Egg-shaped glands in the scrotum. Make a slit into the scrotum and carefully remove one testis. See if you can find a coiled duct attached to the testis **(epididymis)** and a duct that leads from the epididymis into the abdominal cavity **(vas deferens).**

s. *Penis.* Organ of copulation medial to the testes.

When you have finished your dissection, store or dispose of your specimen according to your instructor's directions. Wash your dissecting pan and dissecting instruments with laboratory detergent, dry them, and return them to their storage areas.

ANSWER THE LABORATORY REPORT QUESTIONS AT THE END OF THE EXERCISE.

Introduction to the Human Body

Name _____   Date _____

Laboratory Section _____   Score/Grade _____

PART 1 ■ Multiple Choice

_____ 1. The directional term that best describes the eyes in relation to the nose is (a) distal (b) superficial (c) anterior (d) lateral

_____ 2. Which does *not* belong with the others? (a) right pleural cavity (b) pericardial cavity (c) vertebral cavity (d) left pleural cavity

_____ 3. Which plane divides the brain into an anterior and a posterior portion? (a) frontal (b) median (c) sagittal (d) transverse

_____ 4. The urinary bladder lies in which region? (a) umbilical (b) hypogastric (c) epigastric (d) left iliac

_____ 5. Which is *not* a characteristic of the anatomical position? (a) the subject is erect (b) the subject faces the observer (c) the palms face backward (d) the upper limbs are at the sides

_____ 6. The abdominopelvic region that is bordered by all four imaginary lines is the (a) hypogastric (b) epigastric (c) left hypochondriac (d) umbilical

_____ 7. Which directional term best describes the position of the phalanges with respect to the carpals? (a) lateral (b) distal (c) anterior (d) proximal

_____ 8. The pancreas is found in which body cavity? (a) abdominal (b) pericardial (c) pelvic (d) vertebral

_____ 9. The anatomical term for the leg is (a) brachial (b) tarsal (c) crural (d) sural

_____ 10. In which abdominopelvic region is the spleen located? (a) left lumbar (b) right lumbar (c) epigastric (d) left hypochondriac

_____ 11. Which of the following represents the most complex level of structural organization? (a) organ (b) cellular (c) tissue (d) chemical

_____ 12. Which body system is concerned with support, protection, leverage, blood-cell production, and mineral storage? (a) cardiovascular (b) integumentary (c) skeletal (d) digestive

_____ 13. The skin and structures derived from it, such as nails, hair, sweat glands, and oil glands, are components of which system? (a) respiratory (b) integumentary (c) muscular (d) digestive

_____ 14. Hormone-producing glands belong to which body system? (a) cardiovascular (b) lymphatic and immune (c) endocrine (d) digestive

_____ 15. Which body system brings about movement, maintains posture, and produces heat? (a) skeletal (b) respiratory (c) reproductive (d) muscular

_____ **16.** Which abdominopelvic quadrant contains most of the liver? (a) RUQ (b) RLQ (c) LUQ (d) LLQ

_____ **17.** The physical and chemical breakdown of food for use by body cells and the elimination of solid wastes are accomplished by which body system? (a) respiratory (b) urinary (c) cardiovascular (d) digestive

_____ **18.** The ability of an organism to detect and respond to environmental changes is called (a) metabolism (b) differentiation (c) responsiveness (d) respiration

_____ **19.** In a feedback system, the component that produces a response is the (a) effector (b) receptor (c) input (d) output

PART 2 ▪ Completion

20. The tibia is _____ to the fibula.

21. The ovaries are found in the _____ body cavity.

22. The upper horizontal line that helps divide the abdominopelvic cavity into nine regions is the _____ line.

23. The anatomical term for the hollow behind the knee is _____.

24. A plane that divides the stomach into a superior and an inferior portion is a(n) _____ plane.

25. The wrist is divided as _____ to the elbow.

26. The heart is located in the _____ cavity within the thoracic cavity.

27. The abdominopelvic region that contains the rectum is the _____ region.

28. A plane that divides the body into unequal left and right sides is the _____ plane.

29. The spinal cord is located within the _____ cavity.

30. The body system that removes carbon dioxide from body cells, delivers oxygen to body cells, helps maintain acid-base balance, helps protect against disease, helps regulate body temperature, and prevents hemorrhage by forming clots is the _____ system.

31. The _____ abdominopelvic quadrant contains the descending colon of the large intestine.

32. Which body system returns proteins and plasma to the cardiovascular system, transports lipids from the digestive system to the cardiovascular system, filters blood, protects against disease, and produces white blood cells? _____.

33. The sum of all chemical processes that occur in the body is called _____.

34. A structure that monitors changes in a controlled condition and sends the information to the control center is the _____.

PART 3 ▪ Matching

_____ 35. Right hypochondriac region

_____ 36. Hypogastric region

_____ 37. Left iliac region

_____ 38. Right lumbar region

_____ 39. Epigastric region

_____ 40. Left hypochondriac region

_____ 41. Right iliac region

_____ 42. Umbilical region

_____ 43. Left lumbar region

A. Junction of descending and sigmoid colons of large intestine

B. Descending colon of large intestine

C. Spleen

D. Most of right lobe of liver

E. Appendix

F. Ascending colon of large intestine

G. Middle of transverse colon of large intestine

H. Adrenal (suprarenal) glands

I. Sigmoid colon of large intestine

PART 4 ▪ Matching

_____ 44. Anterior

_____ 45. Skull

_____ 46. Transtubercular line

_____ 47. Armpit

_____ 48. Umbilical region

_____ 49. Medial

_____ 50. Cranial cavity

_____ 51. Anterior surface of knee

_____ 52. Breast

_____ 53. Chest

_____ 54. Buttock

_____ 55. Superior

_____ 56. Groin

_____ 57. Vertebral canal

_____ 58. Cheek

_____ 59. Front of neck

_____ 60. Distal

_____ 61. Pericardial cavity

_____ 62. Forearm

_____ 63. Plantar

_____ 64. Mouth

A. Passes through iliac crests

B. Contains spinal cord

C. Nearer the midline

D. Thoracic

E. Cervical

F. Axillary

G. Contains the heart

H. Cranial

I. Antebrachial

J. Gluteal

K. Mammary

L. Sole

M. Contains navel

N. Buccal

O. Farther from the attachment of a limb

P. Patellar

Q. Toward the head

R. Nearer to or at the front of the body

S. Oral

T. Contains brain

U. Inguinal

Cells

Objectives

At the completion of this exercise you should understand

A The structural and functional characteristics of the various components of the cells of the body.

B The various processes involved in the movement of substances across plasma membranes.

C The stages and processes involved in cell division.

A cell is the basic living structural and functional unit of the body. **Cell biology** is the study of cellular structure and function. The different kinds of cells—blood, nerve, bone, muscle, epithelial, and others—perform specific functions and differ from one another in shape, size, and structure. You will start your study of cells by learning the important components of a theoretical, generalized cell.

A. Cell Parts

Refer to Figure 3.1, which provides an overview of the typical structures found in body cells. Many of the structures shown in this diagram are found in most cells, nevertheless no one cell has them all. With the aid of your textbook and any other items made available by your instructor, label the parts of the cell indicated. In the spaces that follow, describe the three main parts of a cell: the plasma membrane, cytoplasm, and nucleus.

1. *Plasma membrane* _____

    _____

    _____

    _____

    _____

2. *Cytoplasm* (SĪ-tō-plasm′) _____

    _____

_____

_____

_____

_____

a. Cytosol (SĪ-tō-sōl) _____

    _____

    _____

    _____

b. Organelles (or-ga-NELZ = *little organs*) ___

    _____

    _____

    _____

3. *Nucleus* (NOO-klē-us) _____

    _____

    _____

    _____

    _____

    _____

    _____

B. Organelles

1. *Cytoskeleton* _____

    _____

_____

_____

_____

_____

a. Microfilaments _____

b. Intermediate filaments _____

_____

_____

_____

_____

FIGURE 3.1 Generalized animal cell.

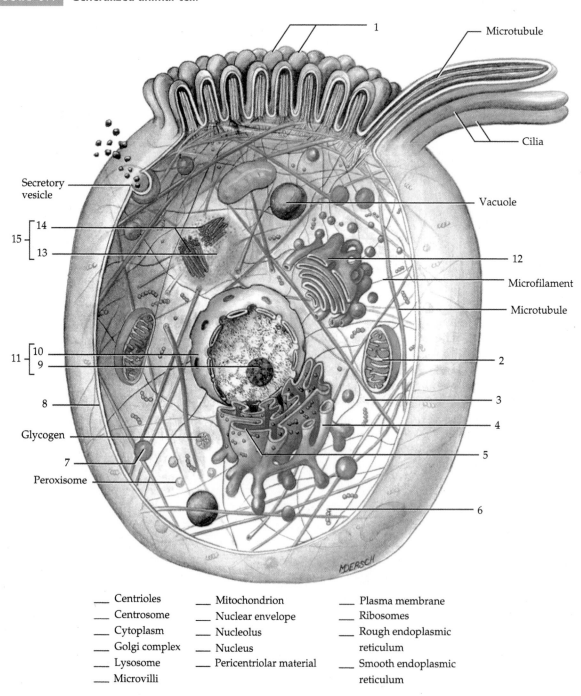

___ Centrioles   ___ Mitochondrion   ___ Plasma membrane

___ Centrosome   ___ Nuclear envelope   ___ Ribosomes

___ Cytoplasm   ___ Nucleolus   ___ Rough endoplasmic reticulum

___ Golgi complex   ___ Nucleus

___ Lysosome   ___ Pericentriolar material   ___ Smooth endoplasmic reticulum

___ Microvilli

c. Microtubules _____

_____

_____

_____

2. *Centrosome* (SEN-trō-sōm′) _____

_____

_____

_____

_____

3. *Cilia* (SIL-ē-a) _____

_____

_____

_____

_____

4. *Flagella* (fla-JEL-a) _____

_____

_____

_____

_____

5. *Ribosome* (RĪ-bō-sōm) _____

_____

_____

_____

_____

6. *Endoplasmic reticulum* (en′-dō-PLAS-mik re-
TIK-ū-lum) or *ER* _____

_____

_____

_____

_____

7. *Golgi* (GOL-jē) *complex* _____

_____

_____

_____

_____

8. *Lysosomes* (LĪ-sō-sōms) _____

_____

_____

_____

_____

9. *Peroxisomes* (pe-ROKS-i-sōms) _____

_____

_____

_____

_____

10. *Proteasomes* (PRO-te-a-sōmes) _____

_____

_____

_____

_____

11. *Mitochondria* (mī′-tō-KON-drē-a) _____

_____

_____

_____

_____

C. Diversity of Cells

Now obtain prepared slides of the following types of cells and examine them under the magnifications suggested:

1. Ciliated columnar epithelial cells (high power)
2. Sperm cells (oil immersion)

Ciliated columnar epithelial cell

Sperm cell

Nerve cell

Muscle cell

3.   Nerve cells (high power)

4.   Muscle cells (high power)

After you have made your examination, draw an example of each of these kinds of cells in the spaces provided and under each cell indicate how each is adapted to its particular function.

D. Movement of Substances Across and Through Plasma Membranes

Biological membranes serve as permeability barriers. They maintain both the internal integrity of the cell and the solute concentrations within the cell that are considerably different from those found in the extracellular environment. In general, most substances in the body have a high solubility in polar liquids, such as water, and low solubility in nonpolar liquids, such as alcohol. Because cellular membranes are composed of a combination of polar and nonpolar components, they present a formidable barrier to the movement of water-soluble substances into and out of a cell. Transport of materials across the plasma membrane is essential to the life of a cell. In general, substances move across cellular membranes by two principal kinds of transport processes—**passive transport processes,** which do not require the expenditure of cellular energy, and **active transport processes,** which do require the expenditure of cellular energy. Energy obtained from hydrolysis of ATP is the source in *primary active transport* and energy stored in an ionic concentration gradient is the source in *secondary active transport.* In passive transport processes, which include diffusion through the lipid bilayer, diffusion through channels, and facilitated diffusion, substances move because of differences in concentration (or pressure) from areas of higher concentration (or pressure) to areas of lower concentration (or pressure). The movement continues until the concentration of substances (or pressure) reaches equilibrium. Passive transport processes are the result of the **kinetic energy** (energy of motion) of the substances themselves, and the cell does not expend energy to move the substances. Examples of passive transport processes are diffusion, facilitated diffusion, osmosis, filtration, and dialysis.

By contrast, in active transport processes, substances move against their concentration gradient, from areas of lower concentration to areas of higher concentration. Moreover, cells must expend energy to drive the substance "uphill" against its concentration gradient. Examples of active transport processes are active transport, endocytosis and exocytosis (phagocytosis and pinocytosis), and transcytosis.

▲   **CAUTION!** *Please reread Section A, "General Safety Precautions and Procedures," on page xi, and Section C, "Precautions Related to Working with Reagents," on page xii, at the beginning of the laboratory manual before* *you begin any of the following experiments. Read the experiments before you perform them to be sure that you understand all the procedures and safety cautions.*

1. Kinetic Energy Transport

Before looking at examples of passive transport processes, we will first examine how molecules move.

a. Brownian Movement

At temperatures above absolute zero ($-273°$C or $-460°$F), all molecules are in constant random motion because of their inherent **kinetic energy.** This phenomenon is called **Brownian movement.** Less energy is required to move small molecules or particles than large ones. Because all molecules are constantly bombarded by the molecules surrounding them, the smaller the particle, the greater its random motion in terms of speed and distance.

PROCEDURE

1. With a medicine dropper, place a drop of dilute detergent solution in a depression (concave) slide.

2. Using another medicine dropper, add a drop of dilute India ink to the first drop in the depression slide.

3. Stir the two solutions with a toothpick and cover the depression with a cover slip.

4. Let the slide stand for 10 min and then place it on your microscope stage and observe it under high power.

5. Using forceps, place the slide on a hot plate and warm it for 15 sec. Then remove the slide with forceps and observe it again under high power.

6. Record your observations in Section D.1.a of the LABORATORY REPORT RESULTS at the end of the exercise.

b. Diffusion

Diffusion is a passive process in which there is the net (greater) movement of molecules or ions from a region of higher concentration to a region of lower concentration until they are evenly distributed (in equilibrium). An example in the human body is the movement of oxygen and carbon dioxide between body cells and blood.

The following two experiments illustrate simple diffusion. Either or both may be performed.

PROCEDURE

1. To demonstrate simple diffusion of a solid in a liquid, *using forceps, carefully* place a large crystal of potassium permanganate ($KMnO_4$) into a test tube filled with water.

▲ **CAUTION!** *Avoid contact of $KMnO_4$ with your skin by using acid- or caustic-resistant gloves.*

2. Place the tube in a rack against a white background where it will not be disturbed.

3. Note the diffusion of the crystal material through the water at 15-min intervals for 2 hr.

4. Record the diffusion of the crystal in millimeters (mm) per minute at 15-min intervals in Section D.1.b of the LABORATORY REPORT RESULTS at the end of the exercise. Simply measure the distance of diffusion using a millimeter ruler.

PROCEDURE

1. To demonstrate simple diffusion of a solid in a solid, *using forceps, carefully* place a large crystal of methylene blue on the surface of agar in the center of a petri plate.

2. Note the diffusion of the crystal through the agar at 15-min intervals for 2 hr.

3. Record the diffusion of the crystal in millimeters (mm) per minute at 15-min intervals, using a millimeter ruler, in Section D.1.b of the LABORATORY REPORT RESULTS at the end of the exercise.

c. Osmosis

Osmosis (oz-MŌ-sis) is the net movement of a *solvent* through a selectively permeable membrane. In living systems, the solvent is water, which moves by osmosis across a plasma membrane from an area of higher water (lower solute) concentration to an area of lower water (higher solute) concentration. In the body, fluids move between cells as a result of osmosis.

PROCEDURE

1. Refer to the osmosis apparatus in Figure 3.2.

2. Tie a knot very tightly at one end of a 4-in. piece of cellophane dialysis tubing that has been soaking in water for a few minutes. Fill the dialysis tubing with a 10% sugar (sucrose) solution that has been colored with Congo red (red food coloring can also be used).

3. Close the open end of the dialysis tubing with a one-hole rubber stopper into which a glass tube has already been inserted by a laboratory assistant or your instructor.

FIGURE 3.2   Osmosis apparatus.

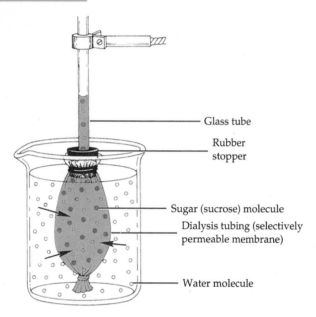

— Glass tube

— Rubber stopper

— Sugar (sucrose) molecule

— Dialysis tubing (selectively permeable membrane)

— Water molecule

⚠ **CAUTION!** *If you are unfamiliar with the procedure, do not attempt to insert the glass tube into the stopper yourself because it may break and result in serious injury.*

4. Tie a piece of string tightly around the dialysis tubing to secure it to the stopper.

5. Secure and suspend the glass tube and dialysis tubing by means of a clamp attached to a ring stand.

6. Insert the dialysis tubing into a beaker or flask of water until the water comes up to the bottom of the rubber stopper.

7. As soon as the sugar solution becomes visible in the glass tube, mark its height with a wax pencil and note the time.

8. Mark the height of liquid in the glass tube after 10-, 20-, and 30-min intervals by using your millimeter (mm) ruler and record your results in Section D.1.c of the LABORATORY REPORT RESULTS at the end of the exercise.

d. Hemolysis and Crenation

Osmosis can be demonstrated by noting the effects of different water concentrations on red blood cells. Red blood cells maintain their normal shape and volume when placed in an **isotonic** (*iso* = same) **solution.** An isotonic solution has the same salt concentration (0.9% NaCl) as that found in red blood cells. If, however, red blood cells are placed in a **hypotonic** (*hypo* = lower) **solution** (a salt solution with less than 0.9% NaCl), a net movement of water

into the cells occurs, causing the red blood cells to swell and eventually to burst. The red blood cells swell because the difference in osmotic pressure inside the cell compared to outside the cell results in a net movement of water into the cell. The rupture of blood cells in this manner is termed **hemolysis** (hē-MOL-i-sis). If, instead, red blood cells are placed in a **hypertonic** (*hyper* = higher) **solution** (a salt solution with more than 0.9% NaCl), a net movement of water out of the red blood cells occurs, causing the cells to shrink. The red blood cells undergo this shrinkage, known as **crenation** (kre-NĀ-shun) because the difference in osmotic pressure inside the cell compared to outside the cell results in the net movement of water out of the cell.

PROCEDURE

⚠ **CAUTION!** *Please reread Section B, "Precautions for Working with Blood, Blood Products, or Other Body Fluids," on page xii, at the beginning of the laboratory manual, before you begin any of the following experiments. Read the experiments before you perform them to be sure that you understand all the procedures and safety precautions. When working with whole blood, take care to avoid any kind of contact with an open sore, cut, or wound. Wear tight-fitting surgical gloves and safety goggles.*
When you finish this part of the exercise, place the reusable items in a fresh bleach solution and the discarded items in a biohazard container.

1. With a wax marking pencil, mark three microscope slides as follows: 0.9%, DW (distilled water), and 3%.

2. Using a medicine dropper, place one drop of fresh (uncoagulated) ox blood on a microscope slide that contains 2 mL of a 0.9% NaCl solution (isotonic solution). Mix gently and thoroughly with a clean toothpick.

3. Using a medicine dropper, place one drop of fresh ox blood on a microscope slide that contains 2 mL of distilled water (hypotonic solution). Mix gently and thoroughly with a clean toothpick.

4. Now, using a medicine dropper, add one drop of fresh ox blood to a microscope slide that contains 2 mL of a 3% NaCl solution (hypertonic solution). Mix gently and thoroughly with a clean toothpick.

5. Using a medicine dropper, place one drop of the red blood cells in the isotonic solution on another microscope slide, cover with a cover slip, and examine the red blood cells under high power. Reduce your illumination.

What is the shape of the cells? _____

_____

Explain their shape. _____

_____

6. Using a medicine dropper, place one drop of the red blood cells in the hypotonic solution on another microscope slide, cover with a cover slip, and examine the red blood cells under high power. Reduce your illumination.

    What is the shape of the cells? _____

    _____

    Explain their shape. _____

    _____

7. Using a medicine dropper, place one drop of the red blood cells in the hypertonic solution on another microscope slide, cover with a cover slip, and examine the red blood cells under high power. Reduce your illumination.

    What is the shape of the cells? _____

    _____

    Explain their shape. _____

    _____

ALTERNATE PROCEDURE

1. Obtain three pieces of raw potato that have an *identical* weight.

2. Immerse one piece in a beaker that contains an isotonic solution; immerse a second piece in a hypotonic solution; immerse the third piece in a hypertonic solution.

3. Continue the experiment for 1 hr. Record the time.

4. At the end of 1 hr, remove the pieces of potato and weigh them separately.

    Weight of potato in isotonic solution _____

    _____

    Weight of potato in hypotonic solution _____

    _____

    Weight of potato in hypertonic solution _____

    _____

    Explain the differences in the weights of the

    three pieces of potato. _____

    _____

    _____

_____

_____

_____

_____

e. Filtration

Filtration is the movement of solvents and dissolved substances across a selectively permeable membrane from regions of higher pressure to regions of lower pressure. Movement of the solvents and dissolved substances occurs under the influence of gravity and the pressure exerted by the solvent, which is termed **hydrostatic pressure.** The selectively permeable membrane prevents molecules with higher molecular weights from passing through the membrane, while the solvent and substances with lower molecular weights easily pass through the selectively permeable membrane. In general, any substance having a molecular weight of less than 100 is filtered, because the pores in the filter paper are larger than the molecules of the substance. Filtration is one mechanism by which the kidneys regulate the chemical composition of the blood.

PROCEDURE

1. Refer to Figure 3.3, which shows the filtration apparatus. In this apparatus, the filter paper represents the selectively permeable membrane of a cell.

FIGURE 3.3 Filtration apparatus.

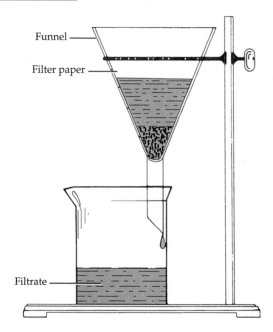

Funnel

Filter paper

Filtrate

2. Fold a piece of filter paper in half and then in half again.

3. Open it into a cone, place it in a funnel, and place the funnel over the beaker.

4. Shake a mixture of a few particles of powdered wood charcoal (black), 1% copper sulfate (blue), boiled starch (white), and water, and slowly pour it into the funnel until the mixture almost reaches the top of the filter paper. Gravity will pull the particles through the pores of the filter paper.

5. Count the number of drops passing through the funnel for the following time intervals: 10, 30, 60, 90, and 120 sec. Record your observations in Section D.1.e of the LABORATORY REPORT RESULTS at the end of the exercise.

6. Observe which substances passed through the filter paper by noting their color in the filtered fluid in the beaker.

7. Examine the filter paper to determine whether any colored particles were not filtered.

8. To determine if any starch is in the liquid (termed the *filtrate*) in the beaker, add several drops of 0.01 M IKI solution. A blue-black color reaction indicates the presence of starch.

f. Dialysis

Dialysis (dī-AL-i-sis) is the separation of smaller molecules from larger ones by a selectively permeable membrane. Such a membrane permits diffusion of the small molecules but not the large ones. Although dialysis does not occur in the human body, it is employed in artificial kidneys.

PROCEDURE

1. Refer to the dialysis apparatus in Figure 3.4.

2. Tie off one end of a piece of dialysis tubing that has been soaking in water. Place a prepared solution containing starch, sodium chloride, 5% glucose, and albumin into the dialysis tubing.

3. Tie off the other end of the dialysis tubing and immerse it in a beaker of distilled water.

4. After 1 hr, test the solution in the beaker for the presence of each of the substances in the tubing, as follows, and record your observations in Section D.1.f of the LABORATORY REPORT RESULTS at the end of the exercise.

⚠ **CAUTION!** *Be extremely careful using nitric acid. It can severely damage your eyes and skin. Use acid- or caustic-resistant gloves.*

a. *Albumin*—*Carefully* add several drops of concentrated nitric acid to a test tube con-

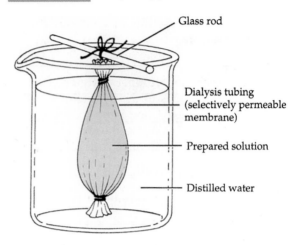

FIGURE 3.4 Dialysis apparatus.

Glass rod

Dialysis tubing (selectively permeable membrane)

Prepared solution

Distilled water

taining 2 mL of the solution in the beaker. Positive reaction = white coagulate.

b. *Sugar*—Test 5 mL of the solution in the beaker in a test tube with 5 mL of Benedict's solution. Place the test tube in a boiling water bath for 3 min.

⚠ **CAUTION!** *Make sure that the mouth of the test tube is pointed away from you and everyone else in the area.*

Using a test tube holder, remove the test tube from the water bath. Note the color. Positive reaction = green, yellow, orange, or red precipitate.

c. *Starch*—Add several drops of IKI solution to 2 mL of the solution in the beaker in a test tube. Note the color. Positive reaction = blue-black color.

d. *Sodium chloride*—Place 2 mL of the solution in the beaker in a test tube and add several drops of 1% silver nitrate. Note the color. Positive reaction = white precipitate.

2. Transport by Transporter Proteins

a. Facilitated Diffusion

Facilitated diffusion is the movement of a substance across a selectively permeable membrane from a region of higher concentration to a region of lower concentration with the assistance of integral proteins. The substance being transported binds to a specific transporter (protein) on one side of the membrane and is released on the other side after the transporter undergoes a change in shape. This process does not require the expenditure of cellular energy. Different sugars, especially glucose, cross plasma membranes by facilitated diffusion.

In this experiment, the substance undergoing facilitated diffusion is neutral red, which is red at a pH just below 7 but yellow at a pH between 7 and 8.

PROCEDURE

1. In a 125-mL flask, combine one gram (1 g) of baker's yeast with 25 milliliters (mL) 0.75% Na_2CO_3 (sodium carbonate) solution. Swirl until the yeast is evenly suspended in the solution.

2. Divide the suspension into two large test tubes marked "U1" (unboiled) and "B1" (boiled) with a wax pencil.

3. Place tube "B1" in a boiling water bath for 2 to 3 min. Be sure that all the suspension is below water level so that all the yeast cells will be killed.

⚠ **CAUTION!** *Make sure that the mouth of the test tube is pointed away from you and all other persons in the area.*

4. Using a test tube holder, place both test tubes into a rack and add 7.5 mL of 0.02% neutral red to each.

5. Record the color of the solution in both test tubes in Section D.2.a of the LABORATORY REPORT RESULTS at the end of the exercise.

6. After 15 min, place 1 mL from each test tube into test tubes marked "U2" and "B2."

7. Add enough drops of 0.75% acetic acid to tube "B2" to make its color identical to that of tube "U2."

8. Mark two microscope slides "U" and "B"; examine a sample from both tubes under high power (using a cover slip) and note which cells are stained. _____

_____

9. Filter half of the solution remaining in tube "B1" and examine the filtrate. Repeat for tube "U1."

10. Observe and record the color of the filtrate and that of the yeast cells on the filter paper.

_____

_____

11. To the suspension remaining in tubes "U1" and "B1," add about 9 mL 0.75% acetic acid. Filter again according to the instructions in step 9.

12. Record the color of these "U1" and "B1" filtrates and cells in Section D.2.a of the

LABORATORY REPORT RESULTS at the end of the exercise.

13. How does boiling affect the facilitated diffusion of neutral red by the cells? _____

_____

Describe the membrane's permeability to neutral red. _____

_____

Did acetic acid enter the living cells? _____

_____

Did sodium carbonate enter the living cells?

_____

b. Active Transport

There are two types of **active transport.** In **primary active transport,** energy derived from hydrolysis of ATP *directly* moves, or "pumps," a substance across a plasma membrane. The cell uses energy from ATP to change the shape of integral membrane proteins. An example of primary active transport is the sodium pump, which maintains a low concentration of sodium ions (Na^+) in the cytosol, the fluid portion of the cytoplasm, by pumping sodium ions out against their concentration gradient. The pump also moves potassium ions (K^+) into cells against their concentration gradient. In **secondary active transport,** the energy stored in a Na^+ or H^+ concentration gradient (difference) drives substances across a plasma membrane. Since ion gradients are established by primary active transport, secondary active transport *indirectly* uses energy obtained from the hydrolysis of ATP. An example of secondary active transport is the movement of an amino acid and Na^+ in the same direction across a plasma membrane with the assistance of an integral membrane protein. Another example is the movement of calcium ions (Ca^{2+}) and Na^+ in opposite directions across a membrane with the assistance of an integral membrane protein.

Because reliable results are difficult to demonstrate simply, you will not be asked to demonstrate active transport.

3. Transport in Vesicles

a. Endocytosis and Exocytosis

Endocytosis (*endo* = within) refers to the passage of large molecules and particles across a plasma membrane, in which a segment of the membrane surrounds the substance (*vesicle*), encloses it, and brings it into the cell. In **exocytosis** (*exo* = out), materials move out of a cell by the fusion with the plasma membrane of vesicles formed inside the cell. Here we will consider two types of endocytosis—phagocytosis and pinocytosis.

(1) Phagocytosis

Phagocytosis (fag′-ō-sī-TŌ-sis; *phago* = to eat) is a form of endocytosis in which the cell engulfs large solid particles or organisms by *pseudopods*, projections of the plasma membrane and cytoplasm of the cell. Once the particle is surrounded by the membrane, the membrane folds inward, pinches off from the rest of the plasma membrane, and forms a *phagosome* around the particle. The particle within the phagosome is digested either by the secretion of enzymes into the vesicle or by the combining of the vesicle with an enzyme-containing lysosome. Only a few types of body cells, termed *phagocytes*, are able to carry out phagocytosis. Two main types of phagocytes are *macrophages*, located in many body tissues, and *neutrophils*, a type of white blood cells.

Phagocytosis can be demonstrated by observing the feeding of an amoeba, a unicellular organism whose movement and ingestion are similar to those of human leukocytes (white blood cells).

PROCEDURE

1. Using a medicine dropper, place a drop of culture containing amoebas that have been starved for 48 hr into the well of a depression slide and cover the well with a cover slip. Cultures containing *Chaos chaos* or *Amoeba proteus* should be used. (Your instructor may wish to use the hanging-drop method instead. If so, she or he will give you verbal instructions.)

2. Examine the amoebas under low power, and be sure that your light is reduced considerably.

3. Observe the locomotion of an amoeba for several minutes. Pay particular attention to the pseudopods that appear to flow out of the cell.

4. To observe phagocytosis, use a medicine dropper and add a drop containing small unicellular animals called *Tetrahymena pyriformis* to the culture containing the amoebas.

5. Examine under low power, and observe the ingestion of *Tetrahymena pyriformis* by an amoeba. Note the action of the pseudopods and the formation of the phagosome around the ingested organism.

(2) Pinocytosis

Pinocytosis (pi-nō-sī-TŌ-sis; *pino* = to drink) is a form of endocytosis in which tiny droplets of extracellular fluid are taken up. The liquid is attracted to the surface of the membrane; the membrane folds inward and surrounds the liquid and detaches, or "pinches off," from the rest of the intact membrane forming a *pinocytic vesicle*. Within the cell, the pinocytic vesicle fuses with a lysosome, where enzymes degrade the engulfed solutes.

(3) Transcytosis

In this active process, vesicles undergo endocytosis on one side of a cell, move across the cell, and then undergo exocytosis on the opposite side. It occurs in endothelial cells when moving materials between the blood plasma and interstitial fluid.

E. Extracellular Materials

Substances that lie outside the plasma membranes of body cells are referred to as **extracellular materials.** They include body fluids, such as interstitial fluid and plasma, which provide a medium for dissolving, mixing, and transporting substances. Extracellular materials also include special substances in which some cells are embedded.

Some extracellular materials are produced by certain cells and deposited outside their plasma membranes where they support cells, bind them together, and provide strength and elasticity. They have no definite shape and are referred to as **amorphous.** These include hyaluronic (hī-a-loo-RON-ik) acid and chondroitin (kon-DROY-tin) sulfate. Others are **fibrous** (threadlike). Examples include collagen, reticular, and elastic fibers.

Using your textbook as a reference, indicate the location and function for each of the following extracellular materials:

Hyaluronic acid

Location _____

_____

_____

_____

Function _____

_____

_____

_____

Chondroitin sulfate

Location _____

_____

_____

_____

Function _____

_____

_____

_____

Collagen fibers

Location _____

_____

_____

_____

Function _____

_____

_____

_____

Elastic fibers

Location _____

_____

_____

_____

Function _____

_____

_____

_____

Reticular fibers

Location _____

_____

_____

_____

Function _____

_____

_____

_____

F. Cell Division

Cell division is the basic mechanism by which cells reproduce themselves. It consists of a nuclear division and a cytoplasmic division. Because nuclear division can be of two types, two kinds of cell division are recognized. In the first type, called **somatic cell division,** a single starting cell called a **parent cell** duplicates itself, and the result is two identical cells called **daughter cells.** In somatic cell division a cell undergoes a nuclear division called **mitosis** (mī-TŌ-sis) and a cytoplasmic division called **cytokinesis** (sī-tō-ki-NĒ-sis; *cyto* = cell; *kinesis* = motion). It provides the body with a means of growth and of replacement of dead or injured cells (Figure 3.5). The second type of cell division is called **reproductive cell division** and is the mechanism by which sperm and oocytes are produced (Exercise 25). Reproductive cell division consists of a special two-step nuclear division called **meiosis** and two cytoplasmic divisions (cytokinesis), and it results in the development of four nonidentical daughter cells.

In order to study somatic cell division, obtain a prepared slide of a whitefish blastula and examine it under high power.

When a cell is not dividing it is said to be in **interphase** of the cell cycle. Interphase is the longest part of the cell cycle and is the period of time during which a cell carries on its physiological activities. One of the most important activities of interphase is the replication of DNA so that the two daughter cells that eventually form will each have the same kind and amount of DNA as the parent cell. In addition, the proteins needed to produce

Cell division: mitosis and cytokinesis. Diagrams and photomicrographs (450×) of the various stages of cell division in whitefish eggs.

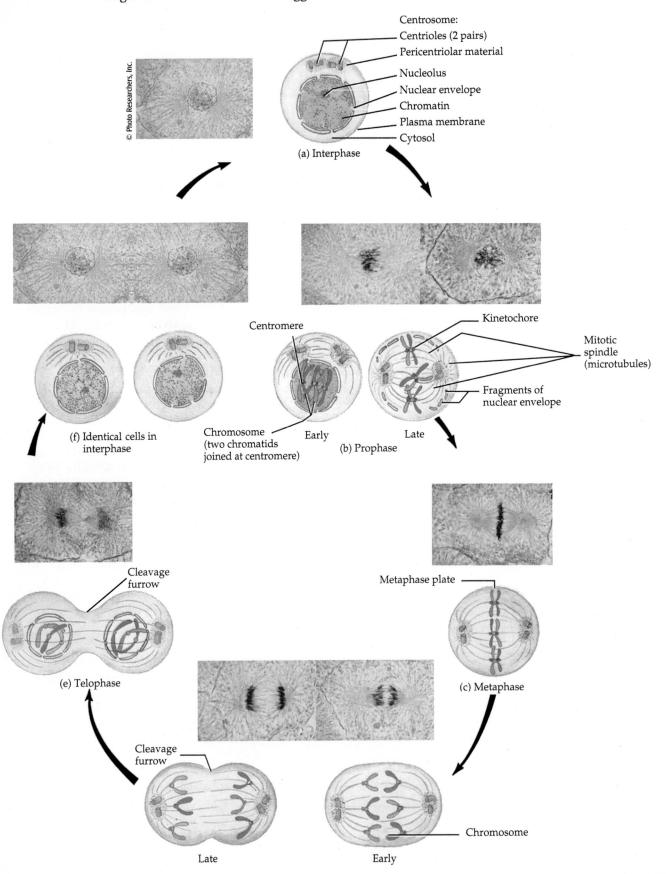

(a) Interphase

Centrosome:
Centrioles (2 pairs)
Pericentriolar material
Nucleolus
Nuclear envelope
Chromatin
Plasma membrane
Cytosol

(f) Identical cells in interphase

Centromere
Chromosome (two chromatids joined at centromere)
Early
Late
Kinetochore
Mitotic spindle (microtubules)
Fragments of nuclear envelope
(b) Prophase

Cleavage furrow
(e) Telophase

Metaphase plate
(c) Metaphase

Cleavage furrow
Late
Early
Chromosome
(d) Anaphase

structures required for doubling all cellular components are manufactured. Scan your slide and find a cell in interphase. Such a parent cell is characterized by a clearly defined nuclear envelope. Within the nucleus, look for the nucleolus (or nucleoli) and **chromatin,** DNA that is associated with protein in the form of a granular substance. Also locate the centrosomes.

Draw a labeled diagram of an interphase cell in the space provided.

Once a cell completes its interphase activities, mitosis begins. Mitosis is the distribution of two sets of chromosomes into two separate and equal nuclei after replication of the chromosomes of the parent cell, an event that takes place in the interphase preceding mitosis. For convenience, biologists divide the process into four stages for purposes of study: prophase, metaphase, anaphase, and telophase.

1. *Prophase*—The first stage of mitosis is called **prophase** (*pro* = before). During early prophase, the chromatin condenses and shortens into visible chromosomes. Because DNA replication took place during interphase, each prophase chromosome contains a pair of identical double-stranded DNA molecules called **chromatids.** Each chromatid pair is held together by a constricted region called a **centromere** that is required for the proper segregation of chromosomes. Attached to the outside of each centromere is a protein complex known as the **kinetochore** (ki-NET-ō-kor). Later in prophase, the nucleolus (or nucleoli) disappears, and the nuclear envelope breaks up. In addition, each centrosome, with its pair of centrioles, moves to an opposite pole (end) of the cell. As they do so, the pericentriolar areas of the centrosomes start to form the **mitotic spindle,** a football-shaped assembly of microtubules that attach to the kinetochore. The spindle is responsible for the separation of chromatids to opposite poles of the cell.

Draw and label a cell in prophase in the space provided.

2. *Metaphase*—During **metaphase** (*meta* = after), the second stage of mitosis, the kinetochore microtubules align the centromeres of the chromatid pairs at the exact center of the mitotic spindle. This midpoint region is called the **metaphase plate.** Draw and label a cell in metaphase in the space provided.

3. *Anaphase*—The third stage of mitosis, **anaphase** (*ana* = upward), is characterized by the splitting and separation of the centromeres (and kinetochores) and the movement of the two sister chromatids of each pair toward opposite poles of the cell. Once separated, the chromatids are referred to as **chromosomes.** As the chromosomes are pulled by the kinetochore microtubules during anaphase, they appear V-shaped because the centromeres lead the way, dragging the trailing arms of the chromosomes toward the pole. Draw and label a cell in anaphase in the space provided.

4. *Telophase*—The final stage of mitosis, **telophase** (*telo* = far or end), begins after chromosomal movement stops. Telophase is essentially the opposite of prophase. During telophase, the identical sets of chromosomes now at opposite poles of the cell uncoil and revert to their threadlike chromatin form. A new nuclear envelope re-forms around each chromatin mass; new nucleoli reappear in the daughter nuclei; and eventually the mitotic spindle disappears.

Draw and label a cell in telophase in the space provided.

Cytokinesis begins during late anaphase or early telophase with the formation of a **cleavage furrow,** a slight indentation of the plasma membrane. The cleavage furrow usually appears midway between the centrosomes and extends around the periphery of the cell. The furrow gradually deepens until opposite surfaces of the cell make contact and the cell is pinched in two. The result is two separated daughter cells, each with separate portions of cytoplasm and organelles and its own set of identical chromosomes.

Following cytokinesis, each daughter cell returns to interphase. Each cell in most tissues of the body eventually grows and undergoes mitosis and cytokinesis, and a new divisional cycle begins. Examine your telophase cell again and be sure that it contains a cleavage furrow.

Using high power and starting at 12 o'clock, move around the blastula and count the number of cells in interphase and in each mitotic phase. It will be easier to do this if you imagine lines dividing the blastula into quadrants. Count the interphase cells in each quadrant, then assign each dividing cell to a specific mitotic stage. It will be hard to assign some cells to a phase—e.g., to distinguish late anaphase from early telophase. If you cannot make a decision, assign one cell to the earlier phase in question and the next cell to the later phase.

Divide the number of cells in each stage by the total number of cells counted and multiply by 100 to determine the percent of the cells in each mitotic stage at a given point in time. Record your results in Section F of the LABORATORY REPORT RESULTS at the end of the exercise.

ANSWER THE LABORATORY REPORT QUESTIONS AT THE END OF THE EXERCISE.

EXERCISE 3 | Cells

Name _____ Date _____

Laboratory Section _____ Score/Grade _____

SECTION D ■ Movement of Substances Across and Through Plasma Membranes

1. Kinetic Energy Transport

a. Brownian Movement

Describe the movement of the India ink particles on the unheated slide. _____

_____

How does this movement differ from that on the heated slide? _____

_____

b. Diffusion

Solid in Liquid		Solid in Solid	
Time, min	Distance, mm	Time, min	Distance, mm
15	_____	15	_____
30	_____	30	_____
45	_____	45	_____
60	_____	60	_____
75	_____	75	_____
90	_____	90	_____
105	_____	105	_____
120	_____	120	_____

c. Osmosis

Time, min	Height of liquid, mm
10	_____
20	_____
30	_____

Explain what happened. _____

_____

e. Filtration

10 sec   _____

30 sec   _____

60 sec   _____

90 sec   _____

120 sec   _____

f. Dialysis
Place a check in the appropriate place to indicate if the following tests are positive (+) or negative (−).

	(+)	(−)
Albumin	_____	_____
Sugar	_____	_____
Starch	_____	_____
Sodium chloride	_____	_____

2. Transport by Transporter Proteins
a. Facilitated Diffusion

Neutral Red		**Acetic Acid**	
Tube "U1"	_____	Tube "U1"	_____
Tube "B1"	_____	Tube "B1"	_____

SECTION F ▪ Cell Division

Percent of cells in interphase   _____

Percent of cells in prophase   _____

Percent of cells in metaphase   _____

Percent of cells in anaphase   _____

Percent of cells in telophase   _____

EXERCISE 3 | # Cells

Name _____ Date _____

Laboratory Section _____ Score/Grade _____

PART 1 ▪ Multiple Choice

_____ 1. The portion of the cell that forms part of the mitotic spindle during division is the (a) endoplasmic reticulum (b) Golgi complex (c) cytoplasm (d) pericentriolar area of the centrosome

_____ 2. Movement of molecules or ions from a region of higher concentration to a region of lower concentration via a process that does not require cellular energy is called (a) phagocytosis (b) diffusion (c) active transport (d) pinocytosis

_____ 3. If red blood cells are placed in a hypertonic solution of sodium chloride, they will (a) swell (b) burst (c) shrink (d) remain the same

_____ 4. The reagent used to test for the presence of sugar is (a) silver nitrate (b) nitric acid (c) IKI (d) Benedict's solution

_____ 5. A cell that carries on a great deal of digestion also contains a large number of (a) lysosomes (b) centrosomes (c) mitochondria (d) nuclei

_____ 6. Which process does *not* belong with the others? (a) active transport (b) dialysis (c) phagocytosis (d) pinocytosis

_____ 7. Movement of oxygen and carbon dioxide between blood and body cells is an example of (a) osmosis (b) active transport (c) diffusion (d) facilitated diffusion

_____ 8. Which type of solution will cause hemolysis? (a) isotonic (b) hypotonic (c) isometric (d) hypertonic

_____ 9. In addition to active transport and simple diffusion, the kidneys regulate the chemical composition of blood by utilizing the process of (a) phagocytosis (b) osmosis (c) filtration (d) pinocytosis

_____ 10. Engulfment of solid particles or organisms by pseudopods is called (a) active transport (b) dialysis (c) phagocytosis (d) filtration

_____ 11. Rupture of red blood cells when exposed to a hypotonic solution is called (a) hemolysis (b) plasmolysis (c) plasmoptysis (d) hemoglobinuria

_____ 12. The area of the cell between the plasma membrane and nucleus where chemical reactions occur is the (a) centrosome (b) vacuole (c) perosisome (d) cytoplasm

_____ 13. The "powerhouses" of the cell where ATP is produced are the (a) ribosomes (b) mitochondria (c) centrosomes (d) lysosomes

_____ 14. The sites of protein synthesis in the cell are (a) peroxisomes (b) flagella (c) ribosomes (d) centrosomes

_____ **15.** Which process does *not* belong with the others? (a) diffusion (b) phagocytosis (c) active transport (d) pinocytosis

_____ **16.** The study of cellular structure and function is called (a) surface anatomy (b) cell biology (c) cytology (d) embryology

_____ **17.** Which extracellular material is found in ligaments and tendons? (a) elastic fibers (b) chondroitin sulfate (c) collagen fibers (d) mucus

_____ **18.** The organelles that contain enzymes for the metabolism of hydrogen peroxide are (a) lysosomes (b) mitochondria (c) Golgi complexes (d) peroxisomes

_____ **19.** The framework of cilia, flagella, centrioles, and the mitotic spindle is formed by (a) endoplasmic reticulum (b) collagen fibers (c) chondroitin sulfate (d) microtubules

_____ **20.** A viscous fluidlike substance that binds cells together, lubricates joints, and maintains the shape of the eyeballs is (a) elastin (b) hyaluronic acid (c) mucus (d) plasmin

PART 2 ■ Completion

21. The flexible outer boundary of the cell through which substances enter and exit is called the

    _____.

22. The cytoskeleton is formed by microtubules, intermediate filaments, and

    _____.

23. The portion of the cell that contains hereditary units called genes is the _____.

24. The tail of a sperm cell is a long whiplike structure called a(n) _____.

25. Cells placed in a(n) _____ solution will undergo crenation.

26. Division of the cytoplasm is referred to as _____.

27. Lipid and protein secretion, formation of transport and secretory vesicles, and assembly of glyco-

    proteins are functions of the _____.

28. Membrane-enclosed vesicles that contain digestive enzymes best describe the

    _____ of a cell.

29. Hairlike projections of cells that move substances along their surfaces are called

    _____.

30. The _____ is the site of phospholipid, fatty acid, and steroid synthesis and a

    temporary storage area for newly synthesized molecules.

31. The _____, located near the nucleus, consists of a pair of centrioles and peri-

    centriolar material.

32. A jellylike substance that supports cartilage, bone, the skin, and blood vessels is

    _____.

33. The framework of many soft organs is formed by _____ fibers.

34. The net movement of water through a selectively permeable membrane from a region of higher concentration of water to a region of lower concentration of water is known as

_____.

35. The principle of _____ is employed in the operation of an artificial kidney.

36. In an interphase cell, DNA is in the form of a granular substance called _____.

37. Distribution of chromosomes into separate and equal nuclei is referred to as

_____.

38. The constant random motion of molecules caused by their inherent kinetic energy is called

_____.

PART 3 ■ Matching

_____ **39.** Anaphase

_____ **40.** Metaphase

_____ **41.** Interphase

_____ **42.** Telophase

_____ **43.** Prophase

A. Mitotic spindle appears

B. Movement of chromosome sets to opposite poles of cell

C. Centromeres line up on metaphase plate

D. Formation of two identical nuclei

E. Phase between divisions

Tissues

Objectives

At the completion of this exercise you should understand

A The functions and anatomical characteristics of the four basic tissues of the human body.

B The types of membranes and their functions.

A tissue (*texere* = to weave) is a group of similar cells that usually have the same embryological origin and function together to carry out specialized activities. The study of tissues is called **histology** (hiss-TOL-ō-jē; *histo* = tissue; *logy* = study of). The body tissues can be classified into four basic types: (1) epithelial, (2) connective, (3) muscle, and (4) nervous. In this exercise you will examine the structure and functions of epithelial and connective tissues, except for bone or blood. Other tissues will be studied later as parts of the systems to which they belong.

A. Epithelial Tissue

Epithelial (ep′-i-THĒ-lē-al) **tissue,** or **epithelium,** may be divided into two types: (1) covering and lining and (2) glandular. Covering and lining epithelium forms the outer covering of the skin and some internal organs; forms the inner lining of blood vessels, ducts, and body cavities; and forms the interior of the respiratory, digestive, urinary, and reproductive systems. In addition, epithelial tissue helps make up special sense organs for smell, hearing, vision, and touch. Glandular epithelium constitutes the secreting portion of glands.

1. Characteristics

Following are the general characteristics of epithelial tissue:

a. Epithelium consists largely or entirely of closely packed cells that are held tightly together by many cell junctions with little intercellular material between cells.

b. Epithelial cells are arranged in continuous sheets, in either single or multiple layers.

c. Epithelial cells have an **apical surface** that is exposed to a body cavity, lining of an internal organ, or the exterior of the body and a **basal surface** that is attached to the basement membrane (described shortly).

d. Cell junctions (points of contact between cells) are plentiful, providing secure attachments among the cells.

e. Epithelial tissue is **avascular** (*a* = without; *vascular* = blood vessels). The vessels that supply nutrients and remove wastes are located in the adjacent connective tissue. The exchange of materials between epithelium and connective tissue occurs by diffusion.

f. Epithelia adhere firmly to nearby connective tissue, which holds the epithelium in position and prevents it from being torn. The attachment between the epithelium and the connective tissue is a thin extracellular layer called the **basement membrane.** It consists of two layers. The **basal lamina** contains collagen fibers, laminins, glycoproteins, and proteoglycans secreted by the epithelium. Cells in the connective tissue secrete the second layer, the **reticular lamina,** which contains reticular fibers, produced by connective tissue cells called fibroblasts.

g. Epithelial tissue has a nerve supply.

h. Epithelial tissue is the only tissue that makes direct contact with the external environment.

i. Since epithelium is subject to physical stress and injury, it has a high rate of cell division allowing epithelial tissue to constantly renew and repair itself.

j. Epithelia are diverse in origin. They are derived from all three primary germ layers (ectoderm, mesoderm, and endoderm).

k. Functions of epithelia include protection, filtration, secretion, absorption, and excretion. In addition, epithelial tissue combines with nervous tissue to form special sense organs.

2. Covering and Lining Epithelium

Before you start your microscopic examination of epithelial tissues, refer to Figure 4.1. Study the tissues carefully to familiarize yourself with their general structural characteristics. For each of the types of epithelium listed, obtain a prepared slide and, unless otherwise specified by your instructor, examine each under high power. In conjunction with your examination, consult a textbook of anatomy and physiology.

a. *Simple squamous* (SKWĀ-mus = flat) *epithelium.* This tissue consists of a single layer of flat cells and is highly adapted for diffusion and filtration because of its thinness. Simple squamous epithelium lines the air sacs of the lungs, glomerular (Bowman's) capsule of the kidneys, and inner surface of the tympanic membrane (eardrum) of the ear. Simple squamous epithelium that lines the heart, blood vessels, and lymphatic vessels is known as **endothelium** (*endo* = within; *thelium* = covering). Simple squamous epithelium that forms the epithelial layer of a serous membrane is known as **mesothelium** (*meso* = middle). Serous membranes line the thoracic and abdominopelvic cavities and cover viscera within the cavities. After you make your microscopic examination, draw several cells in the space that follows and label the plasma membrane, cytoplasm, and nucleus.

Simple squamous epithelium

b. *Simple cuboidal epithelium.* This tissue consists of a single layer of cube-shaped cells. When the tissue is sectioned and viewed from the side, its cuboidal nature is obvious. Highly adapted for secretion and absorption, this tissue covers the surface of the ovaries; lines the anterior surface of the capsule of the lens of eye; forms the pigmented epithelium at the posterior surface of the eye; lines kidney tubules and smaller ducts of many glands; and makes up the secreting portion of some glands, such as the thyroid gland, and ducts of some glands, such as the pancreas.

After you make your microscopic examination, draw several cells in the space that follows and label the plasma membrane, cytoplasm, nucleus, basement membrane, and connective tissue layer.

Simple cuboidal epithelium

c. *Simple columnar (nonciliated) epithelium.* This tissue consists of a single layer of columnar cells, and, when viewed from the side, these cells appear as rectangles. Adapted for secretion and absorption, this tissue lines the gastrointestinal tract from the stomach to the anus, gallbladder, and ducts of many glands. Some columnar cells are modified in that the plasma membranes are folded into microscopic fingerlike cytoplasmic projections called **microvilli** (*micro* = small; *villus* = tuft of hair) that increase the surface area of the plasma membrane for absorption. Other cells are **goblet cells,** modified columnar cells that secrete and store mucus, which serves to lubricate or protect the lining of the digestive, respiratory, reproductive, and most of the urinary tracts. After you make your microscopic examination, draw several cells in the space provided on page 55 and label the plasma membrane, cytoplasm, nucleus, goblet cell, absorptive cell, basement membrane, and connective tissue layer.

FIGURE 4.1 Epithelial tissues. Photomicrographs are unlabeled; the line drawings of the same tissues are labeled.

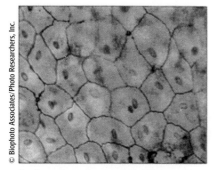

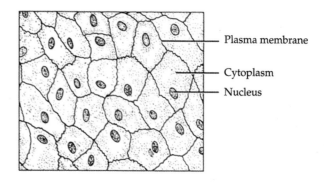

Surface view of mesothelial lining of
peritoneal cavity (240×)

— Plasma membrane

— Cytoplasm

— Nucleus

(a) Simple squamous epithelium

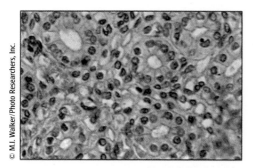

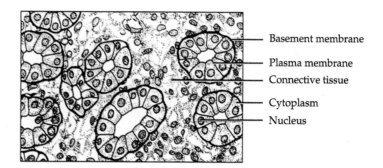

Sectional view of kidney tubules (400×)

— Basement membrane

— Plasma membrane

— Connective tissue

— Cytoplasm

— Nucleus

(b) Simple cuboidal epithelium

Mucus-producing   Connective   Basement   Nucleus of
goblet cell   tissue   membrane   absorptive cell

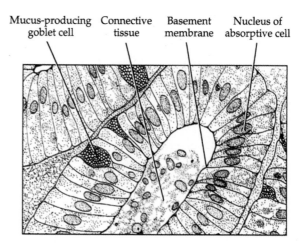

Sectional view of colonic glands (140×)

(c) Simple columnar (nonciliated) epithelium

© Biophoto Associates/Photo Researchers, Inc.

© M.I. Walker/Photo Researchers, Inc.

© Biophoto Associates/Photo Researchers, Inc.

FIGURE 4.1   Epithelial tissues. (Continued)

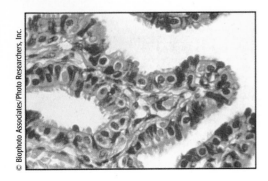

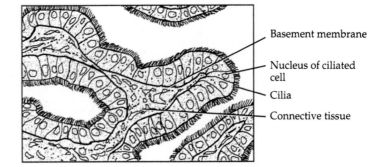

Basement membrane

Nucleus of ciliated cell

Cilia

Connective tissue

Sectional view of uterine (fallopian) tube (175×)

(d) Simple columnar (ciliated) epithelium

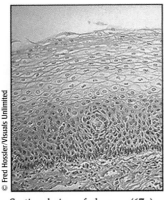

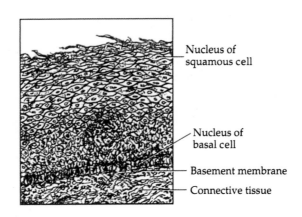

Nucleus of squamous cell

Nucleus of basal cell

Basement membrane

Connective tissue

Sectional view of pharynx (67×)

(e) Stratified squamous epithelium

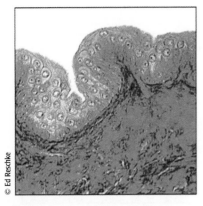

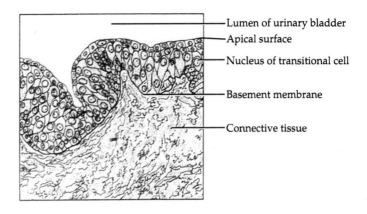

Lumen of urinary bladder

Apical surface

Nucleus of transitional cell

Basement membrane

Connective tissue

Sectional view of urinary bladder in relaxed state (100×)

(f) Relaxed transitional epithelium

FIGURE 4.1 **Epithelial tissues. (Continued)**

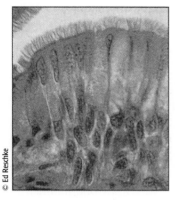

© Ed Reschke

Sectional view of trachea (250×)

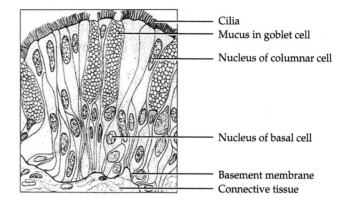

Cilia
Mucus in goblet cell
Nucleus of columnar cell
Nucleus of basal cell
Basement membrane
Connective tissue

(g) Pseudostratified columnar epithelium

Simple columnar (nonciliated) epithelium

Simple columnar (ciliated) epithelium

d. *Simple columnar (ciliated) epithelium.* This type of epithelium consists of a single layer of ciliated rectangular cells. In some locations it contains goblet and absorptive cells. **Cilia** (*cilia* = eyelashes) are hairlike processes that move substances over the surfaces of cells. Simple columnar (ciliated) epithelium lines a few portions of the upper respiratory tract, uterine (fallopian) tubes, uterus, some paranasal sinuses, and the central canal of the spinal cord. Mucus produced by goblet cells forms a thin film over the surface of the tissue, and movements of the cilia propel the mucus and the trapped substances over the surface of the tissue. After you make your microscopic examination, draw several cells in the space that follows and label the plasma membrane, cytoplasm, nucleus, cilia, goblet cell, basement membrane, and connective tissue layer.

e. *Stratified squamous epithelium.* This tissue consists of several layers of cells and affords considerable protection against friction. The superficial cells are flat, whereas cells of the deep layers vary in shape from cuboidal to columnar. The basal (deepest) cells continually multiply by cell division. As new cells grow, the cells of the basal layers are pushed upward toward the apical surface. As the apical surface cells are sloughed off, new cells replace them from the basal layer. The surface cells of **keratinized stratified squamous epithelium** contain a protective protein called **keratin** (*kerato* = horny) that resists friction, heat, microbes, and chemicals. The keratinized variety forms the superficial layer of the skin. Surface cells of **nonkeratinized stratified squamous epithelium** do not contain keratin and remain moist. The nonkeratinized variety lines wet surfaces such as the lining of the mouth, esophagus, vagina, and part of the epiglottis, and it covers the tongue. After you make your microscopic examination, draw several cells in the space that follows and label the plasma membrane, cytoplasm, nucleus, squamous surface cells, basal cells, basement membrane, and connective tissue layer.

Stratified squamous epithelium

f. Stratified squamous epithelium (student prepared). Before examining the next slide, prepare a smear of cheek cells from the epithelial lining of the mouth. As noted previously, epithelium that lines the mouth is nonkeratinized stratified squamous epithelium. However, you will be examining surface cells only, and these will appear similar to simple squamous epithelium.

PROCEDURE

▲ **CAUTION!** *Please reread Section B, "Precautions for Working with Blood, Blood Products, or Other Body Fluids" on page xii at the beginning of the laboratory manual before you begin any of the following experiments. You should also read the experiments before you perform them to be sure that you understand all the procedures and safety precautions. When you finish this part of the exercise, place the reusable items in a fresh bleach solution and the discarded items in a biohazard container.*

a. Using the blunt end of a toothpick, *gently* scrape the lining of your cheek several times to collect some surface cells of the stratified squamous epithelium.

b. Now move the toothpick across a clean glass microscope slide until a thin layer of scrapings is left on the slide.

c. Allow the preparation to air dry.

d. Next, cover the smear with several drops of 1% methylene blue stain. After about 1 min, gently rinse the slide in cold tap water or distilled water to remove excess stain.

e. *Gently* blot the slide dry using a paper towel.

f. Examine the slide under low and high power. See if you can identify the plasma membrane, cytoplasm, nuclear membrane, and nucleoli. Some bacteria are commonly found on the slide and usually appear as very small rods or spheres.

g. **Transitional epithelium.** This kind of stratified epithelium has a variable appearance. When stretched this tissue resembles nonkeratinized stratified squamous epithelium. When stretched, the surface cells are drawn out into squamouslike cells. This drawing out permits the tissue to stretch without the outer cells breaking apart from one another. In its relaxed state, transitional epithelium looks similar to stratified cuboidal epithelium, except that the apical cells tend to be large and round. The tissue lines parts of the urinary system that are subject to expansion from within, such as the urinary bladder, parts of the ureters, and urethra. After you have made your microscopic examination, draw several cells in the space that follows and label the plasma membrane, cytoplasm, nucleus, surface cells, basement membrane, and connective tissue layer.

Transitional epithelium

h. **Pseudostratified columnar epithelium.** Nuclei of cells in this tissue are at varying depths, and, although all the cells are attached to the basement membrane in a single layer, some do not extend to the apical surface. This arrangement gives the false impression of a multilayered tissue when sectioned, thus the name *pseudostratified* (*pseudo* = false). In **pseudostratified ciliated columnar epithelium,** the cells that reach the surface either secrete mucus (goblet cells) or bear cilia. The mucus traps foreign particles and the cilia sweep it away for elimination from the body. This tissue lines most of the upper respiratory tract. In

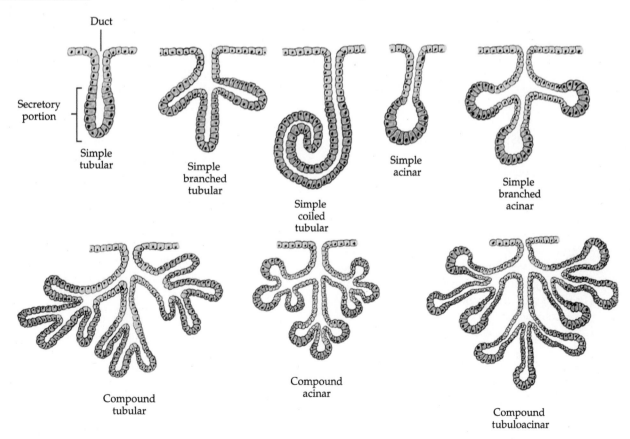

Pseudostratified columnar epithelium

pseudostratified nonciliated columnar epithelium, there are no goblet cells or cilia. This tissue lines the larger ducts of many glands, epididymis, and part of the male urethra. After you have made your microscopic examination, draw and label the basement membrane, a cell that reaches the surface, a cell that does not reach the surface, and nuclei of each cell in the preceding space.

3. Glandular Epithelium

A **gland** may consist of a single cell or a group of cells that secrete various substances. Glands that have no ducts (ductless), secrete hormones, and release their secretions into the blood are called **endocrine** (*endo* = within) **glands.** Examples include the pituitary gland and thyroid gland (Exercise 15). Glands that secrete their products into ducts or onto a surface are called **exocrine** (*exo* = outside) **glands.** Examples include sweat glands and salivary glands.

a. Structural Classification of Exocrine Glands

Based on the shape of the secretory portion and the degree of branching of the duct, exocrine glands can be structurally classified as follows:

1. *Unicellular.* Single-celled glands that secrete mucus. An example is the goblet cell (see Figure 4.1c). These cells line portions of the respiratory and digestive systems.

2. *Multicellular.* Many-celled glands that occur in several different forms (Figure 4.2).

FIGURE 4.2   Structural classification of multicellular exocrine glands.

I. *Simple*

   A. *Tubular*—Secretory portion is straight and attached to a single unbranched duct (intestinal glands).

   B. *Branched tubular*—Secretory portion is branched and attached to a single unbranched duct (gastric glands).

   C. *Coiled tubular*—Secretory portion is coiled and attached to a single unbranched duct (sweat glands).

   D. *Acinar* (AS-i-nar)—Secretory portion is more rounded and attached to a single unbranched duct (glands of penile urethra).

   E. *Branched acinar*—Secretory portion is branched and rounded and attached to a single unbranched duct (sebaceous glands).

II. *Compound*

   A. *Tubular*—Secretory portion is tubular and attached to a branched duct (bulbourethral or Cowper's glands, testes, liver).

   B. *Acinar*—Secretory portion is flasklike and attached to a branched duct (mammary glands).

   C. *Tubuloacinar*—Secretory portion is both tubular and rounded and attached to a branched duct (acinar glands of the pancreas).

Obtain a prepared slide of a representative of each of the types of multicellular exocrine glands just described. As you examine each slide, compare your observations to the diagrams of the glands in Figure 4.2.

b. Functional Classification of Exocrine Glands

Functional classification is based on how their secretion is released. **Holocrine** (HŌ-lō-krin; *holo* = entire) **glands,** such as sebaceous glands, accumulate their secretory product in their cytoplasm. As the secretory cell matures, it ruptures and becomes the secretory product. The sloughed off cell is replaced by a new one. **Merocrine** (MER-ō-krin; *mero* = a part) **glands,** such as the pancreas and salivary glands, produce secretions that are simply formed by the secretory cells and released from the cell in secretory vesicles by exocytosis. **Apocrine** (AP-ō-krin; *apo* = from) **glands,** such as the sudoriferous glands in the axilla, accumulate secretory products at the apical portion of the secreting cells. The margins pinch off as the secretion and the remaining portions of the cells are repaired so that the process can be repeated. Based on electron microscope studies, there is some question as to whether humans have apocrine glands. What were once thought to be apocrine glands (mammary glands) are probably merocrine glands.

B. Connective Tissue

Connective tissue, the most abundant tissue in the body, functions by protecting, supporting, strengthening, and separating structures (e.g., skeletal muscles), and binding structures together. It is also the major transport system within the body, the major site of stored energy reserves (adipose, or fat, tissue), and the main source of immune responses.

1. Characteristics

Following are the general characteristics of connective tissue.

   a. Connective tissue consists of two basic elements: cells and extracellular matrix. The **matrix** fills the wide spaces between its cells. It consists of protein-based fibers and ground substance, the material between the cells and fibers. Unlike epithelial cells, connective tissue cells rarely touch one another; they are separated by a considerable amount of matrix.

   b. In contrast to epithelia, connective tissues do not usually occur on body surfaces, such as the surfaces of a body cavity or the external surface of the body.

   c. Except for cartilage, connective tissue, like epithelia, has a nerve supply.

   d. Unlike epithelium, connective tissue usually is highly vascular (has a rich blood supply). Exceptions include cartilage, which is avascular, and tendons, which have a scanty blood supply.

   e. The matrix consists of a fluid, semifluid, gelatinous, or calcified ground substance and protein fibers. The specific matrix materials determine the tissue's qualities. In blood the matrix, which is not secreted by blood cells, is fluid. In cartilage it is firm but pliable. In bone it is considerably harder and not pliable.

2. Connective Tissue Cells

Following are some of the cells contained in various types of connective tissue. The specific tissues to which they belong will be described shortly.

   a. **Fibroblasts** (FĪ-brō-blasts; *fibro* = fiber) are large, flat cells with branching processes; they secrete the fibers and ground substance of the matrix.

b. **Macrophages** (MAK-rō-fā-jez; *macro* = large; *phages* = eaters) develop from **monocytes,** a type of white blood cell. Macrophages have an irregular shape with short branching projections and are capable of engulfing bacteria and cellular debris by phagocytosis. Thus, they provide a vital defense for the body.

c. **Plasma cells** are small cells that develop from a type of white blood cell called a **B lymphocyte.** Plasma cells secrete antibodies and, accordingly, provide a defense mechanism through immunity.

d. **Mast cells** are abundant alongside blood vessels. They produce *histamine*, a chemical that dilates small blood vessels as part of the inflammatory response, the body's reaction to injury or infection.

e. Other cells in connective tissue include **adipocytes (fat cells)** and **white blood cells (leukocytes).**

3. Connective Tissue Ground Substance

The **ground substance** is the component of a connective tissue between the cells and fibers. It is amorphous, meaning that it has no specific shape and may be a fluid, gel, or solid. Fibroblasts produce the ground substance and deposit it in the space between the cells.

Several examples of ground substance are as follows. **Hyaluronic** (hī-a-loo-RON-ik) acid is a viscous, slippery substance that binds cells together, lubricates joints, and helps maintain the shape of the eyeballs. It also appears to play a role in helping white blood cells migrate through connective tissue during development and wound repair. **Chondroitin** (kon-DROY-tin) **sulfate** provides support and adhesiveness in cartilage, bone, the skin, and blood vessels. The skin, tendons, blood vessels, and heart valves contain **dermatan sulfate,** while bone, cartilage, and the cornea of the eye contain **keratan sulfate.** Also present in the ground substance are **adhesion proteins,** which are responsible for linking components of the ground substance to each other and to the surfaces of cells. The principal adhesion protein is **fibronectin,** which binds to both collagen fibers and ground substance and links them together.

The ground substance supports cells and binds them together and provides a medium through which substances are exchanged between the blood and cells. It plays an active role in tissue development, migration, proliferation, shape, and even metabolic functions.

4. Connective Tissue Fibers

Fibers in the matrix are secreted by fibroblasts and provide strength and support for tissues. Three types of fibers are embedded in the matrix between the cells of connective tissue: collagen, elastic, and reticular fibers.

a. **Collagen** (*colla* = glue) **fibers,** of which there are at least six different types, are very strong and resistant to a pulling force, and promote some flexibility in the tissue because they are not stiff. These fibers often occur in bundles lying parallel to one another. The bundle arrangement affords great strength. Chemically, collagen fibers consist of the protein **collagen.** This is the most abundant protein in your body, representing about 25% of the total protein. Collagen fibers are found in most types of connective tissues, especially bone, cartilage, tendons, and ligaments.

b. **Elastic fibers** are smaller in diameter than collagen fibers and branch and join to form a network within a tissue. They consist of a protein called **elastin** surrounded by a glycoprotein called **fibrillin,** which strengthens and stabilizes the elastic fibers. Like collagen fibers, elastic fibers provide strength. In addition, they can be stretched 150% of their relaxed length without breaking. Elastic fibers are plentiful in the skin, blood vessel walls, and lung tissue.

c. **Reticular** (*reticul* = net) **fibers,** consisting of collagen arranged in fine bundles and a coating of glycoprotein, provide support in the walls of blood vessels and form a network around cells in some tissues, such as areolar and adipose tissues, and smooth muscle tissue. They are much thinner than collagen fibers and form branching networks. Like collagen fibers, reticular fibers provide support and strength and also form the **stroma** (framework) of many soft-tissue organs, such as the spleen and lymph nodes. These fibers also help form the basement membrane.

5. Types

Before you start your microscopic examination of connective tissues, refer to Figure 4.3. Study the tissues carefully to familiarize yourself with their general structural characteristics. For each type of connective tissue listed, obtain a prepared slide and, unless otherwise specified by your instructor, examine each under high power.

Here, we will concentrate on various types of **mature connective tissues,** meaning connective

FIGURE 4.3 Connective tissues. Photomicrographs are unlabeled; the line drawings of the same are labeled.

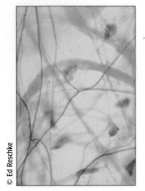

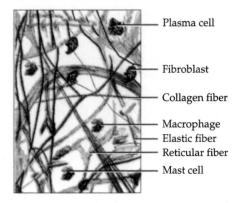

— Plasma cell

— Fibroblast

— Collagen fiber

— Macrophage
— Elastic fiber
— Reticular fiber
— Mast cell

Surface view of subcutaneous tissue (160×)

© Ed Reschke

(a) Areolar connective tissue

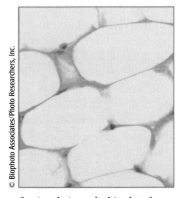

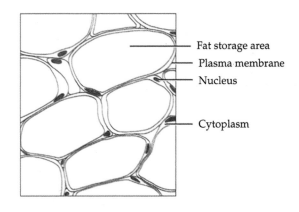

— Fat storage area
— Plasma membrane
— Nucleus

— Cytoplasm

Sectional view of white fat of pancreas (1,600×)

© Biophoto Associates/Photo Researchers, Inc.

(b) Adipose tissue

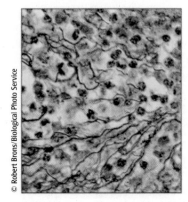

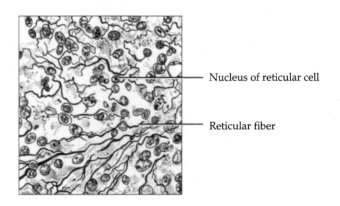

— Nucleus of reticular cell

— Reticular fiber

Sectional view of lymph node (250×)

© Robert Brons/Biological Photo Service

(c) Reticular connective tissue

FIGURE 4.3   Connective tissues. (Continued)

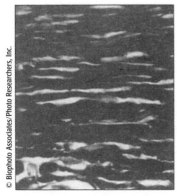

Sectional view of capsule of adrenal gland (250×)

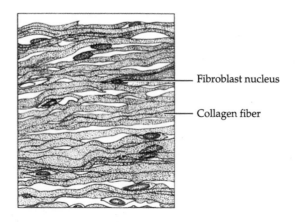

Fibroblast nucleus

Collagen fiber

(d) Dense regular connective tissue

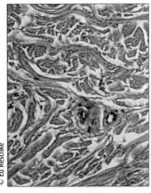

Sectional view of dermis of skin (275×)

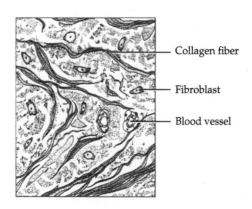

Collagen fiber

Fibroblast

Blood vessel

(e) Dense irregular connective tissue

Sectional view of ligamentum nuchae (400×)

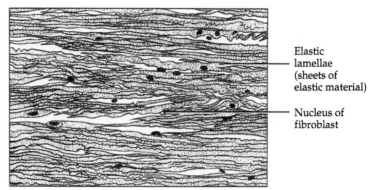

Elastic lamellae (sheets of elastic material)

Nucleus of fibroblast

(f) Elastic connective tissue

FIGURE 4.3   Connective tissues. (Continued)

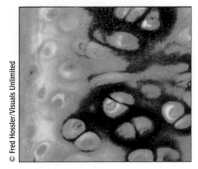

© Fred Hossler/Visuals Unlimited

Sectional view of hyaline cartilage
from trachea (160×)

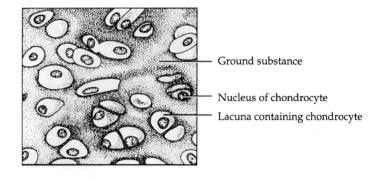

Ground substance

Nucleus of chondrocyte

Lacuna containing chondrocyte

(g) Hyaline cartilage

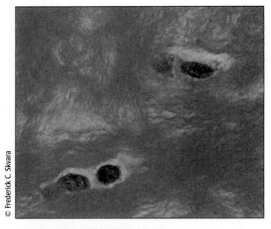

© Frederick C. Skvara

Sectional view of fibrocartilage from medial meniscus
of knee (315×)

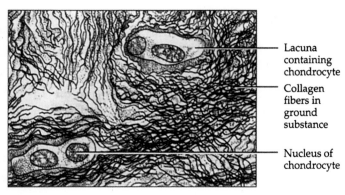

Lacuna
containing
chondrocyte

Collagen
fibers in
ground
substance

Nucleus of
chondrocyte

(h) Fibrocartilage

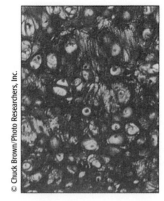

© Chuck Brown/Photo Researchers, Inc.

Sectional view of elastic cartilage
from auricle (pinna) of external ear
(175×)

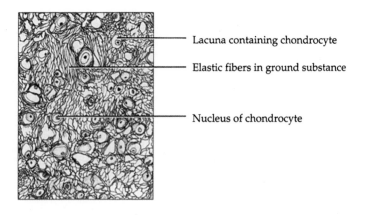

Lacuna containing chondrocyte

Elastic fibers in ground substance

Nucleus of chondrocyte

(i) Elastic cartilage

tissues that are present in the newborn and that do not change afterward. The types of mature connective tissue are **loose connective tissue, dense connective tissue, cartilage, bone tissue,** and **liquid connective tissue.**

a. Loose Connective Tissue

In this general type of connective tissue, the fibers are *loosely* intertwined and there are many cells present.

1. *Areolar* (a-RĒ-ō-lar; *areol* = small space) *connective tissue.* This is one of the most widely distributed connective tissues in the body. It contains several kinds of cells including fibroblasts, macrophages, plasma cells, mast cells, adipocytes, and a few white blood cells. All three types of fibers—collagen, elastic, and reticular—are present and randomly arranged. The ground substance contains hyaluronic acid, chondroitin sulfate, dermatan sulfate, and keratin sulfate. Areolar connective tissue is present in the lamina propria of mucous membranes, the papillary (superficial) region of the dermis of the skin, around blood vessels, nerves, and organs, and, together with adipose tissue, forms the **subcutaneous** (sub'-kyoo-TĀ-nē-us) **layer,** the layer that attaches the skin to underlying tissues and organs. After you make your microscopic examination, draw a small area of the tissue in the space that follows and label the elastic fibers, collagen fibers, fibroblasts, and mast cells.

Areolar connective tissue

2. *Adipose tissue.* This is fat tissue in which cells, called **adipocytes** (*adeps* = fat), are specialized for storage of triglycerides. The cytoplasm and nuclei of the cells are pushed to the periphery of the cell. The tissue is found wherever areolar connective tissue is located and around the kidneys and heart, in the yellow bone marrow of long bones, around joints, and behind the eyeball. It provides insulation, energy reserve, support, and protection. Adipose tissue exists in two forms: **white adipose tissue** and **brown adipose tissue** or **BAT.** BAT is widespread in the fetus and infant; in adults only small amounts are present. After your microscopic examination, draw several cells in the space that follows and label the fat storage area, cytoplasm, nucleus, and plasma membrane.

Adipose tissue

3. *Reticular connective tissue.* This tissue consists of fine interlacing reticular fibers and reticular cells. It forms the stroma (supporting framework) of the liver, spleen, and lymph nodes. It is found in a portion of the basement membrane, in red bone marrow, and around blood vessels and muscle. After you make your microscopic examination, draw a sample of the tissue in the space that follows and label the reticular fibers and cells of the organ.

Reticular connective tissue

b. Dense Connective Tissue

In this general type of connective tissue, the fibers are more numerous, *thicker,* and *densely* packed, but there are fewer cells than in loose connective tissue.

1. *Dense regular connective tissue.* In this tissue, bundles of collagen fibers are *regularly* arranged in parallel patterns that confer great strength. The tissue structure withstands pulling along the axis of the fibers. Fibroblasts, which produce the fibers and ground substance, appear in rows between the fibers. The tissue is silvery white and tough, yet somewhat pliable. Because of its great strength, it is the principal component of **tendons,** which attach muscles to bones; **aponeuroses** (ap′-ō-noo-RŌ-sēz), which are sheetlike tendons that attach muscle to muscle or muscle to bone; and most **ligaments** (collagen ligaments), which attach bone to bone. After you make your microscopic examination, draw a sample of the tissue in the space that follows and label the collagen fibers and fibroblasts.

Dense regular connective tissue

2. *Dense irregular connective tissue.* This tissue contains collagen fibers that are packed more closely together than in loose connective tissue and are usually *irregularly* arranged. It is found in parts of the body where pulling forces are exerted in various directions. The tissue often occurs in sheets. It forms some fasciae, the reticular (deeper) region of the dermis of the skin, the pericardium of the heart, the periosteum of bone, the perichondrium of cartilage, joint capsules, heart valves, and the membrane capsules around organs, such as the kidneys, liver, testes, and lymph nodes. After you make your microscopic examination, draw a sample of the tissue in the space that follows and label the collagen fibers.

3. *Elastic connective tissue.* This tissue has a predominance of branching elastic fibers. These fibers give the unstained tissue a yellowish color. Fibroblasts are present in the

Dense irregular connective tissue

spaces between fibers. Elastic connective tissue can be stretched and can recoil to its original shape after being stretched. It is a component of the walls of elastic arteries, the trachea, bronchial tubes, and the lungs themselves. Elastic connective tissue provides stretch and strength, allowing structures to perform their functions efficiently. Yellow elastic ligaments, as contrasted with collagen ligaments, are composed mostly of elastic fibers; they form the ligaments between vertebrae, the suspensory ligament of the penis, and the true vocal cords. After you make your microscopic examination, draw a sample of the tissue in the space that follows and label the elastic fibers and fibroblasts.

Elastic connective tissue

c. Cartilage

Cartilage is capable of enduring considerably more stress than loose and dense connective tissues. Unlike other connective tissues, cartilage has no blood vessels or nerves, except for those in the perichondrium (membranous covering). Cartilage consists of a dense network of collagen fibers and elastic fibers firmly embedded in chondroitin sulfate, a gel-like component of the ground substance. Whereas the strength of cartilage is due to its collagen fibers, its resilience (ability to assume its original shape after deformation) is due to chondroitin sulfate.

The cells of mature cartilage, called **chondrocytes** (KON-drō-sīts; *chondro* = cartilage), occur singly or in groups within spaces called **lacunae** (la-KOO-nē; = little lake) in the extracellular matrix. The surface of most cartilage is covered by dense irregular connective tissue called the **perichondrium** (per′-i-KON-drē-um; *peri* = around). Unlike other connective tissues, cartilage has no blood vessels or nerves, except in the perichondrium. Three kinds of cartilage are recognized: hyaline cartilage, fibrocartilage, and elastic cartilage.

1. *Hyaline cartilage.* This cartilage contains a resilient gel as its ground substance and appears in the body as a bluish-white, shiny substance. The fine collagen fibers are not visible with ordinary staining techniques, and the prominent chondrocytes are found in lacunae. Most hyaline cartilage is surrounded by a perichondrium. Hyaline cartilage is the most abundant cartilage in the body. It is found at joints over the ends of the long bones (articular cartilage) and at the anterior ends of the ribs (costal cartilage). Hyaline cartilage also helps to support the nose, larynx, trachea, bronchi, and bronchial tubes. Most of the embryonic and fetal skeleton consists of hyaline cartilage, which gradually becomes calcified and develops into bone. Hyaline cartilage affords flexibility and support and, at joints, reduces friction and absorbs shock. Hyaline cartilage is the weakest of the three types of cartilage. After you make your microscopic examination, draw a sample of the tissue in the space that follows and label the perichondrium, chondrocytes, lacunae, and ground substance.

Hyaline cartilage

2. *Fibrocartilage.* Chondrocytes are scattered among clearly visible bundles of collagen fibers within the matrix of this type of cartilage. Fibrocartilage lacks a perichondrium. Fibrocartilage forms a pubic symphysis, the point where the hip bones fuse anteriorly at the midline. It is also found in the intervertebral discs between vertebrae, and the menisci (cartilage pads) of the knee. This tissue combines strength and rigidity and is the strongest of the three types of cartilages. After you make your microscopic examination, draw a sample of the tissue in the space that follows and label the chondrocytes, lacunae, ground substance, and collagen fibers.

Fibrocartilage

3. *Elastic cartilage.* In this tissue, chondrocytes are located in a threadlike network of elastic fibers within the matrix. Elastic cartilage has a perichondrium and provides strength and elasticity and maintains the shape of certain structures—the epiglottis of the larynx, the external part of the ear (auricle), and the auditory (eustachian) tubes. After you make your microscopic examination, draw a sample of the tissue in the space that follows and label the perichondrium, chondrocytes, lacunae, ground substance, and elastic fibers.

C. Membranes

Membranes are flat sheets of pliable tissue that cover or line a part of the body. The combination of an epithelial layer and an underlying layer of connective tissue constitutes an **epithelial membrane.** The principal epithelial membranes of the body are mucous, serous, and cutaneous membranes (skin).

Elastic cartilage

Another kind of membrane, a synovial membrane, has no epithelium. It contains only connective tissue. **Mucous membranes,** also called the **mucosa,** line body cavities that open directly to the exterior, such as the gastrointestinal, respiratory, urinary, and reproductive tracts. The surface tissue of a mucous membrane consists of epithelium and has a variety of functions, depending on location. Accordingly, the epithelial layer secretes mucus but may also secrete enzymes needed for digestion, filter dust, and have a protective and absorbent action. The underlying connective tissue layer of a mucous membrane is areolar connective tissue, called the *lamina propria* (LAM-i-na PRŌ-prē-a). The lamina propria binds the epithelial layer in place, protects underlying tissues, provides the epithelium with nutrients and oxygen and removes wastes, and holds blood vessels in place.

Serous (*serous* = watery) **membranes,** also called the **serosa,** line body cavities that do not open directly to the exterior and cover organs that lie within the cavities. Serous membranes consist of a surface layer of mesothelium (simple squamous epithelium) and an underlying layer of areolar connective tissue. The mesothelium secretes a lubricating fluid. Serous membranes consist of two layers. The layer attached to the cavity wall is called the **parietal** (pa-RĪ-e-tal; *paries* = wall) **layer;** the layer that covers the organs in the cavity is called the **visceral** (*viscer* = body organ) **layer.** The mesothelium of a serous membrane secretes **serous fluid,** a watery lubricating fluid that allows organs to easily glide over one another or to slide against the walls of cavities. Examples of serous membranes are the pleurae, pericardium, and peritoneum.

The **cutaneous membrane,** or skin, is the principal component of the integumentary system, which will be considered in the next exercise.

Synovial (sin-Ō-vē-al) **membranes** line the cavities of freely movable joints. They do not contain epithelium but rather consist of areolar connective tissue, adipose tissue, and synoviocytes. Synovial membranes produce **synovial fluid,** which lubricates the ends of bones as they move at joints and nourishes the articular cartilage around the ends of bones.

ANSWER THE LABORATORY REPORT QUESTIONS AT THE END OF THE EXERCISE.

Tissues

Name _____   Date _____

Laboratory Section _____   Score/Grade _____

PART 1 ■ Multiple Choice

_____ 1. In parts of the body such as the urinary bladder, where considerable distention (stretching) occurs, you can expect to find which epithelial tissue? (a) pseudostratified columnar (b) cuboidal (c) columnar (d) transitional

_____ 2. Stratified epithelium is usually found in areas of the body where the principal activity is (a) filtration (b) absorption (c) protection (d) diffusion

_____ 3. Ciliated epithelium destroyed by disease would cause malfunction in which system? (a) digestive (b) respiratory (c) skeletal (d) cardiovascular

_____ 4. The tissue that provides the skin with resistance to wear and tear and serves to waterproof it is (a) keratinized stratified squamous (b) pseudostratified columnar (c) transitional (d) simple columnar

_____ 5. The connective tissue cell that would most likely increase its activity during an infection is the (a) melanocyte (b) macrophage (c) adipocyte (d) fibroblast

_____ 6. Torn ligaments would involve damage to which tissue? (a) dense regular (b) reticular (c) elastic (d) areolar

_____ 7. Simple squamous epithelial tissue that lines the heart, blood vessels, and lymphatic vessels is called (a) transitional (b) adipose (c) endothelium (d) mesothelium

_____ 8. Microvilli and goblet cells are associated with which tissue? (a) hyaline cartilage (b) simple columnar nonciliated (c) transitional (d) stratified squamous

_____ 9. Superficial fascia contains which tissue? (a) elastic (b) reticular (c) fibrocartilage (d) areolar connective tissue

_____ 10. Which tissue forms articular cartilage and costal cartilage? (a) fibrocartilage (b) elastic cartilage (c) adipose (d) hyaline cartilage

_____ 11. Because the sublingual gland contains a branched duct and flasklike secretory portions, it is classified as (a) simple coiled tubular (b) compound acinar (c) simple acinar (d) compound tubular

_____ 12. Which glands produce secretions that are simply formed by the secretory cells and then discharged into a duct? (a) merocrine (b) apocrine (c) endocrine (d) holocrine

_____ 13. Membranes that line cavities that open directly to the exterior are called (a) synovial (b) serous (c) mucous (d) cutaneous

_____ 14. Which statement about connective tissue is false? (a) Cells are always very closely packed together. (b) Connective tissue always has an abundant blood supply. (c) Matrix is always present in large amounts. (d) It is the most abundant tissue in the body.

_____ **15.** A group of similar cells that have a similar embryological origin and operate together to perform a specialized activity is called a(n) (a) organ (b) tissue (c) system (d) organ system

_____ **16.** Which statement best describes covering and lining epithelium? (a) It is always arranged in a single layer of cells. (b) It contains large amounts of intercellular substance. (c) It has an abundant blood supply. (d) Its free surface is exposed to the exterior of the body or to the interior of a hollow structure.

_____ **17.** Which statement best describes connective tissue? (a) It usually contains a large amount of matrix. (b) It's always arranged in a single layer of cells. (c) It's primarily concerned with secretion. (d) It usually lines a body cavity.

_____ **18.** A gland (a) is either exocrine or endocrine (b) may be single celled or multicellular (c) consists of epithelial tissue (d) is described by all of the preceding statements.

_____ **19.** Which of the following statements is not correct? (a) Simple squamous epithelium lines blood vessels. (b) Endothelium is composed of cuboidal cells. (c) Ciliated epithelium is found in the respiratory system. (d) Transitional epithelium is found in the urinary bladder.

PART 2 ▪ Completion

20. Cells found in epithelium that secrete mucus are called _____ cells.

21. A type of epithelium that appears to consist of several layers but actually contains only one layer of cells is _____.

22. The cell in connective tissue that forms new fibers is the _____.

23. Histamine, a substance that dilates small blood vessels during inflammation, is secreted by _____ cells.

24. Cartilage cells called _____ are found in lacunae.

25. The simple squamous epithelium of a serous membrane that covers viscera is called _____.

26. The tissue that provides insulation, support, and protection, and serves as a food reserve is _____.

27. _____ tissue forms the stroma of organs such as the liver and spleen.

28. The cartilage that provides support for the larynx and external ear is _____.

29. The ground substance that helps lubricate joints and binds cells together is _____.

30. Ductless glands that secrete hormones are called _____ glands.

31. Multicellular exocrine glands that contain branching ducts are classified as _____ glands.

32. _____ glands accumulate their secretory products in their cytoplasm, the cell dies, and is discharged with its contents as the glandular secretions.

33. _____ membranes consist of parietal and visceral layers and line cavities that do not open to the exterior.

34. If the secretory portion of a gland is rounded, it is classified as a(n) _____ gland.

35. Membranes that line joint cavities are called _____ membranes.

36. An example of a simple branched acinar gland is a(n) _____ gland.

37. The structure that attaches epithelium to underlying connective tissue is called the

   _____.

PART 3 ■ Matching

_____ **38.** Lines inner surface of the stomach and intestine

_____ **39.** Lines urinary tract, as in urinary bladder, permitting distention

_____ **40.** Lines mouth; present on outer surface of skin

_____ **41.** Single layer of cube-shaped cells; found in kidney tubules and ducts of some glands

_____ **42.** Lines air sacs of lungs where thin cells are required for diffusion of gases into blood

_____ **43.** Not a true stratified tissue; all cells on basement membrane, but some do not reach surface

_____ **44.** Derived from lymphocytes, gives rise to antibodies and so is helpful in defense

_____ **45.** Phagocytic cell; engulfs bacteria and cleans up debris; important during infection

_____ **46.** Believed to form collagen and elastic fibers in injured tissue

_____ **47.** Abundant along walls of blood vessels; produces histamine, which dilates blood vessels

_____ **48.** Contains lacunae and chondrocytes

_____ **49.** Forms fasciae and dermis of skin

_____ **50.** Stores fat and provides insulation

A. Transitional epithelium

B. Fibroblast

C. Pseudostratified columnar epithelium

D. Dense irregular connective tissue

E. Simple columnar epithelium

F. Macrophage

G. Stratified squamous epithelium

H. Adipose

I. Simple cuboidal epithelium

J. Plasma cell

K. Simple squamous epithelium

L. Mast cell

M. Cartilage

Integumentary System

Objectives

At the completion of this exercise you should understand

A The anatomy of the skin and its layers.

B The functions of the various layers of the skin.

C The structure and functions of the various accessory structures of the skin, such as hair, skin glands, and nails.

The skin and its accessory structures—hair, nails, various glands, muscles, and nerves—make up the **integumentary** (in-teg-ū-MEN-tar-ē; *integumentum* = covering; *inte* = whole; *gument* = body covering) **system,** which you will study in this exercise. An **organ** is an aggregation of tissues of definite form and usually recognizable shape that performs a specific function; a **system** is a group of organs that operate together to perform specialized functions.

A. Skin

The **skin** or **cutaneous membrane** is one of the largest organs of the body in terms of surface area and weight, occupying a surface area of about 2 square meters (2 m^2) (22 square feet) and weighs 4.5 to 5 kilograms (kg) (10 to 11 lb), about 16% of total body weight. Among the functions performed by the skin are regulation of body temperature **(thermoregulation);** protection of underlying tissues from physical abrasion, microorganisms, dehydration, and ultraviolet (UV) radiation; excretion of salts and several organic compounds; absorption of certain lipid-soluble materials; synthesis of vitamin D; reception of stimuli for touch, pressure, pain, and temperature change sensations; serves as a blood reservoir; and immunity.

The skin consists of two main parts. The superficial, thinner **epidermis** (*epi* = above), which is avascular, and a deeper, thicker **dermis** (*derm* = skin), which is vascular. Deep to the dermis and not part of the skin is the **subcutaneous (subQ)** layer **(hypodermis)** that anchors the skin to underlying tissues and organs. This layer also contains nerve endings sensitive to pressure called **lamellated** or **pacinian** (pa-SIN-ē-an) **corpuscles.**

1. Epidermis

The epidermis is composed of keratinized stratified squamous epithelium. It consists of four principal kinds of cells. **Keratinocytes** (ker-a-TIN-ō-sīts; *keratino* = hornlike) are the most numerous cells. They are arranged in four or five layers and produce a protein called keratin, which helps protect the skin and underlying tissues from heat, microbes, and many chemicals. Keratinocytes undergo keratinization; that is, newly formed cells produced in the basal layers are pushed up to the surface and in the process synthesize keratin. **Melanocytes** (MEL-a-nō-sīts; *melano* = black) produce the brown-black pigment, **melanin,** that contributes to skin color and absorbs damaging UV light. The third type of cell in the epidermis is called a **Langerhans** (LANG-er-hans) **cell.** These cells arise from red bone marrow and migrate to the epidermis. They participate in immune responses mounted against microbes that invade the skin, and are easily damaged by UV light. **Merkel cells** are found in the deepest layer of the epidermis where they contact the flattened process of a sensory neuron (nerve cell), a structure called a **tactile (Merkel) disc.** Merkel cells and tactile discs function in the sensation of touch. At this point, we will concentrate only on keratinocytes.

Obtain a prepared slide of human thick skin and carefully examine the epidermis. Identify the following layers from the outside inward:

a. *Stratum basale* (*basale* = base). Single row of cuboidal to columnar keratinocytes that constantly undergo division. This layer contains tactile (Merkel) discs, receptors sensitive to touch, melanocytes, and Langerhans cells. The stratum basale is also known as the **stratum germinativum.**

b. *Stratum spinosum* (*spinos* = thornlike). 8 to 10 rows of many-sided keratinocytes fit closely together.

c. *Stratum granulosum* (*granulos* = little grain). Consists of 3 to 5 layers of flattened keratinocytes that contain **lamellar granules,** which release a water-repellent sealant.

d. *Stratum lucidum* (*lucid* = clear). Present only in the thick skin of the fingertips, palms, and soles. It consists of 3 to 5 layers of flattened clear, dead keratinocytes.

e. *Stratum corneum* (*corneum* = hornlike). 25 to 30 rows of flat, dead keratinocytes. The interior of the cells contains mostly keratin. These cells are continuously shed and replaced by cells from deeper strata. This layer is an effective water-repellent barrier and protects underlying layers.

Label the epidermal layers in Figure 5.1.

2. Dermis

The dermis is divided into two regions and is composed of connective tissue containing collagen and elastic fibers and a number of other structures. The superficial region of the dermis **(papillary layer)** is composed of areolar connective tissue containing fine elastic fibers. This layer contains small finger-like projections, the **dermal papillae** (pa-PIL-ē; *papilla* = nipple). Some papillae contain capillary loops; others contain **corpuscles of touch (Meissner corpuscles),** nerve endings sensitive to touch; others contain free nerve endings that convey sensations of warmth, coolness, pain, tickling, and itching. The deeper region of the dermis **(reticular layer)** consists of dense irregular connective tissue containing bundles of collagen and some coarse elastic fibers. Spaces between the fibers may be occupied by **adipose cells, hair follicles, sebaceous (oil) glands, sudoriferous (sweat) glands,** and **nerves.** The reticular layer of the dermis is attached to the underlying structures (bones and muscles) by the subcutaneous layer.

Carefully examine the dermis and subcutaneous layer on your microscope slide. Label the following structures in Figure 5.1: papillary layer, reticular layer, corpuscle of touch, blood vessels, nerves, sebaceous gland, sudoriferous glands (apocrine and eccrine), and lamellated corpuscle. Also label the epidermis, dermis, and blood vessels in Figure 5.2.

If a model of the skin and subcutaneous layer is available, examine it to see the three-dimensional relationship of the structures to one another.

3. Skin Color

Three pigments contribute to the skin's wide variety of colors: (1) **hemoglobin in red blood cells in capillaries** of the dermis (beneath the epidermis) creating the pink to red appearance of the skin of white people; (2) **carotene** (KAR-o-tēn; *carot* = carrot), a yellow-orange pigment found in the stratum corneum and fatty areas of the dermis and subcutaneous layer; and (3) **melanin** (MEL-a-nin), a yellow-red to brown-black pigment synthesized by **melanocytes** (MEL-a-nō-sīts). The amount of melanin causes the skin's color to vary from pale yellow to tan to black. Melanin is most abundant in the epidermis of the penis, nipples of the breasts, area just around the nipples (areolae), face, and limbs. Because the number of melanocytes is about the same in all people, most differences in skin color are due to the amount of melanin that the melanocytes synthesize and disperse to keratinocytes. Exposure to UV radiation increases melanin synthesis, resulting in darkening (tanning) of the skin to protect the body against further UV radiation.

An inherited inability of an individual to produce melanin results in **albinism** (AL-bi-nizm). Melanin is missing from the hair and eyes as well as from the skin, and the individual is referred to as an **albino.** In some people, melanin tends to accumulate in patches called **freckles.** Others inherit a partial or complete loss of melanocytes from patches of skin producing irregular white spots, a condition called **vitiligo** (vit-i-LĪ-gō).

B. Hair

Hairs (pili) develop from the epidermis and are variously distributed over the body. Each hair is composed of columns of dead, keratinized cells and consists of a **shaft,** most of which projects from the surface of the skin, and a **root,** the

FIGURE 5.1 Structure of the skin.

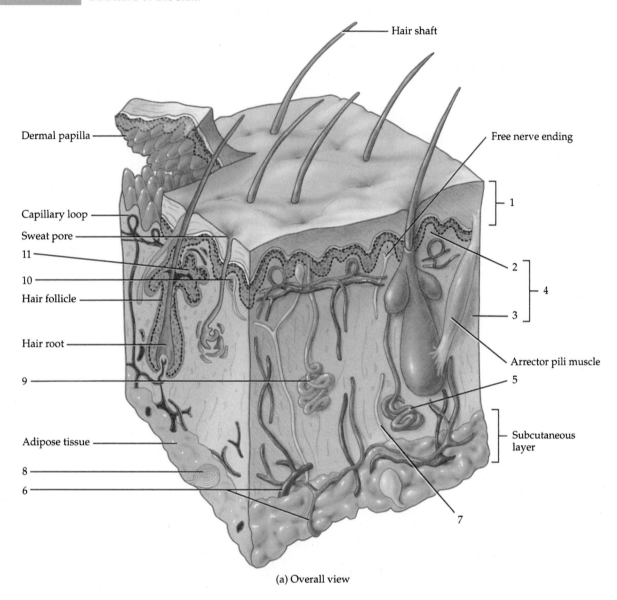

Hair shaft

Dermal papilla

Free nerve ending

1

Capillary loop

Sweat pore

11

2

10

4

Hair follicle

3

Hair root

Arrector pili muscle

9

5

Adipose tissue

Subcutaneous
layer

8

6

7

(a) Overall view

___ Apocrine sweat gland
___ Blood vessels
___ Corpuscle of touch
      (Meissner corpuscle)
___ Dermis
___ Eccrine sweat gland
___ Epidermis
___ Lamellated (pacinian) corpuscle
___ Papillary region
___ Reticular region
___ Sebaceous (oil) gland
___ Sensory nerve
___ Stratum basale
___ Stratum corneum
___ Stratum granulosum
___ Stratum lucidum
___ Stratum spinosum

FIGURE 5.1 Structure of the skin. (Continued)

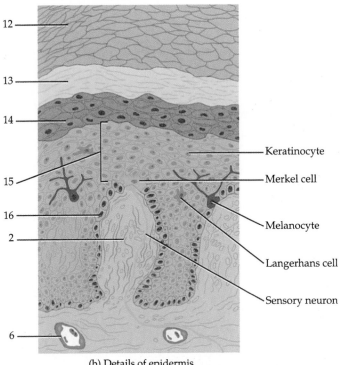

(b) Details of epidermis

FIGURE 5.2 Scanning electron micrograph of the skin and several hairs at a magnification of 260×. (Reproduced by permission from R. G. Kessel and R. H. Kardon, *Tissues and Organs: A Text Atlas of Scanning Electron Microscopy*, W. H. Freeman, 1979.)

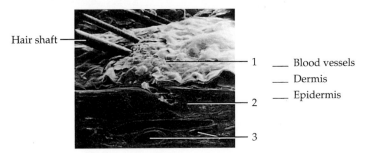

____ Blood vessels
____ Dermis
____ Epidermis

portion deep to the shaft that penetrates deep into the dermis and even into the subcutaneous layer.

The shaft and root consists of the following parts:

1. *Medulla.* Inner region composed of two or three rows of irregularly shaped cells containing pigment and air spaces.

2. *Cortex.* Middle layer; consists of elongated cells that contain pigment granules in dark hair but mostly air in gray or white hair.

3. *Cuticle of the hair.* Outermost layer; consists of a single layer of cells and they are the most heavily keratinized of hair cells. Cuticle cells on the shaft are arranged like shingles on the side of a house.

The root of a hair also contains a medulla, cortex, and cuticle of the hair along with the following associated parts:

1. *Hair follicle.* Structure surrounding the root that consists of an external root sheath and an internal root sheath. These epidermally derived layers are surrounded by a dermal layer of connective tissue.

2. ***External root sheath.*** Downward continuation of the epidermis.

3. ***Internal root sheath.*** Cellular tubular sheath of epithelium that separates the hair from the external root sheath.

4. ***Bulb.*** Enlarged, onion-shaped structure at the base of the hair follicle.

5. ***Papilla of the hair.*** Nipple-shaped indentation into the bulb; contains areolar connective tissue and many blood vessels to nourish the hair.

6. ***Matrix.*** Germinal layer of cells at the base of the bulb derived from the stratum basale and that divides to produce new hair.

7. ***Arrector*** (*arrect* = to raise) ***pili muscle.*** Bundle of smooth muscle extending from the superficial dermis of the skin to the side of the hair follicle; its contraction, under the influ-

ence of fright or cold, causes the hair to move into a vertical position, producing "goose bumps" or "gooseflesh."

8. ***Hair root plexus.*** Nerve endings around each hair follicle that are sensitive to touch and respond when the hair shaft is moved.

Obtain a prepared slide of a transverse (cross) section and a longitudinal section of a hair root and identify as many parts as you can. Using your textbook as a reference, label Figure 5.3. Also label the parts of a hair shown in Figure 5.4.

C. Glands

Several types of exocrine glands are associated with the skin: sebaceous (oil) glands, sudoriferous (sweat) glands, and ceruminous glands. **Seba-**

FIGURE 5.3   **Hair root.**

___ Bulb
___ Cortex
___ Cuticle of hair
___ External root sheath
___ Internal root sheath
___ Matrix
___ Medulla
___ Papilla of hair

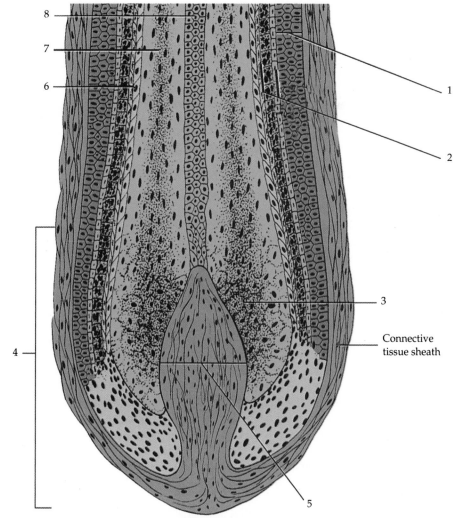

Connective tissue sheath

Longitudinal section

FIGURE 5.4 **Parts of a hair and associated structures.**

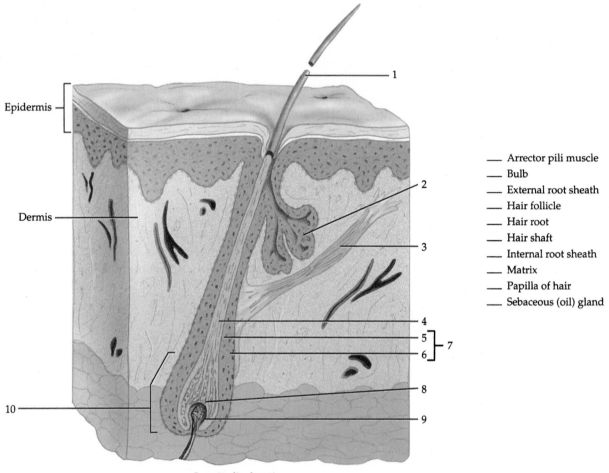

Epidermis

Dermis

— Arrector pili muscle
— Bulb
— External root sheath
— Hair follicle
— Hair root
— Hair shaft
— Internal root sheath
— Matrix
— Papilla of hair
— Sebaceous (oil) gland

1
2
3
4
5
6
7
8
9
10

Longitudinal section

ceous (se-BĀ-shus; *sebace* = grease) or **oil glands,** with few exceptions, are connected to hair folli-cles (see Figures 5.1 and 5.4). They secrete an oily substance called **sebum** (SĒ-bum), a mixture of triglycerides, cholesterol, proteins, and inorganic salts. Sebaceous glands are absent in the palms and soles but are numerous in the skin of the face, neck, upper chest, and breasts. Sebum helps pre-vent hair from drying and forms a protective film over the skin that prevents excessive evaporation and keeps the skin soft and pliable. Sebum also inhibits the growth of certain bacteria.

Sudoriferous (soo'-dor-IF-er-us; *sudori* = sweat; *ferous* = to bear) or **sweat glands** are sepa-rated into two principal types on the basis of struc-ture, location, and secretion. **Apocrine sweat glands** are simple, coiled tubular glands found primarily in the skin of the axilla (armpit), groin, areolae (pigmented areas around the nipples) of the breasts, and bearded regions of the face in adult males. (Recall from Exercise 4 that apocrine glands are probably merocrine glands. For now,

we will still use the term apocrine.) Their secretory portion is located mostly in the subcutaneous layer; the excretory duct opens into hair follicles. Apocrine sweat glands begin to function at pu-berty and produce a more viscous secretion than the other type of sweat gland. Apocrine sweat glands are stimulated during emotional stress and sexual excitement; their secretion is commonly called "cold sweat." **Eccrine sweat glands** are sim-ple, coiled tubular glands found throughout the skin, except for the margins of the lips, nail beds of the fingers and toes, glans penis, glans clitoris, and eardrums. They are most numerous in the skin of the forehead, palms, and soles. The secretory por-tion of these glands is in the deep dermis (some-times in the upper subcutaneous layer); the excre-tory duct projects upward and terminates at a pore at the surface of the epidermis (see Figure 5.1). Eccrine sweat glands function throughout life and produce a more watery secretion than the apocrine glands. Sudoriferous glands produce **perspiration,** a mixture of water, ions, urea, uric acid, amino acids, ammonia, glucose, and lactic acid. The evaporation of perspiration helps to maintain normal body temperature.

Ceruminous (se-ROO-mi-nus; *cer* = wax) **glands** are modified sudoriferous (sweat) glands in the external auditory canal (ear canal). The combined secretion of ceruminous and sudoriferous glands is called **cerumen** (earwax). Cerumen, together with hairs in the external auditory canal, provides a sticky barrier that prevents the entrance of foreign bodies.

D. Nails

Nails are plates of tightly packed, hard, keratinized epidermal cells that form a clear, solid covering over the dorsal surfaces of the terminal portions of the digits. Each nail consists of the following parts:

1. *Nail body.* Portion that is visible.
2. *Free edge.* Part that may project beyond the distal end of the digit.
3. *Nail root.* Portion hidden in nail groove.
4. *Lunula* (LOO-nū-la; *lunula* = little moon). The whitish, cresent-shaped area at proximal end of nail body.
5. *Eponychium* (ep'-ō-NIK-ē-um) or *cuticle.* A narrow band of epidermis that extends from and adheres to the margin (lateral border) of the nail wall.
6. *Hyponychium.* Thickened area of stratum corneum beneath the free edge of the nail.
7. *Nail matrix.* Epithelium deep to the nail root; division of the cells brings about growth of nails.

Using your textbook as a reference, label the parts of a nail shown in Figure 5.5. Also, identify the parts that are visible on your own nails.

FIGURE 5.5 Structure of nails.

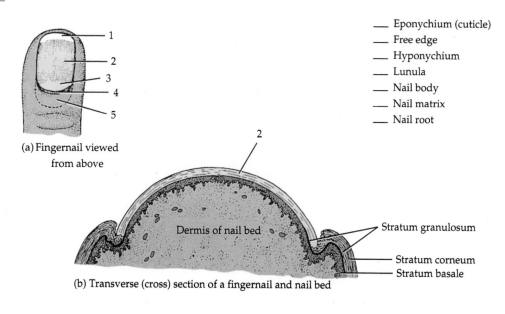

___ Eponychium (cuticle)
___ Free edge
___ Hyponychium
___ Lunula
___ Nail body
___ Nail matrix
___ Nail root

(a) Fingernail viewed from above

Dermis of nail bed

Stratum granulosum

Stratum corneum
Stratum basale

(b) Transverse (cross) section of a fingernail and nail bed

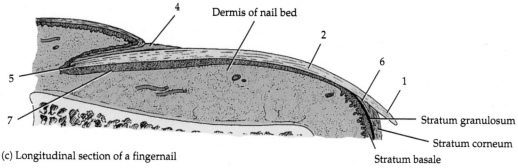

Dermis of nail bed

Stratum granulosum

Stratum corneum

Stratum basale

(c) Longitudinal section of a fingernail

E. Homeostasis of Body Temperature

One of the best examples of homeostasis in humans is the regulation of body temperature by the skin. As warm-blooded animals, we are able to maintain a remarkably constant body temperature of 37°C (98.6°F) even though the environmental temperature varies greatly.

Suppose you are in an environment where the temperature is 38°C (101°F). Heat (the stimulus) continually flows from the environment to your body, raising body temperature. To counteract these changes in a controlled condition, a sequence of events is set into operation. Temperature-sensitive receptors (nerve endings) in the skin called **thermoreceptors** detect the stimulus and send nerve impulses (input) to your brain (control center). A temperature control region of the brain (called the hypothalamus) then sends nerve impulses (output) to the sudoriferous glands (effectors), which produce perspiration more rapidly. As the sweat evaporates from the surface of your skin, heat is lost and your body temperature decreases (response). This cycle continues until body temperature drops to normal (returns to homeostasis). When environmental temperature is low, sweat glands produce less perspiration.

Your brain also sends output to blood vessels (a second set of effectors), dilating (widening) those in the dermis so that skin blood flow increases. As more warm blood flows through capillaries close to the body surface, more heat can be lost to the environment, which lowers body temperature. Thus heat is lost from the body, and body temperature falls to the normal value to restore homeostasis. In response to low environmental temperature, blood vessels in the dermis constrict, blood flow decreases, and heat is lost by radiation.

This temperature regulation involves a *negative feedback system* because the response (cooling) is opposite to the stimulus (heating) that started the cycle. Also, the thermoreceptors continually monitor body temperature and feed this information back to the brain. The brain, in turn, continues to send impulses to the sweat glands and blood vessels until the temperature returns to 37°C (98.6°F).

Regulating the rate of sweating and changing dermal blood flow are only two mechanisms by which body temperature can be adjusted. Other mechanisms include regulating metabolic rate (a slower metabolic rate reduces heat production) and regulating skeletal muscle contractions (decreased muscle tone results in less heat production).

Label Figure 5.6, the role of the skin in regulating the homeostasis of body temperature.

ANSWER THE LABORATORY REPORT QUESTIONS AT THE END OF THE EXERCISE.

FIGURE 5.6 Role of the skin in regulating the homeostasis of body temperature.

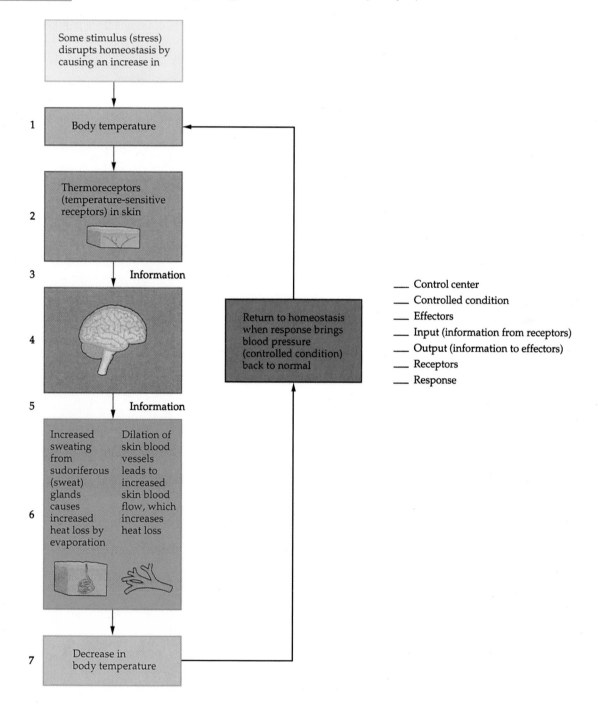

Integumentary System

Name _____   Date _____

Laboratory Section _____   Score/Grade _____

PART 1 ■ Multiple Choice

_____ 1. The waterproofing quality of skin is due to the presence of (a) melanin (b) carotene (c) lamellar granules (d) receptors

_____ 2. Sebaceous glands (a) produce a watery solution called sweat (b) produce an oily substance that prevents excessive water evaporation from the skin (c) are associated with mucous membranes (d) are part of the subcutaneous layer

_____ 3. Which of the following is the proper sequence of layering of the epidermis, going from the free surface toward the underlying tissues? (a) basale, spinosum, granulosum, corneum (b) spinosum, basale, granulosum, corneum (c) corneum, lucidum, granulosum, spinosum, basale (d) corneum, granulosum, lucidum, spinosum

_____ 4. Skin color is *not* determined by the presence or absence of (a) melanin (b) carotene (c) keratin (d) hemoglobin in red blood cells in capillaries in the dermis

_____ 5. Destruction of what part of a single hair would result in its inability to grow? (a) sebaceous gland (b) arrector pili muscle (c) matrix (d) bulb

_____ 6. One would expect to find relatively few, if any, sebaceous glands in the skin of the (a) palms (b) face (c) neck (d) upper chest

_____ 7. Which of the following sequences, from outside to inside, is correct? (a) epidermis, reticular layer, papillary layer, subcutaneous layer (b) epidermis, subcutaneous layer, reticular layer, papillary layer (c) epidermis, reticular layer, subcutaneous layer, papillary layer (d) epidermis, papillary layer, reticular layer, subcutaneous layer

_____ 8. The attached visible portion of a nail is called the (a) nail bed (b) nail root (c) nail fold (d) nail body

_____ 9. Nerve endings sensitive to touch are called (a) corpuscles of touch (Meissner corpuscles) (b) papillae (c) lamellated (pacinian) corpuscles (d) follicles

_____ 10. The cuticle of a nail is referred to as the (a) matrix (b) eponychium (c) hyponychium (d) lunula

_____ 11. One would *not* expect to find sudoriferous glands associated with the (a) forehead (b) axilla (c) palms (d) nail beds

_____ 12. Fingerlike projections of the dermis that contain loops of capillaries and receptors are called (a) dermal papillae (b) nodules (c) polyps (d) pili

_____ 13. Which of the following statements about the function of skin is *not* true? (a) It helps control body temperature. (b) It prevents excessive water loss. (c) It synthesizes several compounds. (d) It absorbs water and salts.

_____ **14.** As it relates to the hair, the downward continuation of the epidermis is called the (a) external root sheath (b) internal root sheath (c) matrix (d) bulb

_____ **15.** Growth in the length of nails is the result of the activity of the (a) eponychium (b) nail matrix (c) hyponychium (d) nail fold

PART 2 ■ **Completion**

16. A group of tissues that performs a definite function is called a(n) _____.

17. The outer, thinner layer of the skin is known as the _____.

18. The skin is attached to underlying structures by the _____.

19. A group of organs that operate together to perform a specialized function is called a(n)

_____.

20. The epidermal layer that is more apparent in the palms and soles is the stratum

_____.

21. The epidermal layer that produces new cells is the stratum _____.

22. The smooth muscle attached to a hair follicle is called the _____ muscle.

23. An inherited inability to produce melanin is called _____.

24. Nerve endings sensitive to deep pressure are referred to as _____ corpuscles.

25. The inner region of a hair shaft and root is the _____.

26. The portion of a hair containing areolar connective tissue and blood vessels is the

_____.

27. Modified sweat glands that line the external auditory canal are called _____

glands.

28. The whitish semilunar area at the proximal end of the nail body is referred to as the

_____.

29. The secretory product of sudoriferous glands is called _____.

30. Melanin is synthesized in cells called _____.

31. In the control of body temperature, the effectors are blood vessels in the dermis and

_____ glands.

Bone Tissue

Objectives

At the completion of this exercise you should understand

A The functions of the skeletal system.

B The parts of a long bone.

C The histological features of bone tissue.

D Bone formation and growth.

E The types of bones and bone surface markings.

Structurally, a bone consists of several different tissues working together: bone or osseous tissue, cartilage, dense connective tissues, epithelium, adipose tissue, and nervous tissue. The entire framework of bones and their cartilages together constitute the **skeletal system.** The microscopic structure of cartilage has been discussed in Exercise 4. In this exercise the gross structure of a typical bone and the histology of **bone (osseous) tissue** will be studied. **Osteology** (os-tē-OL-ō-jē; *osteo* = bone; *logy* = study of) is the study of bone structure and the treatment of bone disorders.

A. Functions of Bone

The skeletal system has the following basic functions:

1. *Support.* It provides the structural framework for the body by supporting the soft tissues and providing points of attachment for the tendons of most skeletal muscles.

2. *Protection.* It protects delicate structures such as the brain, spinal cord, heart, lungs, major blood vessels in the chest, and pelvic viscera.

3. *Assistance in movement.* When skeletal muscles contract, they pull on bones to produce body movements.

4. *Mineral homeostasis.* Bone tissue stores several minerals, especially calcium and phosphorus, which are important in muscle contraction and nerve activity, among other functions. On demand, bone releases minerals into the blood to maintain critical mineral balances (homeostasis) and for distribution to other parts of the body.

5. *Blood cell production.* **Red bone marrow** at the center of certain bones consists of developing blood cells, adipocytes, fibroblasts, and macrophages. It is responsible for producing red blood cells, white blood cells, and platelets, a process called **hemopoiesis** (hē'-mō-poy-Ē-sis).

6. *Triglyceride storage.* **Yellow bone marrow** consists of mostly adipose cells, which store triglycerides. The stored triglycerides are a potential chemical energy reserve.

B. Structure of a Long Bone

Examine the external features of a fresh long bone and locate the following structures:

1. *Diaphysis* (dī-AF-i-sis; *dia* = through; *physis* = growth). Bone's shaft or body of a bone; the long cylindrical main portion of a bone.

2. *Epiphysis* (e-PIF-i-sis; *epi* = above). The distal or proximal ends of the bone.

3. ***Metaphysis*** (me-TAF-i-sis; *meta* = between). In mature bone, the region where the diaphysis joins the epiphysis; in growing bone, the region that includes the **epiphyseal plate,** a layer of hyaline cartilage that allows the diaphysis to grow in length.

4. ***Articular cartilage.*** Thin layer of hyaline cartilage covering the epiphysis where the bone forms an articulation (joint) with another bone. It reduces friction and absorbs shock at freely movable joints.

5. ***Periosteum*** (per′-ē-OS-tē-um; *peri* = around; *osteo* = bone). A tough sheath of dense irregular connective tissue that surrounds the bone surface wherever it is not covered by articular cartilage. The periosteum functions in bone growth, assists in fracture repair, protects the bone, helps nourish bone tissue, and serves as an attachment site for tendons and ligaments.

6. ***Medullary*** (MED-ū-lar-ē; *medulla* = marrow, pith) or ***marrow cavity.*** Space within the diaphysis that contains fatty yellow bone marrow in adults.

7. ***Endosteum*** (end-OS-tē-um; *endo* = within). A thin membrane that lines the medullary cavity. It contains a single layer of bone-forming cells and a small amount of connective tissue.

Label the parts of a long bone indicated in Figure 6.1.

FIGURE 6.1   **Parts of a long bone.**

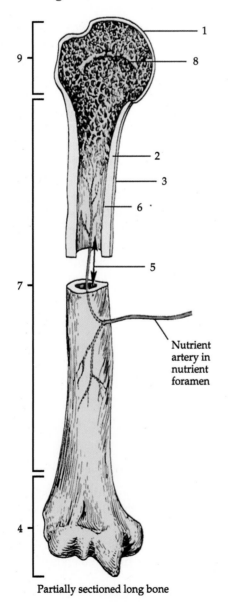

____ Articular cartilage
____ Compact bone tissue
____ Diaphysis
____ Distal epiphysis
____ Endosteum
____ Medullary (marrow) cavity
____ Periosteum
____ Proximal epiphysis
____ Spongy bone tissue

Nutrient artery in nutrient foramen

Partially sectioned long bone

C. Histology of Bone

Bone tissue, like other connective tissues, contains an abundant matrix of intercellular materials that surrounds widely separated cells. Unlike other connective tissues, the matrix of bone is very hard. This hardness results from the presence of inorganic mineral salts, mainly **hydroxyapatite** (calcium phosphate and calcium carbonate). Crystallized mineral salts compose about 50% of the weight of bone. Despite its hardness, bone is also flexible, a characteristic that enables it to resist various forces. The flexibility of bone depends upon its collagen fibers. Collagen fibers compose about 25% of the weight of bone. The remaining 25% of the bone matrix is water. The cells in bone tissue include **osteogenic cells** (os-tē-ō-JEN-ik; *genic* = producing), unspecialized stem cells derived from mesenchyme that differentiate into **osteoblasts** (OS-tē-ō-blasts'; *-blasts* = buds or sprouts), bone-building cells which synthesize and secrete collagen fibers and other organic components needed to build the matrix of bone tissue; **osteocytes** (OS-tē-ō-sīts'; *-cytes* = cell), mature bone cells that maintain the daily metabolism of bone tissue; and **osteoclasts** (OS-tē-ō-clasts'; *-clast* = break), cells that digest the protein and mineral components of bone. This digestion is part of the normal development, growth, maintenance, and repair of bone.

Depending on the size and distribution of spaces between its hard components, the regions of a bone may be categorized as compact or spongy. **Compact bone tissue** contains few spaces and forms the external layer of all bones of the body and the bulk of the diaphyses of long bones. Compact bone tissue provides protection and support and resists the stresses produced by weight and movement. Compact bone tissue is arranged in microscopic units called **osteons (haversian systems).**

Spongy bone tissue, which contains many large spaces, composes most of the bone tissue of short, flat, and irregularly shaped bones and most of the epiphyses of long bones. The spaces within the spongy bone tissue of some bones contain red bone marrow. Spongy bone tissue does not contain osteons. It consists of an irregular latticework of thin columns of bone called **trabeculae** (tra-BEK-ū-lē). See Figure 6.2. Trabeculae are the microscopic units of spongy bone tissue.

Label the spongy and compact bone tissue in Figure 6.1.

Obtain a prepared slide of compact bone tissue in which several osteons (haversian systems) are shown in transverse (cross) section. Observe under high power. Look for the following structures:

1. *Central (haversian) canal.* Circular canal in the center of an osteon (haversian system) that runs longitudinally through the bone; the canal contains blood vessels, lymphatic vessels, and nerves.

2. *Concentric lamellae.* Rings of hard, calcified matrix.

3. *Lacunae* (la-KOO-nē; *lacuna* = little lake). Small spaces between lamellae which contain osteocytes.

4. *Canaliculi* (kan'-a-LIK-ū-lē; *canaliculi* = small channels). Minute channels that radiate in all directions from the lacunae and interconnect with each other and the central canals; contain slender processes of osteocytes; canaliculi provide routes for nutrients and oxygen to reach osteocytes and for wastes to diffuse away.

5. *Osteocyte.* Mature bone cell located within a lacuna.

6. *Osteon (haversian system).* Microscopic structural unit of compact bone made up of a central (haversian) canal plus its surrounding lamellae, lacunae, canaliculi, and osteocytes.

Label the parts of the osteon (haversian system) shown in Figure 6.2.

Now obtain a prepared slide of a longitudinal section of compact bone tissue and examine under high power. Locate the following:

1. *Perforating (Volkmann's) canals.* Canals that extend from the periosteum and contain blood vessels, lymphatic vessels, and nerves; they extend into the central (haversian) canals, medullary cavity, and periosteum.

2. *Endosteum*

3. *Medullary (marrow) cavity*

4. *Concentric lamellae*

5. *Lacunae*

6. *Canaliculi*

7. *Osteocytes*

Label the parts indicated in the microscopic view of bone in Figure 6.2.

FIGURE 6.2 Histology of bone.

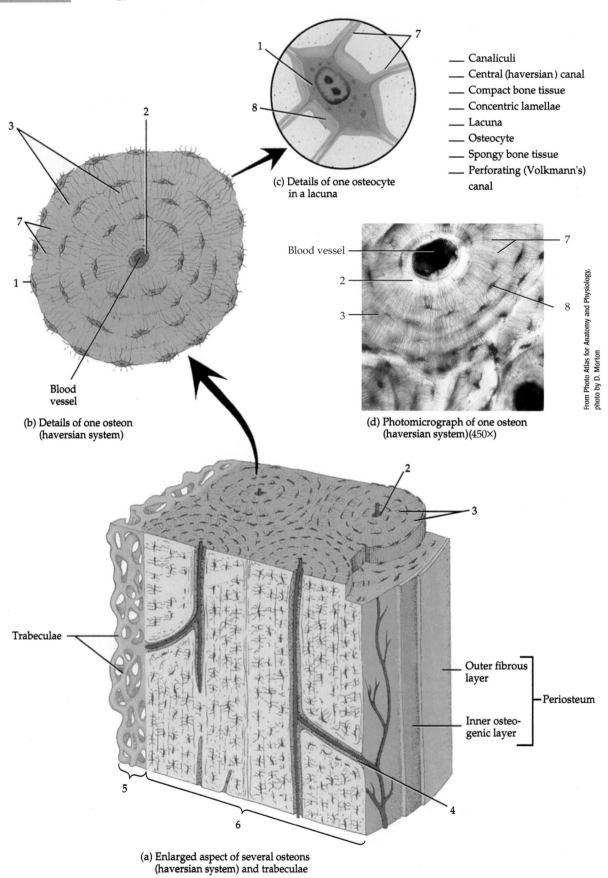

___ Canaliculi
___ Central (haversian) canal
___ Compact bone tissue
___ Concentric lamellae
___ Lacuna
___ Osteocyte
___ Spongy bone tissue
___ Perforating (Volkmann's) canal

(c) Details of one osteocyte in a lacuna

Blood vessel

(b) Details of one osteon (haversian system)

Blood vessel

(d) Photomicrograph of one osteon (haversian system)(450×)

From Photo Atlas for Anatomy and Physiology, photo by D. Morton

Trabeculae

Outer fibrous layer

Inner osteogenic layer

Periosteum

(a) Enlarged aspect of several osteons (haversian system) and trabeculae

D. Chemistry of Bone

PROCEDURE

1. Obtain a bone that has been baked. How does this compare to an untreated one? _____

   _____

   _____

   _____

   What substances does baking remove from the bone (inorganic or organic)? _____

   _____

   _____

   _____

2. Now obtain a bone that has already been soaked in nitric acid by your instructor. How does this bone compare to an untreated one?

   _____

   _____

   _____

   What substances does nitric acid treatment remove from the bone (inorganic or organic)?

   _____

   _____

   _____

E. Bone Formation: Ossification

The process by which bone forms is called **ossification** (os-i-fi-KĀ-shun; *fication* = making). The "skeleton" of a human embryo is composed of loose mesenchymal cells, which are shaped like bones. They provide the supporting structures for ossification. Ossification begins during the sixth week of embryonic life and follows one of two patterns.

1. *Intramembranous* (in'-tra-MEM-bra-nus; *intra* = within; *membran* = membrane) *ossification* refers to the formation of bone directly within mesenchyme arranged in sheetlike layers that resemble membranes. Such bones form *directly* from mesenchyme without first going through a cartilage stage. The flat bones of the skull and mandible (lower jawbone) form by this process. Also, the fontanels ("soft spots") of the fetal skull are replaced by bone via intramembranous ossification.

2. *Endochondral* (en'-dō-KON-dral; *endo* = within; *chondral* = cartilage) *ossification* refers to the formation of bone within hyaline cartilage. In this process, mesenchyme is transformed into chondroblasts which produce a hyaline cartilage matrix that is gradually replaced by bone. Most bones of the body form by this process.

These two kinds of ossification do *not* lead to differences in the gross structure of mature bones. They are simply different methods of bone formation. Both mechanisms involve the replacement of a preexisting connective tissue with bone.

The first stage in the development of bone is the migration of embryonic mesenchymal cells into the area where bone formation is about to begin. These cells increase in number and size and become osteogenic cells. In some skeletal structures where capillaries are lacking, they become chondroblasts; in others where capillaries are present, they become osteoblasts. The **chondroblasts** are responsible for cartilage formation. Osteoblasts form bone tissue by intramembranous or endochondral ossification.

F. Bone Growth

During childhood, bones throughout the body grow in thickness by appositional growth (deposition of matrix on the surface), and long bones lengthen by interstitial growth (the addition of bone material on the diaphyseal side of the epiphyseal plate). Growth in length of bones normally ceases by age 21, although bones may continue to thicken.

1. Growth in Length

To understand how a bone grows in length, you will need to know some of the details of the structure of the epiphyseal plate.

The **epiphyseal** (ep'-i-FIZ-ē-al; *epiphyein* = to grow upon) **plate** is a layer of hyaline cartilage in the metaphysis of growing bone that consists of four zones. The **zone of resting cartilage** is nearest the epiphysis and consists of small, scattered chondrocytes. The cells do not function in bone growth (thus the term "resting"); they anchor the epiphyseal plate to the epiphysis of the bone.

The **zone of proliferating cartilage** consists of slightly larger chondrocytes arranged like stacks of coins. Chondrocytes divide to replace those that die at the diaphyseal surface of the epiphyseal plate.

The zone of **hypertrophic** (hī-per-TRŌF-ik) **cartilage** consists of even larger chondrocytes that are also arranged in columns. The lengthening of the diaphysis is the result of cell divisions in the zone of proliferating cartilage and maturation of the cells in the zone of hypertrophic cartilage.

The **zone of calcified cartilage** is only a few cells thick and consists mostly of dead chondrocytes because the matrix around them has calcified. Osteoclasts dissolve the calcified cartilage, and osteoblasts and capillaries from the diaphysis invade the area. The osteoblasts lay down bone matrix, replacing the calcified cartilage. As a result, the diaphyseal border of the epiphyseal plate is firmly cemented to the bone of the diaphysis.

Label the various zones in the epiphyseal plate in Figure 6.3.

The activity of the epiphyseal plate is the only mechanism by which the diaphysis can increase in length. Unlike cartilage, which can grow by both interstitial and appositional growth, bone can grow in diameter only by appositional growth. Between the ages of 18 and 21, the epi-

physeal plates close; that is, the epiphyseal cartilage cells stop dividing and bone replaces the cartilage. The newly formed bony structure is called the **epiphyseal line.** With the appearance of the epiphyseal line, bone stops growing in length. In general, lengthwise growth in bones in females is completed before that in males.

2. Growth in Thickness

Enlargement of bone thickness or diameter is by appositional growth and occurs as follows. Osteoblasts from the periosteum add new bone tissue to the outer surface. Initially, diaphyseal and epiphyseal ossification produce only spongy bone. Later, the outer region of spongy bone is reorganized into compact bone. At the same time, the bone lining the medullary cavity is destroyed by osteoclasts in the endosteum so that the cavity increases in diameter.

G. Fractures

A **fracture** is any break in a bone. Usually, the fractured ends of a bone can be reduced (aligned to their normal positions) by manipulation without surgery. This procedure of setting a fracture is called **closed reduction.** In other cases, the fracture must be exposed by surgery before the break is rejoined. This procedure is known as **open reduction.**

Using your textbook as a reference, define the fractures listed in Table 6.1 and label the fractures indicated in Figure 6.4.

H. Types of Bones

The 206 named bones of the body may be classified into five principal types based upon shape:

1. *Long.* Have greater length than width, consist of a shaft and a variable number of extremities (ends). Long bones consist mostly of compact bone tissue in their diaphyses but also contain considerable amounts of spongy bone tissue in their epiphyses. They are slightly curved for strength. Examples: thigh (femur), leg (tibia and fibula), arm (humerus), forearm (radius and ulna), and finger and toe (phalanges).

2. *Short.* Somewhat cube-shaped, nearly equal in length and width. They consist of spongy bone tissue except at the surface, where there

FIGURE 6.3 **Histology of the epiphyseal plate.**

Epiphyseal side

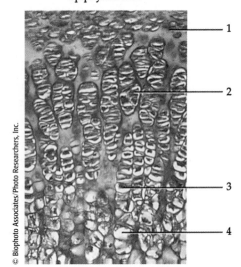

© Biophoto Associates/Photo Researchers, Inc.

Diaphyseal side

Photomicrograph of epiphyseal plate (100x)

____ Zone of calcified cartilage

____ Zone of hypertrophic cartilage

____ Zone of proliferating cartilage

____ Zone of resting cartilage

is a thin layer of compact bone tissue. Examples: wrist bones and ankle bones.

3. *Flat.* Generally thin and composed of two nearly parallel plates of compact bone tissue enclosing a layer of spongy bone tissue. Examples: sternum (breastbone), ribs, and scapulae (shoulder blades).

4. *Irregular.* Very complex shapes; cannot be grouped into any of the three categories just described. Examples: vertebrae, some facial bones, hip bones, and calcaneus.

5. *Sesamoid. Sesamoid* means "shaped like a sesame seed." Small bones that develop in tendons; variable in number; the only constant sesamoid bones are the paired patellae (kneecaps).

An additional type of bone not considered in this structural classification, includes **sutural** (SOO-chur-al) bones, small bones in sutures between certain cranial bones. They are variable in number.

Examine the disarticulated skeleton, Beauchene (disarticulated) skull, and articulated skeleton and find several examples of long, short, flat, and irregular bones. List examples of each type you find.

1. *Long* _____

_____

_____

_____

2. *Short* _____

_____

_____

_____

3. *Flat* _____

_____

_____

_____

4. *Irregular* _____

_____

_____

_____

TABLE 6.1	Summary of selected fractures

Type of fracture	Definition
Closed (simple)	
Open (compound)	
Comminuted (KOM-i-noo'-ted)	
Greenstick	
Impacted	
Stress	
Pott's	
Colles' (KOL-ez)	

FIGURE 6.4 **Types of fractures.**

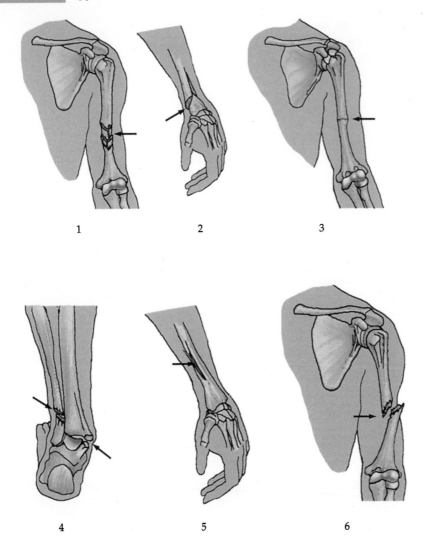

1   2   3

___ Colles' fracture
___ Comminuted fracture
___ Greenstick fracture
___ Impacted fracture
___ Open fracture
___ Pott's fracture

4   5   6

I. Bone Surface Markings

The surfaces of bones contain various structural features that have specific functions. These features are called **bone surface markings** and are listed in Table 6.2. Knowledge of the bone surface markings will be very useful when you learn the bones of the body in Exercise 7.

Next to each marking listed in Table 6.2, write its definition, using your textbook as a reference.

ANSWER THE LABORATORY REPORT QUESTIONS AT THE END OF THE EXERCISE.

TABLE 6.2	Bone surface markings

Marking	Description
Depressions and openings	
Fissure (FISH-ur)	
Foramen (fō-RĀ-men; *foramen* = hole)	
Fossa (*fossa* = basinlike depression)	
Meatus (mē-Ā-tus; *meatus* = canal)	
Sulcus (*sulcus* = ditchlike groove)	
Processes that form joints	
Condyle (KON-dīl; *condylus* = knuckle)	
Facet	
Head	
Processes that form attachment points for connective tissue	
Crest	
Epicondyle (*epi* = above)	
Line	
Spinous process (spine)	
Trochanter (trō-KAN-ter)	
Tubercle (TOO-ber-kul; *tube* = knob)	
Tuberosity	

| EXERCISE 6 | Bone Tissue |

Name _____   Date _____

Laboratory Section _____   Score/Grade _____

PART 1 ■ Completion

1. Small bones located in sutures between certain cranial bones are referred to as

   _____ bones.

2. The technical name for a mature bone cell is a(n) _____.

3. Canals that extend obliquely inward or horizontally from the bone surface and contain blood vessels and lymphatic vessels are called _____ canals.

4. The end, or extremity, of a bone is referred to as the _____.

5. Cube-shaped bones that contain more spongy bone tissue than compact bone tissue are known as

   _____ bones.

6. The cavity within the shaft of a bone that contains fatty yellow bone marrow in the adult is the

   _____ cavity.

7. The thin layer of hyaline cartilage covering the end of a bone where joints are formed is called

   _____ cartilage.

8. Minute canals that connect lacunae are called _____.

9. The membrane around the surface of a bone, except for the areas covered by cartilage, is the

   _____.

10. The shaft of a bone is also referred to as the _____.

11. The _____ are rings of hard calcified matrix.

12. The membrane that lines the medullary cavity and contains a single layer of bone-forming cells is

    the _____.

13. The technical name for bone tissue is _____ tissue.

14. In a mature bone, the region where the shaft joins the extremity is called the

    _____.

15. The microscopic structural unit of compact bone tissue is called a(n) _____.

16. The hardness of bone is primarily due to the mineral salt _____.

17. The process by which bone is formed is called _____.

18. The zone of _____ is closest to the diaphysis of the bone.

19. Growth in thickness of bones occurs by _____ growth.

20. Cells that digest the protein and mineral components of bone are called _____.

21. Most bones of the body form by which type of ossification? _____

22. The lengthening of the diaphysis is the result of cell divisions in the zone of

    _____.

23. Any break in a bone is called a _____.

24. The term _____ refers to the irregular latticework of thin columns of spongy bone.

25. On the basis of shape, vertebrae are classified as _____ bones.

Bones

Objectives

At the completion of this exercise you should understand

A The bones of the human skeleton.

B The major surface markings of the skeletal system.

The 206 named bones of the adult skeleton are grouped into two principal divisions: the 80 bones of the axial skeleton and the 126 bones of the appendicular skeleton. The **axial skeleton** consists of bones that lie around the longitudinal axis of the body. The longitudinal axis is a vertical line that runs through the body's center of gravity, through the head, and down to the space between the feet. The **appendicular skeleton** consists of the bones of the upper and lower limbs (extremities) and the girdles that connect the limbs to the axial skeleton.

In this exercise you will study the names and locations of bones and their markings by examining various regions of the adult skeleton:

Region	Number of bones
Axial skeleton	
Skull	
Cranium	8
Face	14
Hyoid (above the larynx)	1
Auditory ossicles, 3 in each ear	6
Vertebral column	26
Thorax	
Sternum	1
Ribs	24
	Subtotal = 80

Region	Number of bones
Appendicular skeleton	
Pectoral (shoulder) girdles	
Clavicle	2
Scapula	2

Region	Number of bones
Appendicular skeleton *(continued)*	
Upper limbs (extremities)	
Humerus	2
Ulna	2
Radius	2
Carpals	16
Metacarpals	10
Phalanges	28
Pelvic (hip) girdle	
Hip (pelvic or coxal) bone	2
Lower limbs (extremities)	
Femur	2
Fibula	2
Tibia	2
Patella	2
Tarsals	14
Metatarsals	10
Phalanges	28
	Subtotal = 126
	Total = 206

A. Bones of Adult Skull

The **skull,** which contains 22 bones, is composed of two sets of bones—cranial and facial. The 8 **cranial** (*crani* = brain case) **bones** form the cranial cavity and enclose and protect the brain. The cranial bones are 1 **frontal,** 2 **parietals** (pa-RĪ-e-tals), 2 **temporals,** 1 **occipital** (ok-SIP-i-tal), 1 **sphenoid** (SFĒ-noyd), and 1 **ethmoid.** The 14 **facial bones** form the face

FIGURE 7.1 Skull.

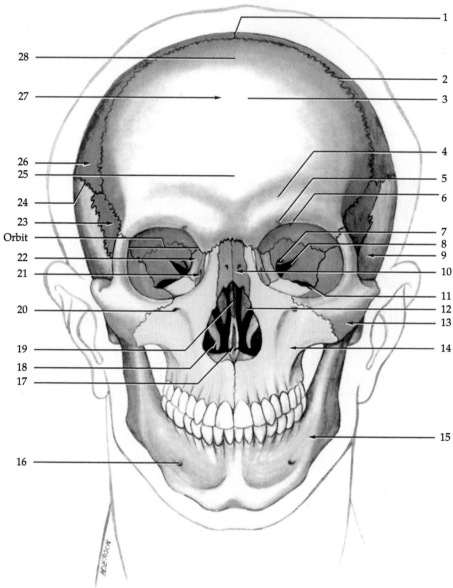

(a) Anterior view

1. Sagittal suture
2. Coronal suture
3. Frontal squama
4. Superciliary arch
5. Supraorbital foramen
6. Supraorbital margin
7. Optic foramen
8. Superior orbital fissure
9. Temporal bone

10. Nasal bone
11. Inferior orbital fissure
12. Middle nasal concha
13. Zygomatic bone
14. Maxilla
15. Mandible
16. Mental foramen
17. Vomer
18. Inferior nasal concha

19. Perpendicular plate
20. Infraorbital foramen
21. Lacrimal bone
22. Ethmoid bone
23. Sphenoid bone
24. Squamous suture
25. Glabella
26. Parietal bone
27. Frontal bone
28. Frontal eminence

FIGURE 7.1   **Skull. (Continued)**

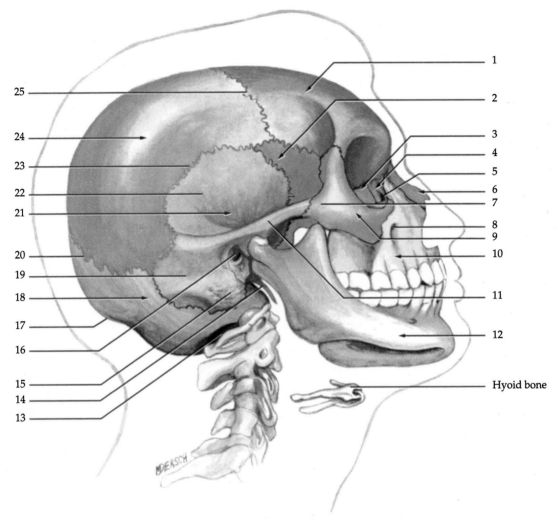

(b) Right lateral view

1. Frontal bone	11. Zygomatic process	18. Occipital bone
2. Sphenoid bone	12. Mandible	19. Mastoid portion
3. Ethmoid bone	13. Foramen magnum	20. Lambdoid suture
4. Lacrimal bone	14. Styloid process	21. Temporal bone
5. Lacrimal fossa	15. Mastoid process	22. Temporal squama
6. Nasal bone	16. External auditory	23. Squamous suture
7. Temporal process	meatus	24. Parietal bone
8. Infraorbital foramen	17. External occipital	25. Coronal suture
9. Zygomatic bone	protuberance	
10. Maxilla		

and include 2 **nasals,** 2 **maxillae** (mak-SIL-ē), 2 **zygomatics,** 1 **mandible,** 2 **lacrimals** (LAK-ri-mals), 2 **palatines** (PAL-a-tīns), 2 **inferior conchae** (KONG-kē), or **turbinates,** and 1 **vomer.** These bones are indicated by *arrows* in Figure 7.1. Using the illustrations in Figure 7.1 for reference, locate the cranial

and facial bones on both a Beauchene (disarticulated) and an articulated skull.

Obtain a Beauchene skull and an articulated skull and observe them in anterior view. Using Figure 7.1a for reference, locate the parts indicated in the figure.

Turn the Beauchene skull and articulated skull so that you are looking at the right side. Using Figure 7.1b for reference, locate the parts indicated in the figure. Note also the **hyoid bone** below the mandible. This is not a bone of the skull; it is noted here because of its proximity to the skull.

If a skull in median section is available, use Figure 7.1c for reference and locate the parts indicated in the figure.

Take an articulated skull and turn it upside down so that you are looking at the inferior surface. Using Figure 7.1d on page 100 for reference, locate the parts indicated in the figure.

Obtain an articulated skull with a removable crown. Using Figure 7.1e on page 101 for reference, locate the parts indicated in the figure.

Examine the right orbit of an articulated skull. Using Figure 7.2 on page 102 for reference, locate the parts indicated in the figure.

Obtain a mandible and, using Figure 7.3 on page 102 for reference, identify the parts indicated in the figure.

Before you move on, refer to Table 7.1, "Summary of foramina of the skull," on page 103. Complete the table by indicating the structures that pass through the foramina listed.

B. Sutures of Skull

A *suture* (SOO-chur; *sutura* = seam) is an immovable joint in an adult found only between skull bones. Sutures hold skull bones together. The four prominent sutures are the **coronal** (*coron* = crown), **sagittal** (SAJ-i-tal; *sagitt* = arrow), **lambdoid** (LAM-doyd), and **squamous** (SKWĀ-mos; *squama* = flat). Using Figures 7.1 and 7.4 on page 104 for reference, locate these sutures on an articulated skull.

C. Fontanels of Skull

At birth, the skull bones are separated by mesenchyme-filled spaces called **fontanels** (fon′-ta-NELZ; *fontanelle* = little fountain). Commonly called "soft spots," fontanels are areas of unossified mesenchyme. Eventually, they will be replaced with bone by intramembranous ossification and become sutures. They (1) enable the fetal skull to compress as it passes through the birth canal, (2) permit rapid growth of the brain during infancy, (3) gauge the degree of brain development by their state of closure, (4) serve as a landmark (anterior fontanel) for withdrawal of blood

for analysis from the superior sagittal sinus, and (5) aid in determining the position of the fetal head prior to birth. The principal fontanels are the **anterior, posterior, anterolateral,** and **posterolateral** fontanels. Using Figure 7.4 for reference, locate the fontanels on the skull of a newborn infant.

D. Paranasal Sinuses of Skull

The **paranasal** (*para* = beside) **sinuses** are cavities in certain cranial and facial bones located near the nasal cavity. Skull bones enclosing the paranasal sinuses are the frontal, sphenoid, ethmoid, and maxillary. Locate the paranasal sinuses on the Beauchene skull or other demonstration models that may be available. Label the paranasal sinuses shown in Figure 7.5 on page 105.

E. Vertebral Column

The **vertebral column** (**backbone** or **spine**) makes up about two-fifths of the total height of the body and is composed of a series of bones called **vertebrae.** The vertebrae of the adult vertebral column are distributed as follows: 7 **cervical** (SER-vi-kal) (neck), 12 **thoracic** (thō-RAS-ik) (chest), 5 **lumbar** (lower back), 5 **sacral** (fused into one bone, the **sacrum** [SĀ-krum] between the hipbones), and usually 4 **coccygeal** (kok-SIJ-ē-al) (fused into one bone, the **coccyx** [KOK-siks] forming the tail of the column). Locate each of these regions on the articulated skeleton. Label the same regions in Figure 7.6a on page 105, the anterior view.

Examine the vertebral column on the articulated skeleton and identify the cervical, thoracic, lumbar, and sacral (sacrococcygeal) curves. Label the curves in Figure 7.6b on page 105, the right lateral view.

F. Vertebrae

A typical **vertebra** consists of the following parts:

1. *Body.* Thick, disc-shaped anterior portion that is the weight-bearing part of a vertebra.

2. *Vertebral arch.* Posterior extension from the body that surrounds the spinal cord and consists of the following parts:

   a. *Pedicles* (PED-i-kuls; *pediculus* = little feet). Two short, thick processes that project

FIGURE 7.1   Skull. (Continued)

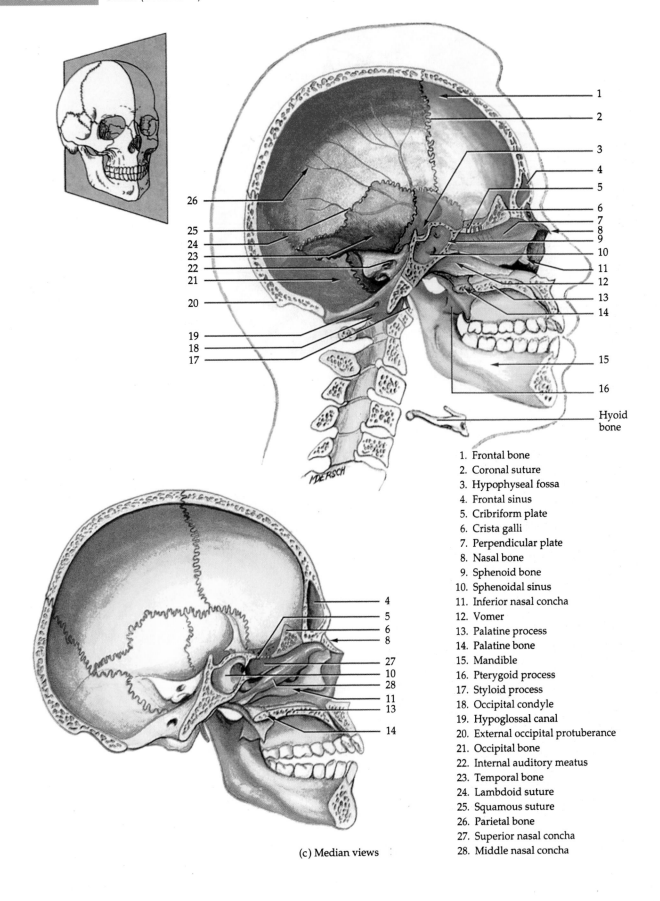

1. Frontal bone
2. Coronal suture
3. Hypophyseal fossa
4. Frontal sinus
5. Cribriform plate
6. Crista galli
7. Perpendicular plate
8. Nasal bone
9. Sphenoid bone
10. Sphenoidal sinus
11. Inferior nasal concha
12. Vomer
13. Palatine process
14. Palatine bone
15. Mandible
16. Pterygoid process
17. Styloid process
18. Occipital condyle
19. Hypoglossal canal
20. External occipital protuberance
21. Occipital bone
22. Internal auditory meatus
23. Temporal bone
24. Lambdoid suture
25. Squamous suture
26. Parietal bone
27. Superior nasal concha
28. Middle nasal concha

Hyoid bone

MDERSCH

(c) Median views

FIGURE 7.1 **Skull. (Continued)**

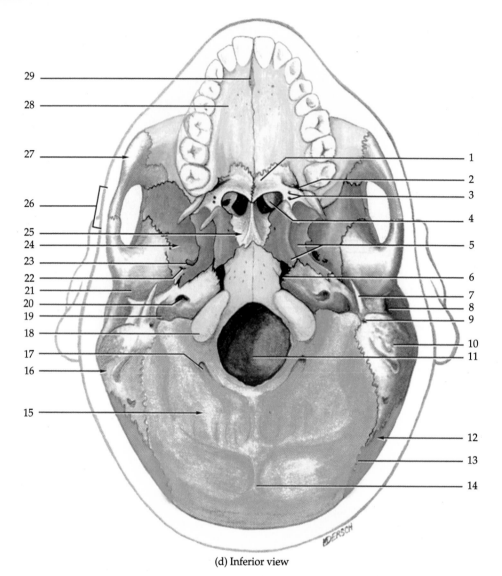

(d) Inferior view

1. Horizontal plate
2. Greater palatine foramen
3. Lesser palatine foramina
4. Middle nasal concha
5. Pterygoid process
6. Foramen lacerum
7. Styloid process
8. External auditory meatus
9. Stylomastoid foramen
10. Mastoid process
11. Foramen magnum
12. Parietal bone
13. Lambdoid suture
14. External occipital protuberance
15. Occipital bone

16. Temporal bone
17. Condylar canal
18. Occipital condyle
19. Jugular foramen
20. Carotid foramen
21. Mandibular fossa
22. Foramen spinosum
23. Foramen ovale
24. Sphenoid bone
25. Vomer
26. Zygomatic arch
27. Zygomatic bone
28. Palatine process
29. Incisive foramen

FIGURE 7.1 **Skull. (Continued)**

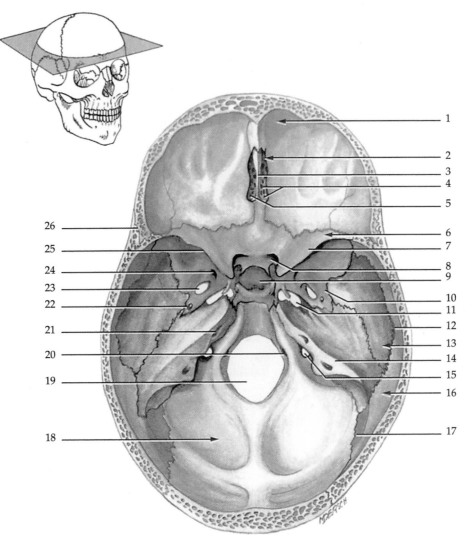

(e) Superior view of floor of cranium

1. Frontal bone
2. Ethmoid bone
3. Crista galli
4. Olfactory foramina
5. Cribriform plate
6. Sphenoid bone
7. Lesser wing
8. Optic foramen
9. Hypophyseal fossa
10. Greater wing
11. Foramen lacerum
12. Squamous suture
13. Temporal bone
14. Petrous portion
15. Jugular foramen
16. Parietal bone
17. Lambdoid suture
18. Occipital bone
19. Foramen magnum
20. Hypoglossal canal
21. Internal auditory meatus
22. Foramen spinosum
23. Foramen ovale
24. Foramen rotundum
25. Superior orbital fissure
26. Coronal suture

FIGURE 7.2 **Right orbit.**

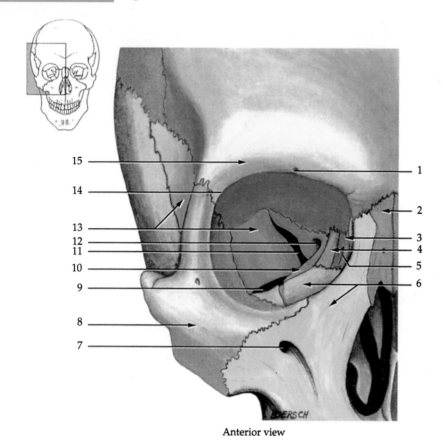

Anterior view

1. Supraorbital foramen
2. Nasal bone
3. Lacrimal bone
4. Ethmoid bone
5. Lacrimal foramen
6. Maxilla
7. Infraorbital foramen
8. Zygomatic bone
9. Inferior orbital fissure
10. Palatine bone
11. Superior orbital fissure
12. Optic foramen
13. Sphenoid bone
14. Supraorbital margin
15. Frontal bone

FIGURE 7.3 **Mandible.**

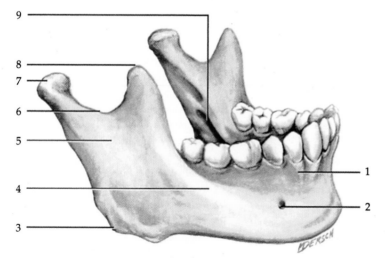

Right lateral view

1. Alveolar process
2. Mental foramen
3. Angle
4. Body
5. Ramus
6. Mandibular notch
7. Condylar process
8. Coronoid process
9. Mandibular foramen

TABLE 7.1	Summary of foramina of the skull

Foramen	Structures passing through
Carotid (relating to carotid artery in neck)	
Hypoglossal (*hypo* = under; *glossus* = tongue)	
Infraorbital (*infra* = below)	
Jugular (*jugul* = the throat)	
Magnum (= large)	
Mandibular (*mand* = to chew)	
Mastoid (= breast-shaped)	
Mental (*ment* = chin)	
Olfactory (*olfact* = to smell)	
Optic (= eye)	
Ovale (= oval)	
Rotundum (= round)	
Stylomastoid (*stylo* = stake or pole)	
Supraorbital (*supra* = above)	

FIGURE 7.4   Fontanels.

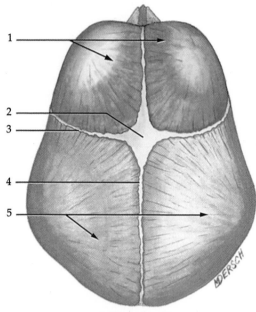

(a) Superior view

1. Frontal bones
2. Anterior fontanel
3. Coronal suture
4. Sagittal suture
5. Parietal bones

(b) Right lateral view

1. Parietal bone
2. Anterior fontanel
3. Coronal suture
4. Frontal bone
5. Anterolateral fontanel
6. Sphenoid bone
7. Temporal bone
8. Squamous suture
9. Posterolateral fontanel
10. Occipital bone
11. Lambdoid suture

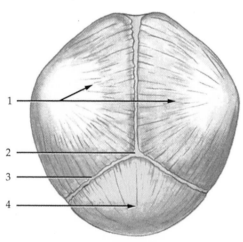

(c) Posterior view

1. Parietal bones
2. Posterior fontanel
3. Lambdoid suture
4. Occipital bone

FIGURE 7.5 Paranasal sinuses.

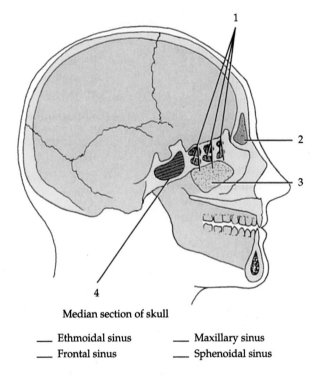

1

2

3

4

Median section of skull

___ Ethmoidal sinus   ___ Maxillary sinus
___ Frontal sinus   ___ Sphenoidal sinus

FIGURE 7.6 Vertebral column.

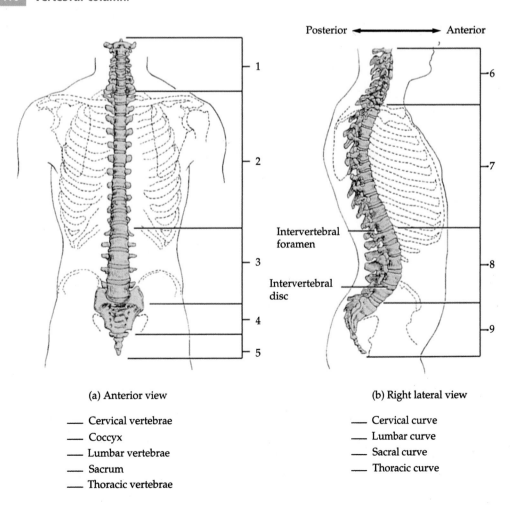

Posterior ←——→ Anterior

1

2

3

4

5

6

7

8

9

Intervertebral foramen

Intervertebral disc

(a) Anterior view

(b) Right lateral view

___ Cervical vertebrae
___ Coccyx
___ Lumbar vertebrae
___ Sacrum
___ Thoracic vertebrae

___ Cervical curve
___ Lumbar curve
___ Sacral curve
___ Thoracic curve

posteriorly; each has superior and inferior indentations **(vertebral notch)** and, when the vertebral notches are stacked on top of one another, they form an opening called an **intervertebral foramen** through which a spinal nerve passes.

b. *Laminae* (LAM-i-nē; *lamina* = thin layer). Flat parts that join to form the posterior wall of the vertebral arch.

c. *Vertebral foramen.* Opening that lies between the vertebral arch and body and contains the spinal cord, adipose tissue, areolar connective tissue, and blood vessels; when all the vertebrae are fitted together, the foramina form a canal, the **vertebral (spinal) cavity.**

3. *Processes.* Seven processes arise from the vertebral arch:

a. Two *transverse processes.* Lateral extensions where the laminae and pedicles join.

b. One *spinous process (spine).* Projects posteriorly from the junction of the laminae.

c. Two *superior articular processes.* Articulate with the inferior articular processes of the vertebra above. Their top surfaces are called **superior articular facets** (*facet* = little face).

d. Two *inferior articular processes.* Articulate with the superior articular processes of the vertebra below. Their bottom surfaces are called **inferior articular facets.**

Obtain a thoracic vertebra and locate each part just described. Now label the vertebra in Figure 7.7a. You should be able to distinguish the general parts on all the different vertebrae that contain them.

Although vertebrae have the same basic design, those of a given region have special distinguishing features. Obtain examples of the following vertebrae and identify their distinguishing features.

1. *Cervical vertebrae (C1-C7)*

a. *Atlas (C1).* First cervical vertebra (Figure 7.7b).

Transverse foramen. Opening in transverse process through which an artery, a vein, and a branch of a spinal nerve pass.

Anterior arch. Anterior wall of vertebral foramen.

Posterior arch. Posterior wall of vertebral foramen.

Lateral mass. Side wall of vertebral foramen.

Label the other indicated parts.

b. *Axis (C2).* Second cervical vertebra (Figure 7.7c).

Dens (*dens* = tooth) or *odontoid process.* Peglike process that projects up through the anterior portion of the vertebral foramina of the atlas.

Label the other indicated parts.

c. *Cervicals 3 through 6.* (Figure 7.7d on page 108).

Bifid spinous process (C3-C6). Cleft in spinous processes of cervical vertebrae 2 through 6.

Label the other indicated parts.

d. *Vertebra prominens (C7).* Seventh cervical vertebra; contains a single long spinous process. It can be seen and felt at the base of the neck.

2. *Thoracic vertebrae (T1-T12).* (Figure 7.7e on page 108).

a. *Facets and demifacets.* For articulation with the tubercle and head of a rib; found on body and transverse processes. Articulations between the thoracic vertebrae and ribs are called **vertebrocostal joints.**

b. *Spinous process.* Usually long, laterally flattened, and directed inferiorly.

Label the other indicated parts.

3. *Lumbar vertebrae (L1-L5).* (Figure 7.7f on page 108).

a. *Spinous processes.* Quadrilateral in shape, are thick and broad, and project posteriorly.

b. *Superior articular processes.* Directed medially, instead of superiorly.

c. *Inferior articular processes.* Directed laterally, instead of inferiorly.

Label the other indicated parts.

4. *Sacrum.* (Figure 7.7g and h on page 109). Formed by the fusion of five sacral vertebrae.

a. *Transverse lines (ridges).* Mark the joining of the sacral vertebrae bodies.

b. *Anterior sacral foramina.* Four pairs of foramina that communicate with posterior sacral foramina; passages for blood vessels and nerves.

c. *Median sacral crest.* Fused spinous processes of upper sacral vertebrae.

FIGURE 7.7 Vertebrae.

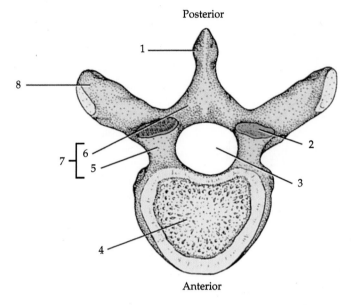

Posterior

Anterior

(a) Superior view of a typical vertebra

___ Body
___ Lamina
___ Pedicle
___ Spinous process
___ Superior articular facet
___ Transverse process
___ Vertebral arch
___ Vertebral foramen

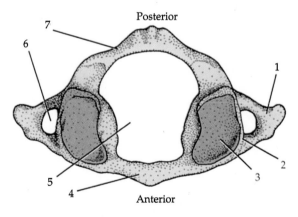

Posterior

Anterior

(b) Superior view of the atlas

___ Anterior arch
___ Lateral mass
___ Posterior arch
___ Superior articular facet
___ Transverse foramen
___ Transverse process
___ Vertebral foramen

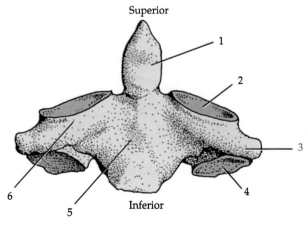

Superior

Inferior

(c) Anterior view of the axis

___ Body
___ Dens
___ Inferior articular facet
___ Lateral mass
___ Superior articular facet
___ Transverse process

FIGURE 7.7   **Vertebrae. (Continued)**

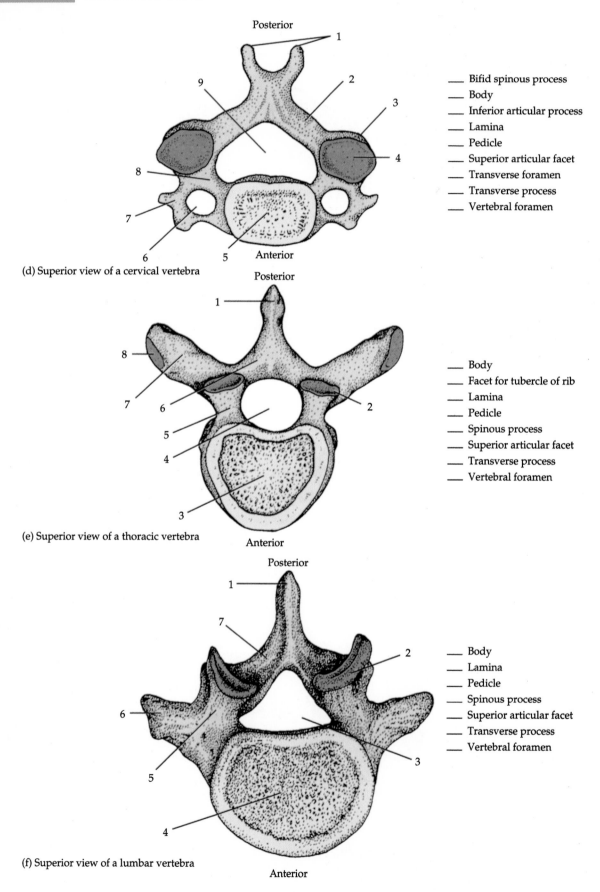

Posterior

1

9

2

3

4

8

7

6   5   Anterior

(d) Superior view of a cervical vertebra

___ Bifid spinous process
___ Body
___ Inferior articular process
___ Lamina
___ Pedicle
___ Superior articular facet
___ Transverse foramen
___ Transverse process
___ Vertebral foramen

Posterior

1

8

7   6

5

4

2

3

(e) Superior view of a thoracic vertebra

Anterior

___ Body
___ Facet for tubercle of rib
___ Lamina
___ Pedicle
___ Spinous process
___ Superior articular facet
___ Transverse process
___ Vertebral foramen

Posterior

1

7

2

6

5   3

4

(f) Superior view of a lumbar vertebra

Anterior

___ Body
___ Lamina
___ Pedicle
___ Spinous process
___ Superior articular facet
___ Transverse process
___ Vertebral foramen

FIGURE 7.7 **Vertebrae. Sacrum and coccyx. (Continued)**

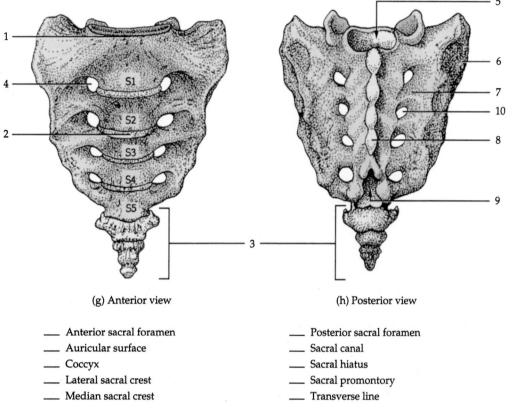

(g) Anterior view

(h) Posterior view

____ Anterior sacral foramen
____ Auricular surface
____ Coccyx
____ Lateral sacral crest
____ Median sacral crest

____ Posterior sacral foramen
____ Sacral canal
____ Sacral hiatus
____ Sacral promontory
____ Transverse line

d. *Lateral sacral crest.* Fused transverse processes of sacral vertebrae.

e. *Posterior sacral foramina.* Four pairs of foramina that communicate with anterior foramina; passages for blood vessels and nerves.

f. *Sacral canal.* Continuation of vertebral canal.

g. *Sacral hiatus* (hi-Ā-tus). Inferior entrance to sacral canal where laminate of S5, and sometimes S4, fail to meet.

h. *Sacral promontory* (PROM-on-tō´-rē). Anteriorly projecting border of the base.

i. *Auricular surface.* Articulates with ilium of each hipbone to form the sacroiliac joint.

5. *Coccyx.* (Figure 7.7g and h). Formed by the fusion of usually four coccygeal vertebrae.

Label the coccyx and the parts of the sacrum in Figure 7.7g and h.

G. Sternum and Ribs

The skeletal part of the **thorax,** the **thoracic cage,** is a bony enclosure formed by the **sternum, costal cartilages, ribs,** and bodies of the **thoracic vertebrae.**

Examine the articulated skeleton and disarticulated bones and identify the following:

1. *Sternum (breastbone).* Flat, narrow bone in the center of the anterior thoracic wall.

a. *Manubrium* (ma-NOO-brē-um; *manubrium* = handlelike). Superior portion.

b. *Body.* Middle, largest portion.

c. *Sternal angle.* Junction of the manubrium and body.

d. *Xiphoid* (ZI-foyd; *xipho* = sword-shaped) *process.* Inferior, smallest portion.

e. *Suprasternal notch.* Depression on the superior surface of the manubrium.

f. *Clavicular notches.* Articular surfaces lateral to suprasternal notches that articulate with the medial ends of the clavicles to form the sternoclavicular joints.

Label the parts of the sternum in Figure 7.8a.

2. *Ribs.* The first through seventh pairs of ribs have a direct attachment to the sternum by a strip of hyaline cartilage called **costal cartilage** (*costa* = rib). These ribs are called **true**

FIGURE 7.8 **Bones of the thorax.**

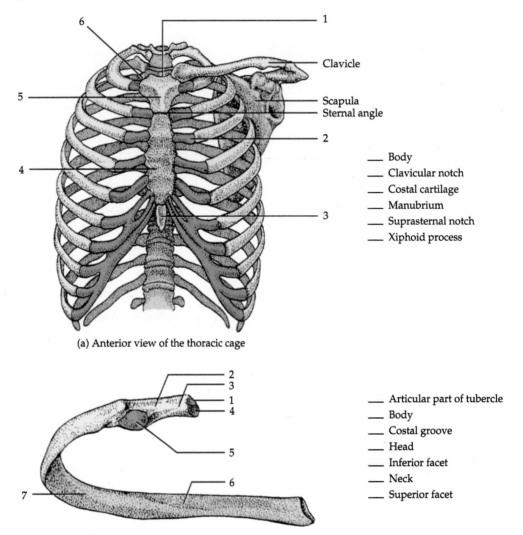

6
1
— Clavicle
5
— Scapula
— Sternal angle
2
4
3

___ Body
___ Clavicular notch
___ Costal cartilage
___ Manubrium
___ Suprasternal notch
___ Xiphoid process

(a) Anterior view of the thoracic cage

2
3
1
4
5
6
7

___ Articular part of tubercle
___ Body
___ Costal groove
___ Head
___ Inferior facet
___ Neck
___ Superior facet

(b) Left rib viewed from behind

(vertebrosternal) ribs. The remaining ribs are called **false ribs** because their costal cartilages either attach indirectly to the sternum or do not attach to the sternum at all. For example, the cartilages of the eighth, ninth, and tenth pairs of ribs attach to each other and then to the cartilages of the seventh pair of ribs. These false ribs are called **vertebrochondral ribs.** The eleventh and twelfth pairs of false ribs are also known as **floating ribs** because their costal cartilage at their anterior end does not attach to the sternum at all. These ribs attach only posteriorly to the thoracic vertebrae.

Parts of a typical rib (third through ninth) include

a. *Body.* Main part of rib.

b. *Head.* Projection at the posterior end of the rib.

c. *Neck.* Constricted portion lateral to head.

d. *Tubercle* (TOO-ber-kul). Knoblike structure on the posterior surface where the neck joins the body; consists of a **nonarticular part** that affords attachment for a ligament and an **articular part** that articulates with the facet of a transverse process of the inferior of the two vertebrae to which the head of the rib is connected.

e. *Costal groove.* Depression on the inner surface that protects blood vessels and a small nerve.

f. *Superior facet.* Articulates with facet on superior vertebra.

g. *Inferior facet.* Articulates with facet on inferior vertebra.

Label the parts of a rib in Figure 7.8b.

H. Pectoral (Shoulder) Girdles

Each **pectoral** (PEK-tō-ral) or **shoulder girdle** consists of two bones—**clavicle** (collar bone) and **scapula** (shoulder blade). Its purpose is to attach the bones of the upper limb to the axial skeleton.

Examine the articulated skeleton and disarticulated bones and identify the following:

1. *Clavicle* (KLAV-i-kul = key). Slender, S-shaped bone with a double curvature; lies horizontally across the anterior part of the thorax superior to the first rib.

    a. *Sternal end.* Rounded, medial end that articulates with manubrium of sternum to form the *sternoclavicular joint.*

    b. *Acromial* (a-KRŌ-mē-al) *end.* Broad, flat, lateral end that articulates with the acromion of the scapula to form acromioclavicular joint.

    c. *Conoid tubercle* (TOO-ber-kul = cone). Projection on the inferior, lateral surface for attachment of the conoid ligament.

Label the parts of the clavicle in Figure 7.9a.

2. *Scapula* (SCAP-ū-la). Large, flat triangular bone in superior part of the posterior thorax between the levels of second and seventh ribs.

    a. *Body.* Flattened, triangular portion.

    b. *Spine.* Ridge across posterior surface of the body of the scapula.

FIGURE 7.9   Pectoral (shoulder) girdle.

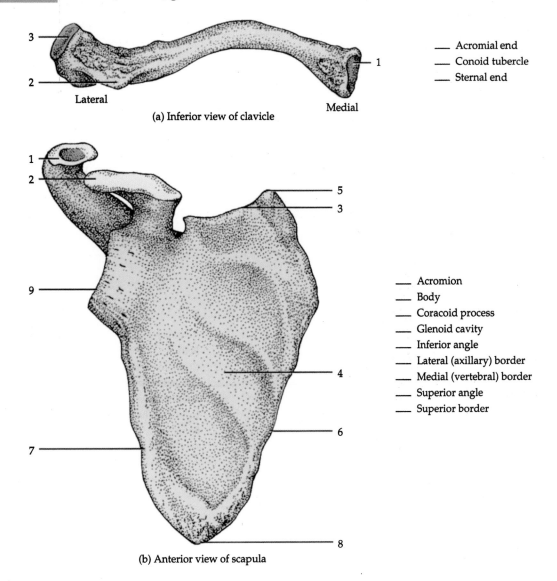

____ Acromial end
____ Conoid tubercle
____ Sternal end

(a) Inferior view of clavicle

____ Acromion
____ Body
____ Coracoid process
____ Glenoid cavity
____ Inferior angle
____ Lateral (axillary) border
____ Medial (vertebral) border
____ Superior angle
____ Superior border

(b) Anterior view of scapula

FIGURE 7.9 Pectoral (shoulder) girdle. (Continued)

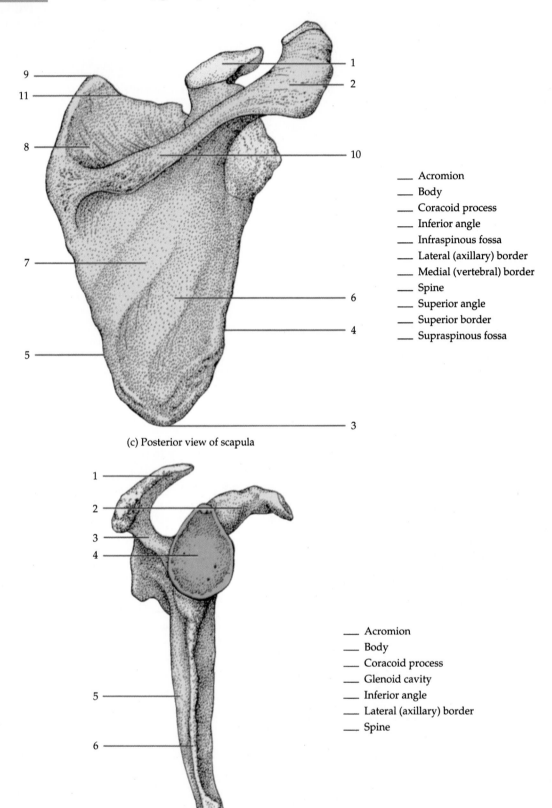

___ Acromion
___ Body
___ Coracoid process
___ Inferior angle
___ Infraspinous fossa
___ Lateral (axillary) border
___ Medial (vertebral) border
___ Spine
___ Superior angle
___ Superior border
___ Supraspinous fossa

(c) Posterior view of scapula

___ Acromion
___ Body
___ Coracoid process
___ Glenoid cavity
___ Inferior angle
___ Lateral (axillary) border
___ Spine

(d) Lateral border view of scapula

c. *Acromion* (a-KRŌ-mē-on; *acrom* = topmost). Flattened, expanded process of the lateral end of the spine.

d. *Medial (vertebral) border.* Edge of body near vertebral column.

e. *Lateral (axillary) border.* Edge of body closer to the arm.

f. *Inferior angle.* Bottom of body where medial and lateral borders join.

g. *Glenoid cavity.* Depression inferior to the acromion that accepts the head of humerus to form the *glenohumeral joint.*

h. *Coracoid* (KOR-a-koyd = like a crow's beak) *process.* Projection at lateral end of superior border, to which tendons of muscles and some ligaments attach.

i. *Supraspinous* (sū'-pra-SPĪ-nus) *fossa.* Surface for muscle attachment above spine.

j. *Infraspinous fossa.* Surface for muscle attachment below spine.

k. *Superior border.* Superior edge of the scapula.

l. *Superior angle.* Top of body where superior and medial borders join.

Label the parts of the scapula in Figure 7.9b, c, and d.

I. Upper Limbs

The skeleton of the **upper limbs** consists of a humerus in each arm, an ulna and radius in each forearm, carpals in each wrist, metacarpals in each palm, and phalanges in the fingers.

Examine the articulated skeleton and disarticulated bones and identify the following:

1. *Humerus* (HŪ-mer-us). Arm bone; longest and largest bone of the upper limb.

   a. *Head.* Proximal end of the humerus that articulates with the glenoid cavity of scapula to form the glenohumeral joint.

   b. *Anatomical neck.* Oblique groove distal to the head; the site of the epiphyseal line.

   c. *Greater tubercle.* Lateral projection distal to the anatomical neck.

   d. *Lesser tubercle.* Anterior projection.

   e. *Intertubercular sulcus.* Between the tubercles.

   f. *Surgical neck.* Constriction in the humerus just distal to the tubercles.

g. *Body.* Shaft.

h. *Deltoid tuberosity.* Roughened, V-shaped area at the middle portion of shaft.

i. *Capitulum* (ka-PIT-ū-lum). Rounded knob on the lateral aspect of the bone that articulates with head of radius.

j. *Radial fossa.* Anterior depression that receives head of radius when forearm is flexed.

k. *Trochlea* (TRŌK-lē-a). Located medial to the capitulum, it is a spool-shaped surface that articulates with the ulna.

l. *Coronoid* (KOR-ō-noyd = crown-shaped) *fossa.* Anterior depression that receives the coronoid process of the ulna when the forearm is flexed.

m. *Olecranon* (ō-LEK-ra-non) *fossa.* Posterior depression that receives the olecranon of the ulna when the forearm is extended (straightened).

n. *Medial epicondyle.* Projection on medial side of distal end.

o. *Lateral epicondyle.* Projection on lateral side of distal end.

Label the parts of the humerus in Figure 7.10a and b.

2. *Ulna.* Medial bone of forearm, it is longer than the radius.

   a. *Olecranon (olecranon process).* Prominence of elbow at proximal end.

   b. *Coronoid process.* Anterior projection that, with olecranon, receives trochlea of humerus.

   c. *Trochlear notch.* Large curved area between olecranon and coronoid process that forms part of the elbow joint.

   d. *Radial notch.* Depression lateral and inferior to trochlear notch that receives the head of the radius to form the proximal radioulnar joint.

   e. *Head.* Rounded portion at distal end.

   f. *Styloid* (*stylo* = stake or pole) *process.* Projection on posterior side of distal end.

Label the parts of the ulna in Figure 7.10c, d, and e.

3. *Radius.* Lateral bone of forearm.

   a. *Head.* Disc-shaped process at proximal end that articulates with the capitulum of the humerus and the radial notch of the ulna.

FIGURE 7.10   Bones of the upper limb.

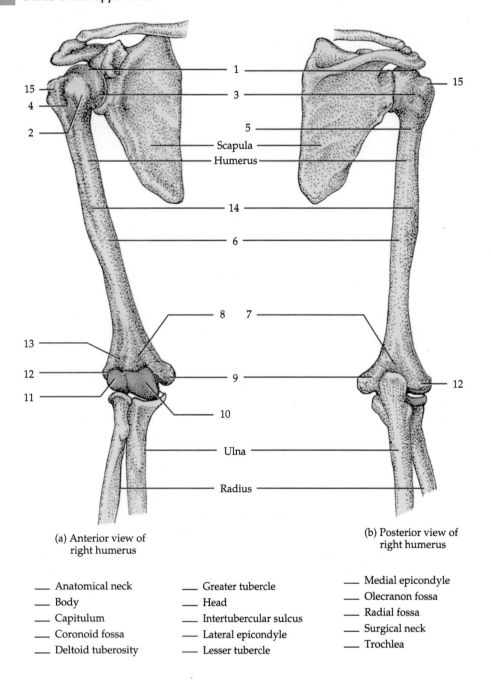

(a) Anterior view of
right humerus

(b) Posterior view of
right humerus

___ Anatomical neck   ___ Greater tubercle   ___ Medial epicondyle
___ Body   ___ Head   ___ Olecranon fossa
___ Capitulum   ___ Intertubercular sulcus   ___ Radial fossa
___ Coronoid fossa   ___ Lateral epicondyle   ___ Surgical neck
___ Deltoid tuberosity   ___ Lesser tubercle   ___ Trochlea

b. ***Radial tuberosity.*** A roughened area inferior to the neck on the medial surface, it
serves as a point of attachment for tendons of the biceps brachii muscle.

c. ***Styloid process.*** Projection on lateral side
of distal end.

d. ***Ulnar notch.*** Medial, concave depression
for articulation with head of ulna to form
distal radioulnar joint.

Label the parts of the radius in Figure 7.10c and d.

4. ***Carpus.*** Wrist, consists of eight small bones
called **carpal bones** *(carpals);* articulations
between carpal bones are called intercarpal
joints.

a. ***Proximal row.*** From lateral to medial are
called **scaphoid, lunate, triquetrum,** and
pisiform.

b. ***Distal row.*** From lateral to medial are
called **trapezium, trapezoid, capitate,** and
hamate.

FIGURE 7.10   Bones of the upper limb. (Continued)

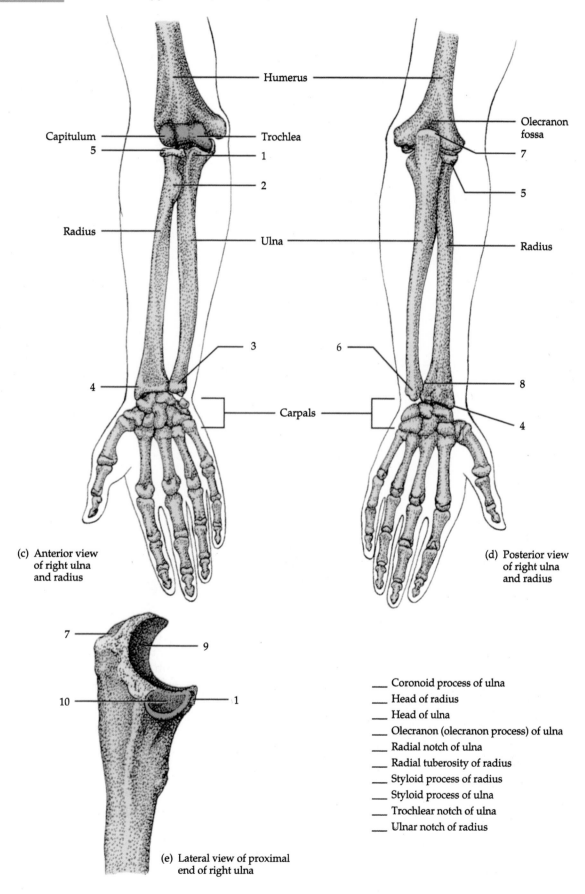

(c) Anterior view
of right ulna
and radius

(d) Posterior view
of right ulna
and radius

(e) Lateral view of proximal
end of right ulna

___ Coronoid process of ulna
___ Head of radius
___ Head of ulna
___ Olecranon (olecranon process) of ulna
___ Radial notch of ulna
___ Radial tuberosity of radius
___ Styloid process of radius
___ Styloid process of ulna
___ Trochlear notch of ulna
___ Ulnar notch of radius

FIGURE 7.10 **Bones of the upper limb. (Continued)**

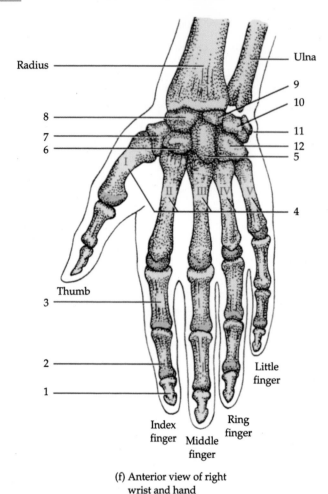

(f) Anterior view of right wrist and hand

— Capitate
— Distal phalanx
— Hamate
— Lunate
— Metacarpal
— Middle phalanx
— Pisiform
— Proximal phalanx
— Scaphoid
— Trapezium
— Trapezoid
— Triquetrum

5. *Metacarpus* (*meta* = beyond). Consists of five bones of the palm called **metacarpal bones (metacarpals).** Each metacarpal bone consists of a proximal *base*, an intermediate *shaft*, and a distal *head*. They are numbered as follows beginning with thumb side: I, II, III, IV, and V (or 1-5) metacarpals.

6. *Phalanges* (fa-LAN-jēz; *phalan* = battle line). Bones of the digits; two in each thumb (proximal and distal) and three in each finger (proximal, middle, and distal). The singular of phalanges is **phalanx.** The metacarpals articulate with proximal phalanges to form the *metacarpophalangeal joints.* Joints between phalanges are called *interphalangeal joints.*

Label the parts of the carpus, metacarpus, and phalanges in Figure 7.10f on this page.

J. Pelvic (Hip) Girdle

The **pelvic (hip) girdle** consists of the two **hip bones** also called **coxal** (KOK-sal) **bones** or **os coxa.** It provides a strong and stable support for the vertebral column and pelvic organs and accepts the bones of the lower limbs, connecting them to the axial skeleton.

Examine the articulated skeleton and disarticulated bones and identify the following parts of the hipbone:

1. *Ilium* (= flank). Superior flattened portion.

   a. *Iliac crest.* Superior border of ilium.

   b. *Anterior superior iliac spine.* Anterior projection of iliac crest.

   c. *Anterior inferior iliac spine.* Projection under anterior superior iliac spine.

   d. *Posterior superior iliac spine.* Posterior projection of iliac crest.

   e. *Posterior inferior iliac spine.* Projection below posterior superior iliac spine.

   f. *Greater sciatic* (sī-AT-ik) *notch.* Concavity below the posterior inferior iliac spine.

   g. *Iliac fossa.* Medial concavity where the tendon of the iliacus muscle attaches.

   h. *Iliac tuberosity.* Point of attachment for sacroiliac ligament posterior to iliac fossa.

   i. *Auricular* (*auric* = little ear) *surface.* Point of articulation with sacrum (sacroiliac joint).

2. *Ischium* (IS-kē-um = hip). Inferior, posterior portion.

   a. *Ischial spine.* Posterior projection of ischium.

   b. *Lesser sciatic notch.* Concavity below ischial spine.

   c. *Ischial tuberosity.* Roughened, thickened projection.

   d. *Ramus.* Portion of ischium that joins the pubis and surrounds the **obturator** (OB-too-rā-ter) **foramen.**

3. *Pubis* or *os pubis* (meaning pubic bone). Anterior, inferior portion.

   a. *Superior ramus.* Upper portion of pubis.

   b. *Inferior ramus.* Lower portion of pubis.

   c. *Pubis symphysis* (SIM-fi-sis). Joint between left and right hip bones.

4. *Acetabulum* (as'-e-TAB-ū-lum). Deep fossa that accepts the rounded head of the femur to form the *hip joint;* formed by the ilium, ischium, and pubis.

Label Figure 7.11.

Again, examine the articulated skeleton. This time compare the male and female pelvis. The **pelvis** consists of the two hip bones, sacrum, and coccyx. Identify the following:

1. *False (greater) pelvis.* Expanded portion situated superior to the pelvic brim; bordered by the lumbar vertebrae posteriorly, the upper portions of the hip bones laterally, and the abdominal well anteriorly.

2. *True (lesser) pelvis.* Inferior to the pelvic brim; bounded by the sacrum, and coccyx posteriorly, inferior portions of the ilium and ischium laterally, and the pubic bones anteriorly, and coccyx; contains an opening above, the **pelvic inlet,** and an opening below, the **pelvic outlet.**

K. Lower Limbs

The bones of the **lower limbs** consist of a femur in each thigh, a patella (kneecap) in front of each knee joint, a tibia and fibula in each leg, tarsals in each ankle, metatarsals in each foot, and phalanges in the toes.

Examine the articulated skeleton and disarticulated bones and identify the following:

1. *Femur.* Thigh bone; longest and heaviest bone in the body.

   a. *Head.* Rounded projection at proximal end that articulates with acetabulum of hipbone to form the hip (coxal) joint.

   b. *Neck.* Constricted region distal to the head.

   c. *Greater trochanter* (trō-KAN-ter). Prominence on lateral side.

   d. *Lesser trochanter.* Prominence inferior and medial to the greater trochanter.

   e. *Intertrochanteric line.* Ridge on anterior surface of the trochanters.

   f. *Intertrochanteric crest.* Ridge on posterior surface of the trochanters.

   g. *Gluteal tuberosity.* Vertical ridge on posterior surface of the body of the femur. It blends into another vertical ridge called the *linea aspera* (LIN-ē-a AS-per-a; *asper* = rough).

   h. *Medial condyle.* Medial posterior projection on distal end that articulates with medial condyle of the tibia to help form tibiofemoral (knee) joint.

   i. *Lateral condyle.* Lateral posterior projection on distal end that articulates with lateral condyle of the tibia to help form tibiofemoral (knee) joint.

   j. *Intercondylar* (in'-ter-KON-di-lar) *fossa.* Depressed area between condyles on posterior surface.

   k. *Medial epicondyle.* Projection above medial condyle.

   l. *Lateral epicondyle.* Projection above lateral condyle.

   m. *Patellar surface.* Between condyles on the anterior surface; articulates with the posterior surface of the patella to help form the tibiofemoral (knee) joint.

Label the parts of the femur in Figure 7.12a and b on page 119.

FIGURE 7.11   Right hipbone.

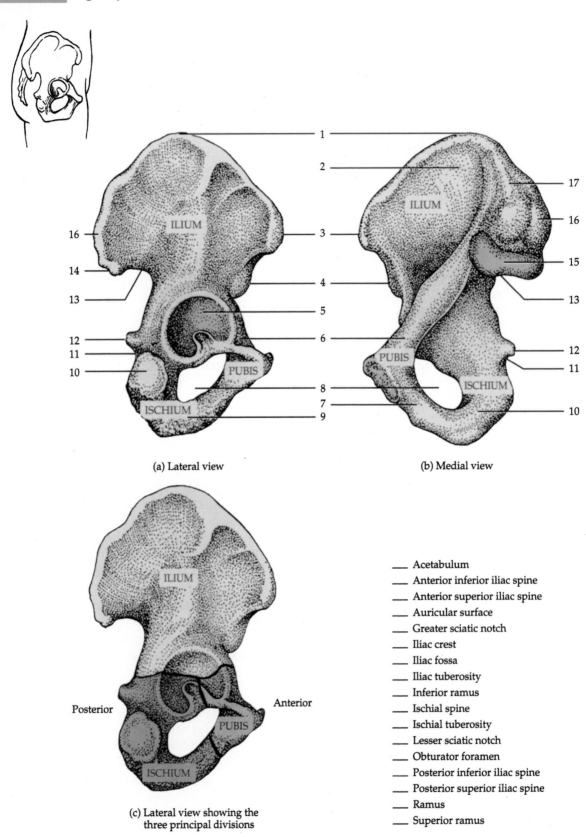

(a) Lateral view

(b) Medial view

(c) Lateral view showing the
three principal divisions

___ Acetabulum
___ Anterior inferior iliac spine
___ Anterior superior iliac spine
___ Auricular surface
___ Greater sciatic notch
___ Iliac crest
___ Iliac fossa
___ Iliac tuberosity
___ Inferior ramus
___ Ischial spine
___ Ischial tuberosity
___ Lesser sciatic notch
___ Obturator foramen
___ Posterior inferior iliac spine
___ Posterior superior iliac spine
___ Ramus
___ Superior ramus

FIGURE 7.12 Bones of the lower limb.

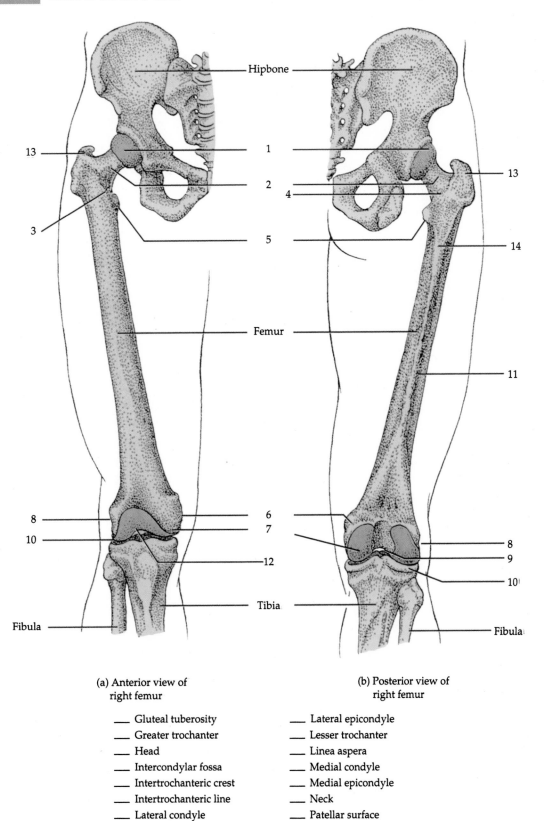

(a) Anterior view of
right femur

(b) Posterior view of
right femur

___ Gluteal tuberosity   ___ Lateral epicondyle

___ Greater trochanter   ___ Lesser trochanter

___ Head   ___ Linea aspera

___ Intercondylar fossa   ___ Medial condyle

___ Intertrochanteric crest   ___ Medial epicondyle

___ Intertrochanteric line   ___ Neck

___ Lateral condyle   ___ Patellar surface

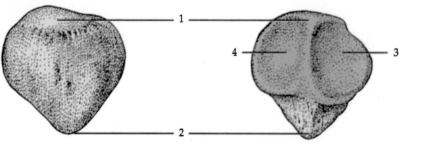

FIGURE 7.12 Bones of the lower limb. (Continued)

(c) Anterior view of
right patella

(d) Posterior view of
right patella

___ Apex
___ Articular facet for
    lateral femoral
    condyle
___ Articular facet for
    medial femoral
    condyle
___ Base

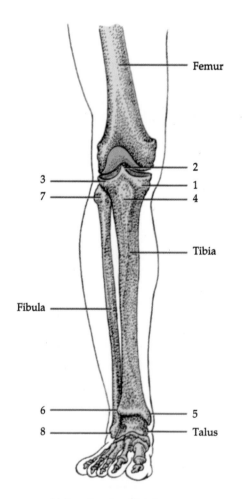

(e) Anterior view of right
fibula and tibia

___ Fibular notch   ___ Lateral malleolus
___ Head of fibula   ___ Medial condyle
___ Intercondylar eminence   ___ Medial malleolus
___ Lateral condyle   ___ Tibial tuberosity

FIGURE 7.12   **Bones of the lower limb. (Continued)**

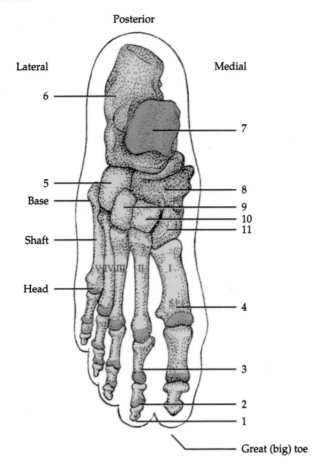

Posterior

Lateral   Medial

6

7

5

Base

8
9
10
11

Shaft

Head

4

3

2

1

____ Calcaneus
____ Cuboid
____ Distal phalanx
____ First (medial) cuneiform
____ Metatarsal
____ Middle phalanx
____ Navicular
____ Proximal phalanx
____ Second (intermediate) cuneiform
____ Talus
____ Third (lateral) cuneiform

Great (big) toe

(f) Superior view of right foot

2. *Patella* (= little dish). Kneecap; a small, triangular bone located anterior to the knee joint. It is a sesamoid bone that develops in tendon of quadriceps femoris muscle.

   a. *Base.* Broad superior portion.

   b. *Apex.* Pointed inferior portion.

   c. *Articular facets.* Articulating surfaces on posterior surface for medial and lateral condyles of femur.

   Label the parts of the patella in Figure 7.12c and d on page 120.

3. *Tibia.* Shinbone; larger, medial, weight-bearing bone of leg.

   a. *Lateral condyle.* Articulates with lateral condyle of femur to help form the tibiofemoral (knee) joint.

   b. *Medial condyle.* Articulates with medial condyle of femur to help form the tibiofemoral (knee) joint.

   c. *Intercondylar eminence.* Upward projection between condyles.

   d. *Tibial tuberosity.* Anterior projection is a point of attachment for the patellar ligament.

   e. *Medial malleolus* (mal-LĒ-ō-lus = hammer). Distal projection that articulates with talus bone of ankle.

   f. *Fibular notch.* Distal depression that articulates with the fibula to form the distal tibiofibular joint.

4. *Fibula.* Lateral, smaller bone of leg.

   a. *Head.* Proximal end that articulates with the inferior surface of the lateral condyle of the tibia to form the proximal tibiofibular joint.

   b. *Lateral malleolus.* Projection at distal end that articulates with the talus bone of ankle.

FIGURE 7.13 **Entire skeleton.**

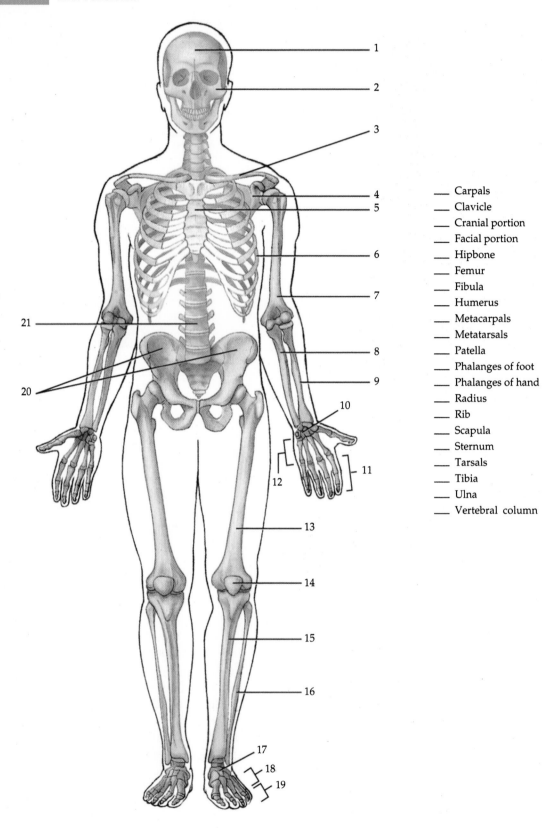

___ Carpals
___ Clavicle
___ Cranial portion
___ Facial portion
___ Hipbone
___ Femur
___ Fibula
___ Humerus
___ Metacarpals
___ Metatarsals
___ Patella
___ Phalanges of foot
___ Phalanges of hand
___ Radius
___ Rib
___ Scapula
___ Sternum
___ Tarsals
___ Tibia
___ Ulna
___ Vertebral column

Anterior view

Label the parts of the tibia and fibula in Figure 7.12e.

5. *Tarsus.* Seven bones of the ankle called **tarsal bones.** Joints between tarsal bones are called *intertarsal joints.*

    a. *Posterior bones*—**talus** (TĀ-lus) and **calcaneus** (kal-KĀ-nē-us) (heel bone).

    b. *Anterior bones*—**cuboid, navicular,** and three **cuneiforms** called the **first (medial), second (intermediate),** and **third (lateral) cuneiforms.**

6. *Metatarsus.* Consists of five *metatarsal bones* of the foot, numbered as follows, beginning on the medial (great or big toe) side: I, II, III, IV, and V (or 1-5) metatarsals.

7. *Phalanges.* Bones of the toes, comparable to phalanges of fingers; two in each great toe (proximal and distal) and three in each small toe (proximal, middle, and distal). The metatarsals articulate with the proximal phalanges to form the *metatarsophalangeal joints.* Joints between phalanges are called *interphalangeal joints.*

Label the parts of the tarsus, metatarsus, and phalanges in Figure 7.12f.

L. Articulated Skeleton

Now that you have studied all the bones of the body, label the entire articulated skeleton in Figure 7.13 on page 122.

ANSWER THE LABORATORY REPORT QUESTIONS AT THE END OF THE EXERCISE.

| EXERCISE 7 | Bones |

Name _____   Date _____

Laboratory Section _____   Score/Grade _____

PART 1 ■ Multiple Choice

_____ 1. The suture between the parietal and temporal bones is the (a) lambdoid (b) coronal (c) squamous (d) sagittal

_____ 2. Which bone does *not* contain a paranasal sinus? (a) ethmoid (b) maxilla (c) sphenoid (d) occipital

_____ 3. Which is the superior, concave curve in the vertebral column? (a) thoracic (b) lumbar (c) cervical (d) sacral

_____ 4. The fontanel between the parietal and occipital bones is the (a) anterolateral (b) anterior (c) posterior (d) posterolateral

_____ 5. Which is *not* a component of the upper limb? (a) radius (b) femur (c) carpus (d) humerus

_____ 6. All are components of the appendicular skeleton *except* the (a) humerus (b) occipital bone (c) calcaneus (d) triquetral

_____ 7. Which bone does *not* belong with the others? (a) occipital (b) frontal (c) parietal (d) mandible

_____ 8. Which region of the vertebral column is closer to the skull? (a) thoracic (b) lumbar (c) cervical (d) sacral

_____ 9. Of the following bones, the one that does *not* help form part of the orbit is the (a) sphenoid (b) frontal (c) occipital (d) lacrimal

_____ 10. Which bone does *not* form a border for a fontanel? (a) maxilla (b) temporal (c) occipital (d) parietal

PART 2 ■ Identification

For each surface marking listed, identify the skull bone to which it belongs:

11. Glabella _____

12. Mastoid process _____

13. Hypophyseal fossa _____

14. Cribriform plate _____

15. Foramen magnum _____

16. Mental foramen _____

17. Infraorbital foramen _____

18. Crista galli _____

19. Foramen ovale _____

20. Horizontal plate _____

21. Optic foramen _____

22. Superior nasal concha _____

23. Zygomatic process _____

24. Styloid process _____

25. Mandibular fossa _____

PART 3 ■ Matching

_____ **26.** Iliac crest

_____ **27.** Capitulum

_____ **28.** Medial malleolus

_____ **29.** Laminae

_____ **30.** Vertebral foramen

_____ **31.** Talus

_____ **32.** Olecranon

_____ **33.** Pisiform

_____ **34.** Acromial extremity

_____ **35.** Pubic symphysis

_____ **36.** Hamate

_____ **37.** Costal groove

_____ **38.** Xiphoid process

_____ **39.** Greater trochanter

_____ **40.** Transverse lines

_____ **41.** Radial tuberosity

_____ **42.** Greater tubercle

_____ **43.** Ischium

_____ **44.** Glenoid cavity

_____ **45.** Lateral malleolus

A. Inferior portion of sternum

B. Medial bone of distal carpals

C. Distal projection of tibia

D. Lateral end of clavicle

E. Points where bodies of sacral vertebrae join

F. Portion of rib that contains blood vessels

G. Prominence of elbow

H. Lateral projection of humerus

I. Medial projection for insertion of biceps brachii muscle

J. Lower posterior portion of hipbone

K. Articulates with head of radius

L. Prominence on lateral side of femur

M. Distal projection of fibula

N. Superior border of ilium

O. Articulates with head of humerus

P. Anterior joint between hipbones

Q. Opening through which spinal cord passes

R. Form posterior wall of vertebral arch

S. Medial bone of proximal carpals

T. Component of tarsus

Joints

Objectives

At the completion of this exercise you should understand

A The kinds of joints on the basis of structure and function.

B The movements that occur at synovial joints.

C The anatomy of the knee joint.

An **articulation** (ar-tik'-ū-LĀ-shun), or **joint,** is a point of contact between two bones, cartilage and bones, or teeth and bones. When we say that one bone **articulates** with another, we mean that the bones form a joint. The scientific study of joints is called **arthrology** (ar-THROL-ō-jē; *arthr* = joint; *logy* = study of).

Some joints permit no movement, others permit a slight degree of movement, and still others permit free movement. In this exercise you will study the structure and action of joints.

A. Kinds of Joints

The joints of the body may be classified structurally, based on their anatomical characteristics, and functionally, based on the type of movement they permit.

The structural classification of joints is based on (1) the presence or absence of a space between articulating bones called a **synovial** or **joint cavity,** and (2) the type of connective tissue that binds the bones together. Structurally, a joint is classified as follows:

1. *Fibrous* (FĪ-brus) *joint.* Lacks a synovial cavity and the bones are held together by fibrous connective tissue.

2. *Cartilaginous* (kar-ti-LAJ-i-nus) *joint.* Lacks a synovial cavity and the bones are held together by cartilage.

3. *Synovial* (si-NŌ-vē-al) *joint.* There is a synovial cavity and the bones forming the joint are united by the dense irregular connective tissue of an articular capsule and often by accessory ligaments (described in detail later).

The functional classification of joints relates to the degree of movement they permit and is as follows:

1. *Synarthroses* (sin'-ar-THRŌ-sēz; *syn* = together; *arthros* = joint). Immovable joints.

2. *Amphiarthroses* (am'-fē-ar-THRŌ-sēz; *amphi* = on both sides). Slightly movable joints.

3. *Diarthroses* (dī-ar-THRŌ-sēz; *diarthros* = movable joint). Freely movable joints.

We will discuss the joints of the body on the basis of their structural classification, referring to their functional classification as well.

B. Fibrous Joints

Allow little or no movement; do not contain synovial cavity; articulating bones held together by fibrous connective tissue.

1. *Sutures* (SOO-churz; *sutur* = seam). A fibrous joint composed of a thin layer of dense fibrous connective tissue that unites only bones of the skull. Sutures are immovable. Example: coronal suture between the frontal and parietal bones.

2. *Syndesmoses* (sin'-dez-MŌ-sēz; *syndesmo* = band or ligament). A fibrous joint in which there is a greater distance between the articulating bones and more fibrous connective tissue than in a suture; the fibrous connective tissue is arranged either as a bundle (ligament) or as a sheet (interosseous membrane) that permits partial movement. Example: distal tibiofibular joint (distal articulation of tibia and fibula).

3. *Gomphoses* (gom-FŌ-sēz; *gomphosis* = a bolt or nail). A type of fibrous joint in which a cone-shaped peg fits into a socket. Gomphoses are immovable. Example: articulations of the roots of the teeth with the sockets (alveoli) of the maxillae and mandible in which the dense connective tissue between the two is the periodontal ligament (membrane).

C. Cartilaginous Joints

Allow little or no movement; lack a synovial cavity; articulating bones held together by hyaline cartilage or fibrocartilage.

1. *Synchondroses* (sin'-kon-DRŌ-sēz; *syn* = together; *chondro* = cartilage). A cartilaginous joint in which the connecting material is hyaline cartilage. Synchondroses are immovable. Example: epiphyseal plate between the epiphysis and diaphysis of a growing bone.

2. *Symphyses* (SIM-fi-sēz = growing together). A cartilaginous joint in which the ends of the articulating bones are covered with hyaline cartilage, but a broad, flat disc of fibrocartilage connects the bones. Symphyses are partially movable. Example: intervertebral joints (between the bodies of vertebrae) and the pubic symphysis (between the anterior surfaces of the hip bones).

D. Synovial Joints

Synovial (si-NŌ-vē-al) **joints** have a variety of shapes and permit several different types of movements. First we will discuss the general structure of a synovial joint and then consider the types of synovial joints and their movements.

1. Structure of a Synovial Joint

The most distinguishing characteristic of a synovial joint is the presence of a space called a **synovial (joint) cavity** between the articulating bones (see Figure 8.1). The synovial cavity in these joints allows them to be freely movable. The bones at a synovial joint are covered by **articular cartilage,** which typically is hyaline cartilage. The cartilage covers the articulating surface of the bones with a smooth, slippery surface; it does not bind them together. Articular cartilage reduces friction between bones in the joint during movement and helps absorb shock.

A sleevelike **articular capsule** surrounds a synovial joint, encloses the synovial cavity, and unites the articulating bones. The articular capsule is composed of two layers, an outer fibrous capsule and an inner synovial membrane. The outer layer of the articular capsule, the **fibrous capsule,** usually consists of dense, irregular connective tissue that attaches to the periosteum of the articulating bones. The fibers of some fibrous capsules are arranged in parallel bundles called **ligaments** (*liga* = to bind) and are often designated by individual names. The strength of the ligaments is one of the principal mechanical factors that hold bones together. The inner layer of the articular capsule, **synovial membrane,** is composed of areolar connective tissue with elastic fibers. At many synovial joints, the synovial membrane includes accumulations of adipose tissue, called **articular fat pads** (see Figure 8.4f).

The synovial membrane secretes **synovial fluid** (*ov* = egg), which forms a thin film over the surfaces within the articular capsule. This viscous, clear or pale yellow fluid was named for its similarity in appearance and consistency to uncooked egg white. Synovial fluid consists of hyaluronic acid and interstitial fluid filtered from blood plasma. Functionally, it lubricates and reduces friction in the joint, absorbs shock, and supplies nutrients to and removes metabolic wastes from the chondrocytes of the articular cartilage. (Recall that cartilage is an avascular tissue.) Synovial fluid also contains phagocytic cells that remove microbes and debris resulting from normal wear and tear in the joint. When there is no joint movement, the synovial fluid is quite viscous (gel-like), but as joint movement increases, the fluid becomes less viscous. One of the benefits of a "warm-up" before exercise is that it stimulates the production and secretion of synovial fluid.

Many synovial joints also contain **accessory ligaments,** called extracapsular ligaments and intracapsular ligaments. **Extracapsular ligaments** lie outside the articular capsule. Examples include the fibular and tibial collateral ligaments of the knee joint (see Figure 8.4b). **Intracapsular lig-**

aments occur within the articular capsule but are isolated from the synovial cavity by folds of the synovial membrane. Examples are the anterior and posterior cruciate ligaments of the knee joint (see Figure 8.4b).

Inside some synovial joints, like the knee, there are pads of fibrocartilage that lie between the articular surfaces of the bones and are attached to the fibrous capsule. These fibrocartilage pads are called **articular discs** or **menisci** (men-IS-ī; singular is *meniscus*). Shown in Figure 8.4b are the medial and lateral menisci in the knee joint. Articular discs modify the shape of the joint surfaces of the articulating bones, allowing bones of different shapes to fit more tightly. Articular discs also help to maintain the stability of the joint and direct the flow of synovial fluid to areas of greatest friction. The tearing of articular discs (menisci) in the knee is called **torn cartilage** and occurs often among athletes. Such damaged cartilage requires surgical removal (meniscectomy) or it will begin to wear and may precipitate arthritis.

The various movements of the body create friction between moving parts. Saclike structures called **bursae** (BER-sē = purse) are strategically situated to alleviate friction in some joints such as the shoulder and knee joints (see Figure 8.4f). Bursae consist of connective tissue lined by a synovial membrane and are filled with a fluid similar to synovial fluid. Bursae are located between the skin and bone in places where skin rubs over bone, between tendons and bones, muscles and bones, ligaments and bones, and within articular capsules. Bursae cushion the movement of one part of the body over another. *Tendon sheaths* are tubelike bursae that wrap around tendons where there is considerable friction. An acute or chronic inflammation of a bursa is called **bursitis.**

Label Figure 8.1, the principal parts of a synovial joint.

2. Axes of Movements at Synovial Joints

When a movement occurs at a joint, one bone remains relatively fixed while the other moves around an axis. If there is movement around a *single* axis, the joint is **monaxial** (mon-AKS-ē-al). An example is bending your lower limb at the knee and then straightening it. If there is movement around two perpendicular axes, the joint is **biaxial** (bī-AKS-ē-al). This can be demonstrated by bending the hand at the wrist and then straightening it (the first axis) and then moving the hand from side to side at the wrist (the second axis). A **multiaxial** or **triaxial** (trī-AKS-ē-al) joint, such as the shoulder joint, permits movement around *three* perpendicular axes and multiple axes in between. For example, at the shoulder joint you can swing your arm forward and backward as occurs during walking (movement around the first axis), move your arm straight out from your side and then back again (movement along the second axis), and turn your palm anteriorly and posteriorly by turning the arm (movement around the third axis).

FIGURE 8.1   **Parts of a synovial joint.**

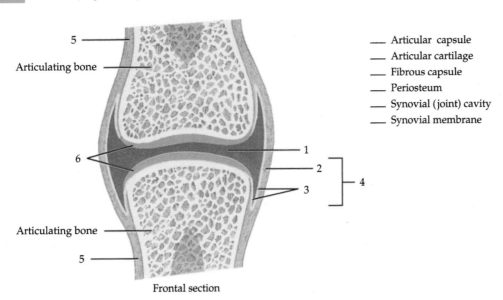

Articulating bone

Articulating bone

Frontal section

___ Articular capsule
___ Articular cartilage
___ Fibrous capsule
___ Periosteum
___ Synovial (joint) cavity
___ Synovial membrane

3. Types of Synovial Joints

Based on the shape of the articulating surfaces and the various types of movements permitted, there are six subtypes of synovial joints: planar, hinge, pivot, condyloid, saddle, and ball-and-socket joints.

a. Planar Joint

In a **planar joint** the articulating surfaces of the bones are flat or slightly curved. Examples of planar joints are the intercarpal joints (between carpal bones at the wrist), intertarsal joints (between tarsal bones at the ankle), sternoclavicular joints (between the manubrium of the sternum and the clavicle), acromioclavicular joints (between the acromion of the scapula and the clavicle), sternocostal joints (between the sternum and ends of the costal cartilage at the tips of the second through seventh pairs of ribs), and vertebrocostal joints (between the heads and tubercles of ribs and transverse processes of thoracic vertebrae). Planar joints permit mainly side-to-side and back-and-forth movements. Because this motion is not around an axis, planar joints are said to be *nonaxial*.

b. Hinge Joint

In a **hinge joint,** the convex surface of one bone fits into the concave surface of another bone. Examples of hinge joints are the knee, elbow, ankle, and interphalangeal joints. Hinge joints produce an angular, opening-and-closing motion like that of a hinged door and they are monaxial because they typically allow motion around a single axis.

c. Pivot Joint

In a **pivot joint,** the articulation is between the rounded or pointed surface of one bone and a ring formed partly by another bone and partly by a ligament. The principal movement permitted at a pivot joint is rotation around its own longitudinal axis and thus the joint is monaxial. Examples of pivot joints are the atlanto-axial joint (between the atlas and axis) in which the atlas rotates around the axis and permits the head to turn from side to side as in signifying "no" and the radioulnar joints (between the radius and ulna) that allow us to turn the palms anteriorly and posteriorly.

d. Condyloid Joint

In a **condyloid joint** (KON-di-loyd; *condyl* = knuckle) or *ellipsoidal* joint, the convex oval-shaped projection of one bone fits into the oval-shaped joint depression of another bone. Examples are the wrist or radiocarpal joint (between radius and carpus) and metacarpophalangeal joints (between the metacarpals and phalanges of the second through fifth digits). Because the movement permitted by a condyloid joint is around two axes, it is biaxial.

e. Saddle Joint

In a **saddle joint,** the articular surface of one bone is saddle-shaped, and the articular surface of the other bone fits into the "saddle" as a sitting rider would. One example of a saddle joint is the carpometacarpal joint (between trapezium of the carpus and metacarpal of the thumb). Movements at a saddle joint are side-to-side and up-and-down movements. Such joints are biaxial.

f. Ball-and-Socket Joint

A **ball-and-socket joint** consist of the ball-like surface of one bone fitting into a cuplike depression of another bone. Examples of ball-and-socket joints are the shoulder and hip joints. Such joints permit movement around three axes, and are thus triaxial (*polyaxial*).

Examine the articulated skeleton and find as many examples as you can of the joints just described. As part of your examination, be sure to note the shapes of the articular surfaces and the movements possible at each joint.

Label Figure 8.2, the types of synovial joints.

4. Types of Movement at Synovial Joints

Anatomists, physical therapists, and kinesiologists use specific terminology to designate movements that occur at synovial joints. These terms may indicate the form of motion, the direction of movement, or the relationship of one body part to another during movement. Movements at synovial joints are grouped into four main categories: (1) gliding, (2) angular movements, (3) rotation, and (4) special movements. This last category includes movements that occur only at certain joints.

a. Gliding

Gliding is a simple movement in which relatively flat bone surfaces move back and forth and from side to side with respect to one another. There is no significant change in the angle between the bones. Gliding occurs at planar joints—the intercarpal, intertarsal, sternoclavicular, acromioclavicular, sternocostal, and vertebrocostal joints.

FIGURE 8.2 Types of synovial joints.

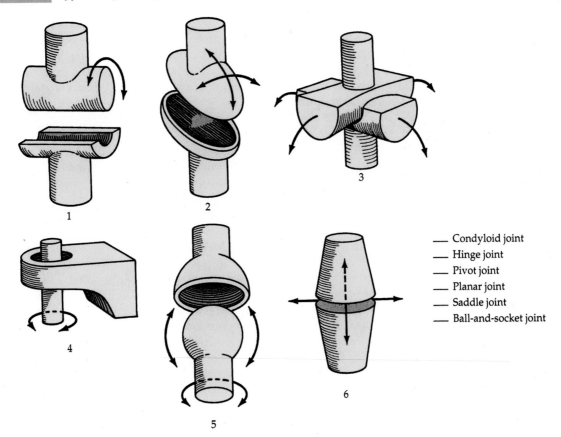

1

2

3

4

5

6

___ Condyloid joint
___ Hinge joint
___ Pivot joint
___ Planar joint
___ Saddle joint
___ Ball-and-socket joint

b. Angular Movements

In **angular movements,** there is an increase or a decrease in the angle between articulating bones. The principal angular movements are flexion, extension, lateral flexion, hyperextension, abduction, adduction, and circumduction. Please remember that as these and all other movements are discussed it is assumed that the body is in the anatomical position.

Flexion (FLEK-shun; *flex* = to bend) involves a decrease in the angle between articulating bones and extension. Following are examples of flexion. **Extension** (eks-TEN-shun; *exten* = to stretch out) involves an increase in the angle between articulating bones, often to restore a body part to the anatomical position after it has been flexed. Flexion and extension usually occur in the sagittal plane. Hinge, pivot, condyloid, saddle, and ball-and-socket joints all permit flexion: (1) bending the head toward the chest at the atlanto-occipital joint (between the atlas and occipital bone) and the cervical intervertebral joints (between the cervical vertebrae) (see Figure 8.3a); (2) moving the humerus forward at the shoulder (glenohumeral) joint as in swinging the arms forward while walking; (3) moving the forearm toward the arm at the elbow joint, between the humerus, ulna, and radius; (4) moving the palm toward the forearm at the wrist (radiocarpal) joint between the radius and carpals (see Figure 8.3h); (5) bending the digits of the hand or foot at the interphalangeal joints; (6) moving the femur forward at the hip joint, as in walking (see Figure 8.3g); and (7) moving the leg toward the thigh at the knee (tibiofemoral) joint as occurs when bending the knee.

Although flexion and extension usually occur in the sagittal plane, there are a few exceptions. For instance, movement of the trunk sideways to the right or left at the waist occurs in the frontal plane. This movement involves the intervertebral joints and is called **lateral flexion** (see Figure 8.3k).

Continuation of extension beyond the anatomical position is called **hyperextension** (*hyper* = beyond or excessive). Examples of hyperextension include the following: (1) bending the head backward at the atlanto-occipital and cervical intervertebral joints (see Figure 8.3a); (2) moving the humerus backward at the shoulder joint as in swinging the arms while walking; (3) moving the palm backward at the wrist joint (see Figure 8.3h); and (4) moving the femur backward at the hip joint as in walking (see Figure 8.3g).

FIGURE 8.3   Movements at synovial joints.

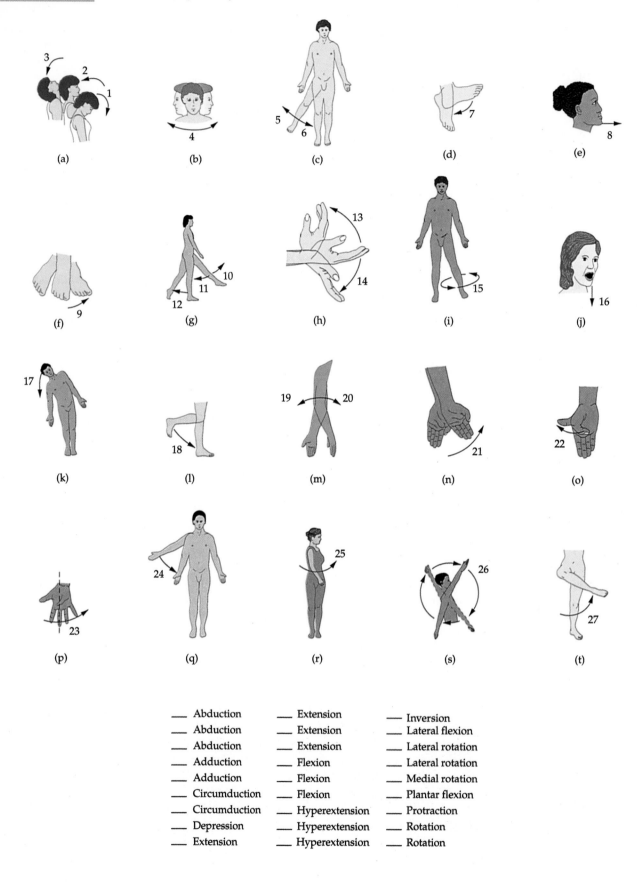

___ Abduction	___ Extension	___ Inversion
___ Abduction	___ Extension	___ Lateral flexion
___ Abduction	___ Extension	___ Lateral rotation
___ Adduction	___ Flexion	___ Lateral rotation
___ Adduction	___ Flexion	___ Medial rotation
___ Circumduction	___ Flexion	___ Plantar flexion
___ Circumduction	___ Hyperextension	___ Protraction
___ Depression	___ Hyperextension	___ Rotation
___ Extension	___ Hyperextension	___ Rotation

Hyperextension of hinge joints, such as the elbow, interphalangeal, and knee joints, is usually prevented by the arrangement of ligaments and the anatomical alignment of the bones.

Abduction (ab-DUK-shun) is the movement of a bone away from the midline, whereas **adduction** (ad-DUK-shun) is the movement of a bone toward the midline. Both movements usually occur in the frontal plane. Condyloid, saddle, and ball-and-socket joints permit abduction and adduction. Examples of abduction include moving the humerus laterally at the shoulder joint (see Figure 8.3q), moving the palm laterally at the wrist joint (see Figure 8.3n), and moving the femur laterally at the hip joint (see Figure 8.3c). The movement that returns each of these body parts to the anatomical position is adduction.

With respect to the digits, the midline of the body is not used as the point of reference for abduction and adduction. When abducting the fingers (but not the thumb), an imaginary line drawn through the longitudinal axis of the middle (longest) finger is the point of reference, and the fingers move away (spread out) from the middle finger (see Figure 8.3p). In abduction of the thumb, the thumb moves away from the palm in the sagittal plane (see Figure 8.3o). When abducting the toes, an imaginary line drawn through the second toe is the point of reference. Adduction of the fingers and toes involves returning them to the anatomical position. Adduction of the thumb moves the thumb toward the palm in the sagittal plane.

Circumduction (ser-kum-DUK-shun) is movement of the distal end of a part of the body in a circle. It occurs as a result of a continuous sequence of flexion, abduction, extension, and adduction. Because circumduction is a combined movement, it is not considered to be a separate axis of movement. Examples of circumduction are moving the humerus in a circle at the shoulder joint (see Figure 8.3s), moving the hand in a circle at the wrist joint, moving the thumb in a circle at the carpometacarpal joint, moving the fingers in a circle at the metacarpophalangeal joints, and moving the femur in a circle at the hip joint (see Figure 8.3i). Both the shoulder and hip joints permit circumduction, but flexion, abduction, extension, and adduction are more limited in the hip joints than the shoulder joints due to the tension on certain ligaments and muscles.

c. Rotation

In **rotation** (rō-TĀ-shun; *rota* = revolve) a bone revolves around its own longitudinal axis. Pivot and ball-and-socket joints permit rotation. One example is turning the head from side to side at the atlanto-axial joint, as in signifying "no" (see Figure 8.3b). Another is turning the trunk from side to side at the intervertebral joints while keeping the hips and lower limbs in the anatomical position (see Figure 8.3r). In the limbs, rotation is defined relative to the midline, and specific qualifying terms are used. If the anterior surface of a bone of the limb is turned toward the midline, the movement is called *medial (internal) rotation* (see Figure 8.3m). You can medially rotate the humerus at the shoulder joint as follows. Starting in the anatomical position, flex your elbow and then draw your palm across the chest. Medial rotation of the forearm at the radioulnar joints (between the radius and ulna) involves turning the palm medially from the anatomical position. You can medially rotate the femur at the hip joint as follows. Lie on your back, bend your knee, and then move your foot laterally from the midline. Although you are moving your foot laterally, the femur is rotating medially. Medial rotation of the leg at the knee joint can be produced by sitting on a chair, bending your knee, raising your lower limb off the floor, and turning your toes medially (see Figure 8.3t). If the anterior surface of the bone of a limb is turned away from the midline, the movement is called *lateral (external) rotation*.

d. Special Movements

As noted previously, **special movements** occur only at certain joints. They include elevation, depression, protraction, retraction, inversion, eversion, dorsiflexion, plantar flexion, supination, pronation, and opposition.

Elevation (el-e-VĀ-shun = to lift up) is an upward movement of a part of the body. Examples are closing the mouth at the temporomandibular joint (between the temporal bone and mandible) to elevate the mandible and shrugging the shoulders at the acromioclavicular joint (between the scapula and clavicle) to elevate the scapula.

Depression (dē-PRESH-un = to press down) is a downward movement of a part of the body. For example, opening the mouth to depress the mandible (see Figure 8.3j), or returning shrugged shoulders to the anatomical position to depress the scapula.

Protraction (prō-TRAK-shun = to draw forth) is a movement of a part of the body anteriorly in the transverse plane. You can protract your mandible at the temporomandibular joint by thrusting it outward (see Figure 8.3e) or protract your clavicles at the acromioclavicular and sternoclavicular joints by crossing your arms.

Retraction (rē-TRAK-shun = to draw back) is a movement of a protracted part of the body back to the anatomical position.

Inversion (in-VER-zhun = to turn inward) is movement of the soles medially at the intertarsal joints so that the soles face each other (see Figure 8.3f).

Eversion (ē-VER-zhun = to turn outward) is a movement of the soles laterally at the intertarsal joints so that the soles face away from each other.

Dorsiflexion (dor-si-FLEK-shun) refers to bending of the foot at the ankle (talocrural) joint in the direction of the dorsum (superior surface). Dorsiflexion occurs when you stand on your heels.

Plantar flexion involves bending of the foot at the ankle joint in the direction of the plantar surface (see Figure 8.3d), as when standing on your toes.

Supination (soo-pi-NĀ-shun) is a movement of the forearm at the proximal and distal radioulnar joints in which the palm is turned anteriorly (see Figure 8.3m). Recall that this position of the palms is one of the defining features of the anatomical position.

Pronation (prō-NĀ-shun) is a movement of the forearm at the proximal and distal radioulnar joints in which the distal end of the radius crosses over the distal end of the ulna and the palm is turned posteriorly (see Figure 8.3m).

Opposition (op-ō-ZISH-un) is the movement of the thumb at the carpometacarpal joint (between the trapezium and metacarpal of the thumb) in which the thumb moves across the palm to touch the tips of the fingers on the same hand. This is the single most distinctive digital movement that gives humans and other primates the ability to grasp and manipulate objects precisely.

Label the various movements illustrated in Figure 8.3.

E. Knee Joint

The knee joint is the largest and most complex joint of the body. It actually consists of three joints within a single synovial cavity: (1) an intermediate patellofemoral joint between the patella and the patellar surface of the femur, which is a planar joint; (2) a lateral tibiofemoral joint between the lateral condyle of the femur, lateral meniscus, and lateral condyle of the tibia, which is a modified hinge joint; and (3) a medial tibiofemoral joint between the medial condyle of the femur, medial meniscus, and medial condyle of the tibia, which is also a modified hinge joint.

The knee joint illustrates the basic structure of a diarthrosis and the limitations on its movement. Some of the structures associated with the knee joint are as follows:

1. *Tendon of quadriceps femoris muscle.* Strengthens joint anteriorly and externally.

2. *Gastrocnemius muscle.* Strengthens joint posteriorly and externally.

3. *Patellar ligament.* Continuation of the common tendon of insertion of the quadriceps femoris muscle that extends from the patella to the tibial tuberosity. The ligament strengthens anterior surface of the joint.

4. *Fibular collateral ligament.* Strong, rounded ligament on the lateral surface of the joint, between femur and fibula; strengthens the lateral aspect of the joint and prohibits side-to-side movement at the joint.

5. *Tibial collateral ligament.* Broad, flat ligament on the medial surface of the joint, between femur and tibia; strengthens the medial aspect of the joint and prohibits side-to-side movement at the joint.

6. *Oblique popliteal ligament.* Broad, flat ligament that extends from the intercondylar fossa of the femur to the head of the tibia; strengthens the posterior surface of the knee.

7. *Anterior cruciate ligament (ACL).* Passes posteriorly and laterally from a point *anterior* to the intercondylar area of the tibia to the posterior part of the medial surface of the lateral condyle of the femur; limits hyperextension of the knee and prevents anterior sliding of the tibia on the femur.

8. *Posterior cruciate ligament (PCL).* Passes anteriorly and medially from a depression on the *posterior* intercondylar area of the tibia and lateral meniscus to the anterior part of the lateral surface of the medial condyle of the femur; prevents posterior sliding of tibis when the knee is flexed.

9. *Articular discs (menisci).* Two fibrocartilage discs between the tibial and femoral condyles that help compensate for the irregular shapes of the bones and circulate synovial fluid.

FIGURE 8.4   Ligaments, tendons, bursae, and menisci of right knee joint.

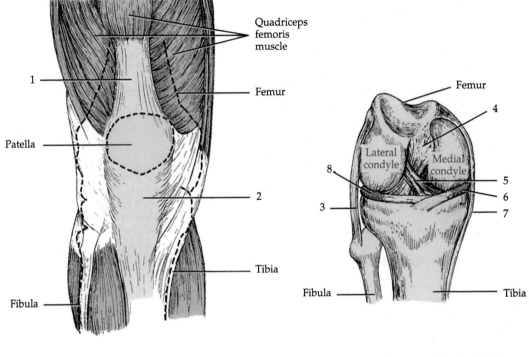

(a) Anterior, superficial view

(b) Anterior deep view (flexed) with many superficial structures removed

___ Anterior cruciate ligament
___ Fibular (lateral) collateral ligament
___ Lateral meniscus
___ Medial meniscus

___ Patellar ligament
___ Posterior cruciate ligament
___ Tendon of quadriceps femoris muscle
___ Tibial (medial) collateral ligament

a. *Medial meniscus.* Semicircular piece of fibrocartilage (C-shaped) attached to the tibia.

b. *Lateral meniscus.* Nearly circular piece of fibrocartilage (approaches an incomplete O in shape) attached to the tibia. The medial and lateral menisci are connected to each other by the *transverse ligament.*

10. The more important **bursae** of the knee include the following:

a. *Prepatellar bursa* between the patella and skin.

b. *Infrapatellar bursa* between superior part of tibia and patellar ligament.

c. *Suprapatellar bursa* between inferior part of femur and deep surface of quadriceps femoris muscle.

Label the structures associated with the knee joint in Figure 8.4.

If a longitudinally sectioned knee joint of a cow or lamb is available, examine it and see how many structures you can identify.

ANSWER THE LABORATORY REPORT QUESTIONS AT THE END OF THE EXERCISE.

FIGURE 8.4 Ligaments, tendons, bursae, and menisci of right knee joint. (Continued)

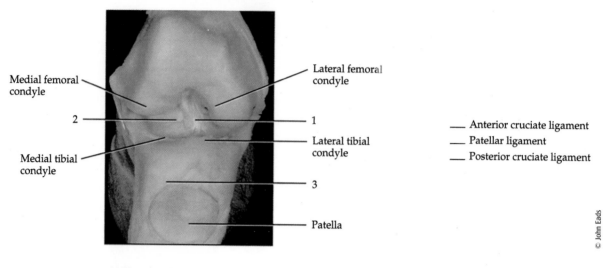

Medial femoral condyle

Lateral femoral condyle

2

Medial tibial condyle

1

Lateral tibial condyle

3

Patella

___ Anterior cruciate ligament
___ Patellar ligament
___ Posterior cruciate ligament

(c) Photograph of internal structure

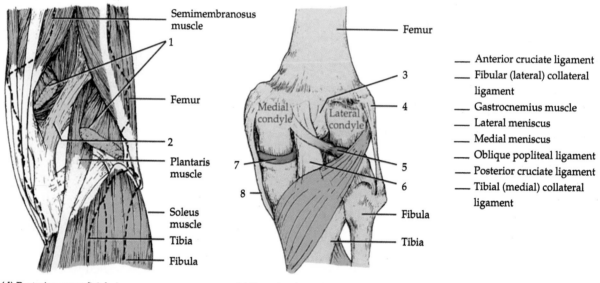

Semimembranosus muscle

1

Femur

2

Plantaris muscle

Soleus muscle

Tibia

Fibula

(d) Posterior, superficial view

Femur

3

Medial condyle

Lateral condyle

4

7

5

8

6

Fibula

Tibia

___ Anterior cruciate ligament
___ Fibular (lateral) collateral ligament
___ Gastrocnemius muscle
___ Lateral meniscus
___ Medial meniscus
___ Oblique popliteal ligament
___ Posterior cruciate ligament
___ Tibial (medial) collateral ligament

(e) Posterior deep view with many superficial structures removed

FIGURE 8.4   Ligaments, tendons, bursae, and menisci of right knee joint. (Continued)

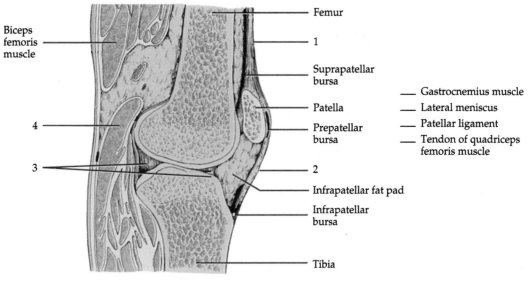

Biceps femoris muscle

Femur

1

Suprapatellar bursa

Patella

Prepatellar bursa

2

Infrapatellar fat pad

Infrapatellar bursa

Tibia

4

3

____ Gastrocnemius muscle
____ Lateral meniscus
____ Patellar ligament
____ Tendon of quadriceps femoris muscle

(f) Sagittal section

EXERCISE 8

Joints

Name _____   Date _____

Laboratory Section _____   Score/Grade _____

PART 1 ■ Multiple Choice

_____ 1. A joint united by dense fibrous tissue that permits a slight degree of movement is a (a) suture (b) syndesmosis (c) symphysis (d) synchondrosis

_____ 2. A joint that contains a broad flat disc of fibrocartilage is classified as a (a) ball-and-socket joint (b) suture (c) symphysis (d) gliding joint

_____ 3. The following characteristics define what type of joint? Presence of a synovial cavity, articular cartilage, synovial membrane, and ligaments. (a) suture (b) synchondrosis (c) syndesmosis (d) hinge

_____ 4. Which joints are slightly movable? (a) diarthroses (b) amphiarthroses (c) synovial (d) synarthroses

_____ 5. Which type of joint is immovable? (a) synarthrosis (b) syndesmosis (c) symphysis (d) diarthrosis

_____ 6. What type of joint provides triaxial movement? (a) hinge (b) ball-and-socket (c) saddle (d) condyloid

_____ 7. Which ligament provides strength on the medial side of the knee joint? (a) oblique popliteal (b) posterior cruciate (c) fibular collateral (d) tibial collateral

_____ 8. On the basis of structure, which joint is fibrous? (a) symphysis (b) synchondrosis (c) pivot (d) syndesmosis

_____ 9. The elbow, knee, and interphalangeal joints are examples of which type of joint? (a) pivot (b) hinge (c) gliding (d) saddle

_____ 10. Functionally, which joint provides the greatest degree of movement? (a) diarthrosis (b) synarthrosis (c) amphiarthrosis (d) syndesmosis

PART 2 ■ Completion

11. The thin layer of hyaline cartilage on articulating surfaces of bones is called

_____ cartilage.

12. The synovial membrane and fibrous capsule together form the _____ capsule.

13. Pads of fibrocartilage between the articular surfaces of bones that maintain stability of the joint are

called _____ .

14. Fluid-filled connective tissue sacs that cushion movements of one body part over another are referred to as _____.

15. The _____ ligament supports the back of the knee and helps to limit hyperextension.

16. Movement of the trunk in the frontal plane is called _____.

17. In extension of the thumb, the thumb moves _____ away from the palm.

18. A joint that moves in two planes is referred to as a _____ joint.

PART 3 ▪ Matching

_____ 19. Circumduction

_____ 20. Adduction

_____ 21. Flexion

_____ 22. Pronation

_____ 23. Elevation

_____ 24. Protraction

_____ 25. Rotation

_____ 26. Plantar flexion

_____ 27. Dorsiflexion

_____ 28. Inversion

_____ 29. Hyperextension

A. Decrease in the angle between articulating bones, usually in the sagittal plane

B. Moving a part superiorly

C. Bending the foot in the direction of the dorsum (upper surface)

D. Anterior movement in the transverse plane

E. Movement of the sole medially

F. Movement toward the midline

G. Bending the foot in the direction of the plantar surface (sole)

H. Turning the palm posteriorly

I. Movement of a bone around its own axis

J. Distal end of a bone moves in a circle while the proximal end remains relatively stable

K. Extension beyond the anatomical position

Muscular Tissue

Objectives

At the completion of this exercise you should understand

A The physiological characteristics of muscular tissue.

B The histological organization of skeletal, cardiac, and smooth muscle tissue.

C The physiological and biochemical processes involved in contraction and relaxation of skeletal muscle.

Although bones provide leverage and form the framework of the body, they cannot move the body by themselves. Motion results from alternating contraction and relaxation of muscles, which constitute 40% to 50% of the total body weight. The prime function of muscle is to change chemical energy (in the form of ATP) into mechanical energy to generate force, perform work, and produce movement. Muscular tissue also stabilizes body positions, regulates organ volume, generates heat, and propels fluids and food matter through various body systems. The scientific study of muscles is known as **myology** (mī-OL-ō-jē; *myo* = muscle; *logy* = study of).

Muscular tissue has four special properties that enable it to function and contribute to homeostasis:

1. *Electrical excitability,* a property of both muscle cells and nerve cells, is the ability to respond to certain stimuli by producing electrical signals such as *action potentials.* For muscle cells, two main types of stimuli can trigger action potentials. One is autorhythmic electrical signals arising in the muscle tissue itself. The other is chemical stimuli, such as neurotransmitters released by neurons, hormones distributed by blood, or even local changes in pH.

2. *Contractility* is the ability of muscle tissue to contract forcefully when stimulated by an action potential. When muscles contract, it generates tension (force of contraction) while pulling on its attachment points.

3. *Extensibility* is the ability of muscle to stretch without being damaged. Extensibility allows a muscle to contract forcefully even if it is already stretched.

4. *Elasticity* is the ability of muscle tissue to return to its original length and shape after contraction or extension.

In this exercise you will examine the histological structure of muscle tissue and conduct exercises on the physiology of muscle.

A. Types of Muscular Tissue

The body has three types of muscular tissue—skeletal, cardiac, and smooth—that differ from one another in their location, microscopic anatomy, and control by the nervous and endocrine systems.

1. *Skeletal muscle tissue* is so named because most skeletal muscles function to move bones of the skeleton. (A few skeletal muscles attach to and move the skin or other muscles.) Skeletal muscle tissue is said to be *striated* because alternating light and dark bands *(striations)* are visible when the tissue is examined under a microscope (see Figure 9.1). Skeletal muscle tissue works in a *voluntary* manner because its activity can be consciously controlled by neurons (nerve cells) that are part of the somatic (voluntary) division of the nervous system.

2. *Cardiac muscle tissue* forms most of the heart wall. It is also *striated* muscle, but its action is

involuntary; that is, its contraction is usually not under conscious control.

3. **Smooth muscle tissue** is located in the walls of hollow internal structures, such as blood vessels, airways, and most organs in the abdominopelvic cavity. It is also found in the skin, attached to hair follicles. Under a microscope, this tissue lacks the striations of skeletal and cardiac muscle tissue. For this reason it looks *nonstriated* or *smooth.* The action of smooth muscle is usually *involuntary.*

B. Structure of Skeletal Muscle Tissue

Each skeletal muscle is a separate organ composed of hundreds or thousands of cells called **muscle fibers** because of their elongated shapes. Thus, the terms *muscle cell* and *muscle fiber* are synonymous (see Figure 9.1). The typical length of a muscle fiber is 100 mm (4 in.) but some are as long as 30 cm (12 in.); the diameter ranges from 10 to 100 μm. There are multiple nuclei in each muscle fiber, and they are located just between the sarcolemma. Within the cytoplasm (sarcoplasm) are threadlike structures called *myofibrils* that extend lengthwise within the fiber. Within myofibrils are two types of even smaller structures called *filaments* that are arranged in compartments called *sarcomeres,* the basic functional units of a myofibril.

Examine a prepared slide of skeletal muscle tissue in longitudinal and transverse (cross) section under high power. Look for the following:

1. **Sarcolemma** (*sarco* = flesh; *lemma* = sheath). Plasma membrane of the muscle fiber.

2. **Sarcoplasm.** Cytoplasm of the muscle fiber.

3. **Nuclei.** Multiple nuclei in each muscle fiber lying close to the sarcolemma.

4. **Striations.** Alternating light and dark bands in each muscle fiber (described on this page).

5. **Epimysium** (ep'-i-MĪZ-ē-um; *epi* = upon). The outermost layer of fibrous connective tissue that encircles the whole muscle.

6. **Perimysium** (per'-i-MĪZ-ē-um; *peri* = around). Fibrous connective tissue surrounding groups of 10 to 100 or more individual muscle fibers, separating them into bundles called **fascicles** (FAS-i-kuls = little bundles).

7. **Endomysium** (en'-dō-MĪZ-ē-um; *endo* = within). A thin sheath of areolar connective tissue penetrating the interior of each fascicle and separating individual muscle fibers (muscle cells) from one another.

Refer to Figure 9.1 and label the structures indicated.

With the use of an electron microscope, additional details of skeletal muscle tissue may be noted. Among these are the following:

1. **Mitochondria.** Organelles that have a smooth outer membrane and folded inner membrane in which ATP is generated.

2. **Sarcoplasmic reticulum** (sar'-kō-PLAZ-mik re-TIK-ū-lum), or **SR.** A fluid-filled system of membranous sacs similar to the smooth endoplasmic reticulum of nonmuscle cells; stores calcium ions in relaxed muscle fibers. The dilated end sacs of sarcoplasmic reticulum on both sides of the T tubule are called **terminal cisterns.**

3. **Transverse (T) tubules.** Thousands of tiny invaginations of the sarcolemma that tunnel in from the surface toward the center of each

FIGURE 9.1   **Histology of skeletal muscle tissue.**

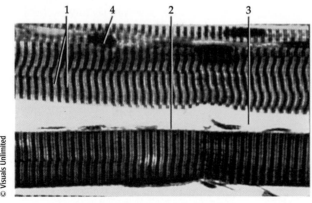

© Visuals Unlimited

____ Endomysium
____ Nucleus
____ Sarcolemma
____ Striations

Photomicrograph of several muscle fibers
in longitudinal section (900X)

muscle fiber and connect with the sarcoplasmic reticulum.

4. **Triad.** Transverse tubule and the terminal cisterns on either side of it.

5. **Myofibrils.** Threadlike structures that extend the entire length of the muscle fiber and consist of **thin filaments** composed of the protein actin and **thick filaments** composed of the protein myosin; the contractile proteins of muscle tissue.

6. **Sarcomere** (*mere* = part). Basic functional units of a myofibril; compartment within a muscle fiber separated from other sarcomeres by dense material called **Z discs.**

7. **A band.** Darker, middle portion of the sarcomere which extends the entire length of the thick filaments and portions of thin filaments where they lie side by side.

8. **I band.** Lighter, less dense area of the sarcomere that contains the rest of the thin filaments but no thick filaments. The combination of alternating dark A bands and light I bands gives the muscle fiber its striated (striped) appearance.

9. **H zone.** Region in the center of the A band of a sarcomere consisting of thick filaments only.

10. **M line.** Supporting proteins that hold the thick filaments together at the center of the H zone.

Figure 9.2 is a diagram of skeletal muscle tissue based on electron micrographic studies. Label the structures shown.

C. Contraction of Skeletal Muscle Tissue

Skeletal muscle contraction is a process of electrical, chemical, and mechanical events. **Excitation-contraction coupling** is the linkage between the electrochemical events and the actual mechanical event we call a skeletal muscle contraction. Muscle

FIGURE 9.2   Enlarged aspect of several myofibrils of skeletal muscle tissue based on an electron micrograph.

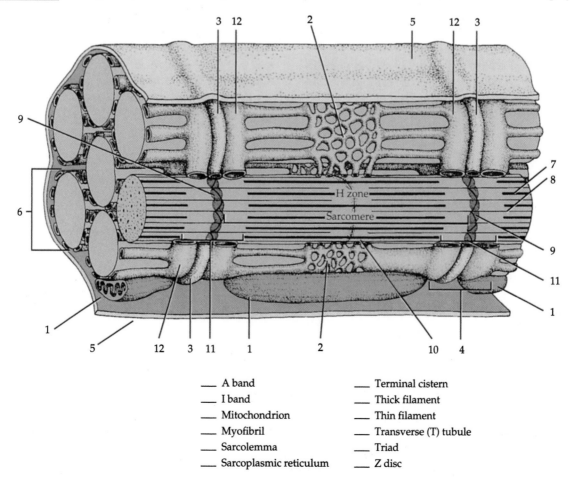

___ A band	___ Terminal cistern
___ I band	___ Thick filament
___ Mitochondrion	___ Thin filament
___ Myofibril	___ Transverse (T) tubule
___ Sarcolemma	___ Triad
___ Sarcoplasmic reticulum	___ Z disc

contraction occurs only if the muscle is stimulated with a stimulus of threshold or suprathreshold intensity. A **threshold stimulus** is defined as a minimum strength stimulus necessary to initiate a contraction. Under normal physiological conditions, this is accomplished by a nerve impulse's being transmitted to the skeletal muscle fiber via a nerve cell called a **motor neuron.** A somatic motor neuron and all of the skeletal muscle fibers it stimulates is termed a **motor unit.**

Each motor neuron possesses a single long process called an **axon,** extending from the nerve cell body to the muscle fiber. As the axon enters the endomysium it branches into fine processes termed **axon terminals.** Each axon terminal enlarges into swellings termed **synaptic end bulbs.** The synapse formed by the invagination of the skeletal muscle sarcolemma and these end bulbs is termed a **neuromuscular (NMJ)** or **myoneural junction** (Figure 9.3). The portion of the skeletal muscle fiber immediately deep to the axon terminal demonstrates a large number of anatomical specializations and is termed the **motor end plate.**

Within each synaptic end bulb are numerous membrane-closed sacs, called *synaptic vesicles.* Each vesicle contains a store of chemicals termed *neurotransmitter substance.* The space between the axon terminal and sarcolemma is termed the *synaptic cleft.* The neurotransmitter substance diffuses across the synaptic cleft and binds to receptors on the motor end plate, initiating the excitation-contraction coupling process.

At the neuromuscular junction, a wave of depolarization (nerve impulse) spreads along the axon opening calcium specific channels. Extracellular calcium ions (Ca^{2+}) enter through these channels and bind to the synaptic vesicles. This binding effect leads to the fusion of the vesicles with the synaptic end bulb membrane, releasing the neurotransmitter substance called **acetylcholine** (as'-ē-til-KŌ-lēn), or **Ach.** Acetylcholine diffuses across the synaptic cleft and binds to receptors on the motor end plate. The binding of Ach with the receptors opens ligand (chemical) gated channels on the sarcolemma, allowing an influx of sodium into the muscle, thus initiating depolarization of the muscle cell membrane or a *muscle action potential.* As the muscle action potential spreads along the sarcolemma it passes into the transverse (T) tubules. The action potential then travels into the sarcoplasmic reticulum (SR) causing the release of intracellular calcium ions into the sarcoplasm. Calcium can now bind to **troponin,** on the **troponin-tropomyosin complex.** This binding effect leads to a changing of the shape of the troponin-tropomyosin complex thereby exposing the *myosin binding sites* on the actin protein molecules of the thin filaments, allowing the myosin cross bridges to attach to the thin filament. The attachment between the myosin cross bridges and the receptors on the actin causes ATP to be split by myosin ATPase, an

FIGURE 9.3 **Neuromuscular junction.**

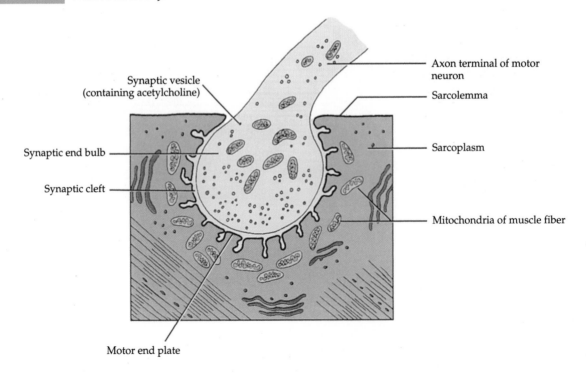

Synaptic vesicle (containing acetylcholine)

Synaptic end bulb

Synaptic cleft

Axon terminal of motor neuron

Sarcolemma

Sarcoplasm

Mitochondria of muscle fiber

Motor end plate

FIGURE 9.4 Sliding filament theory of a skeletal muscle contraction. The position of the various parts of two sarcomeres in relaxed, contracting, and maximally contracted states are shown. Note the movement of thin filaments and relative size of H zone.

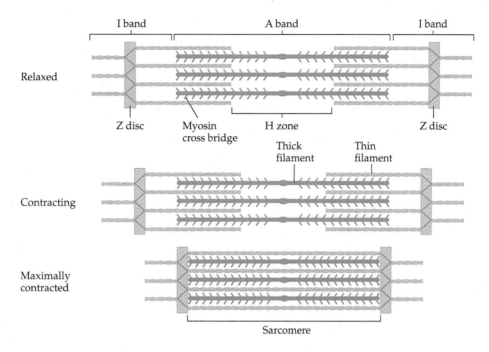

enzyme found on the thick filament. The energy released by the splitting of ATP into ADP and inorganic phosphate causes the myosin cross bridges to move, and the thin filaments slide inward toward the H zone, causing the muscle fiber to shorten (Figure 9.4). This process is called the **power stroke.** As this is occurring, newly synthesized ATP displaces ADP from the myosin molecule. If Ca^{2+} has been taken back into the sarcoplasmic reticulum by the calcium pump, the myosin cross bridges release from the actin receptors and relaxation occurs. However, if Ca^{2+} is still present within the sarcoplasm, the entire process repeats and further contraction occurs. This theory of skeletal muscle contraction is termed the **sliding filament theory of skeletal muscle contraction.**

Two changes permit a muscle fiber to relax after it has contracted. First, ACh is rapidly broken down by an enzyme called **acetylcholinesterase (AChE).** When action potentials cease in a motor neuron, release of ACh stops and AChE rapidly breaks down any ACh already present in the synaptic cleft. Second, active transport pumps rapidly remove Ca^{2+} from the cytosol (sarcoplasm) back into the sarcoplasmic reticulum. As the Ca^{2+} level in the cytosol (sarcoplasm) drops the troponin-tropomyosin

complexes slide back over and cover the myosin-binding sites, and the muscle fiber relaxes.

A contracting skeletal muscle fiber follows the **all-or-none principle.** This principle states that a skeletal muscle fiber will respond maximally or not at all to a stimulus. If the stimulus is of threshold or greater intensity the fiber will respond maximally; if the stimulus is subthreshold in intensity the fiber will not respond. This principle, however, does not imply that the entire skeletal muscle must be either fully relaxed or fully contracted, because individual fibers within a muscle possess varying thresholds for stimulation. Therefore, the muscle as a whole can have graded contractions.

With these facts in mind, you will now perform a few simple lab tests that will illustrate the physiology of skeletal muscle contraction.

D. Laboratory Tests on Skeletal Muscle Contraction

Skeletal muscles produce different kinds of contractions depending on the intensity and frequency of the stimulus applied. Twitch contractions do not occur in the body but are worth demonstrating because they show the different phases of muscle contraction quite clearly. Any record of a muscle contraction is called a **myogram.**

NOTE *Depending on the availability of live animals and the type of laboratory equipment available, you may select from the following procedures related to skeletal muscle contraction.*

a. Procedure Using Biopac

Biopac Student Lab (BSL) system is an interactive hardware and software system that permits students to generate and record data from their bodies and animal or tissue preparations. It permits students to perform experiments, analyze data, draw conclusions, and form hypotheses on the basis of the collected data. The BSL system is available from BIOPAC Systems, Inc., 42 Aero Camino, Goleta, CA 93117. To order call: 1-800-685-0066, Email: info@biopac.com, web page: www.biopac.com.

For activities related to skeletal muscle contraction, select the appropriate experiments from Biopac.

BSL1—Standard and Integrated EMG

BSL2—Motor Unit Recruitment and Fatigue

H06—Finger Twitch

H27—Facial EMG

H36—Muscular Biofeedback

A01—Frog Pith and Prep

b. Procedure Using Physiogrip™

Physiogrip™ is a computer hardware and software system designed to demonstrate the contractile characteristics of human skeletal muscle. Through the use of a specially designed displacement transducer, Physiogrip™ provides an alternative to vivisection for studying classic muscle physiology by allowing students to experience their own flexor digitorum superficialis muscle responding to motor point stimulation. The program is available from INTELITOOL® (1-800-955-7621).

c. Procedure Using Polygraph and Pithed Frogs

You may test muscle contraction through a team exercise or by observing a demonstration prepared by your instructor. Depending on the size of the class and the equipment available, you should work in teams of three to five students: one or two students should be assigned to assist the instructor to set up the physiological apparatus, one or two other students should be assigned as "surgeons" to isolate, remove, and suspend the frog muscle, and another student should act as the recorder and coordinate the work.

E. Biochemistry of Skeletal Muscle Contraction

You will examine the effect of the following solutions on the contraction of glycerinated skeletal muscle fibers:[1] (1) ATP solution, (2) mineral ion solution, and (3) ATP plus mineral ion solution.

PROCEDURE

1. Using 7× or 10× magnification and glass needles or clean stainless steel forceps, gently tease the muscle into very thin groups of myofibers. Single fibers or thin groups must be used because strands thicker than a silk thread curl when they contract.

2. Using a Pasteur pipette or medicine dropper, transfer one strand into a drop of glycerol on a clean glass slide and cover the preparation with a cover slip. Examine the strand under low and high power and note the striations. Also note that each fiber has several nuclei.

3. Transfer one of the thinnest strands to a drop of glycerol on a second microscope slide. Do not add a cover slip. If the amount of glycerol on the slide is more than a small drop, soak the excess into a piece of lens paper held at the edge of the glycerol farthest from the fibers. Using a dissecting microscopic and a millimeter ruler held beneath the slide, measure the length of one of the fibers.

4. Now flood the fibers with the solution containing only ATP and observe their reaction. After 30 sec or more, remeasure the same fiber.

    How much did the fiber contract? _____ mm

5. Using clean slides and medicine droppers, and being especially sure to use clean teasing needles or forceps, transfer other fibers to a drop of glycerol on a slide. Again measure the length of one fiber. Next flood the fibers with a solution containing mineral ions, observe their reaction, and remeasure the fiber.

    How much did the fiber contract? _____ mm

---

[1]Glycerinated muscle preparation and solutions are supplied by the Carolina Biological Supply Company, Burlington, NC 27215.

6. Repeat the exercise, this time using a solution containing a combination of ATP and ions.

7. Observe a contracted fiber under low and high power, and look for differences in appearance between muscle in a contracted state and muscle in a relaxed state (see Figure 9.4).

F. Electromyography

During the contraction of a single muscle fiber, an action potential is generated that lasts between 1 and 4 msec. This electrical activity is dissipated throughout the surrounding tissues. In order to produce a smooth muscle contraction, motor units fire asynchronously, thereby causing muscle fibers to contract at different times. This asynchronous firing of motor units also prolongs the electrical activity resulting from skeletal muscle contraction. By placing two electrodes on the skin or directly within the muscle, an electrical recording, termed an **electromyography,** or **EMG,** of this muscular activity may be obtained when the muscle is stimulated (Figure 9.5). EMGs are utilized clinically to distinguish between peripheral neurological and muscular diseases, and for differentiating abnormalities resulting in reductions of either muscular strength or sensation that may have been caused by either dysfunction or peripheral nervous tissue or central nervous system (CNS) centers. Typically EMGs are recorded under three different activity levels: complete inactivity, slight muscular activity, and extreme (maximal) muscular activity. Ab-normal differences in records obtained under these three conditions may also indicate an abnormality in motor unit recruitment.

1. Subject Preparation

PROCEDURE

1. Locate the **flexor digitorum superficialis** muscle (see Figure 10.14a) on the anterior surface of the forearm and the **triceps brachii** muscle (see Figure 10.13b) on the posterior surface of the subject's arm. EMG electrodes (or ECG electrodes) will be placed on the skin superficial to these muscles.

2. Prior to placing the electrodes on the skin, *gently* scrub the area with a dishwashing pad in order to remove dead epithelial cells and to facilitate recording. *Make sure that the capillaries are not damaged and no bleeding occurs.*

3. Apply a small amount of electrode gel to each electrode and apply them to the proper locations.

4. Connect the two recording electrodes (placed on the anterior surface of the forearm) and the ground electrodes (placed on the posterior surface of the arm) to the preamplifier (high gain coupler) and set the gain to × 100, sensitivity at an appropriate setting between 20 and 100, time constant to 0.03, and paper speed at high.

2. Recording of Spontaneous Muscle Activity

PROCEDURE

1. With the subject completely relaxed and his or her arm placed horizontally on a laboratory table, record any spontaneously occurring electrical activity within the flexor digitorum superficialis over a time period of 1 to 3 min.

2. Record your observations in Section F.1 of the LABORATORY REPORT RESULTS at the end of the exercise.

3. Recruitment of Motor Units

PROCEDURE

1. Have the subject gently flex her or his digits while recording, and note the electrical activity.

2. Have the subject relax his or her digits completely, followed by a more forceful contraction. Note any change in the EMG.

FIGURE 9.5 Diagrammatic representation of normal electromyograms. The single potential in the upper left corner has a measured amplitude of 0.8 mV and a duration of 7 msec.

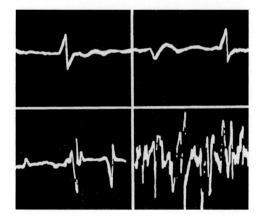

3. Place a tennis ball in your subject's hand and ask him or her to squeeze the ball once or twice as strongly as possible. Note the EMG recording, and then let your subject rest thoroughly for at least 5 min.

4. Record your observations in Section F.2 of the LABORATORY REPORT RESULTS at the end of the exercise.

4. Effect of Fatigue

PROCEDURE

1. *Without recording* an EMG on the polygraph, have your subject repeatedly squeeze the tennis ball as strongly as possible until no longer able to squeeze the ball.

2. When fatigue has been achieved, *quickly* start recording and ask the subject to attempt to squeeze the ball as strongly as possible four or five times in quick succession.

3. Note the EMG recording and record your observations in Section F.3 of the LABORATORY REPORT RESULTS at the end of the exercise.

G. Cardiac Muscle Tissue

In contrast to skeletal muscle fibers, cardiac muscle fibers are shorter in length, larger in diameter, and not quite as circular in transverse section. They also exhibit branching, which gives an individual fiber a Y-shaped appearance (see Figure 9.6). A typical cardiac muscle fiber is 50 to 100 μm long and has a diameter of about 14 μm. Usually there is only one centrally located nucleus. The sarcolemma of cardiac muscle fibers is similar to that of skeletal muscle, but the sarcoplasm is more abundant and the mitochondria are larger and more numerous. Cardiac muscle fibers are arranged in layers and have the same arrangement of actin and myosin and the same bands, zones, and Z discs as skeletal muscle fibers. The transverse (T) tubules of mammalian cardiac muscle tissue are wider but less abundant than those of skeletal muscle: there is only one T tubule per sarcomere, located at the Z disc. Also, the sarcoplasmic reticulum of cardiac muscle fibers is scanty in comparison with that of skeletal muscle fibers.

Cardiac muscle fibers form two separate networks. The muscular walls and partition of the superior chambers of the heart (atria) compose one network. The muscular walls and partition of the inferior chambers of the heart (ventricles) compose the other network. The ends of each cardiac muscle fiber in a network are connected to the ends of neighboring fibers by irregular transverse thickenings of the sarcolemma called **intercalated** (in-TER-ka-lāt-ed; *intercal* = to insert between) **discs** (see Figure 9.6). The discs contain **desmosomes,** which hold the fibers together, and **gap junctions,** which allow muscle action potentials to spread from one cardiac muscle fiber to another. As a consequence, when a single fiber of

FIGURE 9.6 Histology of cardiac muscle tissue.

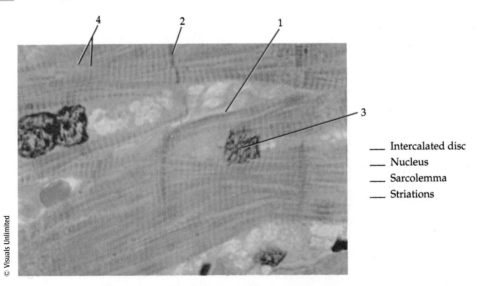

© Visuals Unlimited

____ Intercalated disc
____ Nucleus
____ Sarcolemma
____ Striations

Photomicrograph of several muscle
fibers in longitudinal section (1000×)

either network is stimulated, all the other fibers in the network become stimulated as well. Thus each network contracts as a functional unit. When the fibers of the atria contract as a unit, blood moves into the ventricles. Then, when the ventricular fibers contract as a unit, they pump blood out of the heart into arteries.

Under normal resting conditions, cardiac muscle tissue contracts and relaxes about 75 times a minute. This continuous, rhythmic activity is a major physiological difference between cardiac and skeletal muscle tissue. Another difference is the source of stimulation. Skeletal muscle tissue contracts only when stimulated by acetylcholine released by a nerve impulse in a motor neuron. In contrast, cardiac muscle tissue contracts when stimulated by its own autorhythmic muscle fibers. Its source of stimulation is a conducting network of specialized cardiac muscle fibers within the heart. Stimulation from the nervous system merely causes the conducting fibers to increase or decrease their rate of discharge. Also, cardiac muscle tissue remains contracted 10 to 15 times longer than skeletal muscle tissue. Cardiac muscle tissue also has a long **refractory period** (period of lost ex-

citability), lasting several tenths of a second, that allows time for the heart chambers to relax and fill with blood between beats. The long refractory period permits heart rate to increase significantly but prevents the heart from undergoing **tetanus** (sustained contraction). If heart muscle could undergo tetanus, blood flow would cease.

Examine a prepared slide of cardiac muscle tissue in longitudinal and transverse (cross) section under high power. Locate the following structures: **sarcolemma, endomysium, nuclei, striations,** and **intercalated discs.** Label Figure 9.6.

H. Smooth Muscle Tissue

Smooth muscle fibers are considerably smaller than skeletal muscle fibers. A single, relaxed smooth muscle fiber is 30–200 μm long, thickest in the middle (3–8 μm), and tapered at each end. Within each fiber is a single, oval, centrally located nucleus (see Figure 9.7a). The sarcoplasm of smooth muscle fibers contains both *thick filaments* and *thin filaments,* in ratios between 1:10 and 1:15, but they are not arranged in orderly sarcom-

| FIGURE 9.7 | Histology of smooth muscle tissue. |

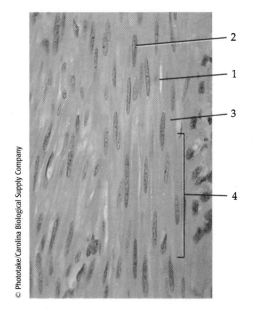

© Phototake/Carolina Biological Supply Company

(a) Photomicrograph of several muscle fibers in longitudinal section (180×)

___ Muscle fiber
___ Nucleus
___ Sarcolemma
___ Sarcoplasm

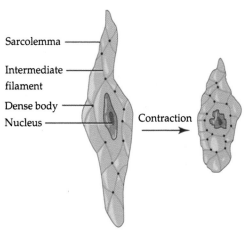

Sarcolemma ———
Intermediate filament ———
Dense body ———
Nucleus ———

Contraction →

(b) Diagram of contraction

eres as in striated muscle. Smooth muscle fibers also contain **intermediate filaments** (described shortly). Because the various filaments have no regular pattern of overlap, smooth muscle fibers do not exhibit striations. This is the reason for the name *smooth*. Smooth muscle fibers also lack transverse tubules and have only a small amount of sarcoplasmic reticulum for storage of calcium ions. Smooth muscle cells have small pouchlike invaginations of the plasma membrane called **caveolae** (kav'-ē-Ō-lē; *cavus* = space) that contain extracellular calcium ions used for muscular contraction.

In smooth muscle fibers, intermediate filaments attach to structures called **dense bodies,** which are functionally similar to Z discs in striated muscle fibers (see Figure 9.7b). Some dense bodies are dispersed throughout the sarcoplasm; others are attached to the sarcolemma. Bundles of intermediate filaments stretch from one dense body to another. During contraction, the sliding-filament mechanism involving thick and thin filaments generates tension that is transmitted to intermediate filaments. These, in turn, pull on the dense bodies attached to the sarcolemma, causing a lengthwise shortening of the muscle fiber. As a smooth muscle fiber contracts, it rotates as a corkscrew turns.

Two kinds of smooth muscle tissue, visceral and multiunit, are recognized. The more common type is **visceral (single-unit) smooth muscle tissue.** It is found in tubular arrangements that form part of the walls of small arteries and veins and hollow viscera such as the stomach, intestines, uterus, and urinary bladder. The fibers in visceral muscle tissue form large networks because they contain gap junctions. These allow muscle action potentials to spread throughout the network so that all the fibers contract in unison as a single unit.

The second kind of smooth muscle tissue is **multiunit smooth muscle tissue.** It consists of individual fibers, each with its own motor neuron terminals and with few gap junctions between neighboring fibers. Whereas stimulation of a single visceral muscle fiber causes contraction of many adjacent fibers, stimulation of a single multiunit fiber causes contraction of only that fiber. Multiunit smooth muscle tissue is found in the walls of large arteries, in large airways to the lungs (bronchioles), in the arrector pili muscles that attach to hair follicles, and in the radial and circular muscles of the iris that adjust pupil diameter of the eye.

In comparison with contraction in a striated muscle fiber, contraction in a smooth muscle fiber starts more slowly and lasts much longer since there are no transverse tubules and only a small amount of sarcoplasmic reticulum. Moreover, smooth muscle can both contract and extend to a greater extent than can striated muscle. Smooth muscle tissue can sustain long-term tone (state of continued partial contraction), which is important in the gastrointestinal tract where the walls maintain a steady pressure on the contents of the tract and in the walls of blood vessels called arterioles that maintain a steady pressure on blood.

Most smooth muscle fibers contract or relax in response to action potentials from the autonomic nervous system. Thus smooth muscle normally is not under voluntary control. In addition, many smooth muscle fibers contract or relax in response to stretching, hormones, or local factors such as changes in pH, oxygen and carbon dioxide levels, temperature, and ion concentrations.

Unlike striated muscle fibers, smooth muscle fibers can stretch considerably and still maintain their contractile function. When smooth muscle fibers are stretched, they initially contract, developing increased tension. Within a minute or so, the tension decreases. This phenomenon, termed the **stress-relaxation response,** allows smooth muscle to undergo great changes in length while still retaining the ability to contract effectively. Thus the smooth muscle in the walls of blood vessels and hollow organs such as the stomach, intestines, and urinary bladder can stretch, but the pressure on the contents within increases very little. After the organ empties, on the other hand, the smooth muscle in the wall rebounds, and the wall is firm, not flabby.

Examine a prepared slide of smooth muscle tissue in longitudinal and transverse (cross) section under high power. Locate the label the following structures in Figure 9.7: **sarcolemma, sarcoplasm, nucleus,** and **muscle fiber.**

ANSWER THE LABORATORY REPORT QUESTIONS AT THE END OF THE EXERCISE.

Muscular Tissue

Name _____   Date _____

Laboratory Section _____   Score/Grade _____

SECTION F ■ Electromyography

1. Include your record of spontaneous electrical activity recorded while the subject was resting. Give a physiological explanation. _____

    _____

2. Include your record of the EMG recorded during the recruitment exercise. Give a physiological explanation and possible function of recruitment. _____

    _____

3. Include your record of the EMG recorded following fatigue, and give a physiological explanation of your observations. _____

    _____

Muscular Tissue

Name _____   Date _____

Laboratory Section _____   Score/Grade _____

PART 1 ■ Multiple Choice

_____ 1. The ability of muscle tissue to return to its original shape after contraction or extension is called (a) excitability (b) elasticity (c) extension (d) tetanus

_____ 2. Which of the following is striated and voluntary? (a) skeletal muscle tissue (b) cardiac muscle tissue (c) visceral muscle tissue (d) smooth muscle tissue

_____ 3. The portion of a sarcomere composed of thin filaments only is the (a) H zone (b) A band (c) I band (d) Z disc

_____ 4. The synapse between a motor axon terminal and its associated skeletal muscle fiber sarcolemma (motor end plate) is called the (a) A band (b) filament (c) transverse tubule (d) neuromuscular junction

_____ 5. The connective tissue layer surrounding bundles (fascicles) of muscle fibers is called the (a) perimysium (b) endomysium (c) ectomysium (d) myomysium

_____ 6. Which of the following is striated and involuntary? (a) smooth muscle tissue (b) skeletal muscle tissue (c) cardiac muscle tissue (d) visceral muscle tissue

_____ 7. The portion of a sarcomere that consists mostly of thick filaments and portions of filaments where thin and thick filaments overlap is called the (a) H zone (b) triad (c) A band (d) I band

PART 2 ■ Completion

8. The ability of muscle tissue to receive and respond to stimuli is called _____.

9. Fibrous connective tissue that separates individual muscle fibers is known as

_____.

10. The regions of a muscle fiber separated by Z discs are called _____.

11. The plasma membrane surrounding a muscle fiber is called the _____.

12. The region in a sarcomere consisting of thick filaments only is known as the

_____.

13. Muscle tissue that is nonstriated and involuntary is _____.

14. The ability of muscle tissue to stretch when pulled is called _____.

15. The phenomenon by which a muscle fiber contracts to its fullest extent or not at all is known as the

    _____.

16. A record of muscle contraction is referred to as a(n) _____.

17. The portion of a skeletal muscle fiber sarcolemma deep to an axon terminal is called a

    _____.

18. The binding of _____ to troponin changes the shape of the troponin-

    tropomyosin complex on the thin filaments to expose myosin binding sites.

19. The energy released from the splitting of _____ causes myosin cross bridges

    to pivot inward to pull the thin filaments inward toward the H zone.

20. In relaxed skeletal muscle fibers, calcium ions are stored in the _____.

21. Whereas thin filaments are composed of the protein actin, thick filaments are composed of the

    protein _____.

22. The combination of alternating dark A bands and light _____ bands gives a

    muscle fiber its striated appearance.

23. A motor neuron and all the muscle fibers it innervates is called a(n) _____.

24. Neurotransmitters are stored in _____, which are located within synaptic end bulbs.

25. Irregular transverse thickenings of the sarcolemma that separate cardiac muscle fibers from each

    other are called _____.

26. Action potentials spread from one muscle fiber to another through structures is called

    _____.

27. Cardiac muscle tissue has a long _____ that allows time for the chambers of

    the heart to relax and fill with blood between beats.

28. Smooth muscle tissue contains thick filaments, thin filaments, and _____ filaments.

29. The two kinds of smooth muscle tissue are visceral and _____.

30. When smooth muscle fibers are extended, they initially contract, developing increased tension, and

    then the tension decreases. This phenomenon is called the _____.

Muscular System

Objectives

At the completion of this exercise you should understand

A   How skeletal muscles produce movement

B   The criteria used for naming skeletal muscles

C   The origin, insertion, and action of the skeletal muscles outlined in this exercise

The term **muscular tissue** refers to all the contractile tissues of the body: skeletal, cardiac, and smooth muscle. The **muscular system,** however, refers to the voluntarily controlled muscles of the body: the skeletal muscle tissue and connective tissues that make up individual muscle organs, such as the biceps brachii muscle. In this exercise you will learn the names, locations, and actions of the principal skeletal muscles of the body.

A. How Skeletal Muscles Produce Movement

1. Origin and Insertion

Skeletal muscles that produce movement do so by exerting force on tendons, which in turn pull on bones or other structures, such as the skin. Most muscles cross at least one joint and typically attach to the articulating bones that form the joint. When a muscle contracts, it draws one of the articulating bones toward the other. Typically the two articulating bones usually do not move equally in response to the contraction. Rather, one bone remains stationary or near its original position, because either contracting and other muscles stabilize that bone by pulling it in the opposite direction or the bone's structure makes it less movable. The site of a muscle's tendon attachment to the more stationary bone is its **origin.** The site of a

muscle's tendon attachment to the more movable bone is its **insertion.** A good analogy is a spring on a door. The part of the spring attached to the door represents the insertion; the part attached to the frame is the origin. The fleshy part of the muscle between the tendons of the origin and insertion is the **belly (gaster).** Often, the origin is proximal to the insertion, especially for muscles in the limbs. In addition, muscles that move a body part generally do not cover the moving part. For example, although contraction of the biceps brachii muscle moves the forearm, the belly of the muscle lies over the humerus.

2. Group Actions

Movements often are the result of several skeletal muscles acting as a group rather than alone. Most skeletal muscles are arranged in opposing (antagonistic) pairs at joints, that is, flexors-extensors, abductors-adductors, and so on. Consider flexing the forearm at the elbow, for example. A muscle that causes a desired action is referred to as the **prime mover** or **agonist** (= leader). In this instance, the biceps brachii is the prime mover (see Figure 10.13a). Simultaneously with the contraction of the biceps brachii, another muscle, called the **antagonist** (*anti* = against), is stretched and yields to the effects of the prime mover. In this movement, the triceps brachii serves as the antagonist (see Figure 10.13b). The antagonist has an action that is opposite to that of the prime mover. You should not assume, however, that the biceps brachii is always the prime mover and the triceps brachii is always the antagonist. For example, when extending the forearm at the elbow, the triceps brachii serves as the prime mover, and the biceps brachii functions as the antagonist; their roles are reversed. If the prime mover and antagonist contracted simultaneously with equal force, there would be no movement.

There are many examples in the body where a prime mover crosses other joints before it reaches the joint at which its primary action occurs. To prevent unwanted movements in intermediate joints, or otherwise aid the movement of the prime mover, muscles called **synergists** (SIN-ergists; *syn* = together; *ergon* = work) contract and stabilize the intermediate joints. As an example, muscles that flex the fingers (prime movers) cross the intercarpal and radiocarpal joints (intermediate joints). If movement at these intermediate joints was unrestrained, you would not be able to flex your fingers without also flexing the wrist at the same time. Synergistic contraction of the wrist extensors stabilizes the wrist joint and prevents it from moving (unwanted movement), while the flexor muscles of the fingers contract to bring about efficient flexion of the fingers (primary action). During flexion of the fingers, extensor muscles of the fingers serve as antagonists (see Figure 10.14c, d). Syngerists are usually located close to the prime mover.

Some muscles in a group also act as **fixators,** which stabilize the origin of the primer mover so that the prime mover can act more efficiently. Fixators steady the proximal end of a limb while movements occur at the distal end. For example, the scapula is a freely movable bone in the pectoral (shoulder) girdle that serves as a firm origin for several muscles that move the arm. However, for the scapula to do this, it must be held steady. Fixator muscles accomplish this by holding the scapula firmly against the back of the chest. In abduction of the arm, the deltoid muscle serves as the prime mover, whereas fixators (pectoralis minor, trapezius, subclavius, serratus anterior muscles, and others) hold the scapula firmly (see Figure 10.11). These fixators stabilize the scapula, which serves as the attachment site for the origin of the deltoid muscle, whereas the insertion of the muscle pulls on the humerus to abduct the arm. Under different conditions and depending on the movement and which point is fixed, many muscles act, at various times, as prime movers, antagonists, synergists, or fixators.

B. Arrangement of Fascicles

Recall from Exercise 9 that skeletal muscle fibers (cells) within the muscle are arranged in bundles called **fascicles.** Within a fascicle, all muscle fibers are parallel to one another. The fascicles, however, may form one of five patterns with respect to the tendons: parallel, fusiform (shaped like a cigar), pinnate (shaped like a feather), circular, or triangular.

Table 10.1 describes the major patterns of fascicles. Using your textbook as a guide, provide an example of each.

C. Naming Skeletal Muscles

The nearly 700 individual muscles of the body are named on the basis of one or more distinctive characteristics. If you understand these characteristics, you will find it much easier to learn and remember the names of individual muscles.

Table 10.2 describes the major characteristics that are used to name skeletal muscles. Using your textbook as a guide, provide an example for each.

D. Connective Tissue Components

Skeletal muscles are protected, strengthened, and attached to other structures by several connective tissue components. For example, the entire muscle is encircled with a dense, irregular connective tissue called the **epimysium** (ep′-i-MĪZ-ē-um; *epi* = upon). When the muscle is cut in transverse (cross) section, invaginations of the epimysium divide the muscle into fascicles. These invaginations of the epimysium are called the **perimysium** (per′-i-MĪZ-ē-um; *peri* = around). In turn, invaginations of the perimysium, called **endomysium** (en′-dō-MĪZ-ē-um; *endo* = within), penetrate into the interior of each fascicle and separate individual muscle fibers from one another. The epimysium, perimysium, and endomysium are all extensions of deep fascia and are all continuous with the connective tissue that attaches the muscle to another structure, such as bone or other muscle. All three connective tissue layers may be extended beyond the muscle fibers to form a **tendon**—a cord of dense regular connective tissue that attaches a muscle to the periosteum of bone. The connective tissue may also extend as a broad, flat layer of tendons called an **aponeurosis** (*apo* = from; *neur* = a tendon). Aponeuroses also attach to the coverings of a bone or another muscle. When a muscle contracts, the tendon and its corresponding bone or muscle are pulled toward the

TABLE 10.1	Arrangements of fascicles		
Arrangement	**Description**		**Example**
PARALLEL	Fascicles are parallel with longitudinal axis of muscle and terminate at either end in flat tendons.		
FUSIFORM	Fascicles are nearly parallel with longitudinal axis of muscle and terminate at either end in flat tendons, but muscle tapers toward tendons where the diameter is less than that of the belly.		
PENNATE	Fascicles are short in relation to muscle length and the tendon extends nearly the entire length of the muscle.		
Unipennate	Fascicles are arranged on only one side of tendon.		
Bipennate	Fascicles are arranged on both sides of a centrally positioned tendon.		
Multipennate	Fascicles attach obliquely from many directions to several tendons.		
CIRCULAR	Fascicles are in concentric circular arrangements forming sphincter muscles that enclose an orifice (opening).		
TRIANGULAR	Fascicles attached to a broad tendon converge at a thick central tendon to give the muscle a triangular appearance.		

contracting muscle. In this way skeletal muscles produce movement.

In Figure 10.1 on page 159, label the epimysium, perimysium, endomysium, fascicles, and muscle fibers.

E. Principal Skeletal Muscles

In the pages that follow, a series of tables has been provided for you to learn the principal skeletal muscles by region.[1] Use each table as follows:

1. First read the **overview.** This information provides a general orientation to the muscles under consideration and emphasizes how the muscles are organized within various regions. The discussion also highlights any distinguishing or interesting features about the muscles.

2. Take each muscle, in sequence, and study the **phonetic pronunciations** and **derivations** that indicate how the muscles are named. They appear in parentheses after the name of the muscle and will help you to pronounce the name of a muscle and to understand the reason for giving a muscle its name.

3. As you learn the name of each muscle, determine its origin, insertion, and action and write these in the spaces provided in the table. Consult your textbook as necessary.

4. Again, using your textbook as a guide, label the diagram referred to in the table.

5. Try to visualize what happens when the muscle contracts so that you will understand its action.

6. Do steps 1 through 5 for each muscle in the table. Before moving to the next table, examine a torso or chart of the skeletal system so that you can compare and approximate the positions of the muscles.

7. When possible, try to feel each muscle on your own body.

Refer to Tables 10.3 through 10.23 and Figures 10.2 through 10.20.

[1]A few of the muscles listed are not illustrated in the diagrams. Please consult your textbook to locate these muscles.

TABLE 10.2	Characteristics used for naming skeletal muscles	
Characteristic	**Description**	**Example**
Direction of muscle fibers	Orientation of muscle fascicles relative to the body's midline. **Rectus** means the fibers run parallel to the midline. **Transverse** means the fibers run perpendicular to the midline. **Oblique** means the fibers run diagonally to the midline.	
Location	Structure near which a muscle is found.	
Size	Relative size of the muscle. **Maximus** means largest. **Medius** means intermediate. **Minimus** means smallest. **Longus** means long. **Brevis** means short. **Latissimus** means widest. **Longissimus** means longest. **Magnus** means large. **Major** means larger. **Minor** means smaller. **Vastus** means huge.	
Number of origins	Number of tendons of origin. **Biceps** means two origins. **Triceps** means three origins. **Quadriceps** means four origins.	
Shape	Relative shape of the muscle. **Deltoid** means triangular. **Trapezius** means trapezoid. **Serratus** means saw-toothed. **Rhomboid** means diamond-shaped. **Orbicularis** means circular. **Pectinate** means comblike. **Piriformis** means pear-shaped. **Platys** means flat. **Quadratus** means square, four-sided. **Gracilis** means slender.	
Origin and insertion	Sites where muscle originates and inserts.	
Action	Principal action of the muscle. **Flexor** (FLEK-sor): decreases the angle at a joint. **Extensor** (eks-TEN-sor): increases the angle at a joint. **Abductor** (ab-DUK-tor): moves a bone away from the midline. **Adductor** (ad-DUK-tor): moves a bone closer to the midline. **Levator** (le-VĀ-tor): raises or elevates a body part. **Depressor** (de-PRES-or): lowers or depresses a body part. **Supinator** (soo'-pi-NĀ-tor): turns the palm anteriorly. **Pronator** (prō-NĀ-tor): turns the palm posteriorly. **Sphincter** (SFINGK-ter): decreases the size of an opening. **Tensor** (TEN-sor): makes a body part rigid. **Rotator** (RŌ-tāt-or): rotates a bone around its longitudinal axis.	

FIGURE 10.1 Connective tissue components of a skeletal muscle.

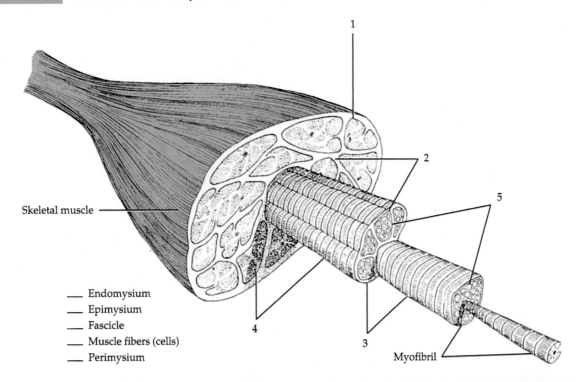

1

2

5

Skeletal muscle

___ Endomysium
___ Epimysium
___ Fascicle
___ Muscle fibers (cells)
___ Perimysium

4

3

Myofibril

TABLE 10.3 Muscles of facial expression (After completing the table, label Figure 10.2.)

Overview: The muscles in this group provide humans with the ability to express a wide variety of emotions, including grief, surprise, fear, and happiness. The muscles themselves lie within the layers of superficial fascia. They arise from the fascia or bones of the skull and insert into the skin. Because of their insertions, the muscles of facial expression move the skin rather than a joint when they contract.

   Among the noteworthy muscles in this group are those that surround the orifices (openings) of the head such as the eyes, nose, and mouth. These muscles function as *sphincters*, which close the orifices, and *dilators*, which open the orifices. For example, the orbicularis oculi muscle closes the eye, while the levator palpebrae superioris muscle opens the eye. As another example, the orbicularis oris muscle closes the mouth, while several other muscles (zygomaticus major, levator labii superioris, depressor labii inferioris, mentalis, and risorius) radiate out from the lips and open the mouth. The occipitofrontalis is an unusual muscle in this group because it is made up of two parts: an anterior part called the frontal belly, which is superficial to the frontal bone, and a posterior part called the occipital belly, which is superficial to the occipital bone. The two muscular portions are held together by a strong aponeurosis (sheetlike tendon), the **galea aponeurotica** or **epicranial aponeurosis,** which covers the superior and lateral surfaces of the skull. The frontalis and occipitalis are treated as separate muscles in this table. The buccinator muscle forms the major muscular portion of the cheek. It functions in whistling, blowing, and sucking and also assists in chewing. The duct of the parotid gland (salivary gland) pierces the buccinator muscle to reach the oral cavity. The buccinator muscle was given its name because it compresses the cheeks (*bucc* = cheek) during blowing, for example, when a musician plays a wind instrument such as a trumpet. Some trumpeters have stretched their buccinator muscles so much that their cheeks bulge out considerably when they blow forcibly.

TABLE 10.3	Muscles of facial expression (After completing the table, label Figure 10.2.) (Continued)

Muscle	Origin	Insertion	Action
Occipitofrontalis (ok-sip-i-tō-fron-TĀ-lis) **Frontal belly** **Occipital belly** (*occipit* = base of head)	This muscle is divisible into two portions: the frontal belly, over the frontal bone, and the occipital belly, over the occipital bone. The two muscles are united by a strong aponeurosis, the galea aponeurotica (epicranial aponeurosis), which covers the superior and lateral surfaces of the skull.		
Orbicularis oris (or-bi'-kū-LAR-is OR-is; *orb* = circular; *oris* = mouth)			
Zygomaticus (zī-gō-MA-ti-kus) **major** (*zygomatic* = cheek bone; *major* = greater)			
Levator labii superioris (le-VĀ-tor LĀ-bē-ī soo-per'-ē-OR-is; *levator* = raises or elevates; *labii* = lip; *superioris* = upper)			
Depressor labii inferioris (de-PRE-sor LĀ-bē-ī in-fer'-ē-OR-is; *depressor* = depresses or lowers; *inferioris* = lower)			
Buccinator (BUK-si-nā'-tor; *bucc* = cheek)			
Mentalis (men-TĀ-lis; *ment* = chin)			
Platysma (pla-TIZ-ma; *platy* = flat, broad)			
Risorius (ri-ZOR-ē-us; *risor* = laughter)			
Orbicularis oculi (or-bi'-kū-LAR-is ok-ū-lī; *oculi* = eye)			
Corrugator supercilii (KOR-a-gā'-tor soo-per-SI-lē-ī; *corrugat* = wrinkle; *supercili* = eyebrow)			
Levator palpebrae superioris (le-VĀ-tor PAL-pe-brē soo-per'-ē-OR-is; *palpebrae* = eyelids) (See Figure 10.3)			

FIGURE 10.2 **Muscles of facial expression.**

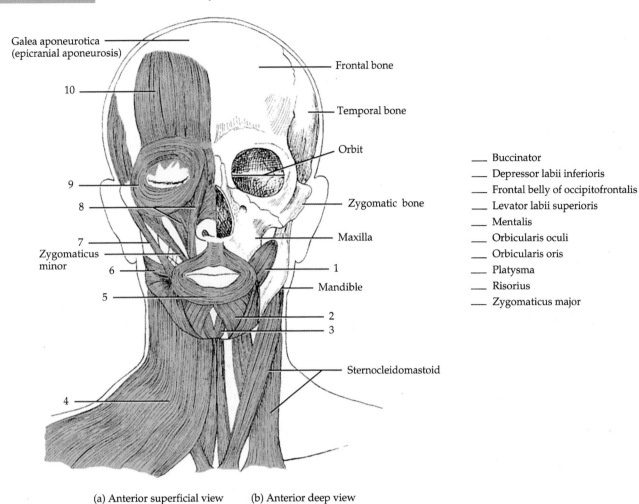

Galea aponeurotica
(epicranial aponeurosis)

10

9

8

7

Zygomaticus
minor

6

5

4

Frontal bone

Temporal bone

Orbit

Zygomatic bone

Maxilla

Mandible

1

2

3

Sternocleidomastoid

(a) Anterior superficial view   (b) Anterior deep view

___ Buccinator
___ Depressor labii inferioris
___ Frontal belly of occipitofrontalis
___ Levator labii superioris
___ Mentalis
___ Orbicularis oculi
___ Orbicularis oris
___ Platysma
___ Risorius
___ Zygomaticus major

Galea aponeurotica
(epicranial aponeurosis)

1

Temporalis

2

3

4

5

9

Masseter

Sternocleidomastoid

6

7

8

(c) Right lateral superficial view

___ Buccinator
___ Frontal belly of occipitofrontalis
___ Levator labii superioris
___ Occipital belly of occipitofrontalis
___ Orbicularis oculi
___ Orbicularis oris
___ Platysma
___ Risorius
___ Zygomaticus major

TABLE 10.4   Muscles that move the eyeballs—the extrinsic muscles* (After completing the table, label Figure 10.3.)

Overview: Muscles associated with the eyeballs are of two principal types: extrinsic and intrinsic. **Extrinsic muscles** originate outside the eyeballs and are inserted on their outer surfaces (sclera). They move the eyeballs in various directions. **Intrinsic muscles** originate and insert entirely within the eyeballs. They move structures within the eyeballs.

Movements of the eyeballs are controlled by three pairs of extrinsic muscles: (1) superior and inferior recti, (2) lateral and medial recti, and (3) superior and inferior oblique. The extrinsic muscles of the eyeballs are among the fastest contracting and most precisely controlled skeletal muscles in the body. The four recti muscles (superior, inferior, lateral, and medial) arise from a tendinous ring in the orbit and insert into the sclera of the eye. Whereas the superior and inferior recti lie in the same vertical plane, the medial and lateral recti lie in the same horizontal plane. The actions of the recti muscles can be deduced from their insertions on the sclera. The superior and inferior recti move the eyeballs superiorly and inferiorly, respectively; the lateral and medial recti move the eyeballs laterally and medially, respectively. It should be noted that neither the superior nor the inferior rectus muscle pulls directly parallel to the long axis of the eyeballs; as a result, both muscles also move the eyeballs medially.

It is not easy to deduce the actions of the oblique muscles (superior and inferior) because of their path through the orbits. For example, the superior oblique muscle originates posteriorly near the tendinous ring and then passes anteriorly and ends in a round tendon, which runs through a pulleylike loop called the *trochlea* (*trochlea* = pulley) in the anterior and medial part of the roof of the orbit. The tendon then turns and inserts on the posterolateral aspect of the eyeball. Accordingly, the superior oblique muscle moves the eyeballs inferiorly and laterally. The inferior oblique muscle originates on the maxilla at the anteromedial aspect of the floor of the orbit. It then passes posteriorly and laterally and inserts on the posterolateral aspect of the eyeballs. Because of this arrangement, the inferior oblique muscles move the eyeballs superiorly and laterally.

Muscle	Origin	Insertion	Action
Superior rectus (REK-tus; *superior* = above; *rectus* = fascicles running parallel to midline)			
Inferior rectus (REK-tus; *inferior* = below)			
Lateral rectus (REK-tus)			
Medial rectus (REK-tus)			
Superior oblique (ō-BLĒK; *oblique* = fascicles running diagonally to midline)			
Inferior oblique (ō-BLĒK)			

*Muscles situated on the outside of the eyeballs.

FIGURE 10.3 Extrinsic muscles of eyeballs.

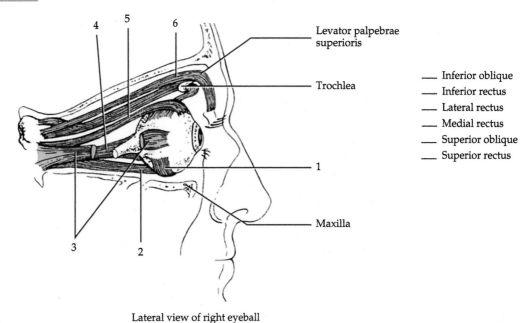

Levator palpebrae superioris

Trochlea

___ Inferior oblique
___ Inferior rectus
___ Lateral rectus
___ Medial rectus
___ Superior oblique
___ Superior rectus

Maxilla

Lateral view of right eyeball

TABLE 10.5 Muscles that move the mandible (lower jaw) (After completing the table, label Figure 10.4.)

Overview: The muscles that move the mandible (lower jaw) at the temporomandibular joint (TMJ) are known as the muscles of mastication because they are involved in chewing (mastication). Of the four pairs of muscles involved in mastication, three are powerful closers of the jaw and account for the strength of the bite: masseter, temporalis, and medial pterygoid. Of these, the masseter is the strongest muscle of mastication. The medial and lateral pterygoid muscles assist in mastication by moving the mandible from side to side to help grind food. Additionally, these muscles protrude the mandible.

In 1996, researchers at the University of Maryland reported that they had identified a new muscle in the skull and named it the **sphenomandibularis muscle.** It extends from the sphenoid bone to the mandible. The muscle is believed to be either a fifth muscle of mastication or a previously unidentified component of an already identified muscle (temporalis or medial pterygoid). The sphenomandibularis muscle is innervated by the maxillary branch of the trigeminal (V) cranial nerve.

Muscle	Origin	Insertion	Action
Masseter (MA-se-ter; *masseter* = chewer)			
Temporalis (tem'-por-A-lis; *tempor* = temples)			
Medial pterygoid (TER-i-goyd; *medial* = closer to midline; *pterygoid* = like a wing)			
Lateral pterygoid (TER-i-goyd; *lateral* = farther from midline)			

FIGURE 10.4 Muscles that move the mandible (lower jaw).

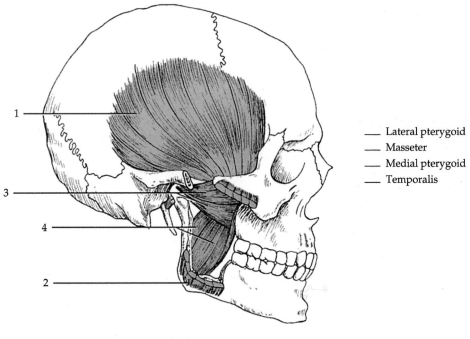

___ Lateral pterygoid
___ Masseter
___ Medial pterygoid
___ Temporalis

Right lateral view

TABLE 10.6   Muscles that move the tongue—the extrinsic muscles (After completing the table, label Figure 10.5.)

Overview: The tongue is a highly mobile structure that is vital to digestive functions such as mastication, perception of taste, and deglutition (swallowing). It is also important in speech. The mobility of the tongue is greatly aided by its suspension from the mandible, styloid process of the temporal bone, and hyoid bone.

The tongue is divided into lateral halves by a median fibrous septum. The septum extends throughout the length of the tongue and is attached inferiorly to the hyoid bone. Like the muscles of the eyeballs, muscles of the tongue are of two principal types—extrinsic and intrinsic. **Extrinsic muscles** originate outside the tongue and insert into it. They move the entire tongue in various directions, such as anteriorly, posteriorly, and laterally. **Intrinsic muscles** originate and insert within the tongue. These muscles alter the shape of the tongue rather than move the entire tongue. The extrinsic and intrinsic muscles of the tongue are arranged in both lateral halves of the tongue.

When you study the extrinsic muscles of the tongue, you will notice that all of the names end in *glossus*, meaning tongue. You will also notice that the actions of the muscles are obvious considering the position of the mandible, styloid process, hyoid bone, and soft palate, which serve as origins for these muscles. For example, the genioglossus (originates on the mandible) pulls the tongue downward and forward, the styloglossus (originates on the styloid process) pulls the tongue upward and backward, the hyoglossus (originates on the hyoid bone) pulls the tongue downward and flattens it, and the palatoglossus (originates on the soft palate) raises the back portion of the tongue.

When general anesthesia is administered during surgery, a total relaxation of the genioglossus muscle results. This will cause the tongue to fall posteriorly, which may obstruct the airway to the lungs. To avoid this, the mandible is either manually thrust forward and held in place, or a tube is inserted from the lips through the laryngopharynx (inferior portion of the throat) into the trachea (endotracheal intubation).

(Continued)

Muscle	Origin	Insertion	Action
Genioglossus (jē′-nē-ō-GLOS-us; *genio* = chin *glossus* = tongue)			
Styloglossus (stī′-lō-GLOS-us; *stylo* = stake or pole)			
Palatoglossus (pal′-a-tō-GLOS-us; *palato* = roof of the mouth or palate)			
Hyoglossus (hī-ō-GLOS-us)			

Muscles that move the tongue.

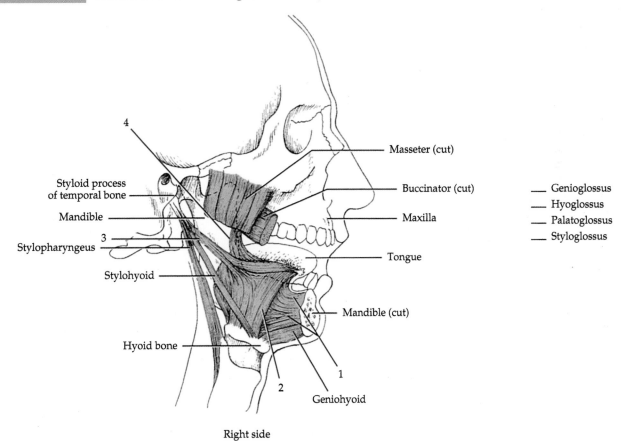

Right side

Genioglossus
Hyoglossus
Palatoglossus
Styloglossus

TABLE 10.7	Muscles of the anterior neck—suprahyoid muscles (After completing the table, label Figure 10.6.)

Overview: There are two groups of muscles associated with the anterior aspect of the neck called (1) **suprahyoid muscles,** because they are located superior to the hyoid bone, and (2) **infrahyoid muscles,** because they lie inferior to the hyoid bone. Acting with the infrahyoid muscles, the suprahyoid muscles stabilize the hyoid bone, thus serving as a firm base upon which the tongue can move. In this table we will consider the suprahyoid muscles, which are associated with the floor of the oral cavity. The next table describes the infrahyoid muscles.

As a group, the suprahyoid muscles elevate the hyoid bone, floor of the oral cavity, and tongue during swallowing. The digastric muscle, as the name suggests, consists of two bellies called an *anterior belly* and *posterior belly,* united by an intermediate tendon that is held in position by a fibrous loop. This muscle elevates the hyoid bone and larynx (voice box) during swallowing and speech and also depresses the mandible. The stylohyoid muscle elevates and draws the hyoid bone posteriorly, thus elongating the floor of the oral cavity during swallowing. The mylohyoid muscle elevates the hyoid bone and helps press the tongue against the roof of the oral cavity during swallowing to move food from the oral cavity into the throat. The geniohyoid muscle elevates and draws the hyoid bone anteriorly to shorten the floor of the oral cavity and to widen the throat to receive food that is being swallowed. It also depresses the mandible.

Muscle	Origin	Insertion	Action
Digastric (dī-GAS-trik; *di* = two; *gastr* = belly)			
Stylohyoid (stī′-lō-HĪ-oid; *stylo* = stake or pole styloid process of temporal bone; *hyo* = U-shaped, pertaining to hyoid bone)			
Mylohyoid (mī′-lō-HĪ-oid; *mylo* = mill)			
Geniohyoid (je′-nē-ō-HĪ-oid; *genio* = chin)			
(See Figure 10.5)			

FIGURE 10.6 **Muscles of the anterior neck—suprahyoid muscles**

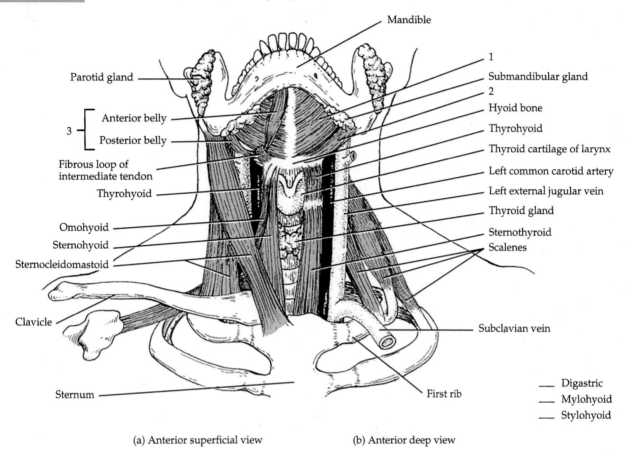

Mandible

Parotid gland

3 — Anterior belly
    Posterior belly

Fibrous loop of
intermediate tendon

Thyrohyoid

Omohyoid

Sternohyoid

Sternocleidomastoid

Clavicle

Sternum

1

Submandibular gland

2

Hyoid bone

Thyrohyoid

Thyroid cartilage of larynx

Left common carotid artery

Left external jugular vein

Thyroid gland

Sternothyroid

Scalenes

Subclavian vein

First rib

____ Digastric
____ Mylohyoid
____ Stylohyoid

(a) Anterior superficial view   (b) Anterior deep view

TABLE 10.8 Muscles of the anterior neck—infrahyoid muscles (After completing the table, label Figure 10.7.)

Overview: The **infrahyoid muscles** are also called "strap" muscles because of their ribbonlike appearance. Most of the infrahyoid muscles depress the hyoid bone and some move the larynx during swallowing and speech.

The omohyoid muscle, like the digastric muscle, is composed of two bellies and an intermediate tendon. In this case, however, the two bellies are referred to as *superior* and *inferior*, rather than anterior and posterior. Together the omohyoid, sternohyoid, and thyrohyoid depress the hyoid bone. In addition, the sternothyroid depresses the thyroid cartilage (Adam's apple) of the larynx, while the thyrohyoid muscle elevates the thyroid cartilage. These actions are necessary during the production of low and high tones, respectively, during phonation.

Muscle	Origin	Insertion	Action
Omohyoid (ō'-mō-HĪ-oid; *omo* = relationship to shoulder; *hyoedes* = U-shaped)			
Sternohyoid (ster'-nō-HĪ-oid; *sterno* = sternum)			
Sternothyroid (ster'-nō-THĪ-roid; *thyro* = thyroid gland)			
Thyrohyoid (thī'-rō-HĪ-oid)			

FIGURE 10.7 Muscles of the anterior neck—infrahyoid muscles.

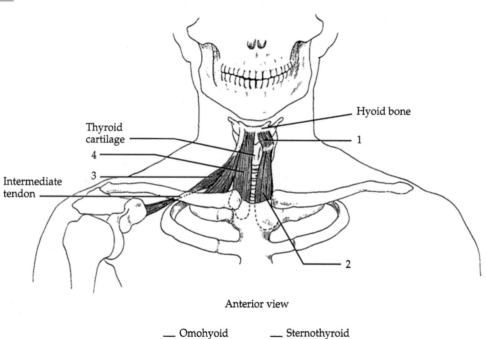

Anterior view

— Omohyoid   — Sternothyroid
— Sternohyoid   — Thyrohyoid

TABLE 10.9	Muscles that move the head

Overview: The head is attached to the vertebral column at the atlanto-occipital joint formed by the atlas and occipital bone. Balance and movement of the head on the vertebral column involve the action of several neck muscles. For example, contraction of the two sternocleidomastoid muscles together (bilaterally) flexes the cervical portion of the vertebral column and head. Acting singly (unilaterally), the muscle laterally flexes and rotates the head. Bilateral contraction of the semispinalis capitis, splenius capitis, and longissimus capitis muscles extends the head. However, when these same muscles contract unilaterally, their actions are quite different, involving primarily rotation of the head.

The cervical (neck) region is divided by the sternocleidomastoid muscle into two principal triangles: anterior and posterior. The **anterior triangle** is bordered superiorly by the mandible, inferiorly by the sternum, medially by the cervical midline, and laterally by the anterior border of the sternocleidomastoid muscle. The anterior triangle is subdivided into an unpaired submental triangle and three paired triangles: submandibular, carotid, and muscular. The **posterior triangle** is bordered inferiorly by the clavicle, anteriorly by the posterior border of the sternocleidomastoid muscle, and posteriorly by the anterior border of the trapezius muscle. The posterior triangle is subdivided into two triangles, occipital and supraclavicular, by the inferior belly of the omohyoid muscle. The triangles are discussed in detail in Exercise 11.

Muscle	Origin	Insertion	Action
Sternocleidomastoid (ster′-nō-klī′-dō-MAS-toid; *sterno* = breastbone; *cleido* = clavicle; *mastoid* = mastoid process of temporal bone) (label this muscle in Figure 10.11)			
Semispinalis capitis (se′-mē-spi-NA-lis KAP-i-tis; *semi* = half; *spine* = spinous process; *capit* = head) (label this muscle in Figure 10.16)			
Splenius capitis (SPLĒ-nē-us KAP-i-tis; *splenon* = bandage) (label this muscle in Figure 10.16)			
Longissimus capitis (lon-JIS-i-mus KAP-i-tis; *longissimus* = longest) (label this muscle in Figure 10.16)			

Overview: The anterolateral abdominal wall is composed of skin, fascia, and four pairs of muscles: the external oblique, the internal oblique, transversus abdominis, and rectus abdominis. The first three muscles are arranged from superficial to deep. The external oblique is the superficial muscle. Its fascicles extend inferiorly and medially. The internal oblique is the intermediate muscle with its fascicles extended at right angles to those of the external oblique. The transversus abdominis is the deepest muscle, with most of its fascicles directed transversely around the abdominal wall. Together, the external oblique, internal oblique, and transversus abdominus form three layers of muscle around the abdomen. The muscle fascicles extend in different directions, a structural arrangement that affords considerable protection to the abdominal viscera, especially when the muscles have good tone.

The rectus abdominis muscle is a long, flat muscle that extends the entire length of the anterior abdominal wall, from the pubic crest and pubic symphysis to the cartilages of ribs 5–7 and the xiphoid process of the sternum. The anterior surface of the muscle is interrupted by three transverse fibrous bands of tissue called **tendinous intersections,** believed to be remnants of septa that separated myotomes during embryological development.

As a group the muscles of the anterolateral abdominal wall help contain and protect the abdominal viscera; flex, laterally flex, and rotate the vertebral column at the intervertebral joints; compress the abdomen during forced expiration; and produce the force required for defecation, urination, and childbirth.

The aponeuroses of the external oblique, internal oblique, and transversus abdominis muscles form the **rectus sheaths,** which enclose the rectus abdominis muscles. The sheaths meet at the midline to form the **linea alba** (= white line), a tough, fibrous band that extends from the xiphoid process of the sternum to the pubic symphysis. In the latter stages of pregnancy, the linea alba stretches to increase the distance between the rectus abdominis muscles. The inferior free border of the external oblique aponeurosis forms the **inguinal ligament,** which runs from the anterior superior iliac spine to the pubic tubercle (see Figure 10.17a). Just superior to the medial end of the inguinal ligament is a triangular slit in the aponeurosis referred to as the **superficial inguinal ring,** the outer opening of the **inguinal canal.** The canal contains the spermatic cord and ilioinguinal nerve in males and round ligament of the uterus and ilioinguinal nerve in females.

The posterior abdominal wall is formed by the lumbar vertebrae, parts of the ilia of the hipbones, psoas major and iliacus muscles (described in Table 10.20), and quadratus lumborum muscle. Whereas the anterolateral abdominal wall can contract and distend, the posterior abdominal wall is bulky and stable by comparison.

Muscle	Origin	Insertion	Action
Rectus abdominis (REK-tus ab-DOM-in-is; *rectus* = fascicles parallel to midline; *abdomin* = abdomen)			
External oblique (ō-BLĒK; *external* = closer to the surface; *oblique* = fascicles diagonal to midline)			

(Continued)

Muscle	Origin	Insertion	Action
Internal oblique (ō-BLĒK; *internal* = farther from the surface)			
Transversus abdominis (tranz-VER-sus ab-DOM-in-is; *transverse* = fascicles perpendicular to midline)			
Quadratus lumborum (kwod-RA-tus lum-BOR-um; *quad* = four; *lumbo* = lumbar region) (See Figure 10.16a)			

FIGURE 10.8 Muscles of the anterior abdominal wall.

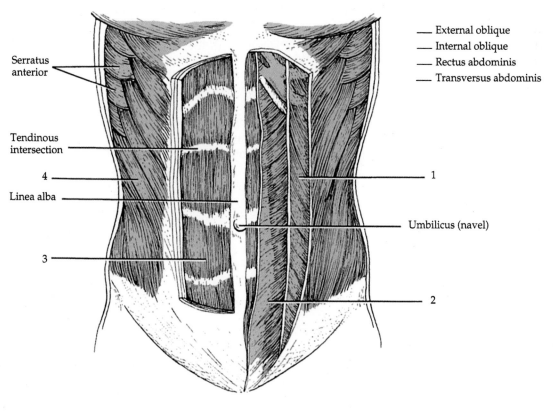

Serratus anterior

Tendinous intersection

4

Linea alba

3

_____ External oblique
_____ Internal oblique
_____ Rectus abdominis
_____ Transversus abdominis

1

Umbilicus (navel)

2

Anterior view

TABLE 10.11   Muscles used in ventilation (breathing) (After completing the table, label Figure 10.9.)

Overview: The muscles described here alter the size of the thoracic cavity so that breathing (ventilation) can occur. Breathing consists of two phases called inhalation and exhalation. Essentially, inhalation occurs when the thoracic cavity increases in size, and exhalation occurs when the thoracic cavity decreases in size.

The dome-shaped diaphragm is the most important muscle that powers breathing. It separates the thoracic and abdominal cavities. The diaphragm is composed of two parts: a peripheral muscular portion and a central portion called the **central tendon.** The central tendon is a strong aponeurosis that serves as the tendon of insertion for all the peripheral muscular fibers of the diaphragm. It fuses with the inferior surface of the fibrous pericardium, the external covering of the heart, and the parietal pleurae, the external coverings of the lungs.

Movements of the diaphragm help return venous blood to the heart as it passes through the abdomen. Together with the anterolateral abdominal muscles, the diaphragm helps to increase intra-abdominal pressure to evacuate the pelvic contents during defecation, urination, and childbirth. This mechanism is further assisted when you take a deep breath and close the rima glottidis (the space between vocal folds). The trapped air in the respiratory system prevents the diaphragm from elevating. The increase in the intra-abdominal pressure as just described will also help support the vertebral column and prevent flexion during weight lifting. This greatly assists the back muscles in lifting a heavy weight.

The diaphragm has three major openings through which various structures pass between the thorax and abdomen. These structures include the aorta along with the thoracic duct and azygos vein, which pass through the **aortic hiatus;** the esophagus with accompanying vagus (X) cranial nerves, which pass through the **esophageal hiatus;** and the inferior vena cava, which passes through the **caval opening (foramen for the vena cava).** In a condition called a hiatus hernia, the stomach protrudes superiorly through the esophageal hiatus.

The other muscles involved in breathing are called intercostal muscles and occupy the intercostal spaces, the spaces between ribs. There are 11 pairs of external intercostal muscles, and their fibers run obliquely inferiorly and anteriorly from the rib above to the rib below. Their role in respiration is to elevate the ribs during inhalation to help expand the thoracic cavity. There are also 11 pairs of internal intercostal muscles, and their fibers run obliquely inferiorly and posteriorly from the rib above to the rib below. Their function is to draw adjacent ribs together during forced exhalation to help decrease the size of the thoracic cavity.

Muscle	Origin	Insertion	Action
Diaphragm (DĪ-a-fram; *dia* = across; *phragm* = wall)			
External intercostals (in'-ter-KOS-tals; *external* = closer to; *inter* = between; *costa* = rib)			
Internal intercostals (in'-ter-KOS-tals; *internal* = farther from surface)			

FIGURE 10.9　Muscles used in ventilation.

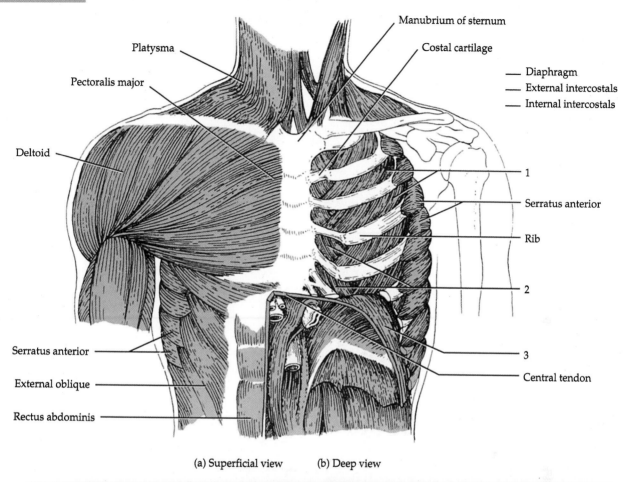

Platysma

Pectoralis major

Deltoid

Serratus anterior

External oblique

Rectus abdominis

Manubrium of sternum

Costal cartilage

— Diaphragm
— External intercostals
— Internal intercostals

1

Serratus anterior

Rib

2

3

Central tendon

(a) Superficial view　　(b) Deep view

TABLE 10.12　Muscles of the pelvic floor (After completing the table, label Figure 10.10.)

Overview: The muscles of the pelvic floor are the levator ani and coccygeus. Together with the fascia covering their internal and external surfaces, these muscles are referred to as the **pelvic diaphragm.** It stretches from the pubis anteriorly to the coccyx posteriorly, and from one lateral wall of the pelvis to the other. This arrangement gives the pelvic diaphragm the appearance of a funnel suspended from its attachments. The pelvic diaphragm is pierced by the anal canal and urethra in both sexes and also by the vagina in the female.

The levator ani muscle is divisible into two muscles called the pubococcygeus and iliococcygeus. The levator ani is the largest and most important muscle of the pelvic floor. It supports the pelvic viscera and resists the inferior thrust that accompanies increases in intra-abdominal pressure during functions such as forced expiration, coughing, vomiting, urination, and defecation. The muscle also functions as a sphincter at the anorectal junction, urethra, and vagina. During childbirth, the levator ani muscles support the head of the fetus, and the muscle may be injured during a difficult childbirth or traumatized during an **episiotomy** (a cut made with surgical scissors to prevent tearing of the perineum during birth of a baby). This may cause urinary stress incontinence, in which there is a leakage of urine whenever intra-abdominal pressure is increased, for example, during coughing. In addition to assisting the levator ani in supporting the pelvic viscera and resisting increases in intra-abdominal pressure, the coccygeus muscle pulls the coccyx anteriorly after it has been pushed posteriorly following defecation or childbirth.

(Continued)

TABLE 10.12	Muscles of the pelvic floor (After completing the table, label Figure 10.10.) (Continued)

Muscle	Origin	Insertion	Action
Levator ani (le-VĀ-tor Ā-nē; *levator* = raises; *ani* = anus)	This muscle is divisible into two parts: the pubococcygeus muscle and the iliococcygeus muscle.		
Pubococcygeus (pū-bō-kok-SIJ-ē-us; *pubo* = pubis; *coccygeus* = coccyx)			
Iliococcygeus (il'-ē-ō-kok-SIJ-ē-us; *ilio* = ilium)			
Coccygeus* (kok-SIJ-ē-us)			

*Not illustrated.

FIGURE 10.10 | Muscles of female pelvic floor and perineum.

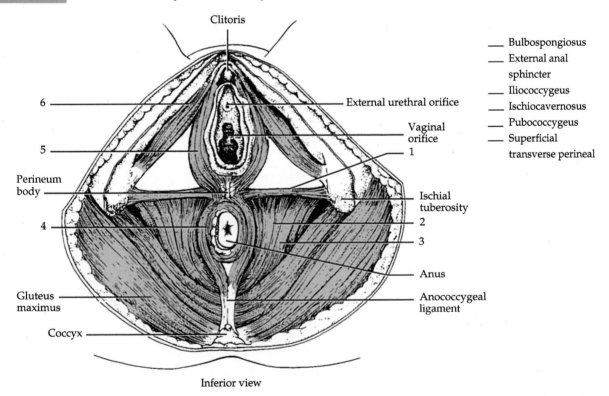

TABLE 10.13	Muscles of the perineum (After completing the table, label Figure 10.10.)

Overview: The **perineum** is the region of the trunk inferior to the pelvic diaphragm. It is a diamond-shaped area that extends from the pubic symphysis anteriorly, to the coccyx posteriorly, and to the ischial tuberosities laterally. A transverse line drawn between the ischial tuberosities divides the perineum into an anterior **urogenital triangle** that contains the external genitals and a posterior **anal triangle** that contains the anus. In the center of the perineum is a wedge-shaped mass of fibrous tissue called the **perineal body.** It is a strong tendon into which several perineal muscles insert.

The muscles of the perineum are arranged in two layers: **superficial** and **deep.** The muscles of the superficial layer are the superficial transverse perineal muscle, bulbospongiosus, and ischiocavernosus. The deep muscles of the perineum are the deep transverse perineus muscle and external urethral sphincter. The muscles of this diaphragm assist in urination and ejaculation in males and urination in females. The external anal sphincter closely adheres to the skin around the margin of the anus and keeps the anal canal and anus closed except during defecation.

Muscle	Origin	Insertion	Action
Superficial transverse perineal (per-i-NĒ-al; *superficial* = near surface; *transverse* = across; *perineus* = perineum)			
Bulbospongiosus (bul'-bō-spon'-jē-O-sus; *bulb* = bulb; *spongio* = sponge)			
Ischiocavernosus (is'-kē-ō-ka'-ver-NŌ-sus; *ischio* = hip)			
Deep transverse perineal* (per-i-NĒ-al; *deep* = farther from surface)			
External urethral sphincter (ū-RĒ-thral SFINGK-ter; *urethral* = pertaining to urethra; *sphincter* = circular muscle that decreases the size of an opening)			
External anal (Ā-nal) **sphincter**			

*Not illustrated.

TABLE 10.14	Muscles that move the pectoral (shoulder) girdle (After completing the table, label Figure 10.11.)

Overview: The main action of the muscles that move the pectoral girdle is to stabilize the scapula so that it can function as a steady origin for most of the muscles that move the humerus. Since scapular movements usually accompany humeral movements in the same direction, the muscles also move the scapula to increase the range of motion of the humerus. For example, it would not be possible to abduct the humerus past the horizontal position if the scapula did not move with the humerus. During abduction, the scapula follows the humerus by rotating upward.

Muscles that move the pectoral girdle can be classified into two groups based on their location in the thorax: **anterior** and **posterior** thoracic muscles. The anterior thoracic muscles are the subclavius, pectoralis minor, and serratus anterior. The subclavius is a small, cylindrical muscle under the clavicle that extends from the clavicle to the first rib. It steadies the clavicle during movements of the pectoral girdle. The pectoralis minor is a thin, flat, triangular muscle that is deep to the pectoralis major. In addition to its role in movements of the scapula, the pectoralis minor muscle also assists in forced inhalation. The serratus anterior is a large, flat, fan-shaped muscle between the ribs and scapula. It is named because of the saw-toothed appearance of its origins on the ribs.

The posterior thoracic muscles are the trapezius, levator scapulae, rhomboid major, and rhomboid minor. The trapezius is a large, flat triangular sheet of muscle extending from the skull and vertebral column medially to the pectoral girdle laterally. It is the most superficial back muscle and covers the posterior neck region and superior portion of the trunk. The two trapezius muscles form a trapezium (diamond-shaped quadrangle), thus its name. The levator scapulae is a narrow, elongated muscle in the posterior portion of the neck. It is deep to the sternocleidomastoid and trapezius muscles. As its name suggests, one of its actions is to elevate the scapula. The rhomboid major and rhomboid minor muscles lie deep to the trapezius and are not always distinct from each other. They appear as parallel bands that pass inferolaterally from the vertebrae to the scapula. They are named on the basis of their shape, that is, a rhomboid (an oblique parallelogram). The rhomboid major is about two times wider than the rhomboid minor. Both muscles are used when forcibly lowering the raised upper limbs, as in driving a stake with a sledgehammer.

In order to understand the action of muscles that move the scapula, it will first be helpful to describe the various movements of the scapula.
Elevation. Superior movement of the scapula, such as shrugging the shoulders or lifting a weight over the head.
Depression. Inferior movement of the scapula, as in doing a "pull-up."
Abduction (protraction). Movement of the scapula laterally and anteriorly, as in doing a "push-up" or punching.
Adduction (retraction). Movement of the scapula medially and posteriorly, as in pulling the oars in a rowboat.
Upward rotation. Movement of the inferior angle of the scapula laterally so that the glenoid cavity is moved upward. This movement is required to abduct the humerus past the horizontal position, as in performing a "jumping jack."
Downward rotation. Movement of the inferior angle of the scapula medially so that the glenoid cavity is moved downward. This movement is seen when a gymnast on parallel bars supports the weight of the body on the hands.

(Continued)

Muscle	Origin	Insertion	Action
Anterior muscles **Subclavius** (sub-KLĀ-vē-us; *sub* = under; *clavius* = clavicle)			
Pectoralis (pek'-tor-A-lis) **minor** (*pector* = breast, chest, thorax; *minor* = lesser)			
Serratus (ser-Ā-tus) **anterior** (*serratus* = sawtoothed; *anterior* = front)			
Posterior muscles **Trapezius** (tra-PĒ-zē-us; *trapezi* = trapezoid-shaped)			
Levator scapulae (le-VĀ-tor SKA-pū-lē; *levator* = raises; *scapulae* = scapula)			
Rhomboid (rom-BOYD) **major** (*rhomboid* = rhomboid- or diamond-shaped)			
Rhomboid (rom-BOYD) **minor**			

FIGURE 10.11 **Muscles that move the pectoral (shoulder) girdle.**

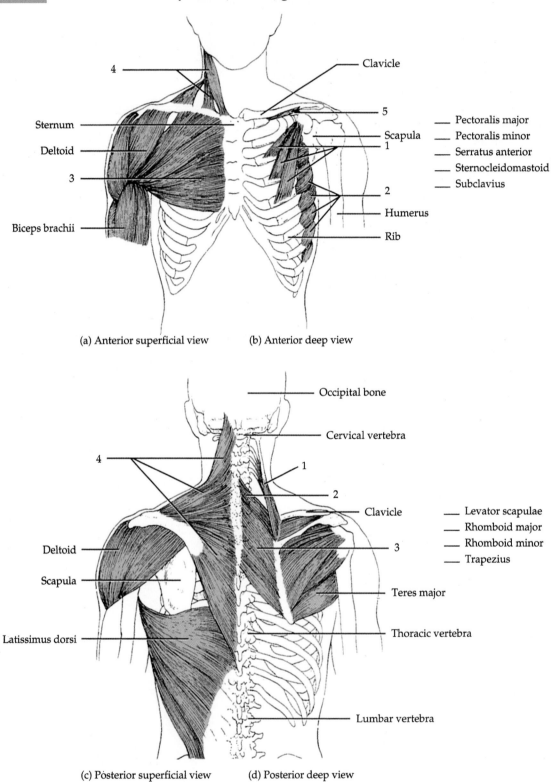

4
Clavicle
Sternum
5
Deltoid
Scapula
1
3
2
Biceps brachii
Humerus
Rib

___ Pectoralis major
___ Pectoralis minor
___ Serratus anterior
___ Sternocleidomastoid
___ Subclavius

(a) Anterior superficial view   (b) Anterior deep view

Occipital bone
Cervical vertebra
4
1
2
Clavicle
Deltoid
3
Scapula
Teres major
Latissimus dorsi
Thoracic vertebra
Lumbar vertebra

___ Levator scapulae
___ Rhomboid major
___ Rhomboid minor
___ Trapezius

(c) Posterior superficial view   (d) Posterior deep view

Overview: The muscles that move the humerus (arm) cross the shoulder joint. Of the nine muscles that cross the shoulder joint, all except the pectoralis major and latissimus dorsi originate on the scapula. These two muscles are thus designated as **axial muscles,** since they originate on the axial skeleton. The remaining seven muscles, the **scapular muscles,** arise from the scapula.

Of the two axial muscles that move the humerus, the pectoralis major is a large, thick, fan-shaped muscle that covers the superior part of the thorax. It has two origins: a smaller clavicular head and a larger sternocostal head. The latissimus dorsi is a broad, triangular muscle located on the inferior part of the back. It is commonly called the "swimmer's muscle" because its many actions are used while swimming.

Among the scapular muscles, the deltoid is a thick, powerful shoulder muscle that covers the shoulder joint and forms the rounded contour of the shoulder. This muscle is a frequent site of intramuscular injections. As you study the deltoid, note that its fibers originate from three different points and that each group of fibers moves the humerus differently. The subscapularis is a large, triangular muscle that fills the subscapular fossa of the scapula and forms part of the posterior wall of the axilla. The supraspinatus is a rounded muscle, named for its location in the supraspinous fossa of the scapula. It lies deep to the trapezius. The infraspinatus is a triangular muscle, also named for its location in the infraspinous fossa of the scapula. The teres minor is a cylindrical, elongated muscle, often inseparable from the infraspinatus, which lies along its inferior border. The teres major is a thick, flattened muscle inferior to the teres minor and also helps to form part of the posterior wall of the axilla. The coracobrachialis is an elongated, narrow muscle in the arm that is pierced by the musculocutaneous nerve.

The strength and stability of the shoulder joint are not provided by the shape of the articulating bones or its ligaments. Instead, four deep muscles of the shoulder—subscapularis, supraspinatus, infraspinatus, and teres minor—strengthen and stabilize the shoulder joint. These muscles join the scapula to the humerus. Their flat tendons fuse together to form a nearly complete circle around the shoulder joint like the cuff on a shirt sleeve. This arrangement is referred to as the **rotator (musculotendinous) cuff.** The supraspinatus muscle is especially predisposed to wear and tear because of its location between the head of the humerus and acromion of the scapula, which compresses its tendon during shoulder movements, especially abduction of the arm.

After you have studied the muscles in this table, arrange them according to the following actions: flexion, extension, abduction, adduction, medial rotation, and lateral rotation. (The same muscle can be used more than once.)

Muscle	Origin	Insertion	Action
Axial muscles **Pectoralis** (pek'-tor-A-lis) **major** (label this muscle in Figure 10.11)			
Latissimus dorsi (la-TIS-i-mus DOR-sī; *latissimus* = widest; *dorsum* = back)			
Scapular muscles **Deltoid** (DEL-toyd; = triangular)			
Subscapularis (sub-scap'-ū-LA-ris; *sub* = below; *scapularis* = scapula)			
Supraspinatus (soo'-pra-spi-NĀ-tus; *supra* = above; *spinatus* = spine of scapula)			
Infraspinatus (in'-fra-spi-NĀ-tus; *infra* = below)			
Teres (TE-rēz) **major** (*teres* = long and round)			
Teres (TE-rēz) **minor**			
Coracobrachialis (kor'-a-kō-brā-kē-Ā-lis; *coraco* = coracoid process; *brachi* = arm)			

FIGURE 10.12 **Muscles that move the humerus (arm).**

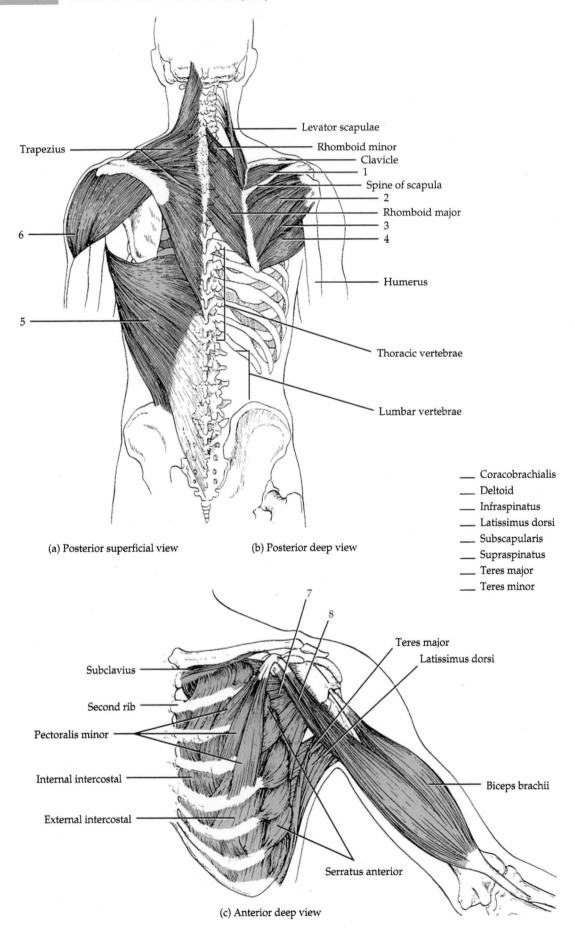

Levator scapulae

Trapezius

Rhomboid minor

Clavicle

1

Spine of scapula

2

Rhomboid major

3

4

6

Humerus

5

Thoracic vertebrae

Lumbar vertebrae

(a) Posterior superficial view   (b) Posterior deep view

___ Coracobrachialis
___ Deltoid
___ Infraspinatus
___ Latissimus dorsi
___ Subscapularis
___ Supraspinatus
___ Teres major
___ Teres minor

7

8

Teres major

Latissimus dorsi

Subclavius

Second rib

Pectoralis minor

Internal intercostal

External intercostal

Biceps brachii

Serratus anterior

(c) Anterior deep view

TABLE 10.16 Muscles that move the radius and ulna (forearm) (After completing the table, label Figure 10.13.)

Overview: Most of the muscles that move the radius and ulna (forearm) cause flexion and extension at the elbow, which is a hinge joint. The biceps brachii, brachialis, and brachioradialis muscles are the flexor muscles. The extensor muscles are the triceps brachii and the anconeus.

The biceps brachii is a large muscle located on the anterior surface of the arm. As indicated by its name, it has two heads of origin (long and short), both from the scapula. The muscle spans both the shoulder and elbow joints. In addition to its role in flexing the forearm at the elbow joint, it also supinates the forearm at the radioulnar joints and flexes the arm at the shoulder joint. The brachialis is deep to the biceps brachii muscle. It is the most powerful flexor of the forearm at the elbow joint. For this reason, it is called the "workhorse" of the elbow flexors. The brachioradialis flexes the forearm at the elbow joint, especially when a quick movement is required or when a weight is lifted slowly during flexion of the forearm.

The triceps brachii is the large muscle located on the posterior surface of the arm. It is the more powerful of the extensors of the forearm at the elbow joint. As its name implies, it has three heads of origin, one from the scapula (long head) and two from the humerus (lateral and medial heads). The long head crosses the shoulder joint; the other heads do not. The anconeus is a small muscle located on the lateral part of the posterior aspect of the elbow that assists the triceps brachii in extending the forearm at the elbow joint.

Some muscles that move the radius and ulna are involved in pronation and supination at the radioulnar joints. The pronators, as suggested by their names, are the pronator teres and pronator quadratus muscles. The supinator of the forearm is aptly named the supinator muscle. You use the powerful action of the supinator when you twist a corkscrew or turn a screw with a screwdriver.

In the limbs, functionally related skeletal muscles and their associated blood vessels and nerves are grouped together by fascia into regions called **compartments.** In the arm, the biceps brachii, brachialis, and coracobrachialis muscles constitute the anterior *(flexor)* compartment; the triceps brachii muscle forms the posterior *(extensor)* compartment.

Muscle	Origin	Insertion	Action
Flexors			
Biceps brachii (BĪ-ceps BRĀ-kē-ī; *biceps* = two heads [of origin]; *brachii* = arm)			
Brachialis (brā′-kē-A-lis)			
Brachioradialis (bra′-kē-ō-rā′-dē-A-lis; *radi* = radius) (see also Figure 10.14a)			
Extensors **Triceps brachii** (TRĪ-ceps BRĀ-kē-ī; *triceps* = three heads [of origin])			

(Continued)

TABLE 10.16	Muscles that move the radius and ulna (forearm) (After completing the table, label Figure 10.13.) (Continued)			
Muscle		**Origin**	**Insertion**	**Action**
Anconeus (an-KŌ-nē-us; *ancon* = the elbow) (see Figure 10.14c)				
Pronators **Pronator teres** (PRŌ-nā′-tor TE-rēz; (*pronator* = turning palm posteriorly)				
Pronator quadratus (PRŌ-nā′-tor kwod-RĀ-tus; *quadratus* = squared, four-sided) (see Figure 10.14a)				
Supinator **Supinator** (SOO-pi-nā-tor; *supinator* = turning palm anteriorly)				

FIGURE 10.13   Muscles that move the radius and ulna (forearm).

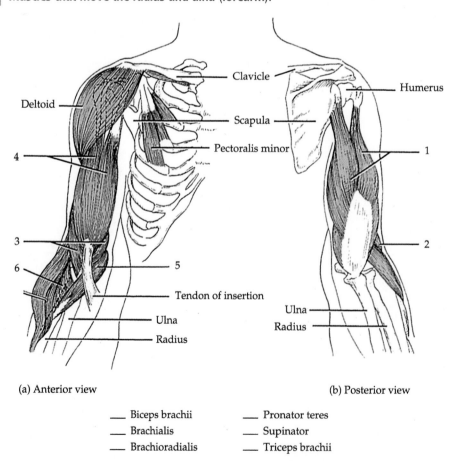

(a) Anterior view   (b) Posterior view

___ Biceps brachii   ___ Pronator teres
___ Brachialis   ___ Supinator
___ Brachioradialis   ___ Triceps brachii

TABLE 10.17	Muscles that move the wrist, hand, and digits (After completing the table, label Figure 10.14)

Overview: Muscles of the forearm that move the wrist, hand, and digits are many and varied. However, as you will see, their names usually give some indication of their origin, insertion, or action. On the basis of location and function, the muscles are divided into two groups—anterior and posterior compartments. The **anterior (flexor) compartment muscles** function as flexors. They originate on the humerus and typically insert on the carpals, metacarpals, and phalanges. The bellies of these muscles form the bulk of the forearm. The **posterior (extensor) compartment muscles** function as extensors. These muscles arise on the humerus and insert on the metacarpals and phalanges. Each of the two principal groups is also divided into superficial and deep muscles.

The superficial anterior compartment muscles are arranged in the following order from lateral to medial: flexor carpi radialis, palmaris longus (absent in about 10% of the population), and flexor carpi ulnaris (the ulnar nerve and artery are just lateral to the tendon of this muscle at the wrist). The flexor digitorum superficialis muscle is actually deep to the other three muscles and is the largest superficial muscle in the forearm.

The deep anterior compartment muscles are arranged in the following order from lateral to medial: flexor pollicis longus (the only flexor of the distal phalanx of the thumb) and flexor digitorum profundus (ends in four tendons that insert into the distal phalanges of the fingers).

The superficial posterior compartment muscles are arranged in the following order from lateral to medial: extensor carpi radialis longus, extensor carpi radialis brevis, extensor digitorum (occupies most of the posterior surface of the forearm and divides into four tendons that insert into the middle and distal phalanges of the fingers), extensor digiti minimi (a slender muscle generally connected to the extensor digitorum), and extensor carpi ulnaris.

The deep posterior compartment muscles are arranged in the following order from lateral to medial: abductor pollicis longus, extensor pollicis brevis, extensor pollicis longus, and extensor indicis.

The tendons of the muscles of the forearm that attach to the wrist or continue into the hand, along with blood vessels and nerves, are held close to bones by strong fasciae. The tendons are also surrounded by tendon sheaths. At the wrist, the deep fascia is thickened into fibrous bands called **retinacula** (*retinacul* = hold fast). The **flexor retinaculum** is located over the palmar surface of the carpal bones. Through it pass the long flexor tendons of the digits and wrist and median nerve. The **extensor retinaculum** is located over the posterior surface of the carpal bones. The extensor tendons of the wrist and digits pass through it.

After you have studied the muscles in the table, arrange them according to the following actions: flexion, extension, abduction, adduction, supination, and pronation. (The same muscles can be used more than once.)

Muscle	Origin	Insertion	Action
Superficial anterior (Flexor) Compartment			
Flexor carpi radialis (FLEK-sor KAR-pē rā′-dē-A-lis; *flexor* = decreases angle at a joint; *carpi* = wrist; *radi* = radius)			
Palmaris longus (pal-MA-ris LON-gus; *palma* = palm *longus* = long)			
Flexor carpi ulnaris (FLEK-sor KAR-pē ul-NAR-is; *ulnar* = ulna)			

(Continued)

TABLE 10.17	Muscles that move the wrist, hand, and digits (Figure 10.14) (Continued)

Muscle	Origin	Insertion	Action
Flexor digitorum superficialis (FLEK-sor di′-ji-TOR-um soo′-per-fish′-ē-A-lis; *digit* = finger or toe; *superficialis* = closer to surface)			
Deep anterior (Flexor) Compartment			
Flexor pollicis longus (FLEK-sor POL-li-sis LON-gus; *pollic* = thumb)			
Flexor digitorum profundus (FLEK-sor di′-ji-TOR-um pro-FUN-dus; *profundus* = deep)			
Superficial posterior (Extensor) Compartment			
Extensor carpi radialis longus (eks-TEN-sor KAR-pē rā′-dē-A-lis LON-gus; *extensor* = increases angle at a joint)			
Extensor carpi radialis brevis (eks-TEN-sor KAR-pē rā′-dē-A-lis BREV-is; *brevis* = short)			
Extensor digitorum (eks-TEN-sor di′-ji-TOR-um)			
Extensor digiti minimi (eks-TEN-sor DIJ-i-tē MIN-i-mē; *digiti* = digit; *minimi* = finger)			
Extensor carpi ulnaris (eks-TEN-sor KAR-pē ul-NAR-is)			
Deep posterior (Extensor) Compartment			
Abductor pollicis longus (ab-DUK-tor POL-li-sis LON-gus; *abductor* = moves a part away from midline)			

(Continued)

Muscle	Origin	Insertion	Action
Extensor pollicis brevis (eks-TEN-sor POL-li-sis BREV-is)			
Extensor pollicis longus (eks-TEN-sor POL-li-sis LON-gus)			
Extensor indicis (eks-TEN-sor IN-di-kis; *indicis* = index)			

FIGURE 10.14 Muscles that move the wrist, hand, and digits.

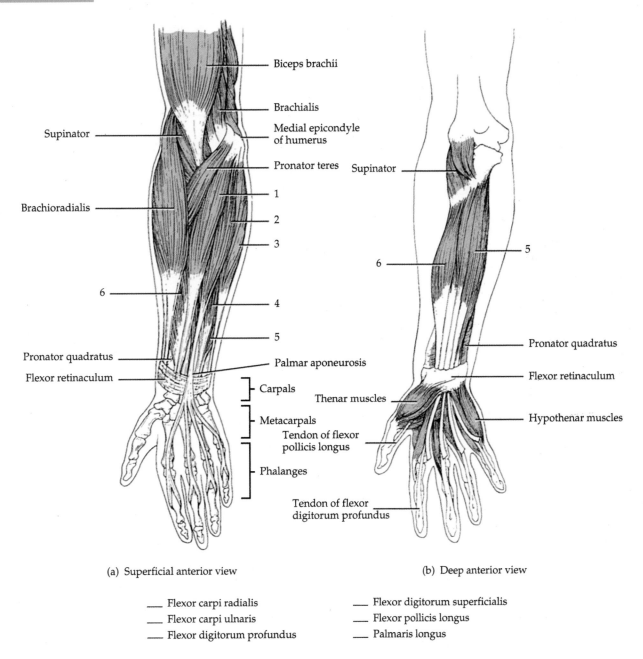

(a) Superficial anterior view

(b) Deep anterior view

___ Flexor carpi radialis   ___ Flexor digitorum superficialis
___ Flexor carpi ulnaris   ___ Flexor pollicis longus
___ Flexor digitorum profundus   ___ Palmaris longus

FIGURE 10.14 **Muscles that move the wrist, hand, and digits. (Continued)**

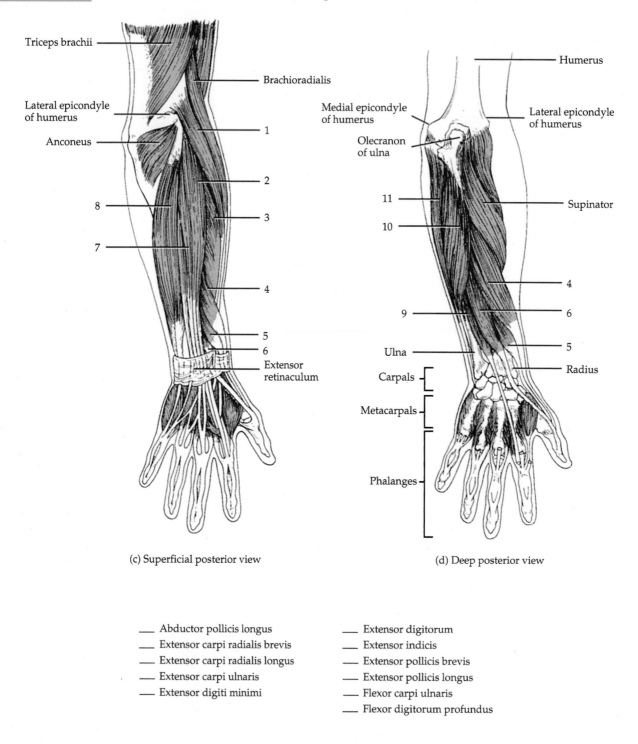

(c) Superficial posterior view

(d) Deep posterior view

___ Abductor pollicis longus
___ Extensor carpi radialis brevis
___ Extensor carpi radialis longus
___ Extensor carpi ulnaris
___ Extensor digiti minimi

___ Extensor digitorum
___ Extensor indicis
___ Extensor pollicis brevis
___ Extensor pollicis longus
___ Flexor carpi ulnaris
___ Flexor digitorum profundus

TABLE 10.18　Intrinsic muscles of the hand (After completing the table, label Figure 10.15.)

Overview: Several of the muscles discussed in Table 10.17 are **extrinsic hand muscles.** They originate outside the hand and insert within the hand. They produce the powerful but crude movements of the digits. The **intrinsic muscles** of the hand are located in the palm. They produce less powerful but intricate and precise movements of the digits. The muscles in this group are so named because their origins and insertions are *within* the hands.

The intrinsic muscles of the hand are divided into three groups: (1) **thenar** (THĒ-nar), (2) **hypothenar** (HĪ-pō-thē-nar), and (3) **intermediate.** The four thenar muscles, shown in Figure 10.14b, act on the thumb and form the **thenar eminence** (see Figure 11.11), the lateral rounded contour on the palm. The thenar muscles include the abductor pollicis brevis, opponens pollicis, flexor pollicis brevis, and adductor pollicis.

The three hypothenar muscles, shown in Figure 10.14b, act on the little finger and form the **hypothenar eminence** (see Figure 11.11), the medial rounded contour on the palm. The hypothenar muscles are the abductor digiti minimi, flexor digiti minimi brevis, and opponens digiti minimi.

The 11 intermediate (midpalmar) muscles act on all the digits except the thumb. The intermediate muscles include the lumbricals, palmar interossei, and dorsal interossei. Both sets of interossei muscles are located between the metacarpals and are important in abduction, adduction, flexion, and extension of the fingers, important movements in skilled activities such as writing, typing, and playing a piano.

The functional importance of the hand is readily apparent when one considers that certain hand injuries can result in permanent disability. Most of the dexterity of the hand depends on the movements of the thumb. The general activities of the hand are free motion, power grip (forcible movement of the fingers and thumb against the palm, as in squeezing), precision handling (a change in position of a handled object that requires exact control of finger and thumb positions, as in winding a watch or threading a needle), and pinch (compression between the thumb and index finger or between the thumb and first two fingers).

Movements of the thumb are very important in the precise activities of the hand, and they are defined in different planes from comparable movements of the fingers because the thumb is positioned at a right angle to the fingers. The five main movements of the thumb, illustrated below, are flexion (movement of the thumb medially across palm), extension (movement of the thumb laterally away from the palm), abduction (movement of the thumb in an anteroposterior plane away from the palm), adduction (movement of the thumb in an anteroposterior plane toward the palm), and opposition (movement of the thumb across the palm so that the tip of the thumb meets the tips of a finger). Opposition is the single most distinctive digital movement that gives humans and other primates the ability to precisely grasp and manipulate objects.

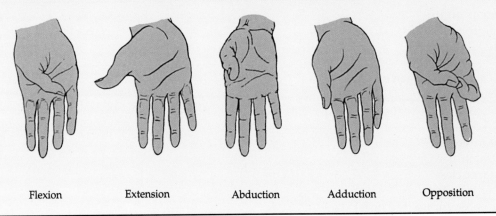

Flexion　　　　Extension　　　　Abduction　　　　Adduction　　　　Opposition

(Continued)

| TABLE 10.18 | Intrinsic muscles of the hand (After completing the table, label Figure 10.15.) (Continued) |

Muscle	Origin	Insertion	Action
Thenar (THĒ-nar) muscles **Abductor pollicis brevis** (ab-DUK-tor POL-li-sis BREV-is; *abductor* = moves part away from middle; *pollic* = thumb; *brevis* = short)			
Opponens pollicis (o-PŌ-nenz POL-li-sis; *opponens* = opposes)			
Flexor pollicis brevis (FLEK-sor POL-li-sis BREV-is; *flexor* = decreases angle at joint)			
Adductor pollicis (ad-DUK-tor POL-li-sis; *adductor* = moves part toward midline)			
Hypothenar (HĪ-pō-thē'-nar) muscles **Abductor digiti minimi** (ab-DUK-tor DIJ-i-tē MIN-i-mē; *digit* = finger or toe; *minimi* = little)			
Flexor digiti minimi brevis (FLEK-sor DIJ-i-tē MIN-i-mē BREV-is)			
Opponens digiti minimi (o-PŌ-nenz DIJ-i-tē MIN-i-mē)			
Intermediate (midpalmar) muscles **Lumbricals** (LUM-bri-kals; *lumbric* = earthworm; four muscles)			
Dorsal interossei (DOR-sal in'-ter-OS-ē-ī; four muscles; *dorsal* = back surface; *inter* = between; *ossei* = bones)			
Palmar interossei (PAL-mar in'-ter-OS-ē-ī; *palmar* = palm) (three muscles)			

FIGURE 10.15   Intrinsic muscles of the hand.

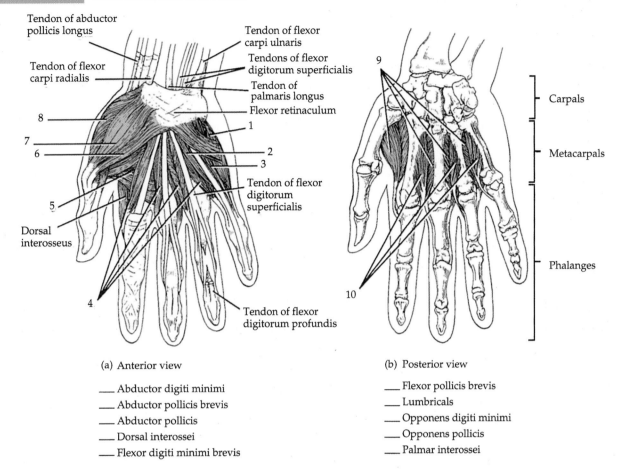

Tendon of abductor
pollicis longus

Tendon of flexor
carpi radialis

Tendon of flexor
carpi ulnaris

Tendons of flexor
digitorum superficialis

Tendon of
palmaris longus

Flexor retinaculum

8

7

6

5

Dorsal
interosseus

4

1

2

3

Tendon of flexor
digitorum
superficialis

Tendon of flexor
digitorum profundis

9

10

Carpals

Metacarpals

Phalanges

(a) Anterior view

____ Abductor digiti minimi
____ Abductor pollicis brevis
____ Abductor pollicis
____ Dorsal interossei
____ Flexor digiti minimi brevis

(b) Posterior view

____ Flexor pollicis brevis
____ Lumbricals
____ Opponens digiti minimi
____ Opponens pollicis
____ Palmar interossei

TABLE 10.19	Muscles that move the vertebral column (backbone) (After completing the table, label Figure 10.16.)

Overview: The muscles that move the vertebral column (backbone) are quite complex because they have multiple origins and insertions and there is considerable overlapping among them. One way to group the muscles is on the basis of the general direction of the muscle bundles and their approximate lengths. For example, the **splenius muscles** arise from the midline and extend laterally and superiorly to their insertions. The **erector spinae muscle** arises from either the midline or more laterally, but usually runs almost longitudinally, with neither a marked lateral nor medial direction as it is traced superiorly. The **transversospinales muscles** arise laterally, but run toward the midline as they are traced superiorly. Deep to these three muscle groups are small **segmental muscles** that run between spinous processes or transverse processes of vertebrae. Since the scalene muscles also assist in moving the vertebral column, they are included in this table.

The bandagelike splenius muscles are attached to the sides and back of the neck. The two muscles in the group are named on the basis of their superior attachments (insertions): splenius capitis (head region) and splenius cervicis (cervical region).

The erector spinae is the largest muscular mass of the back, forming a prominent bulge on either side of the vertebral column. It is the chief extensor of the vertebral column and consists of three groups—iliocostalis (lateral), longissimus (intermediate), and spinalis (medial). These groups, in turn, consist of a series of overlapping muscles, and the muscles within the groups are named according to the regions of the body with which they are associated. The iliocostalis group consists of three muscles named the iliocostalis cervicis (cervical region), iliocostalis thoracis (thoracic region), and iliocostalis lumborum (lumbar region). The longissimus group resembles herringbone and consists of three muscles called the longissimus capitis, longissimus cervicis, and longissimus thoracis. The spinalis group also consists of three muscles, and these are called the spinalis capitis, spinalis cervicis, and spinalis thoracis.

The transversospinales muscles are so named because their fascicles extend from the transverse processes to the spinous processes of the vertebrae. The semispinalis muscles in this group are also named according to the region of the body with which they are associated: semispinalis capitis, semispinalis cervicis, and semispinalis thoracis. The multifidus muscle in this group, as its name implies, is split into several bundles. It extends and laterally flexes the vertebral column. The rotatores muscles of this group are short muscles and run the entire length of the vertebral column. They extend and rotate the vertebral column.

Within the segmental muscle group are the interspinales and intertransversarii muscles, which unite the spinous and transverse processes of consecutive vertebrae.

Within the scalene group, the anterior scalene muscle is anterior to the middle scalene muscle. The middle scalene muscle is intermediate in placement and is the longest and largest of the scalene muscles. The posterior scalene muscle is posterior to the middle scalene muscle and is the smallest of the scalene muscles. These muscles flex, laterally flex, and rotate the head and assist in deep inhalation.

Note in Table 10.10 that the rectus abdominis and quadratus lumborum muscles also assume a role in moving the vertebral column.

(Continued)

Muscle	Origin	Insertion	Action
Splenius (SPLĒ-nē-us) muscles **Splenius capitis** (SPLĒ-nē-us KAP-i-tis; *splenium* = bandage; *capit* = head)			
Splenius cervicis (SPLĒ-nē-us SER-vi-kis; *cervic* = neck)			
Erector spinae This is the largest muscular mass of the back and consists of three groupings—iliocostalis, longissimus, and spinalis. These groups, in turn, consist of overlapping muscles. The iliocostalis group is laterally placed, the longissimus group is intermediate in placement, and the spinalis group is medially placed.			
Iliocostalis (Lateral) Group **Iliocostalis cervicis** (il'-ē-ō-kos-TAL-is SER-vi-kis; *ilio* = flank; *costa* = rib)			
Iliocostalis thoracis (il'-ē-ō-kos-TAL-is thō-RA-kis; *thorac* = chest)			
Iliocostalis lumborum (il'-ē-ō-kos-TAL-is lum-BOR-um)			
Longissimus (Intermediate) Group **Longissimus capitis** (lon-JIS-i-mis KAP-i-tis; *longissimus* = longest)			
Longissimus cervicis (lon-JIS-i-mis SER-vi-kis)			
Longissimus thoracis (lon-JIS-i-mis thō-RA-kis)			
Spinalis (Medial) Group **Spinalis capitis** (spi-NA-lis KAP-i-tis; *spinal* = vertebral column)			
Spinalis cervicis (spi-NA-lis SER-vi-kis)			
Spinalis thoracis (spi-NA-lis thō-RA-kis)			

(Continued)

| TABLE 10.19 | Muscles that move the vertebral column (backbone) (After completing the table, label Figure 10.16.) (Continued) |

Muscle	Origin	Insertion	Action
Transversospinales (trans-ver'-sō-SPĪ-nāl-ez) muscles **Semispinalis capitis** (sem'-ē-spi-NA-lis KAP-i-tis; *semi* = partially or one half)			
Semispinalis cervicis (sem'-ē-spi-NA-lis SER-vi-kis)			
Semispinalis thoracis (sem'-ē-spi-NA-lis thō-RA-kis)			
Multifidus (mul-TIF-i-dus; *multi* = many; *fid* = segmented)			
Rotatores (rō-ta-TO-rēz; singular is rotatore; *rotatore* = to rotate)			
Segmental muscles **Interspinales** (in-ter-SPĪ-nāl-ez; *inter* = between)			
Intertransversarii (in'-ter-trans-vers-AR-ē-ī; singular is intertransversarius)			
Scalene (SKĀ-lēn) muscles **Anterior scalene** (SKĀ-lēn; *anterior* = front *scalene* = uneven)			
Middle scalene (SKĀ-lēn)			
Posterior scalene (SKĀ-lēn; *posterior* = back)			

FIGURE 10.16 Muscles that move the vertebral column (backbone). Many superficial muscles have been removed or cut to show deep muscles.

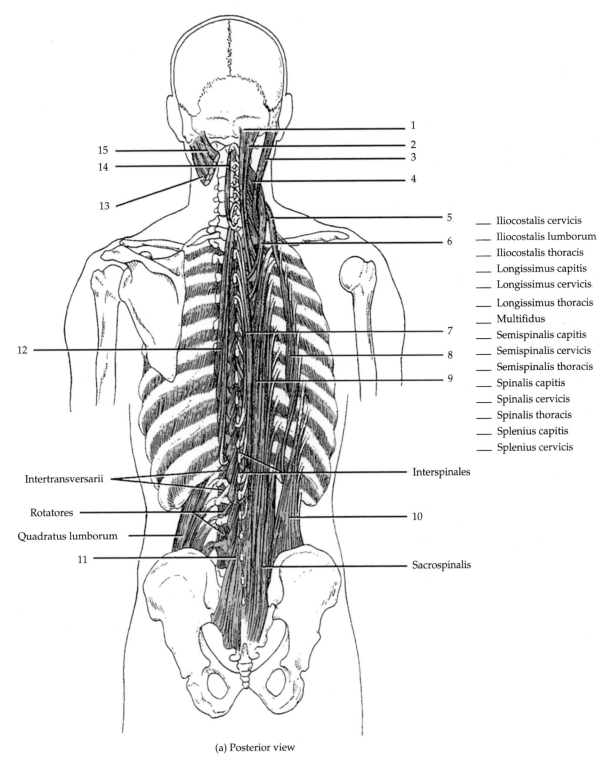

____ Iliocostalis cervicis
____ Iliocostalis lumborum
____ Iliocostalis thoracis
____ Longissimus capitis
____ Longissimus cervicis
____ Longissimus thoracis
____ Multifidus
____ Semispinalis capitis
____ Semispinalis cervicis
____ Semispinalis thoracis
____ Spinalis capitis
____ Spinalis cervicis
____ Spinalis thoracis
____ Splenius capitis
____ Splenius cervicis

Intertransversarii

Rotatores

Quadratus lumborum

Interspinales

Sacrospinalis

(a) Posterior view

FIGURE 10.16 **Muscles that move the vertebral column (backbone). (Continued)**

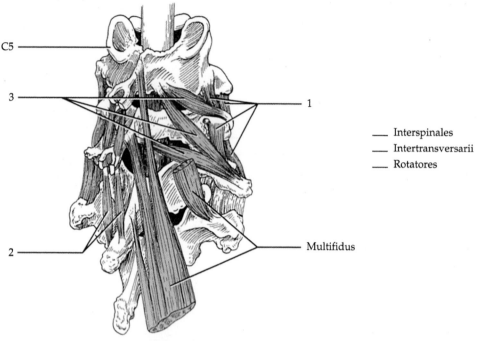

C5

3 ——————— 1

2

Multifidus

___ Interspinales
___ Intertransversarii
___ Rotatores

(b) Posterolateral view of several vertebrae
with their associated muscles

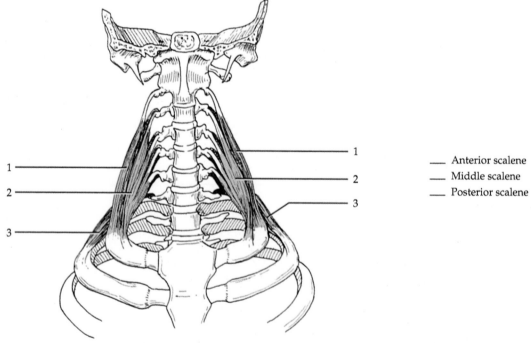

1
2
3

1
2
3

___ Anterior scalene
___ Middle scalene
___ Posterior scalene

(c) Anterior view of scalene muscles

TABLE 10.20	Muscles that move the femur (thigh) (After completing the table, label Figure 10.17.)

Overview: As you will see, muscles of the lower limbs are larger and more powerful than those of the upper limbs since lower limb muscles function in stability, locomotion, and maintenance of posture. Upper limb muscles are characterized by versatility of movement. In addition, muscles of the lower limbs frequently cross two points and act equally on both.

The majority of muscles that move the femur originate from the pelvic girdle and insert on the femur. The psoas major and iliacus muscles are together referred to as the iliopsoas (il'-ē-ō-SŌ-as) muscle because they share a common insertion (lesser trochanter of femur). There are three gluteal muscles: gluteus maximus, gluteus medius, and gluteus minimus. The gluteus maximus is the largest and heaviest of the three muscles and is one of the largest muscles in the body. It is the chief extensor of the femur. The gluteus medius is mostly deep to the gluteus maximus and is a powerful abductor of the femur at the hip joint. This muscle is a common site for an intramuscular injection. The gluteus minimus muscle is the smallest of the gluteal muscles and lies deep to the gluteus medius.

The tensor fasciae latae muscle is located on the lateral surface of the thigh. There is a layer of deep fascia composed of dense connective tissue that encircles the entire thigh and is referred to as the **fascia lata.** It is well developed laterally where, together with the tendons of the tensor fasciae and gluteus maximus muscles, it forms a structure called the **iliotibial tract.** The tract inserts into the lateral condyle of the tibia.

The piriformis, obturator internus, obturator externus, superior gemellus, inferior gemellus, and quadratus femoris muscles are all deep to the gluteus maximus muscle and function as lateral rotators of the femur at the hip joint.

Three muscles on the medial aspect of the thigh are the adductor longus, adductor brevis, and adductor magnus. They originate on the pubic bone and insert on the femur. All three muscles adduct, flex, and medially rotate the femur at the hip joint. The pectineus muscle also adducts and flexes the femur at the hip joint.

Technically, the adductor muscles and pectineus muscles are components of the medial compartment of the thigh and could also be included in Table 10.21. However, they are included here because they act on the femur.

After you have studied the muscles in this table, arrange them according to the following actions: flexion, extension, abduction, adduction, medial rotation, and lateral rotation. (The same muscles can be used more than once.)

Muscle	**Origin**	**Insertion**	**Action**
Psoas (SŌ-as) major (*psoa* = a muscle of loin)			
Iliacus (il'-ē-AK-us; *iliac* = ilium)			
Gluteus maximus (GLOO-tē-us MAK-si-mus; *glutê* = rump or buttock; *maximus* = largest)			
Gluteus medius (GLOO-tē-us MĒ-dē-us; *medî* = middle)			

(Continued)

| TABLE 10.20 | Muscles that move the femur (thigh) (After completing the table, label Figure 10.17.) (Continued) | | | |

Muscle	Origin	Insertion	Action
Gluteus minimus (GLOO-tē-us MIN-i-mus; *minim* = smallest)			
Tensor fasciae latae (TEN-sor FA-shē-ē LĀ-tē; *tensor* = makes tense; *fasciae* = band; *lat* = wide)			
Piriformis (pir-i-FOR-mis; *piri* = pear; *form* = shape)			
Obturator internus* (OB-tū-rā′-tor in-TER-nus; *obturator* = obturator foramen; *intern* = inside)			
Obturator externus (OB-tū-rā′-tor ex-TER-nus; *extern* = outside)			
Superior gemellus (jem-EL-lus; *superior* = above; *gemell* = twins)			
Inferior gemellus (jem-EL-lus; *inferior* = below)			
Quadratus femoris (kwod-RĀ-tus FEM-or-is; *quad* = square, four-sided; *femoris* = femur)			
Adductor longus (LONG-us; *adductor* = moves part closer to midline; *longus* = long)			
Adductor brevis (BREV-is; *brevis* = short)			
Adductor magnus (MAG-nus; *magnus* = large)			
Pectineus (pek-TIN-ē-us; *pectin* = comb)			

*Not illustrated.

FIGURE 10.17 **Muscles that move the femur (thigh).**

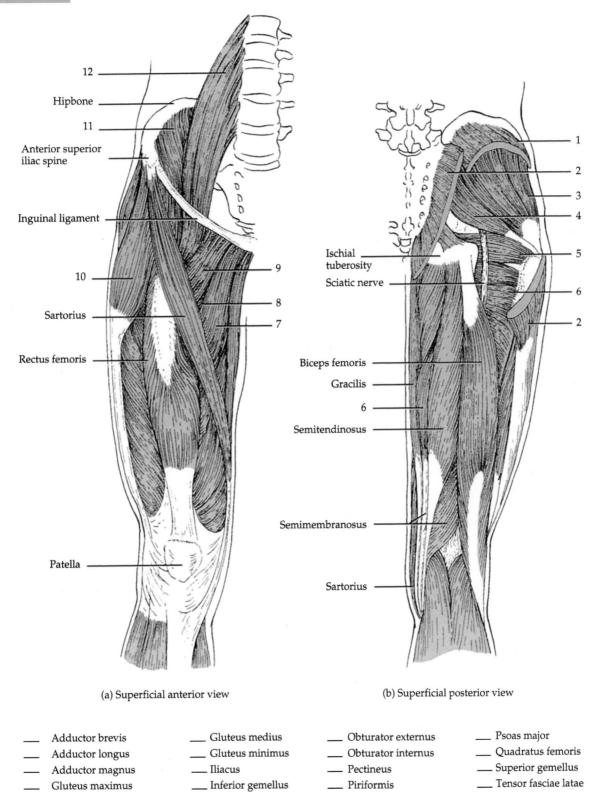

12

Hipbone

11

Anterior superior
iliac spine

Inguinal ligament

10

Sartorius

Rectus femoris

Patella

9

8

7

Ischial
tuberosity

Sciatic nerve

Biceps femoris

Gracilis

6

Semitendinosus

Semimembranosus

Sartorius

1

2

3

4

5

6

2

6

(a) Superficial anterior view

(b) Superficial posterior view

___ Adductor brevis	___ Gluteus medius	___ Obturator externus	___ Psoas major
___ Adductor longus	___ Gluteus minimus	___ Obturator internus	___ Quadratus femoris
___ Adductor magnus	___ Iliacus	___ Pectineus	___ Superior gemellus
___ Gluteus maximus	___ Inferior gemellus	___ Piriformis	___ Tensor fasciae latae

FIGURE 10.17 **Muscles that move the femur (thigh). (Continued)**

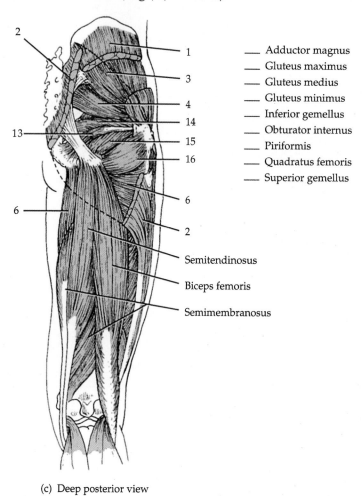

___ Adductor magnus
___ Gluteus maximus
___ Gluteus medius
___ Gluteus minimus
___ Inferior gemellus
___ Obturator internus
___ Piriformis
___ Quadratus femoris
___ Superior gemellus

(c) Deep posterior view

TABLE 10.21 Muscles that act on the femur (thigh bone) and tibia and fibula (leg bones) (After completing the table, label Figure 10.18.)

Overview: The muscles that act on the femur (thigh bone) and tibia and fibula (leg bones) are separated by deep fascia into medial, anterior, and posterior compartments. The **medial (adductor) compartment** is so named because its muscles adduct the femur at the hip joint. As noted earlier, the adductor magnus, adductor longus, adductor brevis, and pectineus, components of the medial compartment, are included in Table 10.20 because they act on the femur. The gracilis, the other muscle in the medial compartment, not only adducts the thigh but also flexes the leg at the knee joint. For this reason, it is included in this table. The gracilis is a long, straplike muscle that lies on the medial aspect of the thigh and knee.

The **anterior (extensor) compartment** extends the leg (and flexes the thigh). This compartment contains the quadriceps femoris and sartorius muscles. The quadriceps femoris muscle is the biggest muscle in the body, covering most of the anterior surface and sides of the thigh. The muscle is actually a composite muscle that includes four distinct parts, usually described as four separate muscles: (1) rectus femoris, on the anterior aspect of the thigh; (2) vastus lateralis, on the lateral aspect of the thigh; (3) vastus medialis, on the medial aspect of the thigh; and (4) vastus intermedius, located deep to the rectus femoris between the vastus lateralis and medialis. The common tendon for the four muscles is known as the **quadriceps tendon,** which inserts into the patella. The tendon continues inferior to the patella as the **patellar ligament**, which attaches to the tibial tuberosity. The quadriceps femoris muscle is the great extensor muscle of the leg. The sartorius muscle is a long, narrow muscle that forms a band across the thigh from the ilium of the hipbone to the medial side of the tibia. The various movements it produces help effect the cross-legged sitting position in which the heel of one limb is placed on the knee of the opposite limb. It is known as the tailor's muscle because tailors often assume this cross-legged sitting position. (Since the major action of the sartorius muscle is to move the thigh rather than the leg, it could also have been included in Table 10.20.)

The **posterior (flexor) compartment** is so named because its muscles flex the leg at the knee joint (and also extend the thigh at the hip joint). This compartment is composed of three muscles collectively called the hamstrings: (1) biceps femoris, (2) semitendinosus, and (3) semimembranosus. The hamstrings are so named because their tendons are long and stringlike in the popliteal area and from an old practice of butchers in which they hung hams for smoking by these long tendons. Since the hamstrings span two joints (hip and knee), they are both extensors of the thigh and flexors of the leg. The **popliteal fossa** is a diamond-shaped space on the posterior aspect of the knee bordered laterally by the tendons of the biceps femoris muscle and medially by the semitendinosus and semimembranosus muscles (see also Figure 10.18b).

Muscle	Origin	Insertion	Action
Medial (adductor) compartment **Adductor magnus** (MAG-nus)			
Adductor longus (LONG-us)			

(Continued)

TABLE 10.21	Muscles that act on the femur (thigh bone) and tibia and fibula (leg bones) (After completing the table, label Figure 10.18.) (Continued)

Muscle	Origin	Insertion	Action
Adductor brevis (BREV-is)			
Pectineus (pek-TIN-ē-us)			
Gracilis (gra-SIL-is; *gracilis* = slender)			
Anterior (extensor) compartment **Quadriceps femoris** (KWOD-ri-ceps FEM-or-is; *quadriceps* = four heads [of origin]; *femoris* = femur)			
Rectus femoris (REK-tus FEM-or-is; *rectus* = fascicles parallel to midline)			
Vastus lateralis (VAS-tus lat'-er-A-lis; *vast* = large; *lateralis* = lateral)			
Vastus medialis (VAS-tus mē-dē-A-lis; *medialis* = medial)			
Vastus intermedius (VAS-tus in'-ter-MĒ-dē-us; *intermedius* = middle)			
Sartorius (sar-TOR-ē-us; *sartor* = tailor; refers to cross-legged position of tailors; longest muscle in body)			
Posterior (flexor) compartment **Hamstrings**—A collective designation for three separate muscles:			
Biceps femoris (BĪ-ceps FEM-or-is; *biceps* = two heads of origin)			
Semitendinosus (sem'-ē-TEN-di-nō-sus; *semi* = half; *tendo* = tendon)			
Semimembranosus (sem'-ē-MEM-bra-nō-sus; *membran* = membrane)			

FIGURE 10.18 Muscles that act on the femur (thigh) and tibia and fibula (leg).

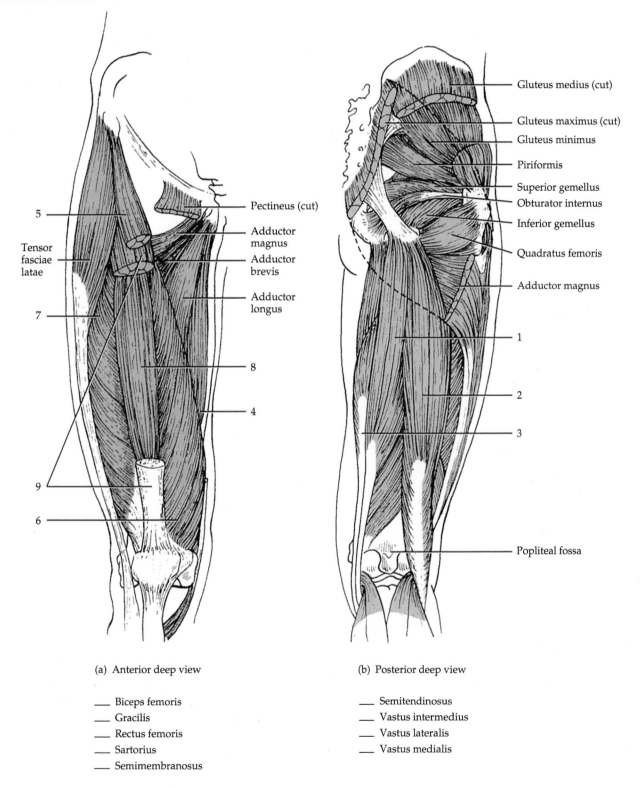

(a) Anterior deep view

(b) Posterior deep view

___ Biceps femoris
___ Gracilis
___ Rectus femoris
___ Sartorius
___ Semimembranosus

___ Semitendinosus
___ Vastus intermedius
___ Vastus lateralis
___ Vastus medialis

TABLE 10.22   Muscles that move the foot and toes (After completing the table, label Figure 10.19.)

Overview: Muscles that move the foot and toes are located in the leg. The muscles of the leg, like those of the thigh, are divided by deep fascia into three compartments: anterior, lateral, and posterior. The **anterior compartment** consists of muscles that dorsiflex the foot. In a situation analogous to the wrist, the tendons of the muscles of the anterior compartment are held firmly to the ankle by thickenings of deep fascia called the **superior extensor retinaculum (transverse ligament of the ankle)** and **inferior extensor retinaculum (cruciate ligament of the ankle).**

Within the anterior compartment, the tibialis anterior muscle is a long, thick muscle against the lateral surface of the tibia, where it is easy to palpate. The extensor hallucis muscle is a thin muscle that lies deep to the tibialis anterior and extensor digitorum longus muscles. This featherlike muscle is lateral to the tibialis anterior muscle, where it can easily be palpated. The fibularis (peroneus) tertius muscle is actually part of the extensor digitorum longus muscle, with which it shares a common origin.

The **lateral (fibular) compartment** contains two muscles that plantar flex and evert the foot: fibularis (peroneus) longus and fibularis (peroneus) brevis.

The **posterior compartment** consists of muscles that are divisible into superficial and deep groups. The superficial muscles share a common tendon of insertion, the **calcaneal (Achilles) tendon,** the strongest tendon of the body, that inserts into the calcaneus bone of the ankle. The superficial muscles and most deep muscles plantar flex the foot at the ankle joint.

The superficial muscles of the posterior compartment are the gastrocnemius, soleus, and plantaris, the so-called calf muscles. The large size of these muscles is directly related to the characteristic upright stance of humans. The gastrocnemius muscle is the most superficial muscle and forms the prominence of the calf. The soleus lies deep to the gastrocnemius and is a broad, flat muscle. It derives its name from its resemblance to a flat fish (sole). The plantaris is a small muscle that may be absent or sometimes there are two of them in each leg. It runs obliquely between the gastrocnemius and soleus muscles.

The deep muscles of the posterior compartment are the popliteus, tibialis posterior, flexor digitorum longus, and flexor hallucis longus. The popliteus muscle is triangular and forms the floor of the popliteal fossa. The tibialis posterior muscle is the deepest muscle in the posterior compartment. It lies between the flexor digitorum longus and flexor hallucis longus muscles. The flexor digitorum muscle is smaller than the flexor hallucis longus muscle, even though the former flexes four toes, while the latter flexes only the great toe at the interphalangeal joints.

Muscle	Origin	Insertion	Action
Anterior compartment **Tibialis** (tib′-ē-A-lis) **anterior** (*tibialis* = tibia; *anterior* = front)			
Extensor hallucis longus (HAL-ū-sis LON-gus; *extensor* = increases angle at joint; *halluc* = hallux or great toe; *longus* = long)			

(Continued)

Muscle	Origin	Insertion	Action
Extensor digitorum longus (di′-ji-TOR-um LON-gus)			
Fibularis (peroneus) tertius (fib-ū-LAR-is TER-shus; *peron* = fibula; *tertius* = third)			
Lateral (fibular) compartment **Fibularis (peroneus) longus** (fib-ū-LAR-is LON-gus)			
Fibularis (peroneus) brevis (fib-ū-LAR-is BREV-is; *brevis* = short)			
Posterior compartment **Superficial** **Gastrocnemius** (gas′-trok-NĒ-mē-us; *gastro* = belly; *cnem* = leg)			
Soleus (SŌ-lē-us; *sole* = a type of flat fish)			
Plantaris (plan-TA-ris; *plantar* = sole of foot)			
Deep **Popliteus** (pop-LIT-ē-us; *poplit* = the back of knee)			
Tibialis (tib′-ē-A-lis) **posterior** (*posterior* = back)			
Flexor digitorum longus (di′-ji-TOR-um LON-gus; *digit* = finger or toe)			
Flexor hallucis longus (HAL-ū-sis LON-gus; *flexor* = decreases angle at joint)			

FIGURE 10.19 **Muscles that move the foot and toes.**

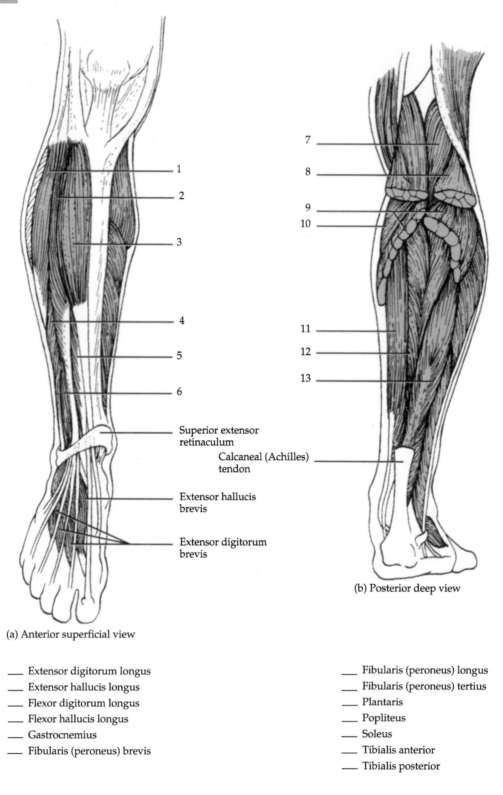

Superior extensor retinaculum

Calcaneal (Achilles) tendon

Extensor hallucis brevis

Extensor digitorum brevis

(a) Anterior superficial view

(b) Posterior deep view

___ Extensor digitorum longus
___ Extensor hallucis longus
___ Flexor digitorum longus
___ Flexor hallucis longus
___ Gastrocnemius
___ Fibularis (peroneus) brevis

___ Fibularis (peroneus) longus
___ Fibularis (peroneus) tertius
___ Plantaris
___ Popliteus
___ Soleus
___ Tibialis anterior
___ Tibialis posterior

TABLE 10.23 Intrinsic muscles of the foot (After completing the table, label Figure 10.20.)

Overview: The muscles discussed here are termed intrinsic muscles because they originate and insert *within* the foot. The **intrinsic muscles** of the foot are, for the most part, comparable to those in the hand. Whereas the muscles of the hand are specialized for precise and intricate movements, those of the foot are limited to support and locomotion. The deep fascia of the foot forms the **plantar aponeurosis (fascia)** that extends from the calcaneus bone to the phalanges of the toes. The aponeurosis supports the longitudinal arch of the foot and transmits the flexor tendons of the foot.

The intrinsic musculature of the foot is divided into two groups: dorsal and plantar. There is only one **dorsal muscle,** the extensor digitorum brevis, which extends toes 2–5 at the metatarsophalangeal joints.

The **plantar muscles** are arranged in four layers, the most superficial layer being referred to as the first layer. The muscles in the first (superficial) layer are the abductor hallucis, which abducts the great toe at the metatarsophalangeal joint, the flexor digitorum brevis, which flexes toes 2–5 at the interphalangeal and metatarsophalangeal joints, and the abductor digiti minimi, which abducts the little toe. The second layer consists of the quadratus plantae, which flexes toes 2–5 at the interphalangeal joints, and the lumbricals, which flex the proximal phalanges and extend the distal phalanges of toes 2–5. The third layer is composed of the flexor hallucis brevis, which flexes the great toe, the abductor hallucis, which adducts the great toe, and the flexor digiti minimi brevis, which flexes the little toe. The fourth (deep) layer consists of the dorsal interossei, four muscles that abduct toes 2–4 and flex the proximal phalanges and extend the distal phalanges, and three plantar interossei, which adduct toes 3–5 and flex the proximal phalanges and extend the distal phalanges.

Muscle	Origin	Insertion	Action
Dorsal muscle **Extensor digitorum brevis** (di'-ji-TOR-um BREV-is; *extensor* = increases angle at joint; *digit* = finger or toe; *brevis* = short) (Illustrated in Figure 10.19a)			
Plantar muscles **First (Most Superficial) Layer** **Abductor hallucis** (HAL-ū-sis; *abductor* = moves part away from midline; *hallucis* = hallux or great toe)			
Flexor digitorum brevis (di'-ji-TOR-um BREV-is; *flexor* = decreases angle at joint)			

(Continued)

| TABLE 10.23 | Intrinsic muscles of the foot (After completing the table, label Figure 10.20.) (Continued) |

Muscle	Origin	Insertion	Action
Abductor digiti minimi (DIJ-i-tē; MIN-i-mē; *minimi* = little)			
Second Layer **Quadratus plantae** (KWOD-RĀ-tus PLAN-tē; *quad* = four; *planta* = sole)			
Lumbricals (LUM-bri-kals; four muscles)			
Third Layer **Flexor hallucis brevis** (HAL-ū-sis BREV-is)			
Adductor hallucis (HAL-ū-sis; *adduct* = moves part closer to midline)			
Flexor digiti minimi brevis (DIJ-i-tē MIN-i-mē BREV-is)			
Fourth (Deepest) Layer **Dorsal interossei** (in'-ter-OS-ē-ī; four muscles; *dorsal* = back surface; *inter* = between; *ossei* = bone)			
Plantar interossei (in'-ter-OS-ē-ī)			

FIGURE 10.20 Intrinsic muscles of the foot.

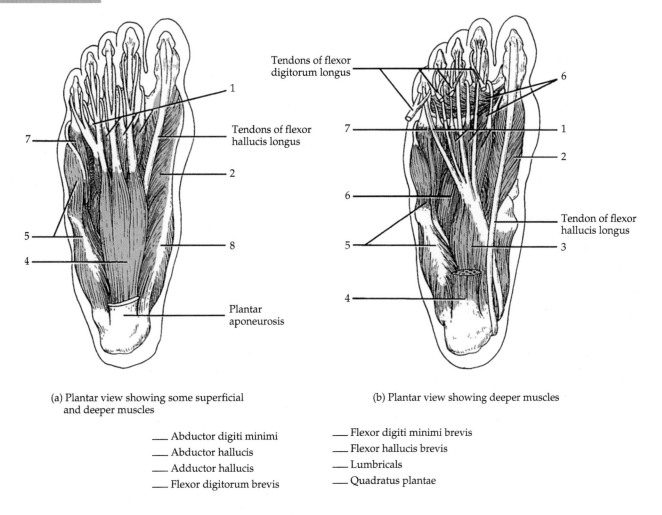

Tendons of flexor
digitorum longus

Tendons of flexor
hallucis longus

Plantar
aponeurosis

Tendon of flexor
hallucis longus

(a) Plantar view showing some superficial
and deeper muscles

(b) Plantar view showing deeper muscles

___ Abductor digiti minimi
___ Abductor hallucis
___ Adductor hallucis
___ Flexor digitorum brevis

___ Flexor digiti minimi brevis
___ Flexor hallucis brevis
___ Lumbricals
___ Quadratus plantae

F. Composite Muscular System

Now that you have studied the muscle of the
body by region, label the composite diagram
shown in Figure 10.21.

**ANSWER THE LABORATORY REPORT QUESTIONS
AT THE END OF THE EXERCISE.**

FIGURE 10.20   **Intrinsic muscles of the foot. (Continued)**

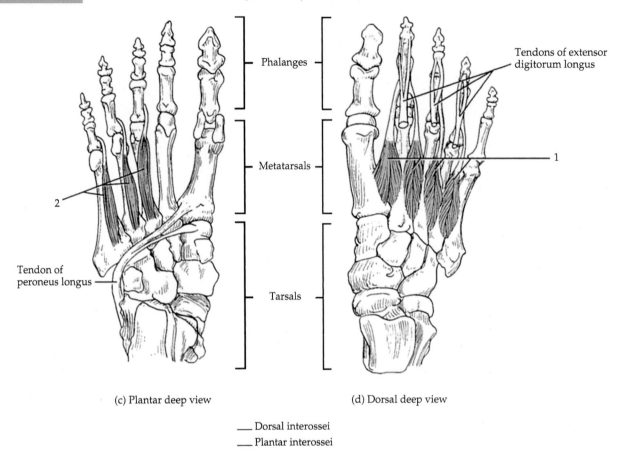

Phalanges

Metatarsals

Tarsals

Tendon of peroneus longus

2

Tendons of extensor digitorum longus

1

(c) Plantar deep view   (d) Dorsal deep view

___ Dorsal interossei
___ Plantar interossei

FIGURE 10.21   Principal superficial muscles.

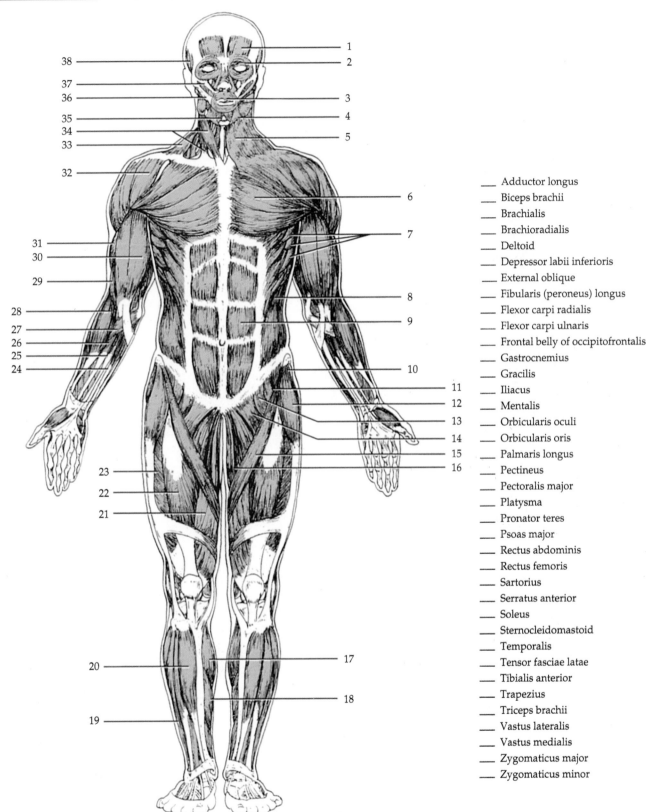

___ Adductor longus
___ Biceps brachii
___ Brachialis
___ Brachioradialis
___ Deltoid
___ Depressor labii inferioris
___ External oblique
___ Fibularis (peroneus) longus
___ Flexor carpi radialis
___ Flexor carpi ulnaris
___ Frontal belly of occipitofrontalis
___ Gastrocnemius
___ Gracilis
___ Iliacus
___ Mentalis
___ Orbicularis oculi
___ Orbicularis oris
___ Palmaris longus
___ Pectineus
___ Pectoralis major
___ Platysma
___ Pronator teres
___ Psoas major
___ Rectus abdominis
___ Rectus femoris
___ Sartorius
___ Serratus anterior
___ Soleus
___ Sternocleidomastoid
___ Temporalis
___ Tensor fasciae latae
___ Tibialis anterior
___ Trapezius
___ Triceps brachii
___ Vastus lateralis
___ Vastus medialis
___ Zygomaticus major
___ Zygomaticus minor

(a) Anterior view

FIGURE 10.21   Principal superficial muscles. (Continued)

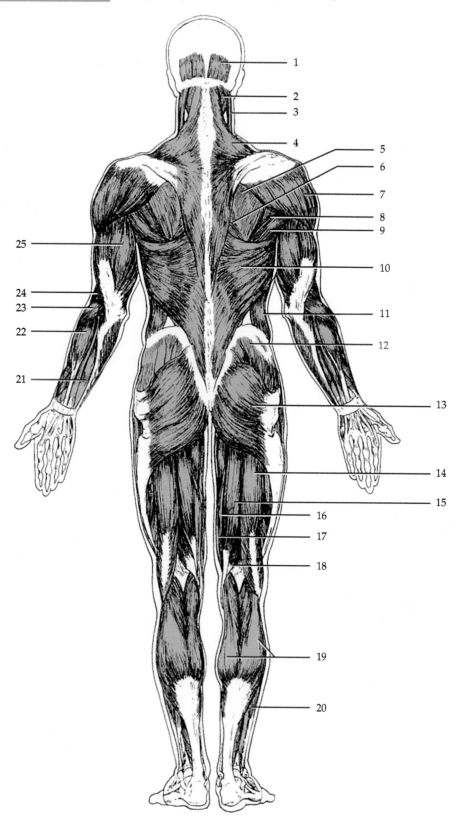

___ Biceps femoris
___ Brachioradialis
___ Deltoid
___ Extensor carpi radialis longus
___ Extensor carpi ulnaris
___ Extensor digitorum
___ External oblique
___ Gastrocnemius
___ Gluteus maximus
___ Gluteus medius
___ Gracilis
___ Infraspinatus
___ Latissimus dorsi
___ Occipital belly of occipitofrontalis
___ Rhomboid major
___ Sartorius
___ Semimembranosus
___ Semispinalis capitis
___ Semitendinosus
___ Soleus
___ Sternocleidomastoid
___ Teres major
___ Teres minor
___ Trapezius
___ Triceps brachii

(b) Posterior view

EXERCISE 10

Muscular System

Name _____   Date _____

Laboratory Section _____   Score/Grade _____

PART 1 ■ Multiple Choice

_____ 1. The connective tissue covering that encloses the entire skeletal muscle is the (a) perimysium (b) endomysium (c) epimysium (d) mesomysium

_____ 2. A cord of connective tissue that attaches a skeletal muscle to the periosteum of bone is called a(n) (a) ligament (b) aponeurosis (c) perichondrium (d) tendon

_____ 3. A skeletal muscle that decreases the angle at a joint is referred to as a(n) (a) flexor (b) abductor (c) pronator (d) evertor

_____ 4. The name *abductor* means that a muscle (a) produces a downward movement (b) moves a part away from the midline (c) elevates a body part (d) increases the angle at a joint

_____ 5. Which connective tissue layer directly encircles the fascicles of skeletal muscles? (a) epimysium (b) endomysium (c) perimysium (d) mesomysium

_____ 6. Which muscle is *not* associated with a movement of the eyeball? (a) superior rectus (b) superior oblique (c) medial rectus (d) external oblique

_____ 7. Of the following, which muscle is involved in compression of the abdomen? (a) external oblique (b) superior oblique (c) medial rectus (d) genioglossus

_____ 8. A muscle directly concerned with breathing is the (a) masseter (b) mentalis (c) brachialis (d) external intercostal

_____ 9. Which muscle is *not* related to movement of the wrist? (a) extensor carpi ulnaris (b) flexor carpi radialis (c) supinator (d) flexor carpi ulnaris

_____ 10. A muscle that helps move the thigh is the (a) piriformis (b) triceps brachii (c) hypoglossus (d) fibularis tertius

_____ 11. Which muscle is *not* related to mastication? (a) temporalis (b) masseter (c) lateral rectus (d) medial pterygoid

_____ 12. Which muscle elevates the tongue? (a) genioglossus (b) styloglossus (c) hyoglossus (d) omohyoid

_____ 13. Which muscle does *not* belong with the others because of its relationship to the hyoid bone? (a) digastric (b) mylohyoid (c) geniohyoid (d) thyrohyoid

_____ 14. Which muscle is *not* a component of the anterolateral abdominal wall? (a) external oblique (b) psoas major (c) rectus abdominis (d) internal oblique

_____ 15. All are components of the pelvic diaphragm *except* the (a) anconeus (b) coccygeus (c) iliococcygeus (d) pubococcygeus

_____ **16.** Which is *not* a flexor of the forearm? (a) biceps brachii (b) brachialis (c) brachioradialis (d) triceps brachii

_____ **17.** The abductor pollicis brevis, opponeus pollicis, flexor pollicis brevis, and adductor pollicis are all components of the (a) thenar eminence (b) midpalmar muscles (c) hypothenar eminence (d) erector spinae

_____ **18.** Which muscle is *not* involved in moving the vertebral column? (a) splenius (b) longissimus (c) spinalis (d) sartorius

_____ **19.** Which muscle flexes the wrist? a) palmaris longus (b) extensor carpi radialis longus (c) supinator (d) extensor indicis

_____ **20.** Which muscle is *not* involved in flexion of the thigh? (a) rectus femoris (b) sartorius (c) biceps femoris (d) vastus intermedius

_____ **21.** Of the muscles that move the foot and toes, the anterior compartment muscles are involved in (a) plantar flexion (b) dorsiflexion (c) abduction (d) adduction

_____ **22.** Which muscle is deepest? (a) plantar interosseous (b) adductor hallucis (c) quadratus plantae (d) abductor hallucis

PART 2 ■ Matching

Identify the characteristic(s) used to name the following muscles:

_____ **23.** Supinator   **A.** Location

_____ **24.** Deltoid   **B.** Shape

_____ **25.** Stylohyoid   **C.** Size

_____ **26.** Flexor carpi radialis   **D.** Direction of fibers

_____ **27.** Gluteus maximus   **E.** Action

_____ **28.** External oblique   **F.** Number of origins

_____ **29.** Triceps brachii   **G.** Insertion and origin

_____ **30.** Adductor longus

_____ **31.** Temporalis

_____ **32.** Trapezius

PART 3 ■ Completion

33. The principal muscle used in compression of the cheek is the _____.

34. The muscle that protracts the tongue is the _____.

35. The eye muscle that moves the eyeball inferiorly, medially, and counterclockwise is the

_____.

36. The _____ muscle flexes the neck on the chest.

37. The abdominal muscle that flexes the vertebral column is the _____.

38. The muscle of the pectoral (shoulder) girdle that depresses the clavicle is the

    _____.

39. Flexion, adduction, and medial rotation of the humerus are accomplished by the
    _____ muscle.

40. The _____ muscle is the most important extensor of the forearm.

41. The muscle that flexes and abducts the wrist is the _____.

42. The four muscles that extend the legs are the vastus lateralis, vastus medialis, vastus intermedius,
    and _____.

43. Muscles that lie inferior to the hyoid bone are referred to as _____ muscles.

44. The neck region is divided into two principal triangles by the _____ muscle.

45. Muscles that move the pectoral (shoulder) girdle originate on the axial skeleton and insert on the
    clavicle or _____.

46. The muscles that move the humerus and do not originate on the scapula are called
    _____ muscles.

47. Together, the subscapularis, supraspinatus, infraspinatus, and teres major muscles form the

    _____.

48. The posterior muscles involved in moving the wrist, hand, and fingers function in extension and

    _____.

49. The posterior muscles of the thigh are involved in _____ of the leg.

50. Together, the biceps femoris, semitendinosus, and semimembranosus are referred to as the
    _____ muscles.

51. _____ fascicles attach obliquely from many directions to several tendons.

52. Movement of the thumb medially across the palm is called _____.

53. The superior oblique muscle moves the eyeball _____ and laterally.

54. The _____ muscles are located superior to the hyoid bone.

55. The _____ muscle depresses the thyroid cartilage (Adam's apple) of the
    larynx.

56. The tendon of insertion for all the peripheral muscle fibers of the diaphragm is called the

    _____.

57. The muscles of the pelvic floor and their fasciae form the _____.

58. A transverse line drawn through the ischial tuberosities divides the perineum into an anal and
    _____ triangle.

59.   Movement of the inferior angle of the scapula laterally is called _____.

60.   The _____ muscle is the "workhorse" of the elbow flexors.

61.   From lateral to medial, the superficial anterior compartment muscles of the forearm are: flexor carpi radialis, _____, and flexor carpi ulnaris.

62.   Whereas the thenar muscles act on the thumb, the _____ muscles act on the little finger.

63.   The chief extensor of the vertebral column is the _____ muscle.

64.   The majority of muscles that move the femur originate from the _____.

65.   The great extensor muscle of the leg is the _____ muscle.

66.   Together, the plantaris, _____, and soleus muscles are referred to as the calf muscles.

67.   The plantar aponeurosis transmits the _____ tendons of the foot.

Surface Anatomy

Objectives

At the completion of this exercise you should understand

A The anatomy of underlying superficial structures as exemplified by surface, anatomical characteristics

B How to locate and identify certain superficial anatomical structures in the head, neck, trunk, and upper and lower limbs by visual inspection and/or palpation through the skin

$\mathbf{N}$ow that you have studied the skeletal and muscular systems, you will be introduced to **surface anatomy,** the study of the anatomical landmarks on the exterior surface of the body.[1] A knowledge of surface anatomy will help you identify certain superficial structures by visual inspection and palpation through the skin. **Palpation** (pal-PĀ-shun) means using the sense of touch to determine the location of an internal part of the body through the skin. Knowledge of surface anatomy is important in health-related activities such as taking a pulse and blood pressure, listening to internal organs, drawing blood, and inserting needles and tubes.

A convenient way to study surface anatomy is to divide the body into its five principal regions: head, neck, trunk, and upper and lower limbs. These may be reviewed in Figure 2.2.

A. Head

A **head** (cephalic region or caput) is divisible into the cranium and face. The **cranium** surrounds and protects the brain; the **face** is the anterior portion of the head. The head also contains the sense organs—eyes, ears, nose, and tongue. Several surface features of the various regions of the head are as follows:

[1]At the discretion of your instructor, surface anatomy may be studied either before or in conjunction with your study of various body systems. This exercise can be used as an excellent review of many topics already studied.

Regions of the **cranium (skull,** or **brain case)** and **face**

a. *Frontal region.* Front of skull that includes the frontal bone.

b. *Parietal region.* Crown of skull that includes the parietal bones.

c. *Temporal region.* Side of skull that includes the temporal bones.

d. *Occipital region.* Base of skull that includes the occipital bone.

e. *Orbital* or *ocular region.* Includes eyeballs, eyebrows, and eyelids.

f. *Nasal region.* Nose.

g. *Infraorbital region.* Inferior to orbit.

h. *Oral region.* Mouth.

i. *Mental region.* Anterior part of mandible.

j. *Buccal region.* Cheek.

k. *Zygomatic region.* Inferolateral to orbit includes the zygomatic bone.

l. *Auricular region.* External ear.

Using your textbook as an aid, label Figure 11.1.

Using a mirror, examine the various features of the head just described. Working with a partner, be sure that you can identify the regions by both common *and* anatomical names.

Within the various regions of the head, it is possible to palpate a number of structures. Read the descriptions of these structures in the following list and then label Figure 11.2. Working with a

FIGURE 11.1 **Regions of the cranium and face.**

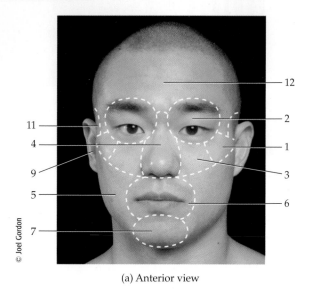

(a) Anterior view

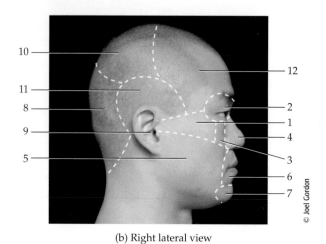

(b) Right lateral view

© Joel Gordon

___ Auricular region
___ Buccal region
___ Frontal region
___ Infraorbital region
___ Mental region
___ Nasal region

___ Occipital region
___ Oral region
___ Orbital (ocular) region
___ Parietal region
___ Temporal region
___ Zygomatic region

partner, be sure that you identify as many structures as you can.

1. **Sagittal suture.** This can be palpated by moving the fingers from side to side over the superior portion of the scalp.

2. **Coronal** and **lambdoid sutures.** These can be palpated using the same side-to-side technique.

3. **External occipital protuberance.** The most prominent bony feature on the posterior aspect of the skull.

4. **Orbit.** Deep to the eyebrow, the superior aspect of the orbit can be palpated. Actually, you can palpate the entire circumference of the orbit.

5. **Nasal bones.** These can be palpated in the nasal region between the orbits on either side of the midline.

6. **Mandible.** The **ramus** (vertical portion), **body** (horizontal portion), and **angle** (area where the ramus meets the body) of the mandible

can be palpated easily at the mental and buccal regions.

7. **Superficial temporal artery.** This can be palpated just anterior to the ear in the temple. A pulse can be detected in this vessel.

8. **Temporomandibular joint.** Inferior to the superficial temporal artery. You can palpate the joint by opening and closing your jaw.

9. **Zygomatic arch.** This can be palpated by moving anteriorly from the temporomandibular joint.

10. **External auditory canal (meatus).** The canal extending from the external ear to the eardrum.

11. **Mastoid process.** Located just posterior to the external ear, it may be palpated as a very prominent projection.

The surface anatomy features of the eyeball and accessory structures are presented in Figure 14.10, of the ear in Figure 14.18, and of the nose in Figure 21.3.

FIGURE 11.2 **Surface anatomy of the head.**

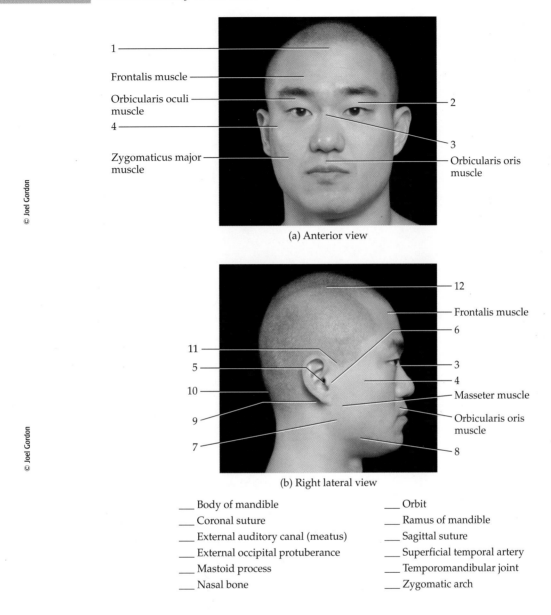

© Joel Gordon

1 ————
Frontalis muscle ————
Orbicularis oculi —— muscle
4 ————
Zygomaticus major —— muscle
———— 2
———— 3
—— Orbicularis oris muscle

(a) Anterior view

© Joel Gordon

———— 12
—— Frontalis muscle
—— 6
11 ————
5 ————
10 ————
9 ————
7 ————
———— 3
———— 4
—— Masseter muscle
—— Orbicularis oris muscle
———— 8

(b) Right lateral view

___ Body of mandible
___ Coronal suture
___ External auditory canal (meatus)
___ External occipital protuberance
___ Mastoid process
___ Nasal bone

___ Orbit
___ Ramus of mandible
___ Sagittal suture
___ Superficial temporal artery
___ Temporomandibular joint
___ Zygomatic arch

B. Neck

The **neck** (cervical) can be divided into an **anterior cervical region,** two **lateral cervical regions,** and a **posterior (nuchal) region.** Within the neck, it is possible to palpate a number of structures. Read the descriptions of these structures in the list that follows and then label Figure 11.3. Working with a partner, be sure that you identify as many structures as you can.

1. ***Thyroid cartilage (Adam's apple).*** Triangular laryngeal cartilage, it is the most prominent structure in the midline of the anterior cervical region.

2. ***Hyoid bone.*** First resistant structure palpated in the midline inferior to the chin, lying just superior to the thyroid cartilage opposite the superior border of C4.

3. ***Cricoid cartilage.*** Inferior laryngeal cartilage that attaches larynx to trachea. This structure can be palpated by running your fingertip down from your chin over the thyroid cartilage. (After you pass the cricoid cartilage, your fingertip sinks in.) This cartilage is used as a landmark in locating the rings of cartilage in the trachea (windpipe) when performing a tracheostomy. The incision is made through the second, third, or fourth tracheal rings, and a tube is inserted to assist breathing.

FIGURE 11.3 Surface anatomy of the neck.

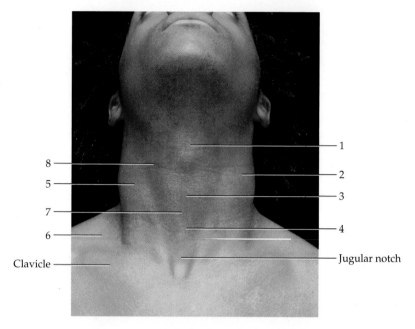

Anterior view

___ Carotid pulse point   ___ Sternocleidomastoid muscle
___ Cricoid cartilage   ___ Thyroid cartilage
___ External jugular vein   ___ Thyroid gland
___ Hyoid bone   ___ Trapezius muscle

4. *Thyroid gland.* Two-lobed gland, one on either side of the trachea.

5. *Sternocleidomastoid muscles.* Form major portion of the lateral aspect of the neck, extending from its insertion on the mastoid process of temporal bone (felt as bump behind ear) to its origin on the sternum and clavicle.

6. *Carotid arteries.* The common carotid artery is deep to the sternocleidomastoid muscle. At the level of the superior margins of the thyroid cartilage, it divides into internal and external carotid arteries. At this point, the carotid pulse can be detected.

7. *External jugular veins.* Prominent veins along lateral cervical regions superficial to the sternocleidomastoid muscles, readily seen when a person is angry or a collar fits too tightly.

8. *Trapezius muscles.* Form portion of lateral cervical region, extending inferiorly and laterally from base of skull. "Stiff neck" is frequently associated with inflammation of these muscles.

9. *Vertebral spines.* The spinous processes of the cervical vertebrae, which may be felt along the midline of the posterior aspect of the neck. Especially prominent at the base of

the neck is the spinous process of the seventh cervical vertebra (see Figure 11.5).

The sternocleidomastoid muscle divides the neck into two major triangles, which contain several important structures (Figure 11.4).

The **anterior triangle** is bordered superiorly by the mandible, inferiorly by the sternum, medially by the cervical midline, and laterally by the anterior border of the sternocleidomastoid muscle. The anterior triangle is subdivided into an unpaired submental triangle and three paired triangles: submandibular, carotid, and muscular. The **submental triangle** is inferior to the chin and contains the submental lymph nodes and beginning of the anterior jugular vein. The **submandibular triangle** is inferior to the mandible and contains the submandibular salivary gland, a portion of the parotid salivary gland, submandibular lymph nodes, the facial artery and vein, the carotid arteries, internal jugular vein, and the glossopharyngeal (IX), vagus (X), and hypoglossal (XII) cranial nerves. The **carotid triangle** is posterior to the hyoid bone. It contains the common carotid artery dividing into internal and external carotid arteries, the internal jugular vein, deep cervical lymph

FIGURE 11.4 Triangles of the neck.

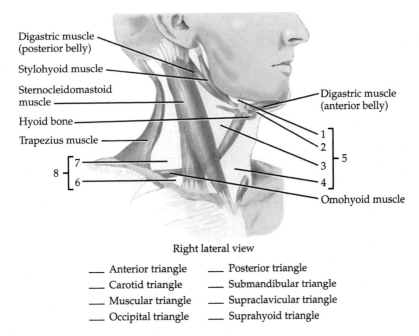

Digastric muscle
(posterior belly)

Stylohyoid muscle

Sternocleidomastoid
muscle

Hyoid bone

Trapezius muscle

Digastric muscle
(anterior belly)

Omohyoid muscle

Right lateral view

____ Anterior triangle   ____ Posterior triangle
____ Carotid triangle   ____ Submandibular triangle
____ Muscular triangle   ____ Supraclavicular triangle
____ Occipital triangle   ____ Suprahyoid triangle

nodes, and the vagus (X), accessory (XI), and hypoglossal (XII) cranial nerves. The **muscular triangle** lies inferior to the hyoid bone. Beneath its floor are located the thyroid gland, larynx, trachea, and esophagus.

The **posterior triangle** is bordered inferiorly by the clavicle, anteriorly by the posterior border of the sternocleidomastoid muscle, and posteriorly by the anterior border of the trapezius muscle. The posterior triangle is subdivided into two triangles, occipital and supraclavicular, by the inferior belly of the omohyoid muscle. The **occipital triangle** lies inferior to the occipital bone. The **supraclavicular triangle** lies superior to the middle third of the clavicle. The posterior triangle as a whole contains part of the subclavian artery, external jugular vein, cervical lymph nodes, brachial plexus, and accessory (XI) cranial nerve.

Label the triangles in Figure 11.4.

C. Trunk

The **trunk** is divided into the back, chest, abdomen, and pelvis.

Read the descriptions of the surface features of the **back** that follow and then label Figure 11.5.

1. *Vertebral spines.* The spinous process of vertebrae that are more prominent when the vertebral column is flexed. The spinous process of C7 (**vertebra prominens**) is the superior of the two prominences found at the base of the neck; the spinous process of T1 is the lower prominence at the base of the neck; the spinous process of T3 is at about the same level as the spine of the scapula; the spinous process of T7 is about opposite the inferior angle of the scapula; a line passing through the highest points of the iliac crests, called the **supracristal line,** passes through the spinous process of L4. See Figure 11.7c.

2. *Scapulae.* Shoulder blades. They lie between ribs 2–7. Depending on how lean a person is, it might be possible to palpate various parts of the scapulae such as the vertebral border, axillary border, inferior angle, spine, and acromion.

3. *Latissimus dorsi muscle.* A broad, flat, triangular muscle of the lumbar region that extends superiorly to the axilla.

4. *Erector spinae (sacrospinalis) muscle.* Located on either side of the vertebral column between the twelfth ribs and iliac crest.

5. *Infraspinatus muscle.* Located inferior to the spine of the scapula.

6. *Trapezius muscle.* Extends from the cervical and thoracic vertebrae to the spine and acromion of the scapula and lateral end of the clavicle. It also occupies a portion of the lateral aspect of the neck.

7. *Teres major muscle.* Located inferior to the infraspinatus and, together with the latissimus

FIGURE 11.5 **Surface anatomy of the back.**

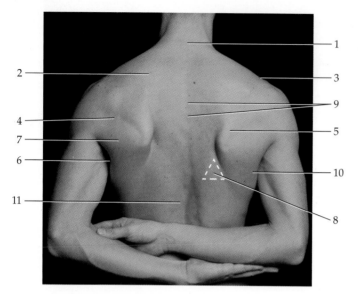

_____ Acromion of scapula
_____ Erector spinae muscle
_____ Infraspinatus muscle
_____ Latissimus dorsi muscle
_____ Posterior axillary fold
_____ Scapula
_____ Teres major muscle
_____ Trapezius muscle
_____ Triangle of ausculation
_____ Vertebra prominens
_____ Vertebral spines (spinous processes)

© Joel Gordon

Posterior view

dorsi, forms the inferior border of the axilla (posterior axillary fold).

8. **Posterior axillary fold.** Formed by the latissimus dorsi and teres major muscles, it can be palpated between the finger and thumb. It forms the posterior wall of the axilla. See Figure 11.9a also.

9. **Triangle of auscultation** (aw-skul-TĀ-shun = listening). A triangular region of the back just medial to the inferior part of the scapula where the rib cage is not covered by superficial muscles. It is bound by the latissimus dorsi, trapezius, and vertebral border of the scapula. The space between the muscles in the region permits respiratory sounds to be heard clearly with a stethoscope.

Read the descriptions of the surface features of the chest (thorax) that follow and then label Figure 11.6.

1. **Clavicles.** Collarbones. These lie in the superior region of thorax and can be palpated along their entire length. Inferior to the clavicle, especially where it articulates with the manubrium (superior portion) of the sternum, the first rib can be palpated.

2. **Sternum.** Breastbone. Lies in midline of chest. The following parts of the sternum are important surface features:

   Suprasternal (jugular) notch. Depression on superior surface of manubrium of sternum between medial ends of clavicles. The trachea can be palpated posterior to the notch.

   Manubrium of sternum. Superior portion of sternum at the same levels as the bodies of the third and fourth thoracic vertebrae and anterior to the arch of the aorta.

   Body of sternum. Midportion of sternum anterior to heart and the vertebral bodies of T5–T8.

   Sternal angle of sternum. Formed by junction of manubrium and body of sternum, about 4 cm (1.5 in.) inferior to suprasternal notch. This is palpable under the skin, located the costal cartilage of the second rib, and is the starting point from which the ribs are counted. At or inferior to the sternal angle and slightly to the right, the trachea divides into right and left primary bronchi.

   Xiphoid process of sternum. Inferior portion of sternum medial to the seventh costal cartilages. The heart lies on the diaphragm deep to the **xiphisternal joint** (joint between the xiphoid process and body of the sternum).

3. **Ribs.** Form bony cage of thoracic cavity. The apex beat of the heart in adults is heard in the left fifth intercostal space, just medial to the left midclavicular line.

4. **Costal margins.** Inferior edges of costal cartilages of ribs 7 through 10. The first costal cartilage lies inferior to the medial end of the

FIGURE 11.6   **Surface anatomy of the chest.**

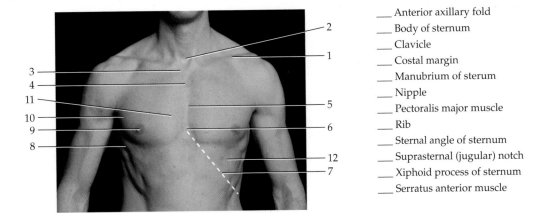

Anterior view

___ Anterior axillary fold
___ Body of sternum
___ Clavicle
___ Costal margin
___ Manubrium of sterum
___ Nipple
___ Pectoralis major muscle
___ Rib
___ Sternal angle of sternum
___ Suprasternal (jugular) notch
___ Xiphoid process of sternum
___ Serratus anterior muscle

© Joel Gordon

clavicle; the seventh costal cartilage is the most inferior to articulate directly with the sternum; the tenth costal cartilage forms the most inferior part of the costal margin, when viewed anteriorly.

5. *Pectoralis major muscle.* Principal upper chest muscle. In the male, the inferior border forms a curved line that serves as a guide to the fifth rib. In the female, the inferior border is mostly covered by the breast.

6. *Serratus anterior muscle.* Inferior and lateral to the pectoralis major muscle.

7. *Mammary glands.* Accessory organs of the female reproductive system located inside the breasts. They overlie the pectoralis major muscle (two-thirds) and serratus anterior muscle (one-third). After puberty, they enlarge to their hemispherical shape, and in young adult females, they extend from the second through sixth ribs and from the lateral margin of the sternum to the **midaxillary line** (an imaginary vertical line that extends downward on the lateral thoracic wall from the center of the axilla).

8. *Nipples.* Superficial to fourth intercostal space or fifth rib about 10 cm (4 in.) from the midline in males and most females. The position of the nipples in females is variable, depending on the size and pendulousness of the breasts. The right dome of the diaphragm is just inferior to the right nipple, the left dome is about 2 to 3 cm (1 in.) inferior to the left nipple, and the central tendon is at the level of the junction of the xiphisternal joint.

9. *Anterior axillary fold.* Formed by the lateral border of the pectoralis major muscle; can be palpated between the fingers and thumb. It

forms the anterior wall of the axilla. See Figure 11.9a also.

Read the descriptions of the surface features of the abdomen and pelvis that follow and then label Figure 11.7.

1. *Umbilicus.* Also called **navel;** previous site of attachment of umbilical cord to fetus. It is level with the intervertebral disc between the bodies of L3 and L4 and is the most obvious surface marking on the abdomen of most individuals. The **abdominal aorta** bifurcates (branches) into the right and left common iliac arteries anterior to the body of vertebra L4. It can be palpated through the upper part of the anterior abdominal wall just to the left of the midline. The **inferior vena cava** lies to the right of the abdominal aorta and is wider; it arises anterior to the body of vertebra L5.

2. *External oblique muscle.* Located inferior to the serratus anterior muscle. The aponeurosis of the muscle on its inferior border is the **inguinal ligament,** a structure along which hernias frequently occur.

3. *Rectus abdominis muscles.* Located just lateral to the midline of the abdomen. They can be seen by raising the shoulders while in the supine position without using the arms.

4. *Linea alba.* Flat, tendinous raphe forming a furrow along the midline between rectus abdominis muscle. The furrow extends from the xiphoid process to the pubic symphysis. It is particularly obvious in thin, muscular individuals. It is broad superior to the umbilicus and narrow inferior to it. The linea alba is a frequently selected site for abdominal surgery

FIGURE 11.7 Surface anatomy of the abdomen and pelvis.

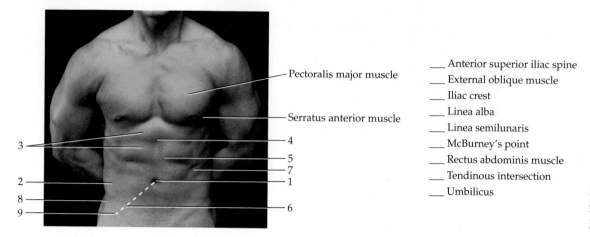

Pectoralis major muscle

Serratus anterior muscle

___ Anterior superior iliac spine
___ External oblique muscle
___ Iliac crest
___ Linea alba
___ Linea semilunaris
___ McBurney's point
___ Rectus abdominis muscle
___ Tendinous intersection
___ Umbilicus

© Joel Gordon

(a) Anterior view of abdomen

since an incision through it severs no muscles and only a few blood vessels and nerves.

5. ***Tendinous intersections.*** Fibrous bands that run transversely or obliquely across the rectus abdominis muscle. Three or more are visible in muscular individuals. One intersection is at the level of the umbilicus, one at the level of the xiphoid process, and one midway between.

6. ***Linea semilunaris.*** The lateral edge of the rectus abdominis muscle that crosses the costal margin at the tip of the ninth costal cartilage.

7. ***McBurney's point.*** Located two-thirds of the way down an imaginary line drawn between the umbilicus and anterior superior iliac spine. An oblique incision through this point is made for an appendectomy (removal of the appendix). Pressure of the finger on this point produces tenderness in acute appendicitis, inflammation of the appendix.

8. ***Iliac crest.*** Superior margin of the ilium of the hipbone that forms the outline of the superior portion of the buttock. When you rest your hands on your hips, they rest on the iliac crests. A horizontal line drawn across the highest point of each iliac crest is called the **supracristal line,** which intersects the spinous process of the fourth lumbar vertebra. This vertebra is a landmark for performing a spinal tap.

9. ***Anterior superior iliac spine.*** The anterior end of the iliac crest that lies at the upper lateral end of the fold of the groin.

10. ***Posterior superior iliac spine.*** The posterior end of the iliac crest that is indicated by a dimple in the skin that coincides with the

middle of the sacroiliac joint where the hipbone attaches to the sacrum.

11. ***Pubic tubercle.*** Projection on the superior border of the pubis of the hipbone. Attached to it is the medial end of the inguinal ligament. The lateral end is attached to the anterior superior iliac spine. The **inguinal ligament** is the inferior free edge of the aponeurosis of the external oblique muscle that forms the **inguinal canal.** Through the canal pass the spermatic cord in males and the round ligament of the uterus in females.

12. ***Pubic symphysis.*** Anterior joint of hipbones. This structure is palpated as a firm resistance in the midline at the inferior portion of the anterior abdominal wall.

13. ***Mons pubis.*** An elevation of adipose tissue covered by skin and pubic hair that is anterior to the pubic symphysis.

14. ***Sacrum.*** The fused spinous processes of the sacrum, the median sacral crest, can be palpated beneath the skin in the superior portion of the **gluteal cleft,** a depression along the midline that separates the buttocks (described shortly).

15. ***Coccyx.*** The inferior surface of the tip of the coccyx can be palpated in the gluteal cleft, about 2.5 cm (1 in.) posterior to the anus.

D. Upper Limb (Extremity)

The **upper limb (extremity)** consists of the shoulder, armpit, arm, elbow, forearm, wrist, and hand (palm and fingers).

Read the descriptions of the surface features of the **shoulder (acromial region)** in the list that

FIGURE 11.7 **Surface anatomy of the abdomen and pelvis. (Continued)**

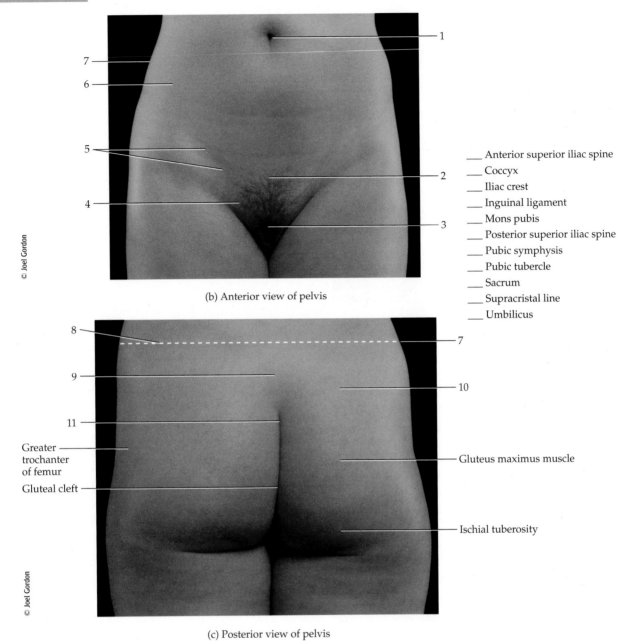

© Joel Gordon

___ Anterior superior iliac spine

___ Coccyx

___ Iliac crest

___ Inguinal ligament

___ Mons pubis

___ Posterior superior iliac spine

___ Pubic symphysis

___ Pubic tubercle

___ Sacrum

___ Supracristal line

___ Umbilicus

(b) Anterior view of pelvis

Greater trochanter of femur

Gluteal cleft

Gluteus maximus muscle

Ischial tuberosity

© Joel Gordon

(c) Posterior view of pelvis

follows and then label Figure 11.8. Working with a partner, be sure that you identify as many structures as you can.

1. *Acromioclavicular joint.* Slight elevation at lateral end of clavicle. It is the joint between the acromion of the scapula and the clavicle.

2. *Acromion.* Expanded lateral end of spine of scapula. This is clearly visible in some individuals and can be palpated about 2.5 cm (1 in.) distal to acromioclavicular joint.

3. *Humerus.* The **greater tubercle** of the humerus may be palpated on the superior aspect of the shoulder. It is the most lateral palpable bony structure.

4. *Deltoid muscle.* Triangular muscle that forms rounded prominence of shoulder. This is a frequent site for intramuscular injections.

Read the descriptions of the surface features of the **arm (brachium)** and **elbow (cubitus)** in the list that follows and then label Figure 11.9. Work-

FIGURE 11.8 **Surface anatomy of the shoulder.**

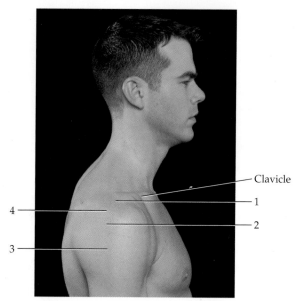

___ Acromioclavicular joint
___ Acromion of scapula
___ Deltoid muscle
___ Greater tubercle of humerus

Clavicle

1

2

4

3

Right lateral view

© Joel Gordon

ing with a partner, be sure that you identify as many structures as you can.

1. *Humerus.* This may be palpated along its entire length, especially at the elbow.

2. *Biceps brachii muscle.* Forms bulk of anterior surface of arm. On the medial side of the muscle is a groove that contains the **brachial artery,** the blood vessel where blood pressure is usually taken. Pressure may be applied to the artery in cases of severe hemorrhage in the forearm and hand.

3. *Triceps brachii muscle.* Forms bulk of posterior surface of arm.

4. *Medial epicondyle.* Medial projection at distal end of humerus near the elbow.

5. *Ulnar nerve.* Can be palpated as a rounded cord in a groove posterior to the medial epicondyle. The "funny bone" is the region where the ulnar nerve rests against the medial epicondyle. Hitting the nerve at this point produces a sharp pain along the medial side of the forearm.

6. *Lateral epicondyle.* Lateral projection at distal end of humerus near the elbow.

7. *Olecranon.* Projection of proximal end of ulna that lies between and slightly superior to the epicondyles when the forearm is extended; it forms the elbow.

8. *Cubital fossa.* Triangular space in anterior region of elbow bounded proximally by an

imaginary line between humeral epicondyles, laterally by the medial border of the brachioradialis muscle, and medially by the lateral border of the pronator teres muscle; contains tendon of biceps brachii muscle, brachial artery and its terminal branches (radial and ulnar arteries), median cubital vein, and parts of median and radial nerves. Pulse can be detected in the brachial artery in the cubital fossa.

9. *Median cubital vein.* Crosses cubital fossa obliquely. This vein is frequently selected for removal of blood or introduction of substances such as medications, contrast media for radiographic procedures, nutrients, and blood cells and/or plasma for transfusions.

10. *Bicipital aponeurosis.* An aponeurotic band that inserts the biceps brachii muscle into the deep fascia in the medial aspect of the forearm. It can be felt when the muscle contracts.

Read the descriptions of the surface features of the **forearm (antebrachium)** and **wrist (carpus)** that follow and then label Figure 11.10. Working with a partner, be sure that you identify as many structures as you can.

1. *Ulna.* Medial bone of the forearm. It can be palpated along its entire length from the olecranon (see Figure 11.9a) to the **styloid process,** a projection on the distal end of the bone at the

FIGURE 11.9   Surface anatomy of the arm and elbow.

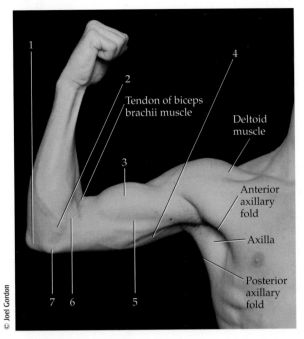

(a) Medial view of arm

Labels in (a): 1, 2, Tendon of biceps brachii muscle, 3, 4, Deltoid muscle, Anterior axillary fold, Axilla, Posterior axillary fold, 7, 6, 5

© Joel Gordon

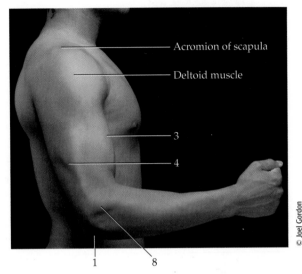

(b) Right lateral view of arm

Labels in (b): Acromion of scapula, Deltoid muscle, 3, 4, 1, 8

© Joel Gordon

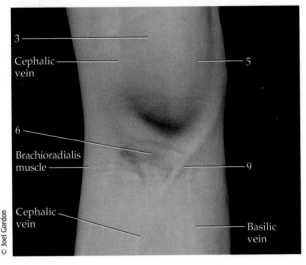

(c) Anterior view of cubital fossa

Labels in (c): 3, Cephalic vein, 5, 6, Brachioradialis muscle, 9, Cephalic vein, Basilic vein

© Joel Gordon

___ Biceps brachii muscle
___ Bicipital aponeurosis
___ Cubital fossa
___ Groove for brachial artery
___ Lateral epicondyle of humerus
___ Medial epicondyle of humerus
___ Median cubital vein
___ Olecranon of ulna
___ Triceps brachii muscle

medial side of the wrist. The **head of the ulna** is a conspicuous enlargement just proximal to the styloid process.

2. *Radius.* When the forearm is rotated, the distal half of the radius can be palpated; the proximal half is covered by muscles. The **styloid process** of the radius is a projection on the distal end of the bone at the lateral side of the wrist.

3. *Muscles.* Because of their close proximity, it is difficult to identify muscles of the forearm. However, it is easy to identify the tendons of some of the muscles as they approach the wrist and then trace them proximally to the muscles.

    Brachioradialis muscle. Located at the superior and lateral aspect of the forearm.

FIGURE 11.10 **Surface anatomy of forearm and wrist.**

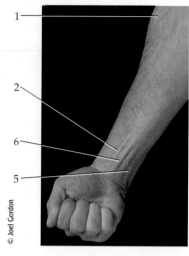

1
2
6
5

© Joel Gordon

(a) Anterior view of forearm and wrist

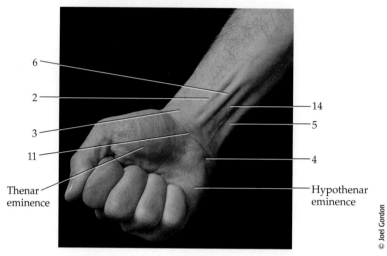

6
2
3
11

Thenar eminence

14
5
4

Hypothenar eminence

© Joel Gordon

(b) Anterior view of forearm and wrist

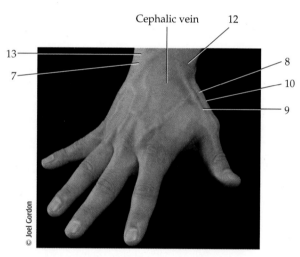

Cephalic vein   12

13
7

8
10
9

© Joel Gordon

(c) Posterolateral view of wrist

___ "Anatomical snuffbox"

___ Brachioradialis muscle

___ Head of ulna

___ Pisiform bone

___ Site for palpation of radial artery

___ Styloid process of radius

___ Styloid process of ulna

___ Tendon of extensor pollicis brevis muscle

___ Tendon of extensor pollicis longus muscle

___ Tendon of flexor carpi radialis muscle

___ Tendon of flexor carpi ulnaris muscle

___ Tendon of flexor digitorum superficialis muscle

___ Tendon of palmaris longus muscle

___ Wrist creases

Flexor carpi radialis muscle. The tendon of this muscle is about 1 cm medial to the styloid process of the radius on the lateral side of the forearm.

Palmaris longus muscle. The tendon of this muscle is medial to the flexor carpi radialis tendon and can be seen if the wrist is slightly flexed and the base of the thumb and little finger are drawn together.

Flexor digitorum superficialis muscle. The tendon of this muscle is medial to the palmaris longus tendon and can be palpated by flexing the fingers at the metacarpophalangeal and proximal interphalangeal joints.

Flexor carpi ulnaris muscle. The tendon of this muscle is on the medial aspect of the forearm.

4. *Radial artery.* Located on the lateral aspect of the wrist between the flexor carpi radialis muscle tendon and styloid process of the radius. It is frequently used to take a pulse.

5. *Pisiform bone.* Medial bone of proximal carpals. The bone is easily palpated as a projection distal and anterior to styloid process of the ulna.

6. *"Anatomical snuffbox."* Triangular depression between tendons of extensor pollicis brevis and extensor pollicis longus muscles. Styloid process of the radius, the base of the first metacarpal, trapezium, scaphoid, and radial artery can all be palpated in the depression.

7. *Wrist creases.* Three more or less constant lines on anterior aspect of wrist (named proximal, middle, and distal) where skin is firmly attached to underlying deep fascia.

Read the descriptions of the surface features of the **hand (manus)** that follow and then label Figure 11.11. Working with a partner, be sure that you identify as many structures as you can.

1. *Knuckles.* Commonly refers to dorsal aspects of the distal ends of metacarpals II–V (or 2–5), but also includes the dorsal aspects of the metacarpophalangeal and interphalangeal joints.

2. *Dorsal venous arch.* Superficial veins on the dorsum of the hand that drain blood into the cephalic vein. It can be displayed by compressing the blood vessels at the wrist for a few moments as the hand is opened and closed.

3. *Tendon of extensor digiti minimi muscle.* This can be seen on the dorsum of the hand in line with the phalanx of the little finger.

4. *Tendons of extensor digitorum muscle.* These can be seen on the dorsum of the hand in line with the phalanges of the ring, middle, and index fingers.

5. *Tendon of the extensor pollicis brevis muscle.* This tendon (described previously) is in line with the phalanx of the thumb (see Figure 11.10c).

6. *Thenar eminence.* Larger, rounded contour on the lateral aspect of the palm formed by muscles that move the thumb. Also called the ball of the thumb.

7. *Hypothenar eminence.* Smaller, rounded contour on the medial aspect of the palm formed by muscles that move the little finger. Also called the ball of the little finger.

8. *Skin creases.* Several more or less constant lines on the palm (**palmar flexion creases**) and anterior aspect of digits (**digital flexion creases**) where skin is firmly attached to underlying deep fascia.

E. Lower Limb (Extremity)

The **lower limb (extremity)** consists of the buttocks, thigh, knee, leg, ankle, and foot.

Read the descriptions of the surface features of the **buttock (gluteal region)** that follow and then label Figure 11.12. Recall that the outline of the buttock is formed by the iliac crests (see Figure 11.7).

1. *Gluteus maximus muscle.* Forms major portion of prominence of buttock. The sciatic nerve is deep to this muscle.

2. *Gluteus medius muscle.* Superior and lateral to gluteus maximus. This is a frequent site for intramuscular injections.

3. *Gluteal (natal) cleft.* Depression along midline that separates the buttocks; it extends as high as the fourth or third sacral vertebra.

4. *Gluteal fold.* Inferior limit of buttock that roughly corresponds to the inferior margin of gluteus maximus muscle.

5. *Ischial tuberosity.* Bony prominence of ischium of hipbone just superior to the medial side of the gluteal fold that bears weight of body when seated.

6. *Greater trochanter.* Projection of proximal end of femur on lateral surface of thigh felt and seen in front of hollow on side of hip. This can be palpated about 20 cm (8 in.) inferior to iliac crest.

FIGURE 11.11 **Surface anatomy of the hand.**

1

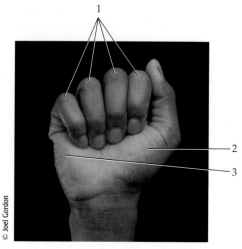

© Joel Gordon

2

3

(a) Anterior and posterior view

1

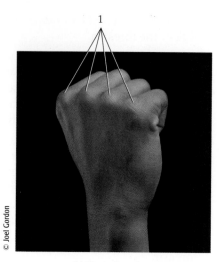

© Joel Gordon

(b) Posterior view

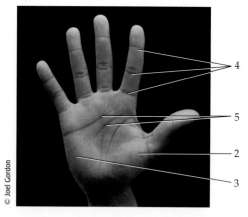

© Joel Gordon

4

5

2

3

(c) Anterior view

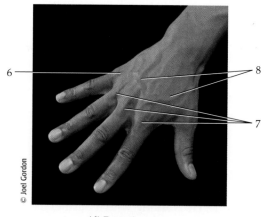

© Joel Gordon

6

8

7

(d) Posterior view

___ Digital flexion creases
___ Dorsal venous arch
___ Hypothenar eminence
___ "Knuckles"

___ Palmer flexion creases
___ Tendon of extensor digiti minimi muscle
___ Tendon of extensor digitorum muscle
___ Thenar eminence

FIGURE 11.12   **Surface anatomy of the buttocks.**

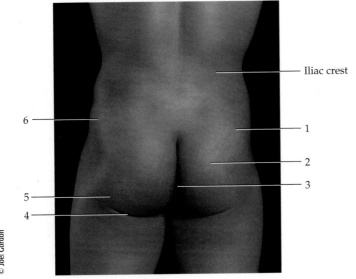

— Iliac crest

6 —
5 —
4 —

1
2
3

© Joel Gordon

Posterior view

___ Gluteal (natal) cleft
___ Guteal fold
___ Gluteus maximus muscle
___ Gluteus medius muscle
___ Greater trochanter of femur
___ Site of ischial tuberosity

Read the descriptions of the surface features of the **thigh (femoral region)** and **knee (genu)** that follow and then label Figure 11.13 on page 230. Be sure that you identify as many structures as you can.

1. **Sartorius muscle.** Superficial anterior muscle that can be traced from the lateral aspect of the thigh to the medial aspect of the knee.

2. **Quadriceps femoris muscle.** Three of the four components of this muscle can be seen.

   Rectus femoris. Located at the midportion of the anterior aspect of the thigh.

   Vastus medialis. Located at the anteromedial aspect of the thigh.

   Vastus lateralis. Located at the anterolateral aspect of the thigh. This muscle is a frequent intramuscular injection site for diabetics.

3. **Adductor longus muscle.** Located at the superior aspect of the thigh, medial to the sartorius.

4. **Hamstring muscles.** Superficial, posterior thigh muscles located inferior to the gluteal folds. They are the

   Biceps femoris. Lies more laterally as it passes inferiorly to the knee.

   Semitendinosus and **semimembranosus.** Lie medially as they pass inferiorly to the knee.

5. **Femoral triangle.** A large space formed by the inguinal ligament superiorly, the sartorius

muscle laterally, and the adductor longus muscle medially. The triangle contains the femoral artery, vein, and nerve; inguinal lymph nodes; and the terminal portion of the great saphenous vein. The triangle is an important arterial pressure point in cases of severe hemorrhage of the lower limb. Hernias occur frequently in this area.

6. **Patella.** Kneecap. A large sesamoid bone located within the tendon of the quadriceps femoris muscle on the anterior surface of the knee along the midline.

7. **Patellar ligament.** Continuation of quadriceps femoris tendon inferior to patella.

8. **Medial condyles of femur and tibia.** Medial projections just inferior to patella. Superior part of projection belongs to distal end of femur; inferior part of projection belongs to proximal end of tibia.

9. **Laberal condyles of femur and tibia.** Lateral projections just inferior to patella. Superior part of projections belongs to distal end of femur. Inferior part of projections belongs to proximal end of tibia.

10. **Popliteal fossa.** Diamond-shaped space on posterior aspect of knee that is clearly visible when knee is flexed. Fossa is bordered superolaterally by the biceps femoris muscle, superomedially by the semimembranosus and semitendinosus muscles, and inferolaterally and inferomedially by the lateral and medial heads of the gastrocnemius muscle, respectively. The **head of the fibula** can easily be

FIGURE 11.13 **Surface anatomy of the thigh and knee.**

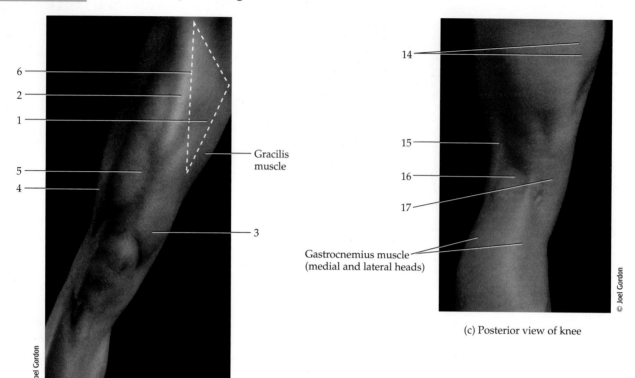

Gracilis muscle

Gastrocnemius muscle
(medial and lateral heads)

(c) Posterior view of knee

© Joel Gordon

(a) Anterior view of thigh

© Joel Gordon

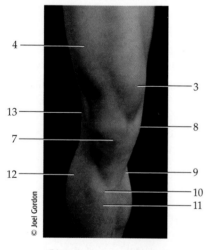

(b) Anterior view of knee

© Joel Gordon

___ Adductor longus muscle
___ Femoral triangle
___ Lateral condyle of femur
___ Lateral condyle of tibia
___ Medial condyle of femur
___ Medial condyle of tibia
___ Patella
___ Patellar ligament
___ Popliteal fossa
___ Rectus femoris muscle
___ Sartorius muscle
___ Site of semitendinosus and
     semimebranosus muscles
___ Tendon of biceps femoris muscle
___ Tendon of semitendinosus
     muscle
___ Tibial tuberosity
___ Vastus lateralis muscle
___ Vastus medialis muscle

palpated on the lateral side of the popliteal fossa. The fossa also contains the popliteal artery and vein. It is sometimes possible to detect a pulse in the popliteal artery.

Read the descriptions of the surface features of the **leg (crus), ankle (tarsus),** and **foot** that follow and then label Figure 11.14. Be sure that you identify as many structures as you can.

1. *Tibial tuberosity.* Bony prominence on the superior, anterior surface of the tibia into which the patellar ligament inserts.

2. *Tibialis anterior muscle.* Lies against the lateral surface of the tibia where it is easy to palpate, particularly when the foot is dorsiflexed.

3. *Tibia.* The medial surface and anterior border (shin) of the tibia are subcutaneous and can be palpated throughout the length of the bone.

4. *Fibularis (peroneus) longus muscle.* A superficial lateral muscle that overlies the fibula.

5. *Gastrocnemius muscle.* Forms the bulk of the midportion and superior portion of the posterior aspect of the leg. The medial and lateral heads can be clearly seen when standing on the toes.

6. *Soleus muscle.* Located deep to the gastrocnemius muscle and together with the gastrocnemius are referred to as the calf muscles.

7. *Calcaneal (Achilles) tendon.* Prominent tendon of the gastrocnemius and soleus muscles on the posterior aspect of the ankle that inserts into the **calcaneus** (heel) bone of the foot.

8. *Lateral malleolus of fibula.* Projection of the distal end of the fibula that forms the lateral prominence of the ankle. The head of the fibula, at the proximal end of the bone, lies at the same level as the tibial tuberosity.

9. *Medial malleolus of tibia.* Projection of the distal end of the tibia that forms the medial prominence of the ankle.

10. *Dorsal venous arch.* Superficial veins on the dorsum of the foot that unite to form the small and great saphenous veins. The great saphenous vein is the longest vein of the body. The location of the great saphenous vein is about 2.5 cm (1 in.) anterior to the medial malleolus of the tibia and is fairly constant. Knowledge of this location may be lifesaving when an urgent transfusion is needed in obese or collapsed patients when other veins cannot be detected.

11. *Tendons of extensor digitorum longus muscle.* Visible in line with phalanges II through V (or 2–5).

12. *Tendon of extensor hallucis longus muscle.* Visible in line with phalanx I (great toe). Pulsations in the dorsalis pedis artery may be felt in most people just lateral to this tendon when the blood vessel passes over the navicular and cuneiform bones of the tarsus.

ANSWER THE LABORATORY REPORT QUESTIONS AT THE END OF THE EXERCISE.

FIGURE 11.14 **Surface anatomy of the leg, ankle, and foot.**

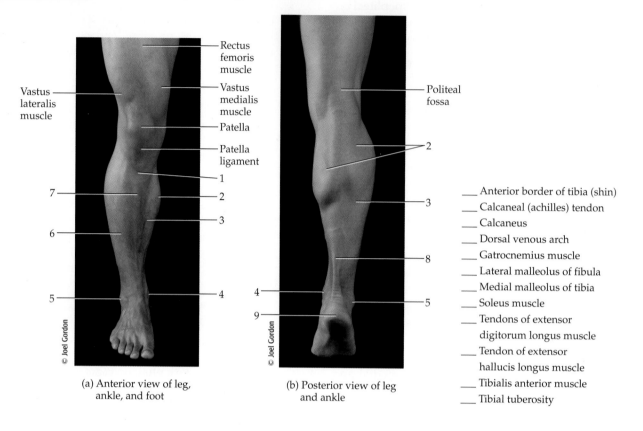

(a) Anterior view of leg, ankle, and foot

(b) Posterior view of leg and ankle

© Joel Gordon

___ Anterior border of tibia (shin)
___ Calcaneal (achilles) tendon
___ Calcaneus
___ Dorsal venous arch
___ Gatrocnemius muscle
___ Lateral malleolus of fibula
___ Medial malleolus of tibia
___ Soleus muscle
___ Tendons of extensor digitorum longus muscle
___ Tendon of extensor hallucis longus muscle
___ Tibialis anterior muscle
___ Tibial tuberosity

FIGURE 11.14   **Surface anatomy of the leg, ankle, and foot. (Continued)**

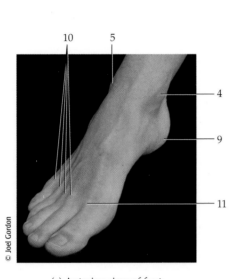

(c) Anterior view of foot

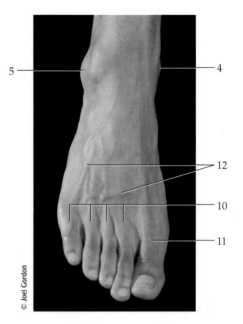

(d) Anterior view of foot

© Joel Gordon

EXERCISE 11

Surface Anatomy

Name _____   Date _____

Laboratory Section _____   Score/Grade _____

PART 1 ■ Multiple Choice

_____ 1. The term used to refer to the crown of the skull is (a) occipital (b) mental (c) parietal (d) zygomatic

_____ 2. The laryngeal cartilage in the midline of the anterior cervical region known as the Adam's apple is the (a) cricoid (b) epiglottis (c) arytenoid (d) thyroid

_____ 3. Inflammation of which muscle is associated with "stiff neck"? (a) teres major (b) trapezius (c) deltoid (d) cervicalis

_____ 4. The skeletal muscle located directly on either side of the vertebral column is the (a) serratus anterior (b) infraspinatus (c) teres major (d) erector spinae

_____ 5. The suprasternal notch and xiphoid process are associated with the (a) sternum (b) scapula (c) clavicle (d) ribs

_____ 6. The expanded end of the spine of the scapula is the (a) acromion (b) linea alba (c) olecranon (d) superior angle

_____ 7. Which nerve can be palpated as a rounded cord in a groove posterior to the medial epicondyle? (a) median (b) radial (c) ulnar (d) brachial

_____ 8. Which carpal bone can be palpated as a projection distal to the styloid process of the ulna? (a) trapezoid (b) trapezium (c) hamate (d) pisiform

_____ 9. The lateral rounded contour on the anterior surface of the hand (at the base of the thumb) formed by the muscles of the thumb is the (a) "anatomical snuffbox" (b) thenar eminence (c) hypothenar eminence (d) dorsal venous arch

_____ 10. The superior margin of the hipbone is the (a) pubic symphysis (b) iliac spine (c) acetabulum (d) iliac crest

_____ 11. Which bony structure bears the weight of the body when a person is seated? (a) greater trochanter (b) iliac crest (c) ischial tuberosity (d) gluteal fold

_____ 12. Which muscle is *not* a component of the quadriceps femoris group? (a) biceps femoris (b) vastus lateralis (c) vastus medialis (d) rectus femoris

_____ 13. The diamond-shaped space on the posterior aspect of the knee is the (a) cubital fossa (b) posterior triangle (c) popliteal fossa (d) nuchal groove

_____ 14. The projection of the distal end of the tibia that forms the prominence on one side of the ankle is the (a) medial condyle (b) medial malleolus (c) lateral condyle (d) lateral malleolus

_____ 15. The tendon that can be seen in line with the great toe belongs to which muscle? (a) extensor digiti minimi (b) extensor digitorum longus (c) extensor hallucis longus (d) extensor carpi radialis

PART 2 ▪ Completion

16. The laryngeal cartilage that connects the larynx to the trachea is the _____ cartilage.

17. The _____ triangle is bordered by the mandible, sternum, cervical midline, and sternocleidomastoid muscle.

18. The depression on the superior surface of the sternum between the medial ends of the clavicles is the _____.

19. The principal upper chest muscle is the _____.

20. Tendinous intersections are associated with the _____ muscle.

21. The muscle that forms the rounded prominence of the shoulder is the _____ muscle.

22. The triangular space in the anterior aspect of the elbow is the _____.

23. The "anatomical snuffbox" is bordered by the tendons of the extensor pollicis brevis muscle and the _____ muscle.

24. The dorsal aspects of the distal ends of metacarpals II through V are commonly referred to as _____.

25. The tendon of the _____ muscle is in line with phalanx of the little finger.

26. The dimple in the skin that coincides with the middle of the sacroiliac joint is a landmark for the _____.

27. The femoral projection that can be palpated about 20 cm (8 in.) inferior to the iliac crest is the _____.

28. The continuation of the quadriceps femoris tendon inferior to the patella is the _____.

29. The tendon of insertion for the gastrocnemius and soleus muscles is the _____ tendon.

30. Superficial veins on the dorsum of the foot that unite to form the small and great saphenous veins belong to the _____.

31. The prominent veins along the lateral cervical regions are the _____ veins.

32. The pronounced vertebral spine of C7 is the _____.

33. The most reliable surface anatomy feature of the chest is the _____.

34. A furrow extending from the xiphoid process to the pubic symphysis is the _____.

35. The vein that crosses the cubital fossa and is frequently used to remove blood is the _____ vein.

36. The submental, submandibular, and carotid triangles are components of the larger _____ triangle.

37. The _____ forms the posterior wall of the axilla.

38. Respiratory sounds are clearly heard in the triangle of _____.

39. The heart lies on the diaphragm deep to the _____ junction.

40. A horizontal line drawn across the highest point of each iliac crest is called the _____ line.

41. The anatomical landmark related to appendectomy and appendicitis is _____.

42. The lateral edge of the rectus abdominis muscle is called the _____.

PART 3 ■ Matching

_____ 43. Mental region

_____ 44. Xiphoid process

_____ 45. Arm

_____ 46. Nucha

_____ 47. Shoulder

_____ 48. Gluteus maximus muscle

_____ 49. Costal margin

_____ 50. Olecranon

_____ 51. Wrist

_____ 52. Auricular region

_____ 53. Semitendinosus muscle

_____ 54. Leg

_____ 55. Cranium

_____ 56. Ankle

_____ 57. Hand

A. Inferior edges of costal cartilages of ribs 7 through 10

B. Forms elbow

C. Inferior portion of sternum

D. Manus

E. Crus

F. Brachium

G. Tarsus

H. Anterior part of mandible

I. Brain case

J. Component of hamstrings

K. Forms main part of prominence of buttock

L. Carpus

M. Posterior neck region

N. Acromial

O. Ear

Nervous Tissue

Objectives

At the completion of this exercise you should understand

A The anatomical and functional subdivisions of the nervous system

B The histological characteristics of the cells of the nervous system

C The types of neuronal circuits and the components of a reflex arc

The diverse activities of the **nervous system** can be grouped into three basic functions: (1) sensory, (2) integrative, and (3) motor.

1. *Sensory function.* Sensory receptors detect internal stimuli such as stretching of your stomach or an increase in blood acidity, and external stimuli such as a raindrop landing on your arm or the aroma of a rose.

2. *Integrative function.* The nervous system *integrates* (processes) sensory information by analyzing and storing some of it and by making *decisions* for appropriate behaviors.

3. *Motor function.* The nervous system's motor function involves responding to integration decisions and by initiating muscular contractions or glandular secretions.

The branch of medical science that deals with the normal functioning and disorders of the nervous system is called **neurology** (noo-ROL-ō-jē; *neuro* = nerve or nervous system; *logy* = study of).

A. Nervous System Divisions

The two main subdivisions of the nervous system are the **central nervous system (CNS)** and the **peripheral** (pe-RIF-er-al) **nervous system (PNS).** The CNS, which consists of the **brain** and **spinal cord,** processes many different kinds of incoming sensory information. The CNS is also the source of thoughts, emotions, and memories. Most nerve impulses that stimulate muscles to contract and glands to secrete originate in the CNS.

The CNS is connected to sensory receptors, muscles, and glands in peripheral parts of the body by the PNS. The PNS consists of **cranial nerves** that arise from the brain, **spinal nerves** that emerge from the spinal cord, **ganglia,** and **sensory receptors**. Portions of these nerves carry nerve impulses into the CNS while other portions carry impulses out of the CNS.

The input component of the PNS consists of nerve cells called **sensory** or **afferent** (AF-er-ent; *af* = toward; *ferrent* = to carry) **neurons.** They conduct nerve impulses from sensory receptors located in various parts of the body *to the CNS* and end within the CNS. The output component consists of nerve cells called **motor** or **efferent** (EF-er-ent; *ef* = away from; *ferrent* = to carry) **neurons.** They originate within the CNS and conduct nerve impulses *from the CNS* to muscles and glands.

The PNS may be subdivided into a **somatic** (*somat* = body) **nervous system (SNS),** an **autonomic** (*auto* = self; *nomic* = law) **nervous system (ANS),** and an **enteric** (*ente* = intestine) **nervous system (ENS).** The SNS consists of sensory neurons that convey information from somatic receptors in the head, body wall, and limbs and from receptors for special senses of vision, hearing, taste, and smell to the CNS and motor neurons that conduct impulses from the CNS to *skeletal muscles* only. Because these motor responses can be consciously controlled, this portion of the SNS is *voluntary.*

The ANS consists of sensory neurons that convey information from autonomic sensory receptors, located primarily in visceral organs such as the stomach and lungs, to the CNS and motor neurons that conduct impulses from the CNS to smooth muscle, cardiac muscle, and glands. Since its motor responses are not normally under conscious control, the ANS is *involuntary.*

The motor portion of the ANS consists of two branches, the **sympathetic division** and the **parasympathetic division.** With few exceptions, the viscera receive instructions from both. Usually, the two divisions have opposing actions. For example, sympathetic neurons speed the heartbeat while parasympathetic neurons slow it down. Processes promoted by sympathetic neurons often involve expenditure of energy while those promoted by parasympathetic neurons restore and conserve body energy.

The ENS is the "brain of the gut" and its operation is involuntary. Once considered part of the ANS, the ENS consists of approximately 100 million neurons in enteric plexuses that extend the entire length of the gastrointestinal (GI) tract. Sensory neurons of the ENS monitor chemical changes within the GI tract and the stretching of its walls. Enteric motor neurons govern contraction of GI tract smooth muscle, secretion of GI tract organs such as acid secretion by the stomach, and activity of GI tract endocrine cells.

In this exercise you will identify the parts of neuron and the components of a reflex arc.

B. Histology of Nervous Tissue

Despite its complexity, the nervous system consists of two types of cells: neurons and neuroglia. **Neurons** (noo-ronz) (nerve cells) constitute the nervous tissue and have the property of **electrical excitability,** the ability to produce action potentials or impulses in response to stimuli. Mature neurons have only limited capacity for replacement or repair. **Neuroglia** (noo-ROG-lē-a; *neuro* = nerve; *glia* = glue) can divide and multiply and support, nurture, and protect neurons and maintain homeostasis of the fluid that bathes neurons. They do not generate or propagate action potentials. Brain tumors are commonly derived from neuroglia. Such tumors, called **gliomas,** are highly malignant and grow rapidly.

A neuron consists of the following parts:

1. **Cell body (perikaryon).** Contains a nucleus, cytoplasm, lysosomes, mitochondria, Golgi complex, *Nissl bodies* (clusters of rough endoplasmic reticulum), and neurofibrils.

2. **Dendrites** (= tree). Usually short, tapered, and highly branched extensions of the cell body that are the receiving or input portions of a neuron. Dendrites conduct nerve impulses toward the cell body.

3. **Axon** (*axon* = axis). Long, thin, cylindrical projection that propagates nerve impulses away from the cell body to another neuron, muscle fiber, or gland cell. An axon, in turn, consists of the following:

   a. **Axon hillock** (= small hill). The origin of an axon from the cell body represented as a small cone-shaped elevation.

   b. **Initial segment.** First part of an axon. Except in sensory neurons, nerve impulses arise at the junction of the axon hillock and initial segment, a region called the **trigger zone.**

   c. **Axoplasm.** Cytoplasm of an axon.

   d. **Axolemma** (*lemma* = sheath or husk). Plasma membrane around the axoplasm.

   e. **Axon collateral.** Side branch of an axon.

   f. **Axon terminals (telodendria).** Fine, branching processes of an axon or axon collateral.

   g. **Synaptic end bulbs.** Bulb-shaped structures at distal end of axon terminals that contain **synaptic vesicles** for neurotransmitters. In some cases, axon terminals exhibit a string of swollen bumps called **varicosities** instead of synaptic end bulbs.

   h. **Myelin sheath.** Multilayered lipid and protein covering of many axons, especially large peripheral ones; the myelin sheath is produced by peripheral nervous system neuroglia called **Schwann cells** and central nervous system neuroglia called **oligodendrocytes** (described shortly).

   i. **Neurolemma (sheath of Schwann).** The outer nucleated cytoplasmic layer of the Schwann cell which encloses the myelin sheath. It is found only around axons in the peripheral nervous system.

   j. **Node of Ranvier** (pronounced RON-vĕ-ā). Unmyelinated gaps in the myelin sheath.

Using your textbook and models of neurons as a guide, label Figure 12.1.

Now obtain a prepared slide of an ox spinal cord (transverse section), human spinal cord

FIGURE 12.1 Structure of a neuron. In (a), arrows indicate the direction in which the nerve impulse travels. (From *Photo Atlas for Anatomy and Physiology*, photo by D. Morton.)

___ Axolemma
___ Axon
___ Axon collateral
___ Axon hillock
___ Axon terminals
___ Cell body
___ Dendrites
___ Initial segment
___ Myelin sheath
___ Node of Ranvier
___ Neurolemma
___ Synaptic end bulbs

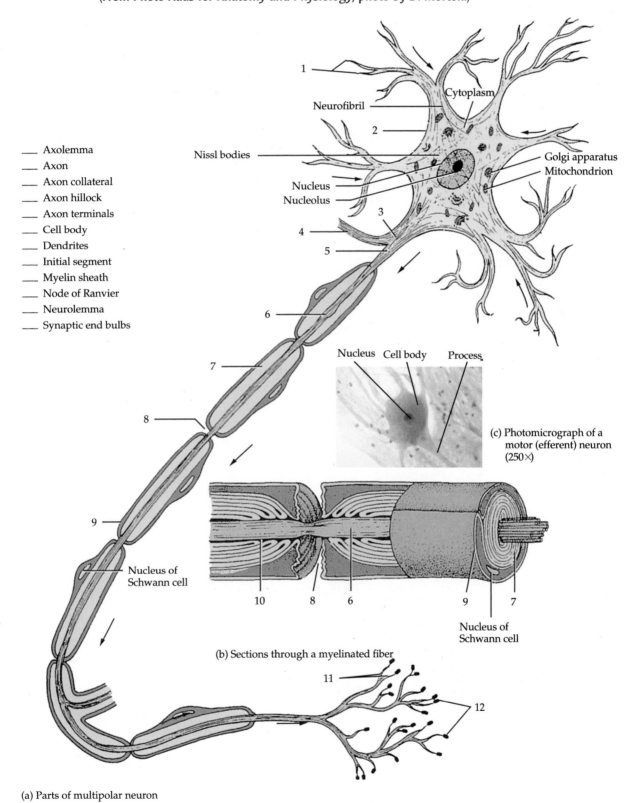

(a) Parts of multipolar neuron

(b) Sections through a myelinated fiber

(c) Photomicrograph of a motor (efferent) neuron (250×)

(transverse and longitudinal sections), nerve endings in skeletal muscle, and a nerve trunk (transverse and longitudinal sections). Examine each under high power and identify as many parts of the neuron as you can.

Neurons may be classified on the basis of structure and function. Structural classification is based on the number of processes extending from the cell body. **Multipolar neurons** usually have several dendrites and one axon. Most neurons in the brain and spinal cord are of this type. **Bipolar neurons** have one main dendrite and one axon. They are found in the retina of the eye, in the inner ear, and in the olfactory area of the brain. **Unipolar neurons** are sensory neurons in which the axon and dendrite are fused into a single process that divides into two branches a short distance from the cell body. Functional classification is based on the type of information carried and the direction in which the information is carried. The neurons that carry sensory information from cranial and spinal nerves into the brain and spinal cord or from a lower to higher level in the spinal cord and brain are **sensory (afferent) neurons. Motor (efferent) neurons** carry information from the brain toward the spinal cord or out of the brain and spinal cord into cranial or spinal nerves. Muscle fibers and glandular cells contacted by motor neurons are termed *effectors*. **Interneurons** are located in the brain, spinal cord, or ganglia, and carry information between the sensory and motor neurons. The neuron you have already labeled in Figure 12.1 is a motor neuron.

Using your textbook and models of neurons as a guide, label Figure 12.2.

C. Histology of Neuroglia

Among the types of neuroglial cells are the following:

1. *Astrocytes* (AS-trō-sīts; *astro* = star; *cyte* = cell). Star-shaped cells with many processes that are the largest and most numerous of the neuroglia. Help maintain appropriate chemical environment for generation of neuron action potentials; provide nutrients to neurons, take up excess neurotransmitters; participate in the metabolism of neurotransmitters and maintain the proper balance of K^+ and Ca^{2+} for generation of nerve impulses by CNS neurons; participate in brain development by assisting migration of neurons; help form the blood-brain barrier, which regulates the passage of substances into the brain; support the neurons; may play a role in learning and memory.

2. *Oligodendrocytes*   (ol'-i-gō-DEN-drō-sīts; *oligo* = few; *dendro* = tree). Smaller than astrocytes, with fewer and shorter processes; they

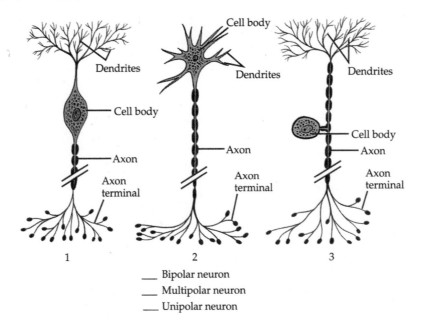

FIGURE 12.2   **Structural classification of neurons.**

1

2

3

___ Bipolar neuron
___ Multipolar neuron
___ Unipolar neuron

form a supporting network around CNS neurons and produce a myelin sheath around several adjacent axons of CNS neurons.

3. *Microglia* (mī-KROG-lē-a; *micro* = small). Small cells with slender processes that derive from mesodermal cells that also give rise to monocytes and macrophage; protect CNS cells from disease by engulfing invading microbes; migrate to areas of injured nerve tissue where they clear away debris of dead cells and may also kill healthy cells.

4. *Ependymal* (ep-EN-de-mal; *epen* = above; *dym* = garment) *cells.* Epithelial cells arranged in a single layer that range from cuboidal to columnar in shape; many possess microvilli and cilia; they line the ventricles of the brain (spaces filled with cerebrospinal fluid [CSF]) and the central canal of the spinal cord; produce CSF and monitor and assist in its circulation.

5. *Schwann cells.* Flattened cells that circle PNS axons. Each cell surrounds multiple unmyelinated axons with a single layer of its plasma membrane or produces part of the myelin sheath around a single axon of a PNS neuron; participates in regeneration of PNS axons.

6. *Satellite* (SAT-i-līt) *cells.* Flattened cells arranged around the cell bodies of neurons of PNS ganglia (collections of neuron cell bodies outside the CNS). They support neurons in PNS ganglia.

Obtain prepared slides of astrocytes (protoplasmic and fibrous), oligodendrocytes, microglia, ependymal cells, Schwann cells, and satellite cells. Using your textbook, Figure 12.3, and models of neuroglia as a guide, identify the various kinds of cells. In the spaces provided, draw each of the cells.

D. Neuronal Circuits

The CNS contains billions of neurons organized into complicated networks called **neuronal circuits,** each of which is a functional group of neurons that processes a specific kind of information.

The site of communication between two neurons or between a neuron and an effector cell is called a **synapse.** At a synapse, the neuron sending the signal is called a **presynaptic neuron,** and the neuron receiving the message is called a **postsynaptic neuron.**

In **simple series circuits** a presynaptic neuron stimulates only a single postsynaptic neuron. The second neuron then stimulates another, and so on. Most circuits, however, are more complex.

A single presynaptic neuron may synapse with several postsynaptic neurons. Such an arrangement, called **divergence,** permits one presynaptic neuron to influence several postsynaptic neurons (or several muscle fibers or gland cells) at the same time. In a **diverging circuit,** the nerve impulse from a single presynaptic neuron causes the stimulation of increasing numbers of cells along the circuit. For example, a small number of neurons in the brain that govern a particular body movement stimulate a much larger number of neurons in the spinal cord. Sensory signals also feed into diverging circuits and are often relayed to several regions of the brain.

In another arrangement, called **convergence,** several presynaptic neurons synapse with a single postsynaptic neuron. This arrangement permits more effective stimulation or inhibition of the postsynaptic neuron. In one type of **converging circuit,** the postsynaptic neuron receives nerve impulses from several different sources. For example, a single motor neuron that synapses with

Astrocyte

Oligodendrocyte

Microglial cell

Ependymal cell

Schwann cell

Satellite cell

FIGURE 12.3 **Neuroglia.**

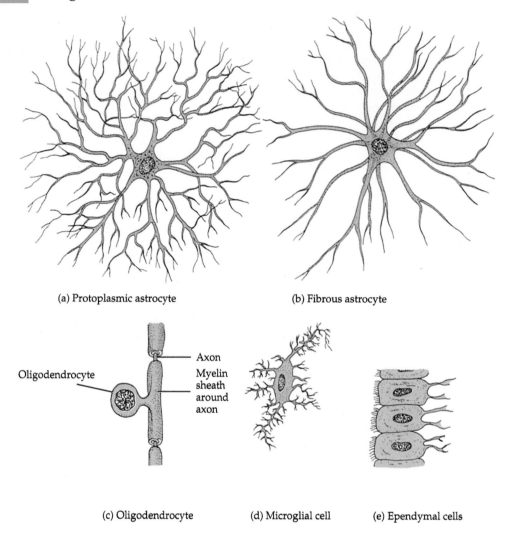

(a) Protoplasmic astrocyte

(b) Fibrous astrocyte

Oligodendrocyte

Axon

Myelin sheath around axon

(c) Oligodendrocyte

(d) Microglial cell

(e) Ependymal cells

skeletal muscle fibers at neuromuscular junctions receives input from several pathways that originate in different brain regions.

Some circuits in your body are constructed so that once the presynaptic cell is stimulated, it will cause the postsynaptic cell to transmit a series of nerve impulses. One such circuit is called a **reverberating circuit.** In this pattern, the incoming impulse stimulates the first neuron, which stimulates the second, which stimulates the third, and so on. Branches from later neurons synapse with earlier ones. This arrangement sends impulses back through the circuit again and again. The output signal may last from a few seconds to many hours, depending on the number of synapses and the arrangement of neurons in the circuit. Inhibitory neurons may turn off a rever-

berating circuit after a period of time. Among the body responses thought to be the result of output signals from reverberating circuits are breathing, coordinated muscular activities, waking up, sleeping (when reverberation stops), and short-term memory.

A fourth type of circuit is the **parallel after-discharge circuit.** In this circuit, a single presynaptic cell stimulates a group of neurons, each of which synapses with a common postsynaptic cell. If the input is excitatory, the postsynaptic neuron then can send out a stream of impulses in quick succession. It is thought that parallel after-discharge circuits may be employed for precise activities such as mathematical calculations.

Using your textbook as a guide, label Figure 12.4.

FIGURE 12.4 Neuronal circuit.

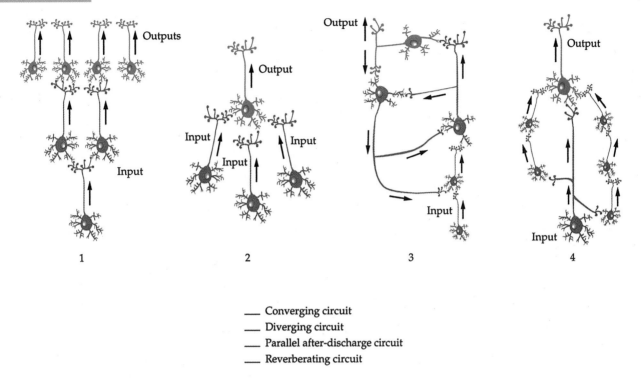

____ Converging circuit
____ Diverging circuit
____ Parallel after-discharge circuit
____ Reverberating circuit

E. Reflex Arc

A **reflex** is a fast, automatic, unplanned sequence of actions that occur in response to a particular stimulus. The stimulus is applied to the periphery and conducted to either the brain or the spinal cord. Reflexes serve to restore functions to homeostasis. Nerve impulses propagating into, through, and out of the CNS follow specific pathways, depending on the kind of information, its origin, and its destination. The pathway followed by nerve impulses that produce a reflex is a **reflex arc** (reflex circuit). A reflex arc includes the following five functional components:

1. *Sensory receptor.* The distal end of a sensory neuron (dendrite) or an associated sensory structure that serves as a receptor. It responds to a specific stimulus—a change in the internal or external environment—by producing a graded potential called a generator (or receptor) potential. If a generator potential reaches the threshold level of depolarization, it will trigger one or more nerve impulses in the sensory neuron.

2. *Sensory neuron.* The nerve impulses propagate from the sensory receptor along the axon of the sensory neuron to the axon terminals, which are located in the gray matter of the spinal cord or brain stem.

3. *Integrating center.* Regions of gray matter within the CNS where the sensory neuron makes a functional connection with one or more neurons. In the simplest type of reflex, the integrating center is a single synapse between a sensory and a motor neuron and is termed a **monosynaptic reflex arc** (*mono* = one). A **polysynaptic reflex arc** (*poly* = many) involves more than two types of neurons and more than one CNS synapse.

4. *Motor neuron.* Impulses triggered by the integrating center propagate out of the CNS along a motor neuron to the part of the body that will respond.

5. *Effector.* The part of the body, either a muscle or a gland, that responds to the motor nerve impulse.

Label the components of a reflex arc in Figure 12.5.

FIGURE 12.5 Components of a reflex arc.

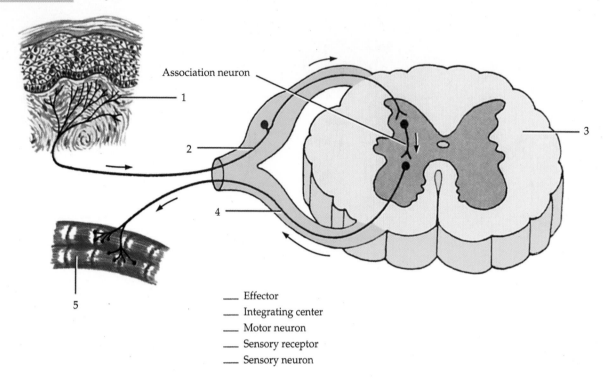

Association neuron

_____ Effector
_____ Integrating center
_____ Motor neuron
_____ Sensory receptor
_____ Sensory neuron

F. Demonstration of Reflex Arc

FLEXICOMP™ is a computer hardware and software system designed to demonstrate characteristics of the reflex arc. It consists of a computer-interfaced goniometer and percussion mallet that permit quantification of the reflex response for various joints of the body and provide visual representation of the reflex arc. The program is available from INTELITOOL® (1–800–955–7621).

ANSWER THE LABORATORY REPORT QUESTIONS AT THE END OF THE EXERCISE.

EXERCISE 12

Nervous Tissue

Name _____   Date _____

Laboratory Section _____   Score/Grade _____

PART 1 ■ Multiple Choice

_____ 1. The portion of a neuron that conducts nerve impulses away from the cell body is the
(a) dendrite (b) axon (c) receptor (d) effector

_____ 2. The fine branching filaments of an axon are called (a) myelin sheaths (b) axolemmas
(c) axon terminals (d) axon hillocks

_____ 3. The component of a reflex arc that responds to a motor impulse is the (a) integrating center (b) receptor (c) sensory neuron (d) effector

_____ 4. Which type of neuron conducts nerve impulses toward the central nervous system?
(a) sensory (b) association (c) connecting (d) motor

_____ 5. In a reflex arc, the nerve impulse is transmitted directly to the effector by the (a) sensory
neuron (b) motor neuron (c) integrating center (d) receptor

_____ 6. Bulblike structures at the distal ends of axon terminals that contain storage sacs for neurotransmitters are called (a) dendrites (b) synaptic end bulbs (c) axon collaterals
(d) neurofibrils

_____ 7. Which neuroglial cell is phagocytic? (a) oligodendrocyte (b) protoplasmic astrocyte
(c) microglial cell (d) fibrous astrocyte

PART 2 ■ Completion

8. A neuron that contains several dendrites and one axon is classified as _____.

9. The portion of a neuron that contains the nucleus and cytoplasm is the _____.

10. The lipid and protein covering around many peripheral axons is called the

_____.

11. The two types of cells that compose the nervous system are neurons and _____.

12. The peripheral, nucleated layer of the Schwann cell that encloses the myelin sheath is the

_____.

13. The side branch of an axon is referred to as the _____.

14.   The part of a neuron that conducts nerve impulses toward the cell body is the

      _____.

15.   Neurons that carry nerve impulses between sensory neurons and motor neurons are called

      _____ neurons.

16.   Neurons with one dendrite and one axon are classified as _____.

17.   Unmyelinated gaps between segments of the myelin sheath are known as

      _____.

18.   The neuroglial cell that produces a myelin sheath around axons of neurons of the central nervous

      system is called a(n) _____.

19.   In the reflex arc, the muscle or gland that responds to a motor impulse is called the

      _____.

20.   The functional contact between two neurons or between a neuron and an effector is called a(n)

      _____.

21.   The circuit that probably plays a role in breathing, waking up, sleeping, and short-term memory is

      a(n) _____ circuit.

Nervous System

Objectives

At the completion of this exercise you should understand

A The principal structural and functional features of the spinal cord

B The components and branches of spinal nerves and how spinal nerves form plexuses

C The main tracts of the spinal cord and their functions

D The principal structural and functional features of the brain

E The names of the cranial nerves and their functions

F How to conduct various experiments for spinal nerve reflexes and cranial nerve functions

G The principal parts of a sheep brain

H The principal structure and function of the autonomic nervous system (ANS)

In this exercise, you will examine the principal structural features of the spinal cord and spinal nerves and perform several experiments on reflexes. Next, you will identify the principal structural features of the brain, trace the course of cerebrospinal fluid, identify the cranial nerves, perform several experiments designed to test for cranial nerve function, and examine the structure and function of the autonomic nervous system. You will also dissect and study the cat and/or sheep brain and cat spinal cord.

A. Spinal Cord and Spinal Nerves

1. Meninges

The **meninges** (me-NIN-jēz) are three connective tissue coverings that encircle the spinal cord and brain (**meninx,** pronounced MĒ-ninks, is singular). They protect the central nervous system. The spinal meninges are:

a. *Dura mater* (DOO-ra MĀ-ter; *dura* = tough; *mater* = mother). The most superficial meninx composed of dense, irregular connective tis-

sue. Between the dura mater and the wall of the vertebral canal is the **epidural space,** which is filled with fat and connective tissue.

b. *Arachnoid mater* (a-RAK-noyd; *arachn* = spider). The middle meninx is an avascular covering composed of a spider's web arrangement of delicate collagen and elastic fibers. Between the dura mater and the arachnoid mater is a space called the **subdural space,** which contains interstitial fluid.

c. *Pia mater* (PĒ-a MĀ-ter; *pia* = delicate). The deep meninx is a thin, transparent connective tissue layer that adheres to the surface of the brain and spinal cord. It consists of interlacing collagen and some fine elastic fibers. Within the pia mater are many blood vessels that supply oxygen and nutrients to the spinal cord. Between the arachnoid mater and the pia mater is a space called the **subarachnoid space** where cerebrospinal fluid circulates. Extensions of the pia mater called **denticulate** (den-TIK-ū-lāt = small tooth) **ligaments** project laterally and fuse with the arachnoid and inner surface of the dura mater along the length of the spinal cord. They protect the spinal cord against shock and sudden displacement.

FIGURE 13.1 Spinal cord.

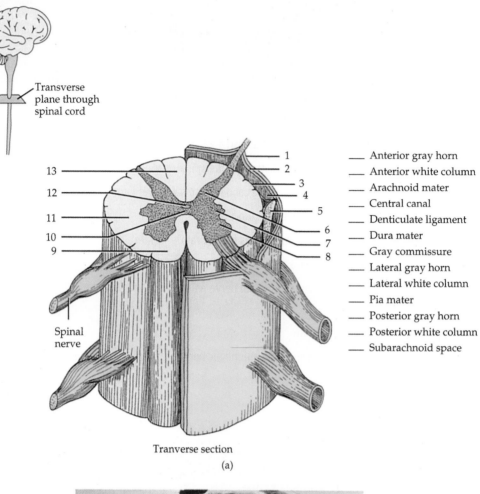

Transverse plane through spinal cord

Spinal nerve

Tranverse section

(a)

__ Anterior gray horn
__ Anterior white column
__ Arachnoid mater
__ Central canal
__ Denticulate ligament
__ Dura mater
__ Gray commissure
__ Lateral gray horn
__ Lateral white column
__ Pia mater
__ Posterior gray horn
__ Posterior white column
__ Subarachnoid space

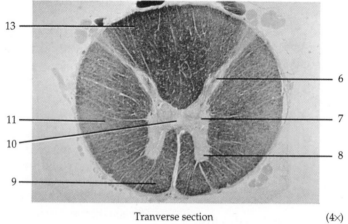

Tranverse section

(b)

(4×)

Label the meninges, subarachnoid space, and denticulate ligament in Figure 13.1.

2. General Features

Obtain a model or preserved specimen of the spinal cord and identify the following general features:

a. **Cervical enlargement.** Between vertebrae C4 and T1; origin of nerves to and from upper limbs.

b. **Lumbar enlargement.** Between vertebrae T9 and T12; origin of nerves to and from lower limbs.

c. **Conus medullaris** (KŌ-nus med-ū-LAR-is; *conus* = cone). Tapered conical portion of spinal cord that ends at the intervertebral disc between L1 and L2.

d. **Filum terminale** (FĪ-lum ter-mi-NAL-ē; *filum* = filament; *terminale* = terminal). An extension of the pia mater that extends inferiorly and anchors the spinal cord to the coccyx.

e. **Cauda equina** (KAW-da ē-KWĪ-na; meaning "horse's tail"). Spinal nerves that angle inferiorly in the vertebral cavity, giving the appearance of wisps of hair.

f. **Anterior median fissure.** Deep, wide groove on the anterior surface of the spinal cord.

g. **Posterior median sulcus.** Shallow, narrow furrow on the posterior surface of the spinal cord.

After you have located the parts on a model or preserved specimen of the spinal cord, label the spinal cord in Figure 13.2.

3. Transverse Section of Spinal Cord

In a freshly dissected section of the brain or spinal cord, some regions look white and glistening whereas others appear gray. **White matter** refers to bundles of myelinated axons of sensory neurons, interneurons, and motor neurons. The whitish color of myelin gives white matter its name. The **gray matter** of the nervous system consists primarily of cell bodies of neurons, neuroglia, unmyelinated axons, axon terminals, and dendrites of interneurons and motor neurons. They look grayish, rather than white, because there is no myelin in these areas.

Obtain a model or specimen of the spinal cord in transverse section and note the gray matter, shaped like a letter H or a butterfly. Identify the following parts:

a. **Gray commissure** (KOM-mi-shur). Cross bar of the letter H.

b. **Central canal.** Small space in the center of the gray commissure that contains cerebrospinal fluid.

c. **Anterior gray horn.** Anterior region of the upright portion of the H. It contains somatic motor nuclei.

d. **Posterior gray horn.** Posterior region of the upright portion of the H. It contains somatic and autonomic sensory nuclei.

e. **Lateral gray horn.** Intermediate region between the anterior and posterior gray horns present in the thoracic, upper lumbar, and sacral segments of the spinal cord. It contains autonomic motor nuclei.

f. **Anterior white column.** Anterior region of white matter.

g. **Posterior white column.** Posterior region of white matter.

h. **Lateral white column.** Intermediate region of white matter between the anterior and posterior white columns.

Each column contains distinct bundles of axons **(tracts)** having a common origin or destination and carrying similar information.

Label these parts of the spinal cord in transverse section in Figure 13.1.

Examine a prepared slide of a spinal cord in transverse section and see how many structures you can identify.

4. Spinal Nerve Attachments

Spinal nerves and their branches are part of the PNS and connect the CNS to sensory receptors, muscles, and glands in all parts of the body. The 31 pairs of spinal nerves are named and numbered according to the region of the spinal cord from which they emerge. The first cervical pair emerges between the atlas and occipital bone; all other spinal nerves leave the vertebral column from intervertebral foramina between adjoining vertebrae. There are 8 pairs of cervical nerves (C1–C8), 12 pairs of thoracic nerves (T1–T12), 5 pairs of lumbar nerves (L1–L5), 5 pairs of sacral nerves (S1–S5), and 1 pair of coccygeal nerves (Co1). Label the spinal nerves in Figure 13.2.

Most spinal nerves are connected to the spinal cord by bundles of axons called roots. The **posterior root** contains sensory nerve fibers only and conducts nerve impulses from sensory receptors in the skin, muscles, and internal organs into the central nervous system. Each posterior root has a swelling, the **posterior (dorsal) root ganglion,** which contains the cell bodies of the sensory neurons. Fibers

FIGURE 13.2   Spinal cord.

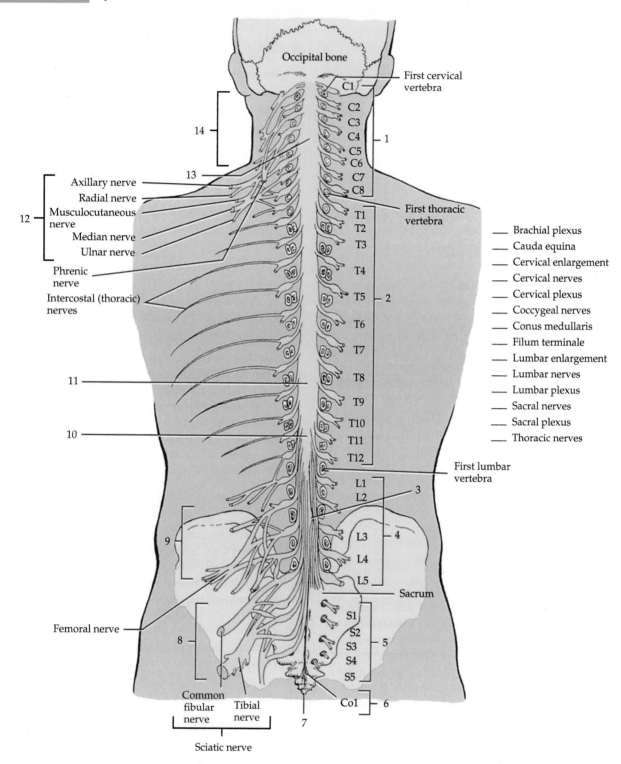

Occipital bone

First cervical vertebra

C1
C2
C3
C4
C5
C6
C7
C8

First thoracic vertebra

T1
T2
T3
T4
T5
T6
T7
T8
T9
T10
T11
T12

First lumbar vertebra

L1
L2
L3
L4
L5

Sacrum

S1
S2
S3
S4
S5

Co1

14

13

Axillary nerve
Radial nerve
Musculocutaneous nerve
Median nerve
Ulnar nerve

12

Phrenic nerve

Intercostal (thoracic) nerves

11

10

9

Femoral nerve

8

Common fibular nerve   Tibial nerve

Sciatic nerve

1

2

3

4

5

6

7

_____ Brachial plexus
_____ Cauda equina
_____ Cervical enlargement
_____ Cervical nerves
_____ Cervical plexus
_____ Coccygeal nerves
_____ Conus medullaris
_____ Filum terminale
_____ Lumbar enlargement
_____ Lumbar nerves
_____ Lumbar plexus
_____ Sacral nerves
_____ Sacral plexus
_____ Thoracic nerves

Posterior view

FIGURE 13.3 Spinal nerve attachments.

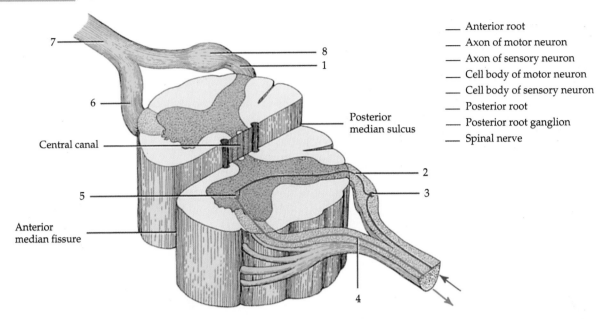

___ Anterior root
___ Axon of motor neuron
___ Axon of sensory neuron
___ Cell body of motor neuron
___ Cell body of sensory neuron
___ Posterior root
___ Posterior root ganglion
___ Spinal nerve

Sections through the thoracic spinal cord

extend from the ganglion into the posterior gray horn. The other point of attachment, the **anterior root,** contains axons of motor neurons, which conduct nerve impulses from the CNS to effector organs and cells. The cell bodies of the motor neurons are located in lateral or anterior gray horns.

Label the posterior root, posterior root ganglion, anterior root, spinal nerve, cell body of sensory neuron, axon of sensory neuron, cell body of motor neuron, and axon of motor neuron in Figure 13.3 on this page.

5. Components and Coverings of Spinal Nerves

The posterior and anterior roots unite to form a spinal nerve at the intervertebral foramen. Because the posterior root contains sensory nerve fibers and the anterior root contains motor nerve fibers, all spinal nerves are **mixed nerves.**

Each cranial nerve and spinal nerve contains layers of protective connective tissue coverings. Individual axons within a nerve, whether myelinated or unmyelinated, are wrapped in **endoneurium** (en'-dō-NOO-rē-um). Groups of axons with their endoneurium are arranged in bundles called **fascicles,** and each bundle is wrapped in a **perineurium** (per'-i-NOO-rē-um). The superficial (outermost) covering over the entire nerve is the **epineurium** (ep'-i-NOO-rē-um).

Obtain a prepared slide of a nerve in transverse section and identify the fibers, endoneurium, perineurium, epineurium, and fascicles. Now label Figure 13.4.

6. Branches of Spinal Nerves

A short distance after passing through its intervertebral foramen, a spinal nerve divides into several branches called **rami** (singular is **ramus** [RĀ-mus]):

a. *Posterior (dorsal) ramus.* Innervates (supplies) deep muscles and skin of the posterior surface of the back.

b. *Anterior (ventral) ramus.* Innervates superficial back muscles and all structures of the upper and lower limbs and the skin of the lateral and anterior trunk; except for thoracic nerves T2–T12, the anterior rami of the other spinal nerves form plexuses before innervating their structures.

c. *Meningeal branch.* Innervates vertebrae, vertebral ligaments, blood vessels of the spinal cord, and meninges.

d. *Rami communicantes* (RĀ-mī ko-mū-nī-KAN-tēz). Gray and white rami communicantes are components of the autonomic nervous system; they connect the anterior rami with sympathetic trunk ganglia.

FIGURE 13.4   Coverings of a spinal nerve.

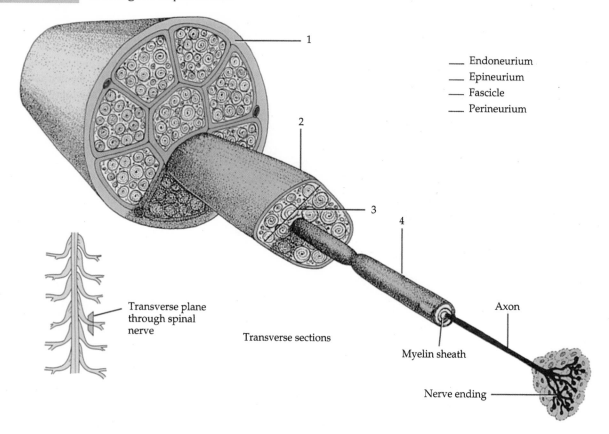

___ Endoneurium
___ Epineurium
___ Fascicle
___ Perineurium

Transverse plane through spinal nerve

Transverse sections

Axon

Myelin sheath

Nerve ending

7. Plexuses

The anterior rami of spinal nerves, except for T2–T12, do not go directly to body structures they supply. Instead, they form networks on both the left and right sides of the body joining with various numbers of axons from anterior rami of adjacent nerves. Such a network of axons is called a **plexus** (= braid or network).

a. *Cervical plexus* (SER-vi-kul PLEK-sus). Formed by the anterior (ventral) rami of the first four cervical nerves (C1–C4) with contributions from C5; one is located on each side of the neck alongside the first four cervical vertebrae; the plexus supplies the skin and muscles of the head, neck, and superior part of shoulders and chest. The phrenic nerves arise from the cervical plexuses and supply motor impulses to the diaphragm.

Using your textbook as a guide, label the nerves of the cervical plexus in Figure 13.5.

b. *Brachial* (BRĀ-kē-al) ***plexus.*** Formed by the anterior (ventral) rami of spinal nerves C5–C8 and T1; each is located on either side of the

last four cervical and first thoracic vertebrae and extends inferiorly and laterally, superior to the first rib posterior the clavicle, and into the axilla; the plexus provides the entire nerve supply for the upper limbs and shoulder region.

Using your textbook as a guide, label the nerves of the brachial plexus in Figure 13.6.

c. *Lumbar* (LUM-bar) ***plexus.*** Formed by the anterior (ventral) rami of spinal nerves L1–L4; each is located on either side of the first four lumbar vertebrae posterior to the psoas major muscle and anterior to the quadratus lumborum muscle; the plexus supplies the anterolateral abdominal wall, external genitals, and part of the lower limbs.

Using your textbook as a guide, label the nerves of the lumbar plexus in Figure 13.7 on page 257.

d. *Sacral* (SĀ-kral) ***plexus.*** Formed by the anterior (ventral) rami of spinal nerves L4–L5 and S1–S4; each is located largely anterior to the sacrum; the plexus supplies the buttocks, perineum, and

FIGURE 13.5 Origin of cervical plexus.

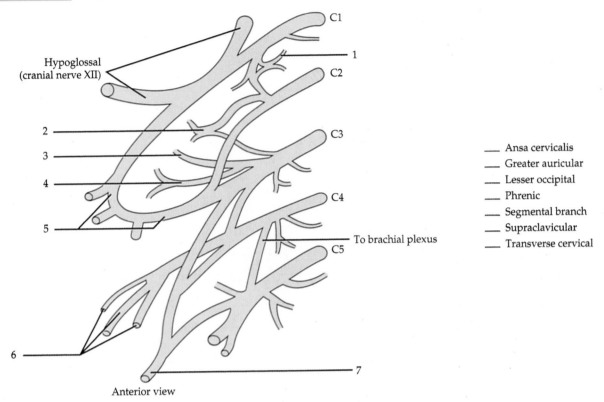

Hypoglossal
(cranial nerve XII)

C1

1

C2

2

3

C3

4

C4

5

To brachial plexus

C5

6

7

Anterior view

___ Ansa cervicalis
___ Greater auricular
___ Lesser occipital
___ Phrenic
___ Segmental branch
___ Supraclavicular
___ Transverse cervical

FIGURE 13.6 Origin of brachial plexus.

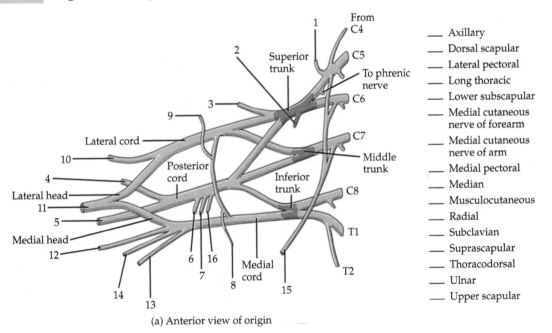

From
C4

1

2

Superior
trunk

C5

To phrenic
nerve

C6

3

9

C7

Lateral cord

Middle
trunk

10

Posterior
cord

Inferior
trunk

C8

4

Lateral head

11

5

Medial head

T1

12

6   16

Medial
cord

T2

7

8

15

14

13

(a) Anterior view of origin

___ Axillary
___ Dorsal scapular
___ Lateral pectoral
___ Long thoracic
___ Lower subscapular
___ Medial cutaneous
        nerve of forearm
___ Medial cutaneous
        nerve of arm
___ Medial pectoral
___ Median
___ Musculocutaneous
___ Radial
___ Subclavian
___ Suprascapular
___ Thoracodorsal
___ Ulnar
___ Upper scapular

FIGURE 13.6 **Distribution of nerves from the brachial plexus. (Continued)**

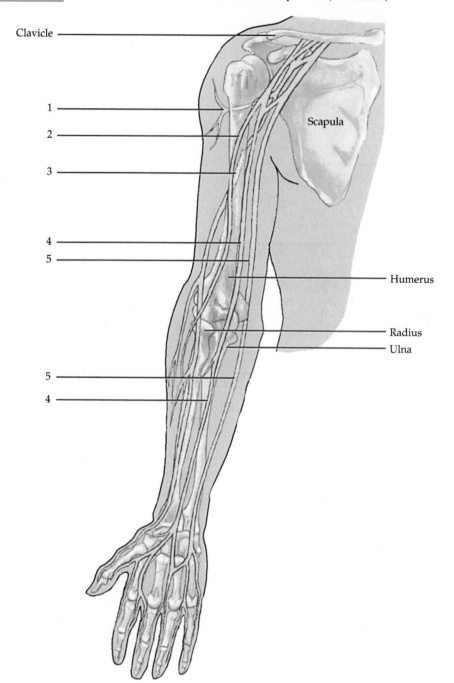

Clavicle

Scapula

1

2

3

4

5

Humerus

Radius

Ulna

5

4

_____ Axillary
_____ Median
_____ Musculocutaneous
_____ Radial
_____ Ulnar

(b) Anterior view of distribution

FIGURE 13.7 Origin of lumbar plexus.

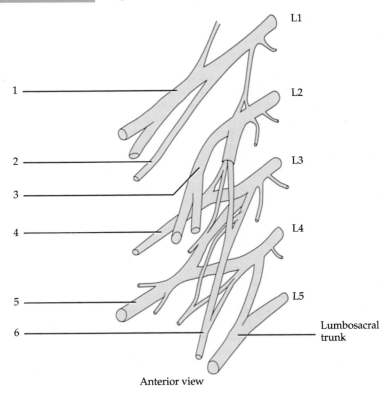

L1
L2
L3
L4
L5

Lumbosacral trunk

Anterior view

___ Femoral
___ Genitofemoral
___ Iliohypogastric
___ Ilioinguinal
___ Lateral cutaneous nerve of thigh
___ Obturator

FIGURE 13.8 Origin of sacral plexus.

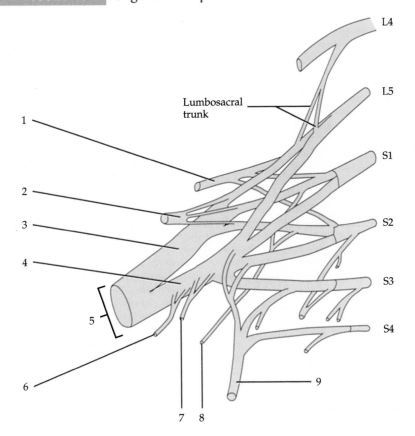

L4
L5
S1
S2
S3
S4

Lumbosacral trunk

___ Common fibular
___ Inferior gluteal
___ Nerve to obturator internus and superior gemellus
___ Nerve to quadratus femoris and inferior gemellus
___ Posterior cutaneous nerve of thigh
___ Pudendal
___ Sciatic
___ Superior gluteal
___ Tibial

(a) Origin in anterior view

FIGURE 13.8   Distribution of sacral plexus. (Continued)

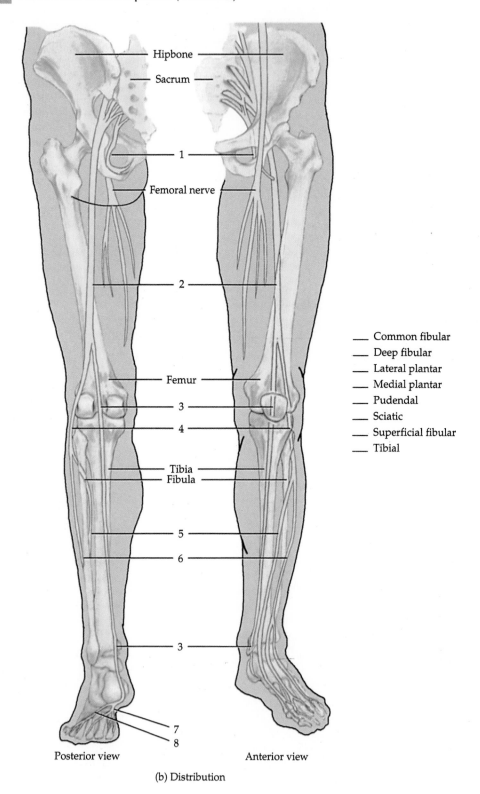

Hipbone

Sacrum

1

Femoral nerve

2

Common fibular
Deep fibular
Lateral plantar
Medial plantar
Pudendal
Sciatic
Superficial fibular
Tibial

Femur

3

4

Tibia
Fibula

5

6

3

7
8

Posterior view   Anterior view

(b) Distribution

lower limbs. The anterior (ventral) rami of spinal nerves S4–S5 and the coccygeal nerves form a small **coccygeal plexus,** which supplies a small area of skin in the coccygeal area.

Using your textbook as a guide, label the nerves of the sacral plexus in Figure 13.8 on page 257.

Label the cervical, brachial, lumbar, and sacral plexuses in Figure 13.2. Also note the names of some of the major peripheral nerves that arise from the plexuses.

For each nerve listed in Table 13.1, indicate the plexus to which it belongs and the structure(s) it innervates.

TABLE 13.1	Nerves of plexuses and innervations		

Nerve	Origin (Plexus)	Distribution (Innervation)
Ansa cervicalis (AN-sa ser-vi-KAL-is)		
Axillary (AK-si-lar-ē)		
Deep (peroneal) fibular		
Femoral (FEM-or-al)		
Iliohypogastric (il′-ē-ō-hī-pō-GAS-trik)		
Long thoracic (thō-RAS-ik)		
Median (MĒ-dē-an)		
Musculocutaneous (mus′-kū-lō-kū-TAN-ē-us)		
Obturator (OB-too-rā-tor)		
Perforating cutaneous (PER-fō-rā-ting kū′-TA-nē-us)		
Phrenic (FREN-ik)		
Pudendal (pū-DEN-dal)		
Radial (RĀ-dē-al)		
Sciatic (sī-AT-ik)		
Thoracodorsal (thō-RA-kō-dor-sal)		
Tibial (TIB-ē-al)		
Transverse cervical (SER-vi-kul)		
Ulnar (UL-nar)		

8. Spinal Cord Tracts

The vital function of conveying sensory and motor information to and from the brain is carried out by sensory (ascending) and motor (descending) tracts and pathways in the spinal cord. The name of a tract or pathway indicates its position in the white matter and where it begins and ends. For example, the anterior spinothalamic tract is located in the **anterior** white column, it originates in the **spinal cord,** and it terminates in the **thalamus** of the brain. Since it conveys nerve impulses from the spinal cord upward to the brain, it is a sensory (ascending) tract.

a. Somatic Sensory Pathways

Somatic sensory pathways from somatic sensory receptors to the cerebral cortex involve three-neuron sets. Axon collaterals (branches) of somatic sensory neurons simultaneously carry signals into the cerebellum and the reticular formation of the brain stem.

1. *First-order neurons.* Conduct impulses from the somatic receptors into the brain stem or the spinal cord. From the face, mouth, teeth, and eyes, somatic sensory impulses propagate along *cranial nerves* into the brain stem. From the back of the head, neck, and rest of the body, somatic sensory impulses propagate along *spinal nerves* into the spinal cord.

2. *Second-order neurons.* Conduct impulses from the spinal cord and brain stem to the thalamus. Axons of second-order neurons *decussate* (cross over to the opposite side) in the spinal cord or brain stem before ascending to the ventral posterior nucleus of the thalamus.

3. *Third-order neurons.* Project from the thalamus to the primary somatosensory area of the cortex on the same side (postcentral gyrus; see Figure 13.15 on page 272), where conscious perception of the sensations results.

There are two general pathways by which somatic sensory signals entering the spinal cord ascend to the cerebral cortex: the **posterior column–medial lemniscus pathway** and the **anterolateral (spinothalamic) pathways.**

1. *Posterior column–medial lemniscus pathway to the cerebral cortex.* Nerve impulses for conscious proprioception and most tactile sensations ascend to the cerebral cortex along the posterior column–medial lemniscus pathway. First-order neurons extend from sensory receptors into the spinal cord and ascend to the medulla oblongata on the same side of the body. The cell bodies of these first-order neurons are in the posterior (dorsal) root ganglia of spinal nerves. Their axons form the **posterior column (gracile fasciculus;** GRAS-ĪL fa-SIK-ū-lus and **cuneate fasciculus;** KŪ-nē-at) in the spinal cord. The axon terminals synapse with second-order neurons in the medulla. The cell body of a second-order neuron is located in the cuneate nucleus (which receives input conducted along axons in the cuneate fasciculus from the neck, upper limbs, and upper trunk) or gracile nucleus (which receives input conducted along axons in the gracile fasciculus from the lower trunk and lower limbs). The axon of the second-order neuron crosses to the opposite side of the medulla and enters the **medial lemniscus,** a thin, ribbonlike projection tract that extends from the medulla to the thalamus. In the thalamus, the axon terminals of second-order neurons synapse with third-order neurons, which project their axons to the somatosensory area of the cerebral cortex.

   Impulses conducted along the posterior column–medial lemniscus pathway give rise to several highly evolved and refined sensations. These are:

   Fine touch. The ability to recognize specific information about a touch sensation, such as what point on the body is touched plus the size, shape, and texture of the source of stimulation and to make two-point discriminations.

   Stereognosis. The ability to recognize the size, shape, and texture of an object by feeling it. Examples are identifying (with closed eyes) a paper clip put into your hand or reading braille.

   Proprioception. The awareness of the precise position of body parts, and **kinesthesia,** the awareness of directions of movement. Proprioceptors also allow **weight discrimination,** the ability to assess the weight of an object.

   Vibratory sensations. The ability to sense rapidly fluctuating touch.

2. *Anterolateral (spinothalamic) pathways to the cerebral cortex.* The **anterolateral (spinothalamic;** spī-nō-THAL-am-ik) **pathways** relay impulses for pain, thermal tickle, and itch sensations. In addition, they relay some tactile impulses which give rise to a very crude, poorly localized touch or pressure sensation. Like the posterior column–medial lemniscus pathway, the anterolateral pathways are also composed of three-neuron sets. The first-order

neuron connects a receptor of the neck, trunk, or limbs with the spinal cord. The cell body of the first-order neuron is in the posterior root ganglion. The axon of the first-order neuron synapses with the second-order neuron, which is located in the posterior gray horn of the spinal cord. The axon of the second-order neuron continues to the opposite side of the spinal cord and passes superiorly to the brain stem in either the **lateral spinothalamic tract** or the **anterior spinothalamic tract.** The axon from the second-order neuron ends in the ventral posterior nucleus of the thalamus. There, it synapses with the third-order neuron. The axon of the third-order neuron projects to the somatosensory area of the cerebral cortex. The lateral spinothalamic tract conveys sensory impulses for pain and temperature whereas the anterior spinothalamic tract conveys impulses for tickle, itch, crude touch, pressure, and vibrations.

3. *Somatic sensory pathways to the cerebellum.* Two tracts in the spinal cord, the **posterior spinocerebellar** (spī′-nō-ser-e-BEL-ar) **tract** and the **anterior spinocerebellar tract,** are major routes for the subconscious proprioceptive input to reach the cerebellum. Sensory input conveyed to the cerebellum along these two pathways is critical for posture, balance, and coordination of skilled movements.

b. Somatic Motor Pathways

The most direct somatic motor pathways extend from the cerebral cortex to skeletal muscles. Other pathways are less direct and include synapses in the basal ganglia, thalamus, reticular formation, and cerebellum.

1. *Direct motor pathways.* Voluntary motor impulses are propagated from the motor cortex to voluntary motor neurons (somatic motor neurons) that innervate skeletal muscles via the **direct** or **pyramidal** (pi-RAM-i-dal) **pathways.** The simplest of these pathways consists of sets of two neurons, upper motor neurons and lower motor neurons. About 1 million pyramidal-shaped cell bodies of direct pathway **upper motor neurons (UMNs)** are in the cortex. Their axons descend through the internal capsule of the cerebrum. In the medulla oblongata, the axon bundles form the ventral bulges known as the **pyramids.** About 90% of the axons of upper motor neurons *decussate* (cross over) to the *contralateral* (opposite) side in the medulla oblongata. They terminate in nuclei of cranial nerves or in the anterior gray horn of the spinal cord. **Lower motor neurons (LMNs)** extend

from the motor nuclei of nine cranial nerves to skeletal muscles of the face and head and emerge from all levels of the spinal cord in the anterior roots of spinal nerves to terminate in skeletal muscles of the limbs and trunk. Close to their termination point, most upper motor neurons synapse with an association neuron, which, in turn, synapses with a lower motor neuron. A few upper motor neurons synapse directly with lower motor neurons.

The direct pathways convey impulses from the cortex that result in precise, voluntary movements. The main parts of the body governed by the direct pathways are the face, vocal cords (for speech), and hands and feet of the limbs. They channel nerve impulses into three tracts:

a. *Lateral corticospinal* (kor′-ti-kō-SPĪ-nal) *tracts.* These pathways begin in the right and left motor cortex and descend through the **internal capsule** of the cerebrum and through the cerebral peduncle of the midbrain and the pons on the same side. The axons of upper motor neurons decussate (cross over to the opposite side) in the medulla oblongata. These axons form the lateral corticospinal tracts in the right and left lateral white columns of the spinal cord. Thus the motor cortex of the right side of the brain controls muscles on the left side of the body, and vice versa. The lower motor neurons then receive input from both upper motor neurons and association neurons. Axons of lower motor neurons (somatic motor neurons) exit all levels of the spinal cord via the anterior roots of spinal nerves and terminate in skeletal muscles. These motor neurons control muscles located in distal parts of the limbs.

b. *Anterior corticospinal tracts.* The axons of upper motor neurons do not cross in the medulla oblongata. They pass through the medulla oblongata, descend on the same side, and form the anterior corticospinal tracts in the right and left anterior white columns. At their level of termination, these upper motor neurons decussate and end in the anterior gray horn on the opposite side. Axons of these lower motor neurons exit the cervical and upper thoracic segments of the cord via the anterior roots of spinal nerves. They terminate in skeletal muscles that control movements of the neck and part of the trunk, thus coordinating movements of the axial skeleton.

c. *Corticobulbar* (kor′-ti-kō-BUL-bar) *tracts.* Some axons of upper motor neurons of these tracts accompany the corticospinal tracts from the motor cortex through the internal capsule to the brain stem. There some decussate, whereas others remain uncrossed. They terminate in the motor nuclei of nine pairs of cranial nerves in the pons and medulla oblongata: cranial nerves III (oculomotor), IV (trochlear), V (trigeminal), VI (abducens), VII (facial), IX (glossopharyngeal), X (vagus), XI (accessory), and XII (hypoglossal). The lower motor neurons of cranial nerves convey impulses that control voluntary movements of the eyes, tongue, and neck; chewing; facial expression; and speech.

2. *Indirect pathways.* The **indirect pathways,** or **extrapyramidal pathways,** include all somatic tracts other than the corticospinal and corticobulbar tracts. Nerve impulses conducted along the indirect pathways follow complex, polysynaptic circuits that involve the motor cortex, basal ganglia, limbic system, thalamus, cerebellum, reticular formation, and nuclei in the brain stem. Axons of upper motor neurons that carry nerve impulses from the indirect pathways descend from various nuclei of the brain stem into five major tracts of the spinal cord and terminate on association neurons or lower motor neurons.

Lower motor neurons receive both excitatory and inhibitory input from many presynaptic neurons in both direct and indirect pathways, an example of convergence. For this reason, lower motor neurons are also called the **final common pathway.** Most nerve impulses from the brain are conveyed to association neurons before being received by lower motor neurons. The sum total of the input from upper motor neurons and association neurons determines the final response of the lower motor neuron. It is not just a simple matter of the brain sending an impulse and the muscle always contracting.

FIGURE 13.9   Selected sensory (ascending) and motor (descending) tracts of the spinal cord. The sensory tracts are indicated in red, and the motor tracts are indicated in blue.

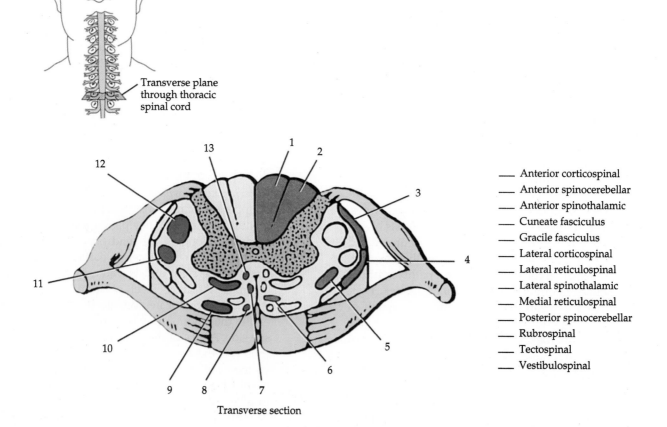

Transverse plane through thoracic spinal cord

___ Anterior corticospinal
___ Anterior spinocerebellar
___ Anterior spinothalamic
___ Cuneate fasciculus
___ Gracile fasciculus
___ Lateral corticospinal
___ Lateral reticulospinal
___ Lateral spinothalamic
___ Medial reticulospinal
___ Posterior spinocerebellar
___ Rubrospinal
___ Tectospinal
___ Vestibulospinal

Transverse section

| **TABLE 13.2** | Sensory (ascending) and motor (descending) tracts and their functions |

Sensory (ascending) tracts	Function
Posterior column	
Lateral spinothalamic	
Anterior spinothalamic	
Posterior spinocerebellar	
Anterior spinocerebellar	

Motor (descending) tracts	Function
Direct	
Lateral corticospinal	
Anterior corticospinal	
Corticobulbar	
Indirect	
Rubrospinal	
Tectospinal	
Vestibulospinal	
Lateral reticulospinal	
Medial reticulospinal	

The five major tracts of the indirect pathways are the **rubrospinal** (ROO-brō-spī-nal), **tectospinal** (TEK-tō-spī-nal), **vestibulospinal** (ves-TIB-ū-lō-spī-nal), **lateral reticulospinal** (re-TIK-ū-lō-spī-nal), and **medial reticulospinal.**

Label the sensory (ascending) and motor (descending) tracts shown in Figure 13.9.

Indicate the function of each sensory and motor tract listed in Table 13.2.

9. Reflex Experiments

In this section, you will determine the responses obtained in various common human reflexes. While performing the procedures it is important for you to consider the various components of a reflex arc, and to try to determine the receptors and effector organs involved.

a. *Patellar reflex.* Have your partner sit on a table so that the knee is off of the table and the leg hangs freely. Strike the quadriceps (patellar) tendon just inferior to the kneecap with the reflex hammer. What is the response?

_____

_____

_____

_____

Test the subject while he or she interlocks fingers and pulls one hand against the other. Is there any change in the level of activity (sensitivity) of the reflex?

_____

_____

_____

_____

Formulate a hypothesis regarding the purpose of the patellar reflex.

_____

_____

_____

_____

b. *Achilles reflex.* Have a subject kneel on a chair and let the feet hang freely over the edge of the chair. Bend one foot to increase the tension on the gastrocnemius muscle. Tap the calcaneal (Achilles) tendon with the reflex hammer.

What is the result?

_____

_____

_____

c. *Plantar reflex.* Scratch the sole of your partner's foot by moving a blunt object along the sole toward the toes.

What is the result?

_____

_____

_____

What is the Babinski sign?

_____

_____

_____

_____

Why is it normal in children under the age of 18 months?

_____

_____

_____

What does the Babinski sign indicate in an adult?

_____

_____

_____

_____

B. Brain

1. Parts

The brain may be divided into four principal parts (1) **brain stem,** which is continuous with the spinal cord and consists of the medulla oblongata, pons, and midbrain; (2) **diencephalon** (dī-en-SEF-a-lon; *di* = through; *encephalos* = brain), which consists of the thalamus, hypothalamus, and epithalamus; (3) **cerebellum** (ser'-e-BEL-um), which is posterior to the brain stem; and (4) **cerebrum** (se-RĒ-brum), which is supported on the brain stem and diencephalons and is the largest part of the brain.

Examine a model and preserved specimen of the brain and identify the parts just described. Then refer to Figure 13.10 and label the parts of the brain.

2. Meninges

As in the spinal cord, the brain is protected by **meninges.** The cranial meninges are continuous with the spinal meninges. The cranial meninges are the superficial **dura mater,** the middle **arachnoid mater,** and the deep **pia mater.** Three extensions of the dura mater separate parts of the brain: (1) the *falx cerebri* (FALKS CER-e-brē) separates the two hemispheres of the cerebrum. (2) The *falx cerebelli* (cer-e-BEL-ī) separates the two hemispheres of the cerebellum. (3) The *tentorium cerebelli* (ten-TŌ-rē-um) separates the cerebrum from the cerebellum.

Refer to Figure 13.11 on page 266 and label all of the meninges.

3. Cerebrospinal Fluid (CSF)

The central nervous system (brain and spinal cord) is nourished and protected by **cerebrospinal fluid (CSF).** The fluid circulates through the subarachnoid space around the brain and spinal cord and through the **ventricles** (VEN-tri-kuls; *ventriculus* = little cavity) of the brain. The ventricles are cavities in the brain that communicate with each other, with the central canal of the spinal cord, and with the subarachnoid space.

FIGURE 13.10 Principal parts of the brain.

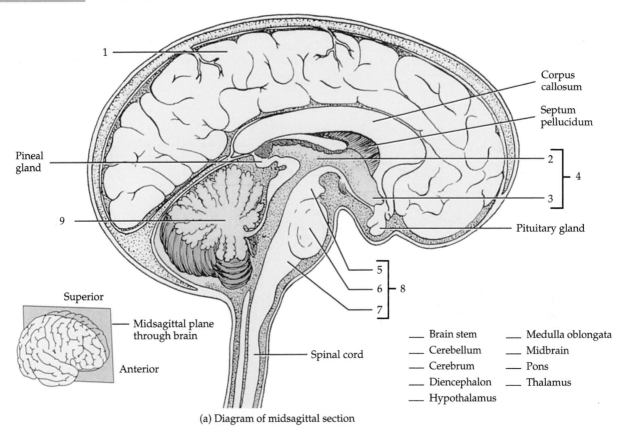

Corpus callosum

Septum pellucidum

Pineal gland

Pituitary gland

Superior

Midsagittal plane through brain

Anterior

Spinal cord

___ Brain stem   ___ Medulla oblongata
___ Cerebellum   ___ Midbrain
___ Cerebrum   ___ Pons
___ Diencephalon   ___ Thalamus
___ Hypothalamus

(a) Diagram of midsagittal section

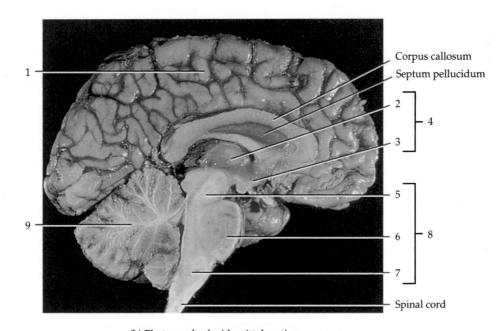

Corpus callosum

Septum pellucidum

Spinal cord

(b) Photograph of midsagittal section

FIGURE 13.11   **Brain and spinal cord.**

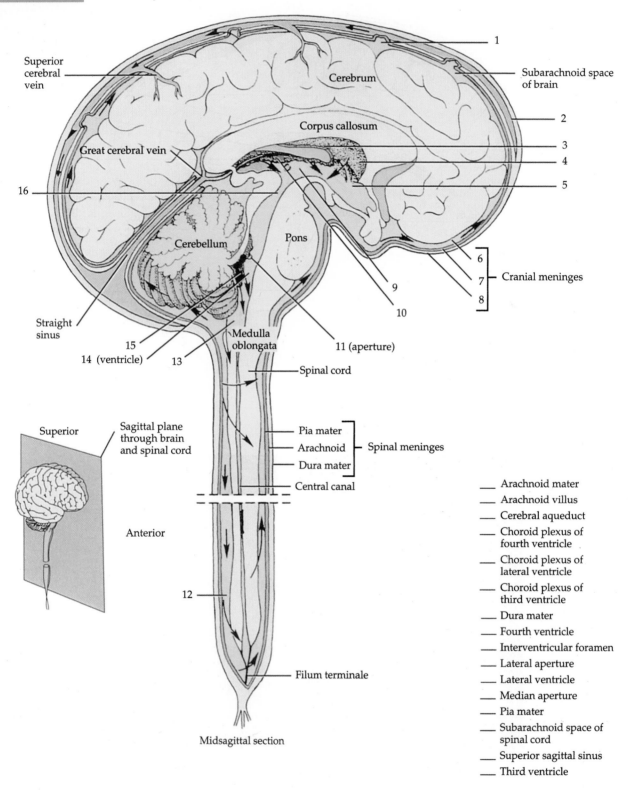

Superior cerebral vein

Cerebrum

Subarachnoid space of brain

1

2

3

4

5

Corpus callosum

Great cerebral vein

16

6

7

8

Cranial meninges

9

10

11 (aperture)

Pons

Cerebellum

Straight sinus

Medulla oblongata

15

14 (ventricle)   13

Spinal cord

Superior

Sagittal plane through brain and spinal cord

Pia mater

Arachnoid

Dura mater

Spinal meninges

Anterior

Central canal

12

Filum terminale

Midsagittal section

___ Arachnoid mater

___ Arachnoid villus

___ Cerebral aqueduct

___ Choroid plexus of fourth ventricle

___ Choroid plexus of lateral ventricle

___ Choroid plexus of third ventricle

___ Dura mater

___ Fourth ventricle

___ Interventricular foramen

___ Lateral aperture

___ Lateral ventricle

___ Median aperture

___ Pia mater

___ Subarachnoid space of spinal cord

___ Superior sagittal sinus

___ Third ventricle

Cerebrospinal fluid is formed primarily by filtration and secretion from networks of capillaries covered by ependymal cells in the ventricles, called **choroid** (KŌ-royd; *chorion* = membrane) **plexuses** (Figure 13.11). Each of the two **lateral ventricles** is located within each hemisphere (side) of the cerebrum under the corpus callosum. Anteriorly, the lateral ventricles are separated by a thin membrane called the **septum pellucidum** (pe-LOO-si-dum; *pellucid* = transparent). The fluid formed in the choroid plexuses of the lateral ventricles circulates through an opening called the **interventicular foramen** into the third ventricle. The **third ventricle** is a narrow cavity along the midline superior to the hypothalamus and between the right and left halves of the thalamus. More fluid is secreted by the choroid plexus of the third ventricle. Then the fluid circulates through a canal-like structure called the **cerebral aqueduct** (AK-we-duct) into the fourth ventricle. The **fourth ventricle** lies between the brain stem and the cerebellum. More fluid is secreted by the choroid plexus of the fourth ventricle. The roof of the fourth ventricle has three openings: one **median aperture** (AP-er-chur) and paired **lateral apertures.** The fluid circulates through the apertures into the subarachnoid space around the posterior portion of the brain and inferiorly through the central canal of the spinal cord to the subarachnoid space around the posterior surface of the spinal cord, up the anterior surface of the spinal cord, and around the anterior part of the brain. Cerebrospinal fluid is gradually reabsorbed through arachnoid villi into a vein called the superior sagital sinus. Normally, cerebrospinal fluid is absorbed as rapidly as it is formed.

Refer to Figure 13.11 and label the choroid plexus of the lateral ventricle, lateral ventricle, interventricular foramen, choroid plexus of third ventricle, third ventricle, cerebral aqueduct, choroid plexus of fourth ventricle, fourth ventricle, median aperture, lateral aperture, subarachnoid space of spinal cord, superior sagittal sinus, and arachnoid villus.

Note the arrows in Figure 13.11, which indicate the path taken by cerebrospinal fluid. With the aid of your textbook, starting at the choroid plexus of the lateral ventricle and ending at the superior sagittal sinus, see if you can follow the remaining path of the fluid.

Now complete Figure 13.12.

FIGURE 13.12 Formation, circulation, and absorption of cerebrospinal fluid (CSF).

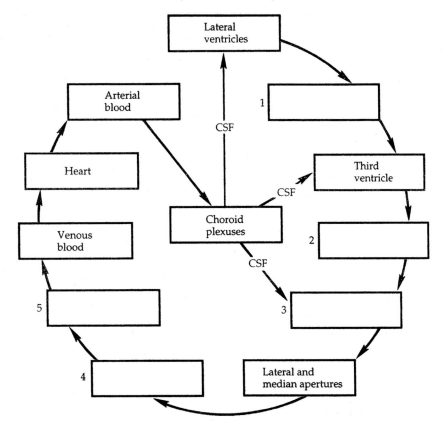

4. Medulla Oblongata

The **medulla oblongata** (me-DOOL-la ob'-long-GA-ta), or more simply **medulla,** is a continuation of the spinal cord and forms the inferior part of the brain stem. The medulla contains all sensory (ascending) and motor (descending) tracts that extend between the spinal cord and various parts of the brain. On the anterior aspect of the medulla are two roughly triangular bulges called **pyramids.** They contain the largest motor tracts that pass from the cerebrum to the spinal cord. Approximately 90% of the fibers in the left pyramid cross to the right side of the spinal cord, and 90% fibers in the right pyramid cross to the left side of the spinal cord. This crossing is called the **decussation** (dē-ku-SĀ-shun) **of pyramids** and explains why motor areas of one side of the cerebral cortex control muscular movements on the opposite side of the body. The medulla contains several nuclei which control vital reflex centers called the **cardiovascular center** (regulates rate and force of the heart beat and blood vessel diameter) and **medullary rhythmicity center** (adjusts basic rhythm of breathing). Other centers in the medulla coordinate vomiting, coughing, swallowing, hiccuping, and sneezing. The posterior part of the medulla contains two pairs of prominent nuclei: **gracile nucleus** and **cuneate nucleus.** Most sensory impulses initiated on one side of the body cross to the thalamus on the opposite side either in the spinal cord or in these medullary nuclei. Just lateral to each pyramid is an oval swelling called the **olive.** It is caused mostly by the **inferior olivary nucleus** within the medulla. Neurons in the olive relay impulses from proprioceptors (monitoring joint and muscle positions) to the cerebellum. They carry signals that ensure the efficiency of precise, voluntary movements and maintain equilibrium and posture. The medulla also contains the nuclei of origin of several cranial nerves. These are the cochlear branch of the vestibulocochlear nerves (cranial nerve VIII), glossopharyngeal nerves (cranial nerve IX), vagus nerves (cranial nerve X), spinal portions of the accessory nerves (cranial nerve XI), and hypoglossal nerves (cranial nerve XII). Examine a model or specimen of the brain and identify the parts of the medulla. Then refer to Figure 13.10 and locate the medulla.

5. Pons

The **pons** (*pons* = bridge) lies directly superior to the medulla oblongata and anterior to the cerebellum. Like the medulla, the pons consists of both nuclei and tracts of white matter. As its name implies, the pons is a bridge connecting the spinal cord with the brain and parts of the brain with each other. These connections are provided by bundles of axons. Some axons of the pons connect the right and left sides of the cerebellum. Others are part of ascending sensory tracts and descending motor tracts. The pons contains the nuclei or origin of the following pairs of cranial nerves: trigeminal nerves (cranial nerve V), abducens nerves (cranial nerve VI), facial nerves (cranial nerve VII), and vestibular branch of the vestibulocochlear nerves (cranial nerve VIII). Other important nuclei in the pons are the **pneumotaxic** (noo-mō-TAK-sik) **area** and **apneustic** (ap-NOO-stik) **area** that help control respiration.

Identify the pons on a model or specimen of the brain. Locate the pons in Figure 13.10.

6. Midbrain

The **midbrain** extends from the pons to the diencephalon. The anterior part of the midbrain contains the paired **cerebral peduncles,** which conduct nerve impulses from the cerebrum to the spinal cord, medulla, and pons. The posterior part of the midbrain, called the **tectum,** contains four rounded elevations. Two of the elevations, the **superior colliculi** (ko-LIK-ū-lī), serve as reflex centers for movements of the eyeballs and the head in response to visual and other stimuli. **Colliculus,** which is singular, means little hill. The other two elevations, the **inferior colliculi,** serve as reflex centers for movements of the head and trunk in response to auditory stimuli. Several nuclei are also found in the midbrain. Among these are the left and right **substantia nigra** (sub-STAN-shē-a NĪ-gra = black substance). These large, darkly pigmented nuclei that control subconscious muscle activities are near the cerebral peduncles. In Parkinson's disease, dopamine-containing neurons in the substantia nigra degenerate. The left and right **red nuclei** are also found in the midbrain. The name derives from their rich blood supply and an iron-containing pigment in their neuronal cell bodies. Fibers from the cerebellum and cerebral cortex form synapses in the red nuclei, which function with the cerebellum to coordinate muscular movements. The midbrain contains the nuclei of origin for two pairs of cranial nerves: oculomotor nerves (cranial nerve III) and trochlear nerves (cranial nerve IV).

Identify the parts of the midbrain on a model or specimen of the brain. Locate the midbrain in Figure 13.10.

7. Cerebellum

The **cerebellum** is posterior to the medulla and pons and inferior to the posterior portion of the cerebrum. It is separated from the cerebrum by the **transverse fissure** and by an extension of the cranial dura mater called the **tentorium cerebelli.** The central constricted area of the cerebellum is called the **vermis** (meaning worm), and the lateral lobes are referred to as **cerebellar hemispheres.** Each hemisphere consists of lobes that are separated by deep and distinct fissures. The **anterior lobe** and **posterior lobe** govern subconscious aspects of skeletal muscle movements. The **flocculonodular** (*flocculus* = wool-like tuft) **lobe** on the inferior surface contributes to equilibrium and balance. Between hemispheres is a sickle-shaped extension of the cranial dura mater, the **falx cerebelli.**

The superficial layer of the cerebellum, called the **cerebellar cortex,** consists of gray matter in a series of slender parallel ridges called **folia** (*leaves*). Deep to the gray matter are **white matter tracts** called **arbor vitae** (*tree of life*) that resemble branches of a tree. Deep within the white matter are masses of gray matter, the **cerebellar nuclei.** The nuclei give rise to nerve fibers that carry impulses from the cerebellum to other parts of the brain and the spinal cord.

The cerebellum is attached to the brain stem by three paired cerebellar peduncles. The **inferior cerebellar peduncles** carry sensory information for equilibrium, balance, and proprioception (joint and muscle sense) into the cerebellum. The **middle cerebellar peduncles** contain axons, which relay motor commands for voluntary movements from the pontine nuclei into the cerebellum. The **superior cerebellar peduncles** contain axons that extend from the cerebellum to the red nuclei of the midbrain.

The cerebellum evaluates how well movements programmed by motor areas in the cerebrum are actually being carried out. It constantly receives sensory input from proprioceptors in muscles, tendons, and joints, receptors for equilibrium, and visual receptors of the eyes. If the intent of the cerebral motor areas is not being attained by skeletal muscles, the cerebellum detects the variation and sends feedback signals to the motor areas to either stimulate or inhibit the activity of skeletal muscles. This interaction helps to smooth and coordinate complex sequences of skeletal muscle contractions. Besides coordinating skilled movements, the cerebellum is the main brain region that regulates posture and balance. These aspects of cerebellar function make possible all skilled motor activities, from catching a baseball to dancing to speaking.

Examine a model or specimen of the brain and locate the parts of the cerebellum. Now refer to Figure 13.13 and label the parts of the cerebellum.

8. Determination of Cerebellar Function

Working with your lab partner, perform the following tests to determine cerebellar function:

1. With your upper limbs down straight at your sides, walk heel to toe for 20 ft without losing your balance.

2. Stand away from any supporting object and place your feet together and upper limbs down straight at your sides; look straight ahead. Now close your eyes and stand for 2 to 3 min. Your lab partner should stand off to one side and then immediately in front of you and observe any and all body movements.

3. Stand away from any supporting object and place your feet together and upper limbs down straight at your sides; look straight ahead. Now, with your eyes open, raise one foot off the ground and run the heel of that foot down the shin of the other leg. Your lab partner should observe the smoothness of the movement and whether your heel remains in contact with the shin at all times.

4. Place your left palm in the supinated (palm up) position with the elbow at a 90° angle from the body. Now place the palm of your right hand into the left palm. Alternately pronate (palm down) and supinate the right hand as quickly as possible, while contacting the left palm only when the movement is completed (i.e., supinated right hand gently slaps left palm, and then pronated right hand gently slaps left palm). Do not let the medial surface of the right hand contact the left palm at any time. Note the ease with which your partner completes this activity. Then have your partner repeat the actions with the right and left hands switching roles.

5. Stand away from any supporting object and place your feet together and upper limbs abducted (away from midline) and palms facing anteriorly; look straight ahead. Now, gently close your eyes and touch the tip of your nose with your right index finger and return the right upper limb to its abducted position. Now touch the tip of your nose with your left index finger and return the left upper limb to its abducted position. Slowly increase the rapidity of the activity.

6. With your eyes closed, touch your nose with the index finger of each hand.

FIGURE 13.13
Cerebellum. The arrows in the insets in (a) and (b) indicate the directions from which the cerebellum is viewed (superior and inferior, respectively).

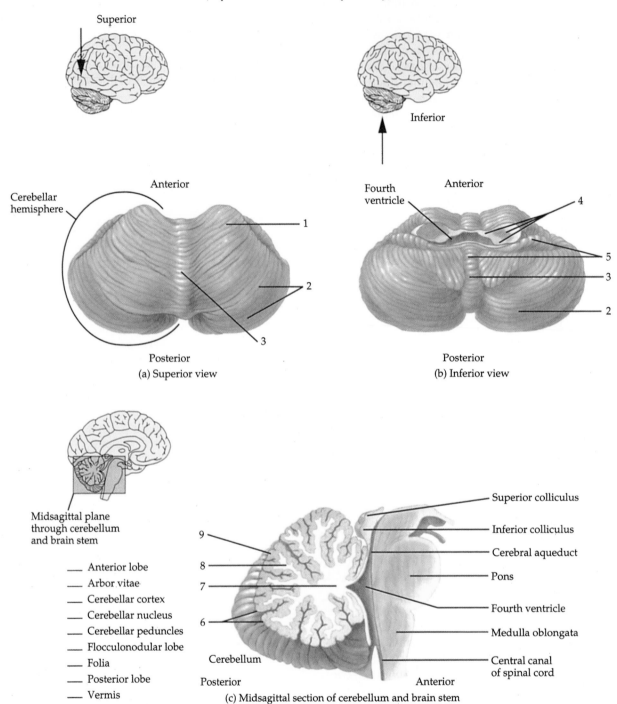

Superior

Anterior

Cerebellar hemisphere

1

2

3

Posterior

(a) Superior view

Inferior

Fourth ventricle

Anterior

4

5

3

2

Posterior

(b) Inferior view

Midsagittal plane through cerebellum and brain stem

___ Anterior lobe
___ Arbor vitae
___ Cerebellar cortex
___ Cerebellar nucleus
___ Cerebellar peduncles
___ Flocculonodular lobe
___ Folia
___ Posterior lobe
___ Vermis

9

8

7

6

Cerebellum

Posterior

Superior colliculus

Inferior colliculus

Cerebral aqueduct

Pons

Fourth ventricle

Medulla oblongata

Central canal of spinal cord

Anterior

(c) Midsagittal section of cerebellum and brain stem

9. Thalamus

The **thalamus** (THAL-a-mus = inner chamber) consists of paired oval masses of gray matter, organized into nuclei, with interspersed tracts of white matter and comprises about 80% of the di- encephalon. The two masses are joined by a bridge of gray matter called the **intermediate mass.** The thalamus is the major relay station for most sensory impulses. The thalamic nuclei are divided into seven major groups on each side of the thalamus. The most prominent nuclei of the ventral group are the **medial geniculate** (je-NIK-ū-lāt) **nuclei** (hearing), **lateral geniculate nuclei** (vision), **ventral posterior nuclei** (touch, pressure, proprioception, vibration, heat, cold, and pain from the face and body), **ventral lateral nuclei** (voluntary motor actions), and **ventral anterior nuclei** (voluntary motor actions and movement planning). The thalamus also allows crude perception of some sensations, such as pain, temperature, and pressure. Precise localization of such sensations depends on nerve impulses being relayed from the thalamus to the cerebral cortex.

Identify the thalamic nuclei on a model or specimen of the brain. Then refer to Figure 13.14 and label the nuclei.

10. Hypothalamus

The **hypothalamus** (*hypo* = under) is located inferior to the thalamus and forms the floor and part of the wall of the third ventricle. Among the functions served by the hypothalamus are the control and integration of the activities of the autonomic nervous system and parts of the endocrine system (pituitary gland) and the control of body temperature. The hypothalamus also assumes a role in feelings of rage and aggression, regulation of eating and drinking, and the waking state and sleep patterns.

Identify the hypothalamus on a model or specimen of the brain. Locate the hypothalamus in Figure 13.10.

11. Cerebrum

The **cerebrum** is the largest portion of the brain and is supported on the brain stem. Its outer rim of gray matter is called the **cerebral cortex** (*cortex* = rind or bark). Beneath the cerebral cortex is the cerebral white matter. The cortical region expands rapidly during embryonic development. As a re-

FIGURE 13.14 **Thalamic nuclei.**

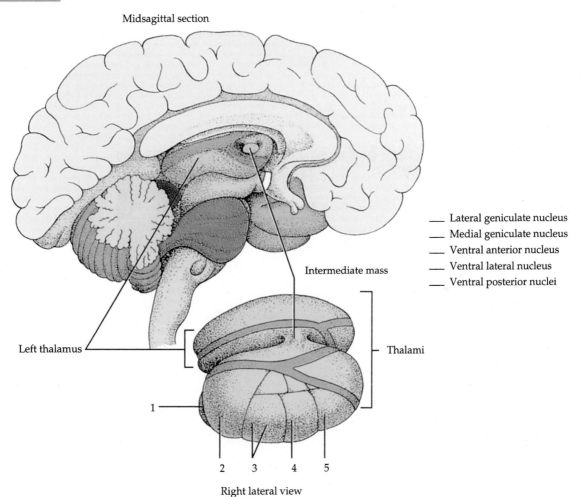

Midsagittal section

_____ Lateral geniculate nucleus
_____ Medial geniculate nucleus
_____ Ventral anterior nucleus
_____ Ventral lateral nucleus
_____ Ventral posterior nuclei

Intermediate mass

Left thalamus

Thalami

1

2   3   4   5

Right lateral view

sult, the cortical region rolls and folds upon itself. The folds of the cerebral cortex are termed **gyri** (JĪ-rī) or **convolutions,** the deep grooves between folds are termed **fissures,** and the shallow grooves between folds are termed **sulci** (SUL-sī).

The most prominent fissure, the **longitudinal fissure,** separates the cerebrum into right and left halves called **hemispheres.** The hemispheres are connected internally by the **corpus callosum,** a broad band of white matter containing axons that extend between the hemispheres. Each hemisphere is further subdivided into four lobes by sulci or fis-

sures. The **central sulcus** (SUL-kus) separates the **frontal lobe** from the **parietal lobe.** The **lateral cerebral sulcus** separates the frontal lobe from the **temporal lobe.** The **parieto-occipital sulcus** separates the **parietal lobe** from the **occipital lobe.** Another prominent fissure, the **transverse fissure,** separates the cerebrum from the cerebellum. Another lobe of the cerebrum, the **insula,** lies deep within the lateral cerebral fissure under the parietal, frontal, and temporal lobes. It cannot be seen in external view. Two important gyri on either side of the **central sulcus** are the **precentral gyrus** and

FIGURE 13.15   Lobes of cerebrum.

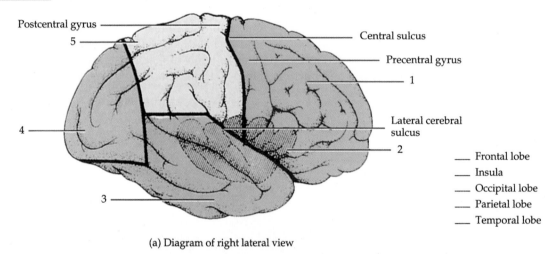

Postcentral gyrus

5

4

3

Central sulcus

Precentral gyrus

1

Lateral cerebral sulcus

2

___ Frontal lobe
___ Insula
___ Occipital lobe
___ Parietal lobe
___ Temporal lobe

(a) Diagram of right lateral view

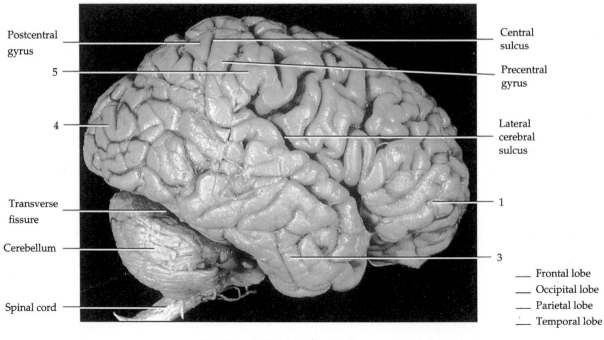

Postcentral gyrus

5

4

Transverse fissure

Cerebellum

Spinal cord

Central sulcus

Precentral gyrus

Lateral cerebral sulcus

1

3

___ Frontal lobe
___ Occipital lobe
___ Parietal lobe
___ Temporal lobe

(b) Photograph of right lateral side

the **postcentral gyrus.** The olfactory nerves (cranial nerve I) and optic nerves (cranial nerve II) are associated with the cerebrum.

Examine a model of the brain and identify the parts of the cerebrum just described. Refer to Figure 13.15 and label the lobes.

12. Cerebral White Matter

The white matter underlying the cerebral cortex consists of myelinated and unmyelinated axons in three types of tracts.

a. **Association tracts** conduct nerve impulses between gyri in the same hemisphere.

b. **Commissural tracts** conduct impulses from the gyri in one cerebral hemisphere to the corresponding gyri in the opposite cerebral hemisphere. Three important groups of commissural tracts are the **corpus callosum, anterior commissure,** and **posterior commissure.**

c. **Projection tracts** conduct impulses from the cerebrum and other parts of the brain to the spinal cord or from the spinal cord to the brain.

Examine a model of the brain and identify the fibers just described. Refer to Figure 13.16 and label the fibers.

13. Basal Ganglia

Basal ganglia (GANG-lē-a) are three nuclei (groups of gray matter) within the cerebral hemispheres. Two of the basal ganglia nuclei are side-by-side, just lateral to the thalamus. They are the **globus pallidus** (*globus* = ball; *pallidus* = pale) and the **putamen** (pū-TĀ-men). Together, the globus pallidus and putamen are referred to as the **lentiform nucleus.** The third member of the basal ganglia is the **caudate nucleus** (*caud* = tail). Together, the caudate and lentiform nuclei are known as the **corpus striatum.**

The basal ganglia are interconnected by many nerve fibers. They also receive input from and provide output to the cerebral cortex, thalamus, and hypothalamus. The caudate nucleus and the putamen control large automatic movements of skeletal muscles, such as swinging the arms while walking. The globus pallidus is concerned with the regulation of muscle tone required for specific body movements.

Examine a model or specimen of the brain and identify the basal ganglia described. Now refer to Figure 13.17 and label the basal ganglia shown.

14. Functional Areas of Cerebral Cortex

The functions of the cerebrum are numerous and complex. In a general way, the cerebral cortex can be divided into sensory, motor, and association areas. The **sensory areas** receive and interpret sensory impulses, the **motor areas** control muscular movement, and the **association areas** are concerned with emotional and intellectual processes.

The principal sensory and motor areas of the cerebral cortex are indicated by numbers based on K. Brodmann's map of the cerebral cortex. His map, first published in 1909, attempts to correlate

FIGURE 13.16 **Cerebral white matter tracts.**

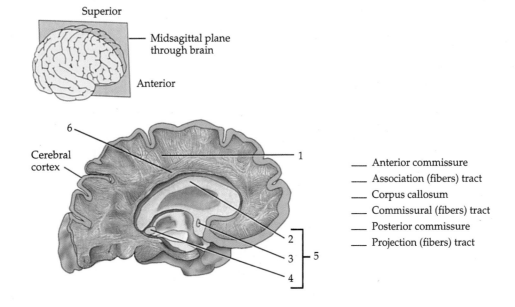

___ Anterior commissure
___ Association (fibers) tract
___ Corpus callosum
___ Commissural (fibers) tract
___ Posterior commissure
___ Projection (fibers) tract

FIGURE 13.17 **Basal ganglia.**

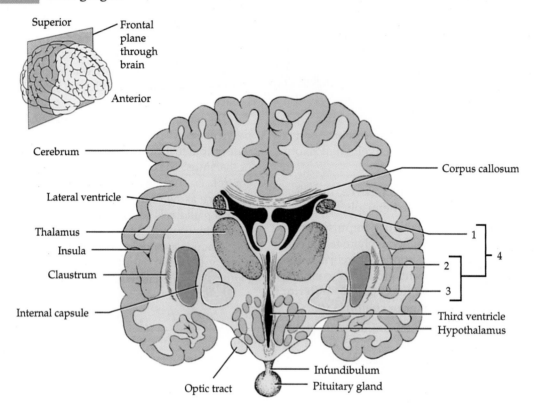

Frontal section of cerebrum

___ Caudate nucleus   ___ Globus pallidus
___ Corpus striatum   ___ Putamen

structure and function. Refer to Figure 13.18 and match the name of the sensory or motor area next to the appropriate number.

C. Cranial Nerves: Names and Components

Of the 12 pairs of **cranial nerves,** 10 originate from the brain stem, but all emerge through foramina in the skull. The cranial nerves are designated by Roman numerals and names (see Figure 13.19). The Roman numerals indicate the order in which the nerves arise from the brain, from anterior to posterior. The names indicate the distribution or function of the nerves.

Obtain a model of the brain and using your text and any other aids available identify the 12 pairs of cranial nerves. Now refer to Figure 13.19 and label the cranial nerves.

D. Tests of Cranial Nerve Function

The following simple tests may be performed to determine cranial nerve function. Although they provide only superficial information, they will help you to understand how the various cranial nerves function. Perform each of the tests with your partner.

FIGURE 13.18 Functional areas of the cerebrum.

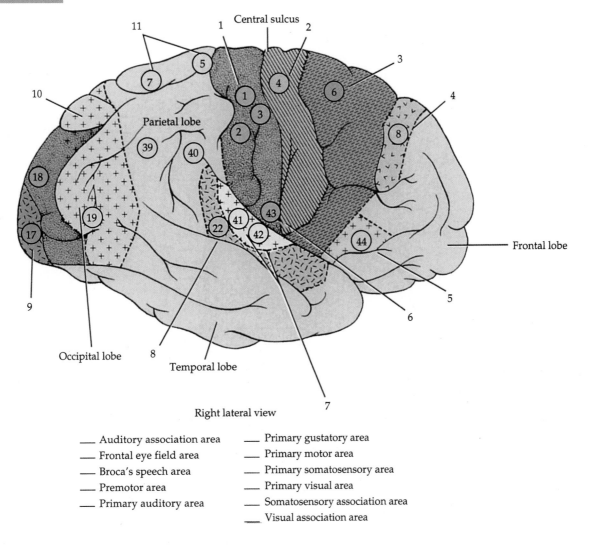

Right lateral view

___ Auditory association area
___ Frontal eye field area
___ Broca's speech area
___ Premotor area
___ Primary auditory area

___ Primary gustatory area
___ Primary motor area
___ Primary somatosensory area
___ Primary visual area
___ Somatosensory association area
___ Visual association area

1. ***Olfactory nerve (cranial nerve I).*** Have your partner smell several familiar substances such as spices, first using one nostril and then the other. Your partner should be able to distinguish the odors equally with each nostril.

   What structures could be malfunctioning if the ability to detect odors is lost? _____

   _____

   _____

   _____

2. ***Optic nerve (cranial nerve II).*** Have your partner read a portion of a printed page using each eye. Do the same while using a Snellen chart at a distance of 6.10 m (20 ft).

   Describe the visual pathway from the optic

   nerve to the cerebral cortex. _____

   _____

   _____

   _____

   _____

3. ***Oculomotor (cranial nerve III), trochlear*** (TROK-lē-ar) ***(cranial nerve IV), and abducens*** (ab-DOO-senz) ***nerves (cranial nerve VI).*** To test the motor responses of these nerves, have your partner follow your finger with his or her eyes without moving the head. Move your finger up, down, medially, and laterally.

FIGURE 13.19   Cranial nerves of human brain. The arrow in the inset indicates the direction from which the brain is viewed (inferior).

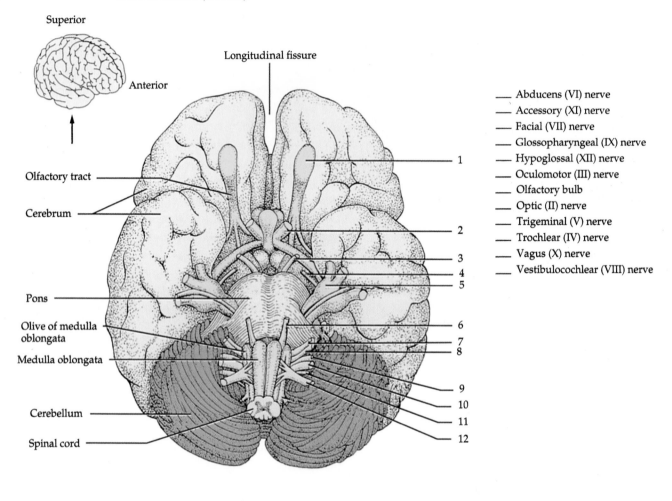

Superior

Anterior

Longitudinal fissure

Olfactory tract

Cerebrum

Pons

Olive of medulla oblongata

Medulla oblongata

Cerebellum

Spinal cord

1
2
3
4
5
6
7
8
9
10
11
12

____ Abducens (VI) nerve
____ Accessory (XI) nerve
____ Facial (VII) nerve
____ Glossopharyngeal (IX) nerve
____ Hypoglossal (XII) nerve
____ Oculomotor (III) nerve
____ Olfactory bulb
____ Optic (II) nerve
____ Trigeminal (V) nerve
____ Trochlear (IV) nerve
____ Vagus (X) nerve
____ Vestibulocochlear (VIII) nerve

(a) Diagram of inferior surface

Which nerves control which movements of the eyeball? _____

_____

Look for signs of ptosis (drooping of one or both eyelids).

Which cranial nerve innervates the upper eye lid? _____

▲ **CAUTION!** *Do not make contact with the eyes.*

To test for the papillary light reflex, shine a small flashlight into each eye from the side. Observe the pupil.

What cranial nerve controls this reflex? _____

_____

4. *Trigeminal* (trī-JEM-i-nal) *nerve (cranial nerve V).* To test the motor response of this nerve, have your partner close his or her jaws tightly.

What muscles are used? _____

_____

Now, while holding your hand under your partner's lower jaw to provide resistance, ask your partner to open his or her mouth. To test the sensory responses of this nerve, have your partner close his or her eyes and lightly whisk a piece of dry cotton over the mandibular, maxillary, and ophthalmic areas on each side of the face. Do the same with cotton that has been moistened with cold water.

What cranial nerves bring about this response?

_____

FIGURE 13.19   **Cranial nerves of human brain. (Continued)**

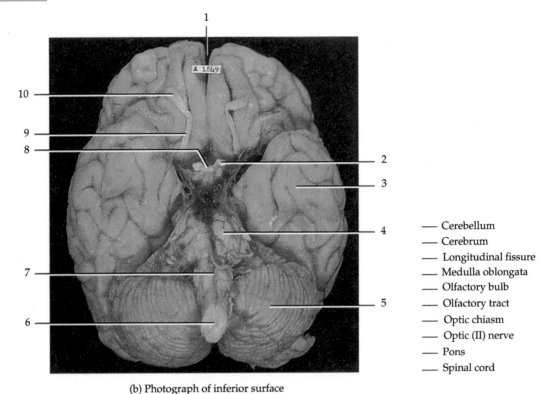

© Martin M. Rotker

(b) Photograph of inferior surface

— Cerebellum
— Cerebrum
— Longitudinal fissure
— Medulla oblongata
— Olfactory bulb
— Olfactory tract
— Optic chiasm
— Optic (II) nerve
— Pons
— Spinal cord

---

5. *Facial nerve (cranial nerve VII).* To test the motor responses of this nerve, ask your partner to bring the corners of the mouth straight back, smile while showing his or her teeth, whistle, puff his or her cheeks, frown, raise his or her eyebrows, and wrinkle his or her forehead.

Define Bell's palsy. _____

_____

To test the sensory responses of this nerve, touch the tip of your partner's tongue with an applicator that has been dipped into a salt solution. Rinse the mouth out with water and now place an applicator dipped into a sugar solution along the anterior surface of the tongue.

What are the five basic tastes that can be distinguished? _____

_____

6. *Vestibulocochlear* (vest-ib'-ū-lō-KŌK-lē-ar) *nerve (cranial nerve VIII).*

a. To test for the functioning of the vestibular portion of this nerve, have your partner sit on a swivel stool with his or her head slightly bent forward toward the chest. Now have a student (the operator) turn your partner *slowly and very carefully* to the right about 10 times (about one turn every 2 sec). Stop the stool suddenly and look for nystagmus (nis-TAG-mus), which is rapid movement or quivering of the eyeballs. Note the direction of the nystagmus.

⚠ **CAUTION!** *Do not allow the subject to stand or walk until dizziness is completely gone.*

b. When the nystagmus movements have ceased, have the subject sit on the chair and, with his or her eyes open, reach forward and touch the operator's index finger with his or her finger. Now spin the subject to the right about 10 times (as before) and then suddenly stop the stool. *Immediately* after stopping the stool have the subject again, with his or her eyes open, try to reach forward and touch the operator's index finger with his or her finger. Note the degree of ease or difficulty with which this is accomplished.

c. Again, when the nystagmus movements have ceased, repeat procedure b. This time, however, *immediately* after stopping the chair have the operator extend an index finger and have the subject open his or her eyes, locate the finger visually, and then *close his or her eyes and try to touch the operator's index finger with his or her eyes closed.* Note the degree of ease or difficulty with which this is accomplished, as well as the direction (right or left) in which the subject missed the operator's finger.

d. Now repeat the procedure in part a with the subject resting his or her head on their right shoulder. Note any changes that might occur with the nystagmus response.

_____

Formulate a hypothesis as to the differences in results obtained in procedures a through d.

_____

_____

_____

_____

To test the functioning of the cochlear portion of the nerve, have your subject sit on a chair with his or her eyes closed with six members of the class standing in a wide circle around the individual. Now, one at a time, and in a random order, have the students click two coins together and have the subject point in the direction from which the sound originated. Then, with the students still standing in a circle of equal radius around the chair, and with the subject's eyes still closed, hand a ticking watch to one of the individuals in the circle and have him or her slowly move the watch closer to the subject until the subject first hears the sound and can correctly indicate from which direction the sound is originating. Repeat this procedure until the distance at which the sound is first heard and the direction correctly identified have been determined for all six directions. Measure the distances between the subject and each encircling student. Are all six distances equal?

_____

7. *Glossopharyngeal* (glos-ō-fa-RIN-jē-al) *(cranial nerve IX)* and *vagus nerves (cranial nerve X)*. The palatal (gag) reflex can be used to test the functioning of both of these nerves. Using a cotton-tipped applicator, *very slowly and gently* tough your partner's uvula.

Develop a hypothesis concerning the purpose

of this reflex. _____

_____

_____

To test the sensory responses of these nerves, have your partner swallow. Does swallowing occur easily? Now *gently* hold your partner's tongue down with a tongue depressor and ask him or her to say "ah." Does the uvula move? Are the movements on both sides of the soft palate the same? The sensory function of the glossopharyngeal nerve may be tested by *lightly* applying a cotton-tipped applicator dipped in quinine to the top, sides, and back of the tongue.

In which area of the tongue was the quinine

tasted? _____

What taste sensation is located there? _____

_____

8. *Accessory nerve (cranial nerve XI).* The strength and muscle tone of the sternocleidomastoid and trapezius muscles indicate the proper functioning of the accessory nerve. To ascertain the strength of the sternocleidomastoid muscle, have your partner turn his or her head from side to side against *slight* resistance that you supply by placing your hands on either side of your partner's head. To ascertain the strength of the trapezius muscle, place your hands on your partner's shoulders and while *gently* pressing down firmly ask him or her to shrug his or her shoulders.

Do both muscles appear to be reasonably

strong? _____

9. *Hypoglossal nerve (cranial nerve XII).* Have your partner protrude his or her tongue. It should protrude without deviation. Now have your partner protrude his or her tongue and move it from side to side while you attempt to *gently* resist the movements with a tongue depressor.

Expose the roots of several spinal nerves and identify the parts just described.

E. Dissection of Sheep Brain

▲ **CAUTION!** *Please reread Section D, "Precautions Related to Dissection," at the beginning of the laboratory manual on page xiii before you begin your dissection.*

The brains of the fetal pig, sheep, and human show many similarities. They possess the protective membranes called the **meninges,** which can easily be seen as you proceed with the dissections. The outermost layer is the **dura mater.** It is the toughest, most protective one and may be missing in the preserved sheep brain. The middle membrane is the **arachnoid mater,** and the inner one containing blood vessels and adhering closely to the surface of the brain itself is the **pia mater.**

PROCEDURE

1. Use Figures 13.20 through 13.23 as references for this dissection.

2. The most prominent external parts of the brain are the pair of large **cerebral hemispheres** and the posterior **cerebellum** on the dorsal surface. These large hemispheres are separated from each other by the **longitudi-**

FIGURE 13.20 Sheep brain in which cerebellum has been spread apart from the cerebrum.

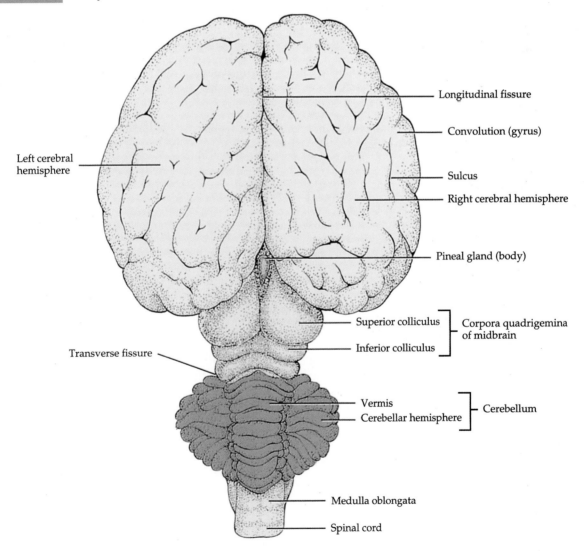

Longitudinal fissure

Convolution (gyrus)

Sulcus

Right cerebral hemisphere

Pineal gland (body)

Left cerebral hemisphere

Superior colliculus

Inferior colliculus

Corpora quadrigemina of midbrain

Transverse fissure

Vermis

Cerebellar hemisphere

Cerebellum

Medulla oblongata

Spinal cord

Dorsal view

FIGURE 13.21 Sheep brain.

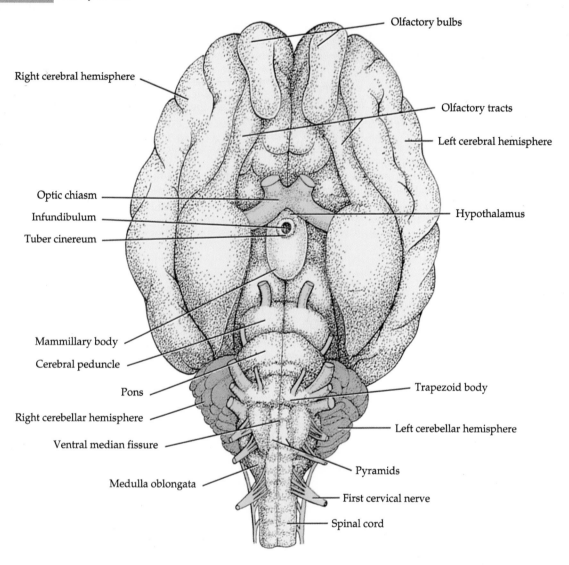

Ventral view

nal fissure. The **transverse fissure** separates the hemispheres from the cerebellum. The surfaces of these hemispheres form many **gyri,** or raised ridges, that are separated by grooves, or **sulci.**

3. If you spread the hemispheres gently apart, you can see, deep in the longitudinal fissure, thick bundles of white transverse fibers. These bundles form the **corpus callosum,** which connects the hemispheres.

4. Most of the following structures can be identified by examining a midsagittal section of the sheep brain, or by cutting an intact brain along the longitudinal fissure completely through the corpus callosum.

5. If you break through the thin ventral wall, the **septum pellucidum** of the corpus callosum, you can see part of a large chamber, the **lateral ventricle,** inside the hemisphere.

6. Each hemisphere has one of these ventricles. Ventral to the septum pellucidum, locate a smaller band of white fibers called the **fornix.** Close by where the fornix disappears is a small, round bundle of fibers called the **anterior commissure.**

7. The **third ventricle** and the **thalamus** are located ventral to the fornix. The third ventricle is outlined by its shiny epithelial lining, and the thalamus forms the lateral walls of this ventricle. This ventricle is crossed by a large

FIGURE 13.22 Sheep brain.

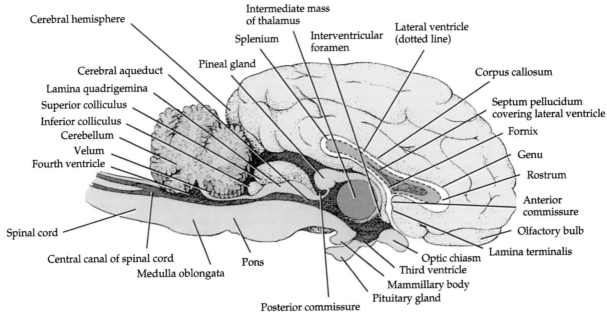

Midsagittal section

circular mass of tissue, the **intermediate mass,** which connects the two sides of the thalamus. Each lateral ventricle communicates with the third ventricle through an opening, the **interventricular foramen,** which lies in a depression anterior to the intermediate mass and can be located with a dull probe.

8. Spreading the cerebral hemispheres and the cerebellum apart reveals the roof of the midbrain (mesencephalon), which is seen as two pairs of round swellings collectively called the **corpora quadrigemina.** The larger, more anterior pair are the **superior colliculi.** The smaller posterior pair are the **inferior colliculi.** The **pineal gland (body)** is seen directly between the superior colliculi. Just posterior to the inferior colliculi, appearing as a thin white strand, is the **trochlear (IV) nerve.**

9. The cerebellum is connected to the brain stem by three prominent fiber tracts called **peduncles.** The **superior cerebellar peduncle** connects the cerebellum with the midbrain, the **inferior cerebellar peduncle** connects the cerebellum with the medulla, and the **cerebellar peduncle** connects the cerebellum with the pons.

10. Most of the following parts can be located on the ventral surface of the intact brain.

11. Just beneath the cerebral hemispheres are two **olfactory bulbs,** which continue posteriorly as two **olfactory tracts.** Posterior to these tracts, the **optic (II) nerves** undergo a crossing (decussation) known as the **optic chiasm.**

12. Locate the **pituitary gland (hypophysis)** just posterior to the chiasm. This gland is connected to the **hypothalamus** portion of the diencephalon by a stalk called the **infundibulum.** The **mammillary body** appears immediately posterior to the infundibulum.

13. Just posterior to this body are the paired **cerebral peduncles,** from which arise the large **oculomotor (III) nerves.** They may be partially covered by the pituitary gland.

14. The **pons** is a posterior extension of the hypothalamus, and the **medulla oblongata** is a posterior extension of the pons.

15. The **cerebral aqueduct** dorsal to the peduncles runs posteriorly and connects the third ventricle with the **fourth ventricle,** which is located dorsal to the medulla and ventral to the cerebellum.

16. The medulla merges with the **spinal cord** and is separated by the **ventral median fissure.** The **pyramids** are the longitudinal bands of tissue on either side of this fissure.

FIGURE 13.23   Cranial nerves of a sheep brain.

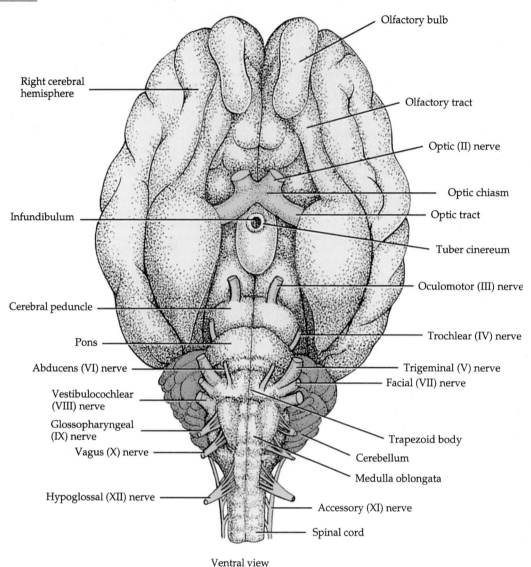

Ventral view

17. Identify the remaining **cranial nerves** on the ventral surface of the brain. They are the trigeminal (V), abducens (VI), facial (VII), vestibulocochlear (VIII), glossopharyngeal (IX), vagus (X), accessory (XI), and hypoglossal (XII). The previously identified cranial nerves are the olfactory (I), optic (II), oculomotor (III), and trochlear (IV), for a total of 12.

18. A transverse section through a cerebral hemisphere reveals **gray matter** near the surface of the **cerebral cortex** and **white matter** beneath this layer.

19. A transverse section through the spinal cord reveals the **central canal,** which is connected to the fourth ventricle and contains **cerebrospinal fluid.**

20. A midsagittal section through the cerebellum reveals a treelike arrangement of gray and white matter called the **arbor vitae** (tree of life).

F. Autonomic Nervous System

The **autonomic nervous system (ANS)** regulates the activities of smooth muscle, cardiac muscle, and certain glands, usually without conscious control. In the somatic nervous system (SNS), which is voluntary, the cell bodies of the motor neurons are in the CNS, and their axons extend all the way to skeletal muscles in spinal nerves. The ANS always has two motor neurons in the pathway. The first of the two motor neurons, the **preganglionic neuron,** has its cell body in the brain or spinal cord. Its axon leaves the CNS and synapses in an autonomic ganglion with the second neuron called the **postganglionic neuron.** The cell body of the postganglionic neuron lies entirely outside the CNS, located inside an autonomic ganglion, and its axon terminates in a **visceral effector** (muscle or gland).

Label the components of the autonomic pathway shown in Figure 13.24.

The output (motor) part of the ANS consists of two divisions: sympathetic and parasympathetic (Figure 13.25 on page 284). Most organs have **dual innervation:** they receive impulses from both sympathetic and parasympathetic neurons. In general, nerve impulses from one division stimulate the organ to increase its activity (excitation), whereas nerve impulses from the other division decrease its activity (inhibition).

In the **sympathetic division,** the cell bodies of the preganglionic neurons are located in the lateral gray horns of the gray matter in the 12 thoracic and first two (and sometimes three) lumbar segments of the spinal cord. The axons of preganglionic neurons are myelinated and leave the spinal cord through the anterior root of a spinal nerve. Each axon travels briefly in an anterior (ventral) ramus and then through a small branch called a **white ramus communicans** to enter a

sympathetic trunk ganglion. These ganglia lie in a vertical row, on either side of the vertebral column, from the base of the skull to the coccyx. In the ganglion, the axon may synapse with a postganglionic neuron, travel upward or downward through the sympathetic trunk ganglia to synapse with postganglionic neurons at different levels, or pass through the ganglion without synapsing to form part of the splanchnic nerves. If the preganglionic axon synapses in a sympathetic trunk ganglion, it reenters the anterior or posterior ramus of a spinal nerve via a small branch called a **gray ramus communicans.** If the preganglionic axon forms part of the splanchnic nerves, it passes through the sympathetic trunk ganglion but synapses with a postganglionic neuron in a **prevertebral (collateral) ganglion.** These ganglia are anterior to the vertebral column close to large abdominal arteries from which their names are derived (celiac, superior mesenteric, and inferior mesenteric).

In the **parasympathetic division,** the cell bodies of the preganglionic neurons are located in nuclei of four cranial nerves in the brain stem and lateral gray horn of the second through fourth sacral segments of the spinal cord. The axons emerge as part of cranial or spinal nerves. The preganglionic axons synapse with postganglionic neurons in **terminal ganglia,** close to or within the walls of visceral effectors.

FIGURE 13.24   Components of an autonomic pathway.

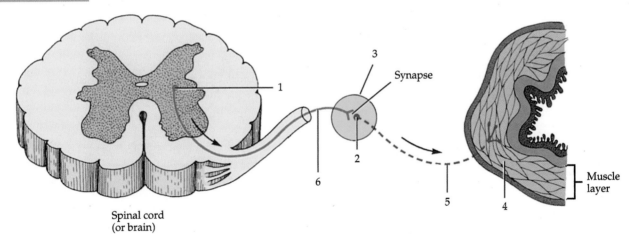

Spinal cord
(or brain)

___ Autonomic ganglion

___ Cell body of postganglionic neuron

___ Cell body of preganglionic neuron

___ Axon of postganglionic neuron

___ Axon of preganglionic neuron

___ Visceral effector

FIGURE 13.25 **Structure of the autonomic nervous system.**

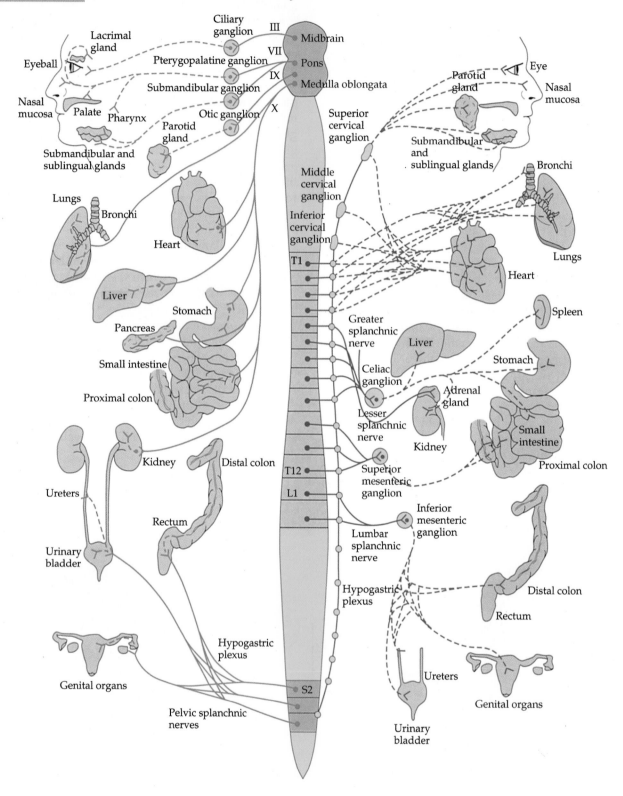

The sympathetic division is primarily concerned with processes that expend energy. During stress, the sympathetic division sets into operation a series of reactions collectively called the **fight-or-flight response,** designed to help the body counteract the stress and return to homeostasis. During the fight-or-flight response, the heart and breathing rates increase and the blood sugar level rises, among other things.

The parasympathetic division is primarily concerned with activities that restore and conserve energy. The parasympathetic division enhances **"rest-and-digest"** activities. Under normal conditions, the parasympathetic division dominates the sympathetic division in order to maintain homeostasis.

Autonomic fibers, like other axons of the nervous system, release neurotransmitters at synapses as well as at points of contact with visceral effectors **(neuroeffector junctions).** On the basis of the neurotransmitter produced, autonomic fibers may be classified as either cholinergic or adrenergic. **Cholinergic** (kō-lin-ER-jik) **neurons** release **acetylcholine (ACh)** and include the following: (1) all sympathetic and parasympathetic preganglionic neurons, (2) all parasympathetic postganglionic neurons, and (3) a few sympathetic postganglionic neurons that innervate most sweat glands. **Adrenergic** (ad'-ren-ER-jik) **neurons** produce **norepinephrine (NE)** or **noradrenalin.** Most sympathetic postganglionic axons are adrenergic.

The actual effects produced by ACh are determined by the type of receptor with which it interacts. The two types of cholinergic receptors are known as nicotinic receptors and muscarinic receptors. **Nicotinic receptors** are present in the plasma membrane of dendrites and cell bodies of both sympathetic and parasympathetic postganglionic neurons and in the motor endplate at the neuromuscular junction. These receptors are so named because nicotines mimic the actions of ACh by binding to these receptors. **Muscarinic receptors** are present in the plasma membranes of all effectors (smooth muscle, cardiac muscle, and glands) innervated by parasympathetic postganglionic axons. These receptors are so named because a mushroom poison called muscarine mimics the actions of ACh by binding to them. The effects of NE, like those of ACh, are also determined by the type of receptor with which they interact. Such receptors are found on visceral effectors innervated by most sympathetic postganglionic axons and are referred to as **alpha receptors** and **beta receptors.** In general, alpha receptors are excitatory. Some beta receptors are excitatory, while others are inhibitory. Although cells of most effectors contain either alpha or beta receptors, some effector cells contain both. NE, in general, stimulates alpha receptors to a greater extent than beta receptors, and epinephrine, in general, stimulates both alpha and beta receptors about equally.

Using your textbook as a reference, write in the effects of sympathetic and parasympathetic stimulation for the visceral effectors listed in Table 13.3.

ANSWER THE LABORATORY REPORT QUESTIONS AT THE END OF THE EXERCISE.

TABLE 13.3   Activities of the autonomic nervous system

Visceral effector	Effect of sympathetic stimulation	Effect of parasympathetic stimulation
Glands Adrenal medullae		
Lacrimal (tear)		
Pancreas		
Posterior pituitary		
Pineal		
Sweat		
Adipose tissue		
Liver		
Kidney, juxtaglomerular cells		
Cardiac (heart) muscle **Smooth muscle** Iris, radial muscle		
Iris, circular muscle		
Ciliary muscle of eye		
Lungs, bronchial muscles		
Gallbladder and ducts		
Stomach and intestines		

(Continued)

Visceral effector	Effect of sympathetic stimulation	Effect of parasympathetic stimulation
Spleen		
Ureters		
Urinary bladder		
Uterus		
Sex organs		
Hair follicles, arrector pili muscle		
Vascular smooth muscle Salivary gland arterioles		
Gastric gland arterioles		
Intestinal gland arterioles		
Coronary (heart) arterioles		
Skin and mucosal arterioles		
Skeletal muscle arterioles		
Abdominal viscera arterioles		
Brain arterioles		
Kidney arterioles		
Systemic veins		

EXERCISE 13

Nervous System

Name _____ Date _____

Laboratory Section _____ Score/Grade _____

PART 1 ■ **Multiple Choice**

_____ 1. The tapered, conical portion of the spinal cord is the (a) filum terminale (b) conus medullaris (c) cauda equine (d) lumbar enlargement

_____ 2. The superficial meninx composed of dense fibrous connective tissue is the (a) pia mater (b) arachnoid mater (c) dura mater (d) denticulate

_____ 3. The portion of a spinal nerve that contains axons of motor neurons only is the (a) posterior root (b) posterior root ganglion (c) lateral root (d) anterior root

_____ 4. The connective tissue covering around individual axons is the (a) endoneurium (b) epineurium (c) perineurium (d) ectoneurium

_____ 5. On the basis of organization, which does *not* belong with the others? (a) pons (b) medulla oblongata (c) thalamus (d) midbrain

_____ 6. The lateral ventricles are connected to the third ventricle by the (a) interventricular foramen (b) cerebral aqueduct (c) median aperture (d) lateral aperture

_____ 7. The vital centers for heartbeat, respiration, and blood vessel diameter regulation are found in the (a) pons (b) cerebrum (c) cerebellum (d) medulla oblongata

_____ 8. The reflex centers for movements of the head and trunk in response to auditory stimuli are located in the (a) inferior colliculi (b) medial geniculate nucleus (c) superior colliculi (d) ventral posterior nucleus

_____ 9. Which thalamic nucleus relays sensory impulses for vision? (a) medial geniculate (b) ventral posterior (c) ventral lateral (d) ventral anterior

_____ 10. Integration of the autonomic nervous system, control of body temperature, and the regulation of eating and drinking are functions of the (a) pons (b) thalamus (c) cerebrum (d) hypothalamus

_____ 11. The left and right cerebral hemispheres are separated from each other by the (a) central sulcus (b) transverse fissure (c) longitudinal fissure (d) insula

_____ 12. Which structure does *not* belong with the others? (a) putamen (b) caudate nucleus (c) insula (d) globus pallidus

_____ 13. Which peduncles connect the cerebellum with the midbrain? (a) superior (b) inferior (c) middle (d) lateral

_____ 14. Which cranial nerve has the most anterior origin? (a) XI (b) IX (c) VII (d) IV

—————— **15.** Extensions of the pia mater that protect the spinal cord against shock and sudden displacement are the (a) choroid plexuses (b) pyramids (c) denticulate ligaments (d) superior colliculi

—————— **16.** Which branch of a spinal nerve enters into formation of plexuses? (a) meningeal (b) posterior (c) rami communicantes (d) anterior

—————— **17.** Which plexus innervates the upper limbs and superior part of the shoulders? (a) sacral (b) brachial (c) lumbar (d) cervical

—————— **18.** How many pairs of thoracic spinal nerves are there? (a) 1 (b) 5 (c) 7 (d) 12

PART 2 ▪ Completion

19. The narrow, shallow furrow on the posterior surface of the spinal cord is the ——————————————.

20. The space between the dura mater and wall of the vertebral canal is called the ——————————————.

21. In a spinal nerve, the cell bodies of sensory neurons are found in the ——————————————.

22. The superficial connective tissue covering over the entire spinal nerve is the ——————————————.

23. The middle meninx is referred to as the ——————————————.

24. The nuclei of origin for cranial nerves IX, X, XI, and XII are found in the ——————————————.

25. The portion of the brain containing the cerebral peduncles is the ——————————————.

26. Cranial nerves V, VI, VII, and VIII have their nuclei of origin in the ——————————————.

27. A shallow groove of the cerebral cortex is called a(n) ——————————————.

28. The —————————————— separates the frontal lobe of the cerebrum from the parietal lobe.

29. White matter tracts of the cerebellum are called ——————————————.

30. The space between the dura mater and the arachnoid is referred to as the ——————————————.

31. Together, the thalamus and hypothalamus constitute the ——————————————.

32. Cerebrospinal fluid passes from the third ventricle into the fourth ventricle through the ——————————————.

33. The cerebrum is separated from the cerebellum by the —————————————— fissure.

34. The branches of a spinal nerve that are components of the autonomic nervous system are known as ——————————————.

35. The plexus that innervates the buttocks, perineum, and lower limbs is the —————————————— plexus.

36. There are _____ pairs of spinal nerves.

37. The part of the brain that coordinates subconscious movements in skeletal muscles is the

    _____.

38. The cell bodies of _____ neurons of the ANS are found inside autonomic

    ganglia.

39. The portion of the ANS concerned with the fight-or-flight response is the

    _____.

40. The autonomic ganglia that are anterior to the vertebral column and close to large abdominal arter-

    ies are called _____ ganglia.

41. The _____ nerve innervates the diaphragm.

42. The _____ tract conveys nerve impulses for crude touch and pressure.

43. The flexor muscles of the thigh and extensor muscles of the leg are innervated by the

    _____ nerve.

44. The _____ tract conveys nerve impulses related to muscle tone and posture.

45. The _____ nerve supplies the extensor muscles of the arm and forearm.

46. The _____ area of the cerebral cortex receives sensations from cutaneous,

    muscular, and visceral receptors in various parts of the body.

47. The portion of the cerebral cortex that translates thoughts into speech is the

    _____ area.

48. The term _____ refers to bundles of myelinated axons from many neurons.

49. The pneumotaxic and apneustic areas are two important nuclei located in the

    _____.

50. _____ -order neurons carry sensory information from the spinal cord and

    brain stem to the thalamus.

51. The part of the brain that allows crude perception of some sensations, such as pain, temperature,

    and pressure, is the _____.

52. Autonomic fibers are classified as either cholinergic or _____.

PART 3 ▪ **Matching**

Using B for brachial, C for cervical, L for lumbar, and S for sacral, indicate to which plexus the following nerves belong.

_____ 53. Sciatic

_____ 54. Femoral

_____ 55. Radial

_____ 56. Ansa cervicalis

_____ 57. Median

_____ 58. Obturator

_____ 59. Pudendal

_____ 60. Tibial

_____ 61. Ilioinguinal

_____ 62. Ulnar

_____ 63. Phrenic

PART 4 ▪ **Matching**

_____ 64. Posterior column–medial meniscus pathway

_____ 65. Spinocerebellar tracts

_____ 66. Lateral corticospinal tracts

_____ 67. Lateral spinothalamic tract

_____ 68. Anterior corticospinal tracts

_____ 69. Tectospinal tract

_____ 70. Lateral reticulospinal tract

A. Conduct subconscious proprioceptive input to the cerebellum

B. Convey impulses that control movements of the neck and part of the trunk

C. Conducts impulses for conscious proprioception and most tactile sensations

D. Conveys impulses that facilitate flexor reflexes, inhibit extensor reflexes, and decrease muscle tone in muscles of the axial skeleton and proximal limbs

E. Control skilled movements of the distal limbs

F. Conduct mainly pain and temperature impulses

G. Conveys impulses that move the head and eyes in response to visual stimuli

Sensory Receptors and Sensory and Motor Pathways

Objectives

At the completion of this exercise you should understand

A The definition and characteristics of a sensation

B The classification of receptors

C The anatomy and physiology of somatic receptors

D How to conduct tests for somatic senses

E The anatomy and physiology of the special senses

F The principal somatic sensory and motor pathways and their functions

---

Now that you have studied the nervous system, you will study sensations. In its broadest definition, **sensation** is the conscious or subconscious awareness of changes in the external or internal environment. If a stimulus is strong enough, one or more nerve impulses arise in sensory nerve fibers. After the nerve impulses propagate to a region of the spinal cord or brain, they are translated into a sensation. The nature of the sensation and the type of reaction generated vary according to the destination of sensory impulses in the CNS. Each unique type of sensation—for example, touch or hearing—is called a **sensory modality.**

A. Characteristics of Sensations

Conscious sensations may be classified into two types. **General senses** include **somatic senses,** such as touch, pressure, vibration, temperature, pain, itch, tickle, and proprioception, and **visceral senses. Special senses** include the modalities of smell, taste, vision, hearing, and equilibrium (balance).

For a sensation to arise (either consciously or subconsciously), four events typically occur:

1. *Stimulation of the sensory receptor.* An appropriate **stimulus** must occur within a sensory receptor's *receptive field,* that is, the body region where stimulation produces a response.

2. *Transduction of the stimulus.* A **sensory receptor** or **sense organ transduces** (converts) energy in a stimulus into a graded potential. A sensory receptor or sense organ is a specialized peripheral structure or a specialized type of neuron that is sensitive to certain types of stimuli. A **graded potential** is the electrical response of a sensory receptor to a stimulus. It has a variable amplitude (size) that varies with the strength of the stimulus and is not propagated like a nerve impulse (action potential). When a generator potential is large enough to reach threshold, it generates one or more nerve impulses in a sensory neuron.

3. *Generation of nerve impulses and conduction.* The graded potential triggers nerve impulses that are propagated along a sensory neural pathway to the CNS. The sensory neuron that

conveys impulses from the PNS to the CNS is termed a **first-order neuron.** Additional neurons (termed **second-order neurons** and **third-order neurons**) carry the action potential to a particular region within the brain. The number of neurons (two or three) involved in the transmission of information to the brain varies depending upon the sensory tract utilized to transmit the information to the brain.

4. *Integration of sensory input.* A particular region of the CNS must receive and integrate the sensory nerve impulses. Most conscious sensations or perceptions are integrated in the cerebral cortex of the brain after passing through the thalamus. In other words, you see, hear, and feel in the brain. You seem to see with your eyes, hear with your ears, and feel pain in an injured part of your body because sensory impulses from each part of the body arrive in a specific region in the cerebral cortex, which interprets the sensation as coming from the stimulated sensory receptors.

A **sensory unit** consists of a single peripheral neuron and its terminal ending. The neuron's cell body is typically located in either the posterior root ganglion or within a cranial nerve ganglion. The peripheral processes of this neuron may terminate as free nerve endings, encapsulated nerve endings, or synapse with specialized, separate cells. Sensory receptors may be either very simple or quite complex, containing highly specialized neurons, epithelium, and connective tissue components. All sensory receptors contain the terminal processes of a neuron, exhibit a high degree of excitability, and possess a *selectivity* for each different type of sensation, which is termed a **sensory modality.** In addition, a sensory unit possesses a peripheral **receptive field,** which is the body region where stimulation elicits a response. The majority of sensory impulses are conducted to the sensory areas of the cerebral cortex, for this is the region of the brain that initiates the processing of sensory information that ultimately results in the conscious feeling of the sensory stimulus. Different sets of sensory nerve fibers, when activated, will elicit different sensations by virtue of their unique CNS connections. Therefore, a particular sensory neuron will provoke an identical sensation *regardless of how it is excited.* This interpretation of sensory perception is termed **Muller's Doctrine of Specific Energies.**

One characteristic of most sensory receptors, that of **projection,** describes the process by which the brain refers sensations to their point of **learned origin** of the stimulation. This process accounts for a phenomenon termed **phantom pain** that is quite common in amputees. When phantom pain is experienced, the individual feels pain (or one of many other sensations, such as itching or tickling) in the part of the body that was amputated because of the irritation of a nerve ending in the healing wound surface of the amputation. The action potential is carried to the brain and projected back to the portion of the limb that is no longer intact.

A second characteristic of most sensory receptors is **adaptation,** that is, a change in sensitivity, usually a decrease, even though a stimulus is still being applied. For example, when you first get into a tub of hot water, you might feel an intense burning sensation. However, after a brief period of time the sensation decreases to one of comfortable warmth, even though the stimulus (hot water) is still present and has not diminished in intensity.

A third characteristic is that of **afterimages,** that is, the persistence of a sensation after the stimulus has been removed. One common example of an afterimage occurs when you look at a bright light and then look away. You will still see the light for several seconds afterward.

A fourth characteristic is that of **sensory modality.** Modality is the possession of distinct properties by which one sensation may be distinguished from another. For example, pain, pressure, touch, body position, equilibrium, hearing, vision, smell, and taste are all distinctive because the brain perceives each differently.

B. Classification of Receptors

Receptors vary in their complexity. The **somatic (general) senses** include touch, pressure, vibration, itch, tickle, warmth, cold, and pain plus proprioception (perception of both static [nonmoving] positions of limbs and body parts and movements of the limbs and head). **Visceral senses** provide information about conditions within internal organs. Somatic receptors are relatively simple and utilize relatively simple neural pathways.

Somatic sensations arise in receptors located in the skin or subcutaneous layer; in mucous membranes of the mouth, vagina, and anus; in muscles, tendons, joints, and the inner ear (proprioceptive sensations). **Cutaneous** (kū-TĀ-nē-us; *cutane* = skin) **sensations** include:

a. **Tactile** (TAK-tīl; *tact* = touch) **sensations** (touch, pressure, vibration, itch, and tickle).

b. **Thermal sensations** (cold and warmth).

c. **Pain sensations.**

Proprioceptive (prō-prē-ō-SEP-tiv) **sensations** provide us with an awareness of the activities of muscles, tendons, and joints and equilibrium.

The receptors for the **special senses**—smell, taste, vision, hearing, and equilibrium—are located in sense organs such as the eye and ear. Special sense receptors are relatively complex and utilize relatively complex neural pathways.

Special sensations arise in receptors located in organs in different parts of the body. The special sensations are as follows:

a. **Olfactory** (ōl-FAK-tō-rē) **sensations** (smell).

b. **Gustatory** (GUS-ta-tō-rē) **sensations** (taste).

c. **Visual sensations** (sight).

d. **Auditory sensations** (hearing).

e. **Equilibrium** (ē′-kwi-LIB-rē-um) **sensations** (orientation of the body).

Several structural and functional characteristics of sensory receptors can be used to group them into different classes.

On a microscopic level, sensory receptors may be classified as:

1. **Free nerve endings,** bare dendrites lacking any structural specializations that can be seen under a light microscope. They serve as receptors for pain, thermal sensations, tickle, itch, and some touch sensations.

2. **Encapsulated nerve endings,** whose dendrites are enclosed in a connective tissue capsule with a distinctive microscopic structure—for example, lamellated (pacinian) corpuscles. They provide information such as touch, pressure, and vibration.

3. **Separate cells** that synapse with sensory (first order) neurons. They include *hair cells* for hearing and equilibrium in the inner ear, *gustatory receptor cells* in the taste buds, and *photoreceptor cells* in the retina of the eye.

The classification according to the location of the receptors and the origin of the stimuli that activate them is as follows:

1. **Exteroceptors** (EKS′-ter-ō-sep′-tors), located at or near the external surface of the body, provide information about the *external* environment. They transmit sensations such as hearing, vision, smell, taste, touch, pressure, vibration, temperature, and pain.

2. **Interoceptors** or **visceroceptors** (VIS-er-ō-sep′-tors), located in blood vessels, visceral organs, muscles, and the nervous system, provide information about the *internal* environment. Sensations from these receptors usually do not reach conscious perception but may be felt as pain or pressure.

3. **Proprioceptors** (PRŌ-prē-ō-sep′-tors; *proprio* = one's own), located in muscles, tendons, joints, and the inner ear. They provide information about body position, muscle length and tension, and the position and movement of our joints.

The classification of sensations based on the type of stimuli they detect is as follows:

1. **Mechanoreceptors** detect mechanical stimuli such as deformation, stretching, or bending of cells. They provide sensations of touch, pressure, vibration, proprioception (awareness of the location and movement of body parts), hearing, and equilibrium.

2. **Thermoreceptors** detect changes in temperature.

3. **Nociceptors** (NŌ-sē-sep′-tors) respond to stimuli resulting from physical or chemical damage to tissues, giving rise to the sensation of pain.

4. **Photoreceptors** detect light that strikes the retina of the eye.

5. **Chemoreceptors** detect chemicals in the mouth (taste), nose (smell), and body fluids such as water, oxygen, carbon dioxide, hydrogen ions, certain other electrolytes, hormones, and glucose.

6. **Osmoreceptors** detect the osmotic pressure of body fluids.

C. Receptors for Somatic Senses

1. Tactile Receptors

Although touch, pressure, vibration, itch, and tickle are classified as separate sensations, all are detected by mechanoreceptors.

Tactile sensations generally result from stimulation of tactile receptors in the skin or in tissues immediately deep to the skin. *Crude touch* refers to the ability to perceive that something has touched the skin, even though the exact location, shape, size, or texture cannot be determined. *Fine touch* provides specific information about a touch sensation, such as exactly what point on the body is touched plus the shape, size, and texture of the source of stimulation.

Touch receptors include corpuscles of touch, hair root plexuses, and type I and type II cutaneous mechanoreceptors. **Corpuscles of touch,** or **Meissner's** (MĪS-ners) **corpuscles,** are receptors for fine touch that are located in dermal papillae. They

Corpuscles of touch (Meissner's corpuscles)

Hair root plexuses

Type I cutaneous mechanoreceptors (Merkel discs)

Type II cutaneous mechanoreceptors (Ruffini corpuscles)

have already been discussed in Exercise 5. **Hair root plexuses** are free nerve endings wrapped around hair follicles that detect movements on the skin surface that disturb hairs. **Type I cutaneous mechanoreceptors,** also called **Merkel** (MER-kel) **discs,** are saucer-shaped, flattened nerve endings that make contact with epidermal cells of the stratum basale called Merkel cells. They are distributed in many of the same locations as corpuscles of touch and also function in fine touch. **Type II cutaneous mechanoreceptors,** or **Ruffini corpuscles,**

Lamellated (pacinian) corpuscles

are elongated, encapsulated receptors located deep in the dermis and in ligaments and tendons. They are most sensitive to stretching that occurs as digits or limbs are moved.

Receptors for touch, like other cutaneous receptors, are not randomly distributed; some areas contain many receptors, others contain few. Such a clustering of receptors is called **punctuate distribution.** Touch receptors are most numerous in the fingertips, palms, and soles. They are also abundant in the eyelids, tip of the tongue, lips, nipples, clitoris, and tip of the penis.

Examine prepared slides of corpuscles of touch (Meissner's corpuscles), hair root plexuses, type I cutaneous mechanoreceptors (Merkel discs), and type II cutaneous mechanoreceptors (Ruffini corpuscles). With the aid of your textbook, draw each of the receptors in the spaces provided in the box to the left.

Pressure sensations generally result from stimulation of tactile receptors in deeper tissues. **Pressure** is a sustained sensation that is felt over a larger area than touch. It occurs with deformation of deeper tissues. Receptors for pressure sensations include lamellated corpuscles, type I cutaneous mechanoreceptors, and corpuscles of touch. **Lamellated (pacinian) corpuscles** are found in the subcutaneous layer and have already been discussed in Exercise 5.

Pressure receptors are found in the dermis and subcutaneous tissue under the skin, in the deep subcutaneous tissues that lie under mucous and serous membranes, around joints, tendons, and muscles in the mammary glands, in the external genitals of both sexes, and in some viscera, such as the urinary bladder and pancreas.

Examine a prepared slide of lamellated (pacinian) corpuscles. With the aid of your textbook, draw the receptors in the box to the left.

Vibration sensations result from rapidly repetitive sensory signals from tactile receptors. Receptors for vibration include corpuscles of touch (Meissner's corpuscles) that detect low-frequency vibration and lamellated (pacinian) corpuscles that detect high-frequency vibration.

The **itch sensation** results from stimulation of free nerve endings by certain chemicals, such as bradykinin, often as a result of a local inflammatory response. Free nerve endings and lamellated corpuscles also are thought to mediate the **tickle sensation.** This unusual sensation is the only one that you may not be able to elicit on yourself. The explanation seems to lie in the impulses that conduct to and from the cerebellum when you are moving your fingers and touching yourself that don't occur when someone else is tickling you.

2. Thermoreceptors

The cutaneous receptors for the sensation of warmth and coolness are free nerve endings that are located in the stratum basale of the epidermis and the dermis. They are also located in the cornea of the eye, tip of the tongue, and external genitals. Separate thermoreceptors respond to warm and cold stimuli.

3. Pain Receptors

Receptors for **pain,** called **nociceptors** (NŌ-sē-sep'-tors; *noci* = harmful), are free nerve endings (see Figure 14.1 on page 300). Pain receptors are found in every tissue of the body except the brain and adapt only slightly or not at all. Intense thermal, mechanical, or chemical stimuli can activate nociceptors. For example, when stimuli for other sensations such as touch, pressure, heat, and cold reach a certain threshold, they stimulate pain receptors as well. Pain receptors, because of their sensitivity to all stimuli, have a general protective function of informing us of changes that could be potentially dangerous to health or life. Adaptation to pain does not readily occur. This low level of adaptation is important, because pain indicates disorder or disease. If we became used to it and ignored it, irreparable damage could result.

4. Proprioceptive Receptors

An awareness of the activities of muscles, tendons, and joints and inner ear is provided by the **proprioceptive (kinesthetic) sense.** It informs us of the degree to which tendons are tensed and muscles are contracted. The proprioceptive sense enables us to recognize the location and rate of movement of one part of the body in relation to other parts. It also allows us to estimate weight and to determine the muscular effort necessary to perform a task. With the proprioceptive sense, we can judge the position and movements of our limbs without using our eyes when we walk, type, play a musical instrument, or dress in the dark.

    Proprioceptors, the receptors for proprioception, are located in skeletal muscles and tendons, in and around joints, and in the internal ear. Proprioceptors adapt slowly and only slightly. This slight adaptation is beneficial because the brain continually receives nerve impulses related to the position of different body parts and makes adjustments to ensure coordination.

    Receptors for proprioception are as follows. The **joint kinesthetic** (kin'-es-THET-ik) **receptors** are located within and around the articular cap-

Joint kinesthetic receptors

Muscle spindles

Tendon organs

sules of synovial joints. These receptors provide feedback information on the degree and rate of angulation (change of position) of a joint. **Muscle spindles** consist of several slowly adapting sensory endings that wrap around 3 to 10 specialized muscle fibers. They are located in nearly all skeletal muscles and are more numerous in the muscles of the limbs. Muscle spindles provide feedback information on the degree of muscle stretch. This information is relayed to the CNS to assist in the coordination and efficiency of muscle contraction. **Tendon organs** are located at the junction of a muscle and tendon. They function by sensing the tension applied to a tendon and the force of contraction of associated muscles. The information is translated by the CNS.

    Proprioceptors in the internal ear are the maculae and cristae that function in equilibrium. These are discussed at the end of the exercise.

    Examine prepared slides of joint kinesthetic receptors, muscle spindles, and tendon organs. With the aid of your textbook, draw the receptors in the preceding spaces.

D. Tests for Somatic Senses

Areas of the body that have few cutaneous receptors are relatively insensitive, whereas those regions that contain large numbers of cutaneous receptors are quite sensitive. This difference can be demonstrated by the **two-point discrimination**

test for touch. In the following tests, students can work in pairs, with one acting as subject and the other as experimenter. The subject will keep his or her eyes closed during the experiments.

1. Two-Point Discrimination Test

In this test, the two points of a measuring compass are applied to the skin and the distance in millimeters (mm) between the two points is varied. The subject indicates when he or she feels two points and when he or she feels only one.

PROCEDURE

▲ **CAUTION!** *Wash the compass in a fresh bleach solution before using it on another subject to prevent transfer of saliva.*

1. *Very gently,* place the compass on the tip of the tongue, an area where receptors are very densely packed.

2. Narrow the distance between the two points to 1.4 mm. At this distance, the points are able to stimulate two different receptors, and the subject feels that he or she is being touched by two objects.

3. Decrease the distance to less than 1.4 mm. The subject feels only one point, even though both points are touching the tongue, because the points are so close together that they reach only one receptor.

4. Now *gently* place the compass on the back of the neck, where receptors are relatively few and far between. Here the subject feels two distinctly different points only if the distance between them is 36.2 mm or more.

5. The two-point discrimination test shows that the more sensitive the area, the closer the compass points can be placed and still be felt separately.

6. The following order, from greatest to least sensitivity, has been established from the test: tip of tongue, tip of finger, side of nose, back of hand, and back of neck.

7. Test the tip of finger, side of nose, and back of hand and record your results in Section D.1 of the LABORATORY REPORT RESULTS at the end of the exercise.

2. Identifying Touch Receptors

PROCEDURE

1. Using a water-soluble colored felt marking pen, draw a 1-in. square on the back of the forearm and divide the square into 16 smaller squares.

2. With the subject's eyes closed, press a Von Frey hair or bristle against the skin, just enough to cause the hair to bend, once in each of the 16 squares. The pressure should be applied in the same manner each time.

3. The subject should indicate when he or she experiences the sensation of touch, and the experimenter should make dots in square 1 at the places corresponding to the points at which the subject feels the sensations.

4. The subject and the experimenter should switch roles and repeat the test.

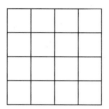

The pair of students working as a team should examine their 1-in. squares after the test is done. They should compare the number of positive and negative responses in each of the 16 small squares, and see how uniformly the touch receptors are distributed throughout the entire 1-in. square. Other general areas used for locating touch receptors are the arm and the back of the hand.

3. Identifying Pressure Receptors

PROCEDURE

1. The experimenter touches the skin of the subject (whose eyes are closed) with the point of a piece of colored chalk.

2. With eyes still closed, the subject then tries to touch the same spot with a piece of differently colored chalk. The distance between the two points is then measured.

3. Proceed using various parts of the body, such as the palm, arm, forearm, and back of neck.

4. Record your results in Section D.3 of the LABORATORY REPORT RESULTS at the end of the exercise.

4. Identifying Thermoreceptors

PROCEDURE

1. Draw a 1-in. square on the back of the wrist.

2. Place a forceps or other metal probe in ice-cold water for a minute, dry it quickly, and, with the *dull* point, explore the area in the square for the presence of cold spots.

3. Keep the probe cold and, using ink, mark the position of each spot that you find.

4. Mark each corresponding place in square 2 with the letter "c."

5. Immerse the forceps in hot water so that it will give a sensation of warmth when removed and applied to the skin, but *avoid having it so hot that it causes pain.*

6. Proceeding as before, locate the position of the warm spots in the same area of the skin.

7. Mark these spots with ink of a different color, and then mark each corresponding place in square 2 with the letter "h."

8. Repeat the entire procedure, using both cold forceps and warm forceps on the back of the hand and the palm, respectively, and mark squares 3 and 4 as you did square 2.

9. In order to demonstrate adaptation of thermoreceptors, fill three 1-liter beakers with 700 mL of (1) ice water, (2) water at room temperature, and (3) water at 45°C. Immerse your left hand in the ice water and your right hand in the water at 45°C for 1 min. Now move your left hand to the beaker with water at room temperature and record the sensation.

_____

Move your right hand to the beaker with the water at room temperature and record the

sensation. _____

How do you explain the experienced difference in the temperature of the water in the beaker with the water at room temperature?

_____

_____

5. Identifying Pain Receptors

PROCEDURE

1. Using the same 1-in. square of the forearm that was previously used for the touch test in Exercise D.2, perform the following experiment.

2. Apply a piece of absorbent cotton soaked with water to the area of the forearm for 5 min to soften the skin.

3. Add water to the cotton as needed.

4. Place the blunt end of a probe to the surface of the skin and press enough to produce a sensation of pain. Explore the marked area systematically.

5. Using dots, mark the places in square 5 that correspond to the points that give pain sensation when stimulated.

6. Distinguish between sensations of pain and touch. Are the areas for touch and pain identical?

_____

7. At the end of the test, compare your squares as you did in Exercise D.2.

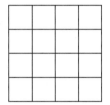

Perform the following test to demonstrate the phenomenon of **referred pain.** Referred pain is felt in or just deep to the skin that overlies a stimulated organ or in a surface area far from the stimulated organ. Place your elbow in a large shallow pan of ice water, and note the progression of sensation that you experience. At first, you will feel some discomfort in the region of the elbow. Later, pain sensations will be felt elsewhere.

Where do you feel the referred pain? _____

_____

Label the areas of referred pain indicated in Figure 14.1.

6. Identifying Proprioceptors

PROCEDURE

1. Face a blackboard close enough so that you can easily reach to mark it. Mark a small X on the board in front of you and keep the chalk on the X for a moment. Now close your eyes, raise your right hand above your head, and then, with your eyes still closed, mark a dot as near as possible to the X. Repeat the procedure by placing your chalk on the X, closing your eyes, raising your arm above your head, and then marking another dot as close as possible to the

FIGURE 14.1 **Referred pain.**

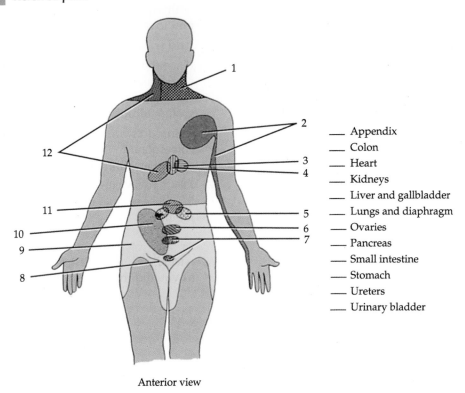

___	Appendix
___	Colon
___	Heart
___	Kidneys
___	Liver and gallbladder
___	Lungs and diaphragm
___	Ovaries
___	Pancreas
___	Small intestine
___	Stomach
___	Ureters
___	Urinary bladder

Anterior view

X. Repeat the procedure a third time. Record your results by estimating or measuring how far you missed the X for each trial in Section D.6 of the LABORATORY REPORT RESULTS at the end of the exercise.

2. Write the word "physiology" on the left line that follows. Now, with your *eyes closed*, write the same word immediately to the right. How do the samples of writing compare?

_____

Explain your results. _____

_____

_____

3. The following experiments demonstrate that kinesthetic sensations facilitate repetition of certain acts involving muscular coordination.

Students should work in pairs for these experiments.

a. The experimenter asks the subject to carry out certain movements with his or her eyes closed, for example, point to the middle finger of the subject's left hand with the index finger of the subject's right hand.

b. With his or her eyes closed, the subject extends the right arm as far as possible behind the body, and then brings the index finger quickly to the tip of his or her nose.

How accurate is the subject in doing this?

_____

c. Ask the subject, with eyes shut, to touch the named fingers of one hand with the index finger of the other hand.

How well does the subject carry out the

directions? _____

_____

_____

E. Somatic Sensory Pathways

Most input from somatic receptors on one side of the body crosses over to the opposite side in the spinal cord or brain stem before ascending to the thalamus. It then projects from the thalamus to the somatosensory area of the cerebral cortex,

FIGURE 14.2 The posterior column–medial lemniscus pathway.

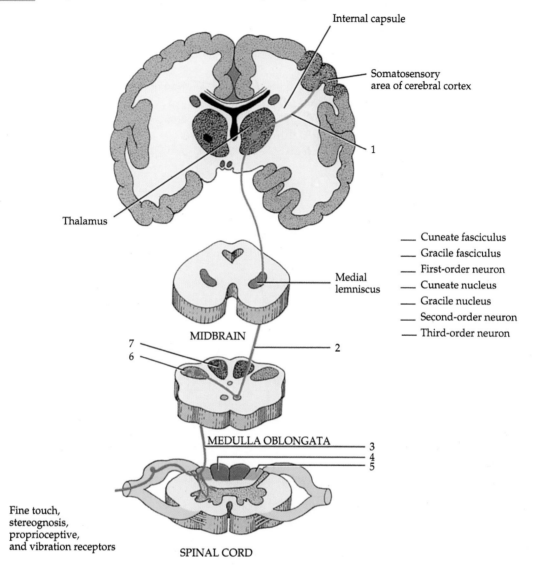

Internal capsule

Somatosensory
area of cerebral cortex

1

Thalamus

___ Cuneate fasciculus
___ Gracile fasciculus
___ First-order neuron
___ Cuneate nucleus
___ Gracile nucleus
___ Second-order neuron
___ Third-order neuron

Medial
lemniscus

MIDBRAIN

7
6

2

MEDULLA OBLONGATA

3
4
5

Fine touch,
stereognosis,
proprioceptive,
and vibration receptors

SPINAL CORD

where conscious sensations result. Axon collaterals (branches) of somatic sensory neurons also carry signals into the cerebellum and the reticular formation of the brain stem. Two general pathways lead from sensory receptors to the cortex: the posterior column–medial lemniscus pathway and the anterolateral (spinothalamic) pathway.

1. Posterior Column–Medial Lemniscus Pathway

Nerve impulses for conscious proprioception and most tactile sensations ascend to the cerebral cor-

tex along a common pathway formed by three-neuron sets.

Based on the description provided on page 260, label the components of the posterior column–medial lemniscus pathway in Figure 14.2.

2. Anterolateral (Spinothalamic) Pathways

The **anterolateral (spinothalamic) pathways** relay impulses for pain, thermal, tickle, and itch sensations. In addition, they carry some tactile impulses, which give rise to a very crude, poorly localized touch or pressure sensation. Like the

FIGURE 14.3 The lateral spinothalamic pathway.

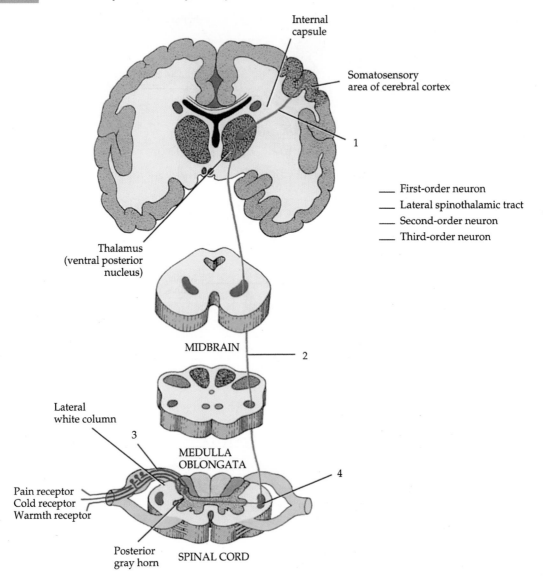

Internal capsule

Somatosensory area of cerebral cortex

1

____ First-order neuron
____ Lateral spinothalamic tract
____ Second-order neuron
____ Third-order neuron

Thalamus (ventral posterior nucleus)

MIDBRAIN

2

Lateral white column

3

MEDULLA OBLONGATA

4

Pain receptor
Cold receptor
Warmth receptor

Posterior gray horn

SPINAL CORD

posterior column–medial lemniscus pathway, the anterolateral pathways are also composed of three-neuron sets. The axon of the second-order neuron extends to the opposite side of the spinal cord and passes superiorly to the brain stem in either the **lateral spinothalamic tract** or the **anterior spinothalamic tract.** The lateral spinothalamic tract conveys sensory impulses for pain and temperature, whereas the anterior spinothalamic tract conveys tickle, itch, crude touch, pressure, and vibration.

Based on the descriptions provided, label the components of the lateral spinothalamic tract in Figure 14.3 and the anterior spinothalamic tract in Figure 14.4 on page 303.

We will consider **special senses** that involve complex receptors and neural pathways. These include **olfaction** (ol-FAK-shun; = smell), **gustation** (gus-TĀ-shun; = taste), **vision, hearing,** and **equilibrium** (ē'-kwi-LIB-rē-um).

F. Olfaction

1. Olfactory Receptors

The receptors for the sense of smell or **olfaction** (*olfact* = smell) are found in the olfactory epithelium in the superior portion of the nasal cavity, covering the inferior surface of the cribriform

FIGURE 14.4 The anterior spinothalamic pathway.

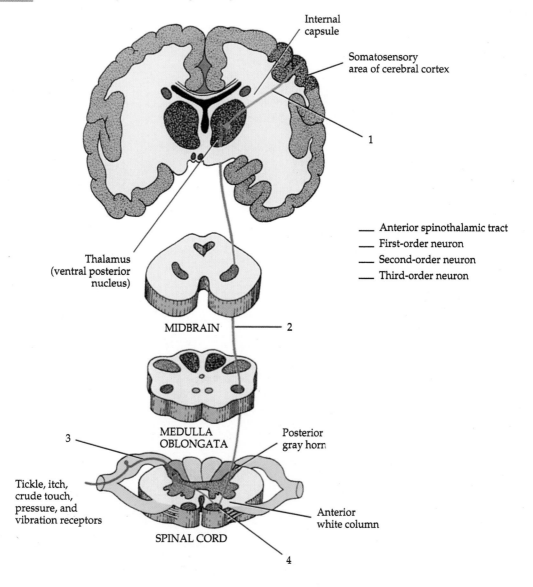

Internal capsule

Somatosensory area of cerebral cortex

1

Thalamus (ventral posterior nucleus)

___ Anterior spinothalamic tract
___ First-order neuron
___ Second-order neuron
___ Third-order neuron

MIDBRAIN — 2

MEDULLA OBLONGATA

Posterior gray horn

3

Tickle, itch, crude touch, pressure, and vibration receptors

Anterior white column

SPINAL CORD

4

plate and extending along the superior nasal conchae and upper part of the middle nasal conchae. The olfactory epithelium consists of three principal kinds of cells: olfactory receptors, supporting cells, and basal stem cells. The **olfactory receptors** are bipolar neurons. Their cell bodies lie between the supporting cells. The distal (free) end of each olfactory cell contains a knob-shaped **dendrite** from which cilia, called **olfactory hairs,** protrude. The **supporting cells** are columnar epithelial cells of the mucous membrane that line the nose. **Basal (stem) cells** lie between the bases of the supporting cells and produce new olfactory receptors, which live for only a month or so before being replaced. Within the connective tissue deep to the

olfactory epithelium are **olfactory (Bowman's) glands** that secrete mucus.

The unmyelinated axons of the olfactory receptors unite to form the **olfactory nerves (cranial nerve I),** which pass through about 20 olfactory foramina in the cribriform plate of the ethmoid bone. The olfactory nerves terminate in paired masses of gray matter, the **olfactory bulbs,** which lie inferior to the frontal lobes of the cerebrum and lateral to the crista galli of the ethmoid bone. The first synapse of the olfactory neural pathway occurs in the olfactory bulbs between the axons of the olfactory (I) nerves and the dendrites of neurons inside the olfactory bulbs. Axons of these neurons extend posteriorly and form the **olfactory**

FIGURE 14.5 **Olfactory receptors.**

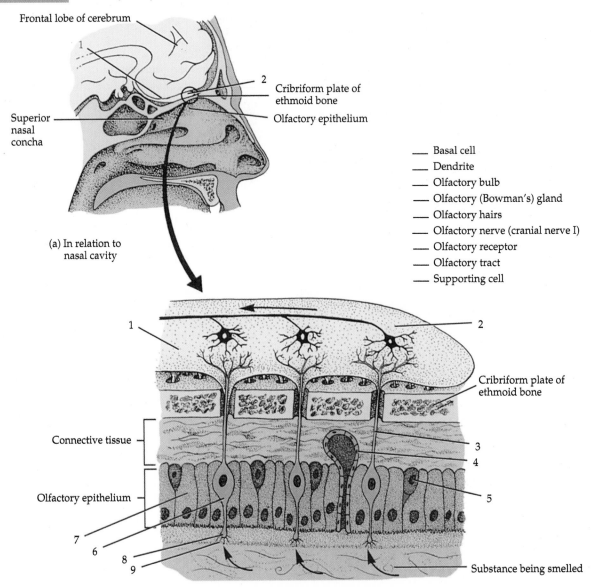

Frontal lobe of cerebrum

Cribriform plate of
ethmoid bone

Olfactory epithelium

Superior
nasal
concha

___ Basal cell
___ Dendrite
___ Olfactory bulb
___ Olfactory (Bowman's) gland
___ Olfactory hairs
___ Olfactory nerve (cranial nerve I)
___ Olfactory receptor
___ Olfactory tract
___ Supporting cell

(a) In relation to
nasal cavity

Cribriform plate of
ethmoid bone

Connective tissue

Olfactory epithelium

Substance being smelled

(b) Enlarged aspect

tract. From here, nerve impulses are conveyed to the lateral olfactory area in the temporal lobe of the cerebral cortex. In the cortex, the nerve impulses are interpreted as odor and give rise to the sensation of smell.

Adaptation (decreasing sensitivity) happens quickly, especially adaptation to odors. For this reason, we become accustomed to some odors and are also able to endure unpleasant ones.

Rapid adaptation also accounts for the failure of a person to detect gas that accumulates slowly in a room.

Label the structures associated with olfaction in Figure 14.5.

Now examine a slide of the olfactory epithelium under high power. Identify the olfactory receptors and supporting cells and label Figure 14.6 on page 305.

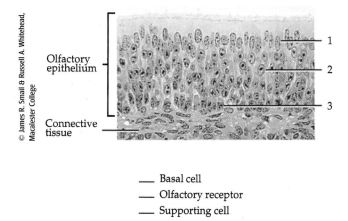

FIGURE 14.6 Photomicrograph of the olfactory epithelium (350×).

Olfactory epithelium

Connective tissue

—— Basal cell
—— Olfactory receptor
—— Supporting cell

2. Olfactory Adaptation

PROCEDURE

1. The subject should close his or her eyes after plugging one nostril with cotton.

2. Hold a bottle of oil of cloves, or other substance having a distinct odor, under the open nostril.

3. The subject breathes in through the open nostril and exhales through the mouth. Note the time required for the odor to disappear, and repeat with the other nostril.

4. As soon as olfactory adaptation has occurred, test an entirely different substance.

5. Compare results for the various materials tested.

Olfactory stimuli such as pepper, onions, ammonia, ether, and chloroform are irritating and may cause tearing because they stimulate the receptors of the trigeminal nerve (cranial nerve V) as well as the olfactory neurons.

G. Gustation

1. Gustatory Receptors

The receptors for taste, or **gustation** (*gust* = taste), are located in the taste buds. Although taste buds are most numerous on the tongue, they are also found on the soft palate, pharynx, and epiglottis. **Taste buds** are oval bodies consisting of three kinds of epithelial cells: supporting cells, gustatory receptor cells, and basal cells. The **supporting cells** are specialized epithelial cells that form a capsule. Inside each capsule are

about 50 **gustatory receptor cells.** Each gustatory receptor has a single, long microvillus called a **gustatory hair** that projects to the external surface through an opening in the taste bud called the **taste pore.** Gustatory cells make contact with taste stimuli through the taste pore. **Basal cells** are found at the periphery of the taste bud and produce new supporting cells, which then develop into gustatory receptors cells with a lifespan of about 10 days.

Examine a slide of taste buds and label the structures associated with gestation in Figure 14.7.

Taste buds are found in elevations on the tongue called **papillae** (pa-PIL-ē). They give the upper surface of the tongue its rough texture and appearance. **Vallate (circumvallate)** (VAL-āt) **papillae** are circular and form an inverted V-shaped row at the back of the tongue. **Fungiform** (FUN-ji-form) **papillae** are mushroom-shaped elevations scattered over the entire surface of the tongue. All circumvallate and most fungiform papillae contain taste buds. **Foliate papillae** (FŌ-lē-āt = leaflike) are located in small trenches on the lateral margins of tongue but their taste buds degenerate in early childhood. **Filiform** (FIL-i-form) **papillae** are pointed, threadlike structures that are also distributed over the entire surface of the tongue. They contain tactile receptors but no taste buds.

Have your partner protrude his or her tongue and examine its surface with a hand lens to identify the shape and position of the papillae.

2. Identifying Taste Zones

For gustatory receptor cells to be stimulated, substances tasted must be in solution in the saliva in order to enter the taste pores in the taste buds. Despite the many substances tasted, there are basically only five primary taste sensations: sour, salty, bitter, sweet, and umami (ū-MAM-ē). The umami taste, recently described by Japanese scientists, is described as "meaty" or "savory." Each taste is due to a different response to different chemicals. Some regions of the tongue react more strongly than others to particular taste sensations.

To identify the taste zones for four of the taste sensations, perform the following steps and record the results in Section G.2 of the LABORATORY REPORT RESULTS at the end of the exercise by inserting a plus sign (taste detected) or a minus sign (taste not detected) where appropriate.

PROCEDURE

1. The subject thoroughly dries his or her tongue (use a clean paper towel). The experimenter

FIGURE 14.7 **Structure of a taste bud.**

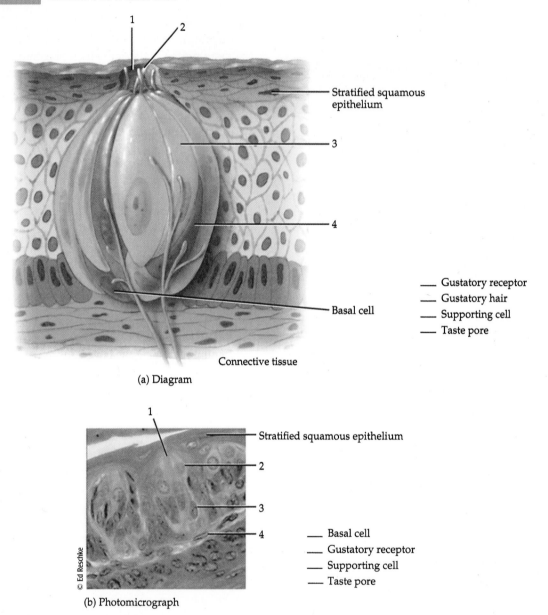

Stratified squamous epithelium

3

4

Basal cell

___ Gustatory receptor
___ Gustatory hair
___ Supporting cell
___ Taste pore

Connective tissue

(a) Diagram

1

Stratified squamous epithelium

2

3

4

© Ed Reschke

___ Basal cell
___ Gustatory receptor
___ Supporting cell
___ Taste pore

(b) Photomicrograph

places some granulated sugar on the tip of the tongue and notes the time. The subject indicates when he or she tastes sugar by raising his or her hand. The experimenter notes the time again and records how long it takes for the subject to taste the sugar.

2. Repeat the experiment, but this time use a drop of sugar solution. Again record how long it takes for the subject to taste the sugar. How do you explain the difference in time periods?

3. The subject rinses his or her mouth again. The experiment is then repeated using the quinine solution (bitter taste), and then the salt solution.

4. After rinsing yet again, the experiment is repeated using the acetic acid solution or vinegar (sour) placed on the tip and *sides* of the tongue.

3. Taste and Inheritance

Taste for certain substances is inherited, and geneticists for many years have been using the chemical **phenylthiocarbamide (PTC)** to test taste. To some individuals this substance tastes bitter; to others it is sweet; and some cannot taste it at all.

PROCEDURE

1. Place a few crystals of PTC on the subject's tongue. Does he or she taste it? If so, describe the taste.

2. Special paper that is flavored with this chemical may be chewed and mixed with saliva and tested in the same manner.

3. Record on the blackboard your response to the PTC test. Usually about 70% of the people tested can taste this compound; 30% cannot. Compare this percentage with the class results.

4. Taste and Smell

This test combines the effect of smell on the sense of taste.

PROCEDURE

1. Obtain small cubes of carrot, onion, potato, and apple.

2. The subject dries the tongue with a clean paper towel, closes the eyes, and pinches the nostrils shut. The experimenter places the cubes, one by one, on the subject's tongue.

3. The subject attempts to identify each cube in the following sequences: (1) immediately, (2) after chewing (nostrils closed), and (3) after opening the nostrils.

4. Record your results in Section G.4 of the LABORATORY REPORT RESULTS at the end of the exercise.

H. Vision

Structures related to **vision** are the eyeball (which is the receptor organ for visual sensations), optic nerve (cranial nerve II), brain, and accessory structures. The extrinsic muscles of the eyeball may be reviewed in Figure 10.3.

1. Accessory Structures

The **accessory structures** of the eye are the eyebrows, eyelids, eyelashes, lacrimal (tearing) apparatus, and extrinsic eye muscles.

Eyebrows protect the eyeball from foreign objects, prevent perspiration from getting into the eye, and shade the eye from the direct rays of the sun. **Eyelids,** or **palpebrae** (PAL-pe-brē), consist primarily of skeletal muscle covered externally by skin. The inner aspect of the eyelid is lined by a thin,

protective mucous membrane called the **palpebral conjunctiva** (kon-junk-TĪ-va). The **bulbar (ocular) conjunctiva** passes from the eyelids onto the anterior surface of the eyeball where it covers the sclera (the white of the eye) but not the cornea. Also within the eyelids are **tarsal (Meibomian) glands,** modified sebaceous glands whose secretion keeps the eyelids from adhering to each other. Infection of these glands produces a **chalazion** (cyst) in the eyelid. Eyelids shade the eyes during sleep, protect the eyes from excessive light and foreign objects, and spread lubricating secretions over the surface of the eyeballs. Projecting from the border of each eyelid is a row of short, thick hairs, the **eyelashes.** Sebaceous glands at the base of the hair follicles of the eyelashes, called **sebaceous ciliary glands,** release a lubricating fluid into the follicles. An infection of these glands is called a **sty.**

The **lacrimal** (LAK-ri-mal; *lacrima* = tear) **apparatus** is a group of structures that produces and drains **lacrimal fluids** (tears). Each **lacrimal gland** is located at the superior lateral portion of both orbits. Leading from the lacrimal glands are 6 to 12 **excretory lacrimal ducts** that empty tears onto the surface of the conjunctiva of the upper lid. From here, the tears pass medially over the anterior surface of the eyeball and enter two small openings called **lacrimal puncta** that appear as two small pores, one in each papilla of the eyelid, at the medial commissure of the eye. The tears then pass into two ducts, the **lacrimal canals,** and are next conveyed into the lacrimal sac. The **lacrimal sac** is the superior expanded portion of the **nasolacrimal duct,** a canal that carries the tears into the nasal cavity. Tears clean, lubricate, and moisten the external surface of the eyeball.

Label the parts of the lacrimal apparatus in Figure 14.8.

2. Structure of the Eyeball

The adult eyeball measures about 2.5 cm (1 in.) in diameter. Of its total surface area, only anterior one-sixth is exposed. The eyeball can be divided into three principal layers: (1) fibrous tunic, (2) vascular tunic, and (3) retina. See Figure 14.9 on page 310.

a. Fibrous Tunic

The **fibrous tunic** is the superficial coat of the eyeball. It is divided into the posterior sclera and the anterior cornea. The **sclera** (SKLE-ra; *skleros* = hard), the "white of the eye," is a layer of dense connective tissue that covers the entire eyeball except the cornea. The sclera gives shape to the eyeball and protects its inner parts. The anterior

FIGURE 14.8 Lacrimal apparatus.

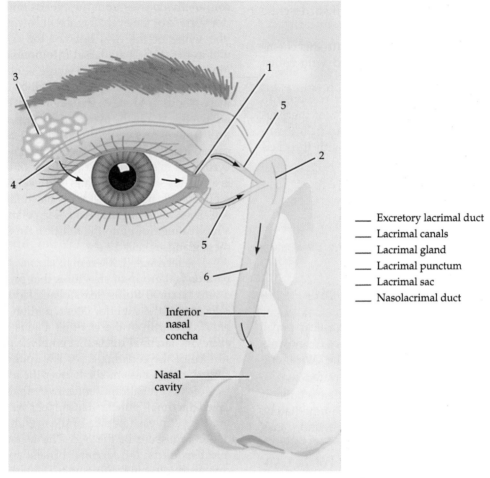

____ Excretory lacrimal duct
____ Lacrimal canals
____ Lacrimal gland
____ Lacrimal punctum
____ Lacrimal sac
____ Nasolacrimal duct

Inferior nasal concha

Nasal cavity

Anterior view

portion of the fibrous tunic is known as the **cornea** (KOR-nē-a). This nonvascular, transparent coat covers the colored iris. Because it is cured, the cornea helps focus light. The outer surface of the cornea consists of nonkeratinized stratified squamous epithelium that is continuous with the epithelium of the bulbar conjunctiva. At the junction of the sclera and cornea is the **scleral venous sinus (canal of Schlemm),** through which circulates a fluid called aqueous humor.

b. Vascular Tunic

The **vascular tunic** is the middle layer of the eyeball and consists of three portions: choroids, ciliary body, and iris. The highly vascularized **choroid** (KŌ-royd) is the posterior portion of the vascular tunic. It lines most of the internal surface of the

sclera and maintains the nutrition of the retina. The anterior portion of the vascular tunic is the **ciliary** (SIL-ē-ar′-ē) **body.** It extends from the **ora serrata** (Ō-ra ser-RĀ-ta), the jagged anterior margin of the retina (inner tunic) to a point just posterior to the junction of the sclera and cornea. The ciliary body consists of the **ciliary processes** (folds of the ciliary body that secrete aqueous humor) and the **ciliary muscle** (a circular band of smooth muscle that alters the shape of the lens for near or far vision). The **iris** (*irid* = colored circle), the third portion of the vascular tunic, is the colored portion of the eyeball and consists of circular and radial smooth muscle fibers arranged to form a flattened doughnut-shaped structure. The hole in the center of the iris is the **pupil,** through which light enters the eyeball. One function of the iris is to regulate the amount of light entering the eyeball.

c. Retina

The third and inner coat of the eyeball, the **retina** lines the posterior three-quarters of the eyeball. Its primary function is image formation. The optic part of the retina consists of a pigment layer and a neural layer. The outer **pigment layer** (nonvisual portion) is a sheet of melanin-containing epithelial cells located between the choroids and the neural part of the retina. The inner **neural layer** (visual portion) is composed of three distinct layers of neurons. Named in the order in which they conduct nerve impulses, these are the **photoreceptor layer, bipolar cell layer,** and **ganglion cell layer.** Structurally, the photoreceptor layer is just internal to the pigment layer, which lies adjacent to the choroids. The ganglion cell layer is the innermost zone of the neural portion.

The two types of photoreceptors are called rods and cones because of their respective shapes. **Rods** are specialized for vision in dim light. In addition, they allow discrimination between different shades of dark and light and permit discernment of shapes and movement. **Cones** are specialized for color vision and for sharpness of vision, that is, **visual acuity** or **resolution.** Cones are stimulated only by bright light and are most densely concentrated in the **central fovea,** a small depression in the center of the macula lutea. The **macula lutea** (MAK-ū-la LOO-tē-a) is situated in the exact center of the posterior portion of the retina and corresponds to the visual axis of the eye. The fovea is the area of sharpest vision because it contains only cones. Rods are absent from the fovea and macula but increase in density toward the periphery of the retina.

When light stimulates photoreceptors, impulses are conducted across synapses to the bipolar neurons in the intermediate zone of the neural portion of the retina. From there, the impulses pass to the ganglion cell layer. Axons of the ganglion cells extend posteriorly to a small area of the retina called the **optic disc (blind spot).** This region contains openings through which fibers of the ganglion neurons exit as the **optic nerve (cranial nerve II).** Because this area contains neither rods nor cones, and only nerve fibers, we cannot see an image that strikes it. For this reason it is called the blind spot.

d. Lens

The eyeball itself also contains the lens, just behind the pupil and iris. The **lens** is constructed of numerous layers of protein called **crystallins,** arranged like the layers of an onion. Normally, the lens is perfectly transparent due to its lack of blood vessels and is enclosed by a clear capsule and held in position by encircling **zonular fibers (suspensory ligaments).** A loss of transparency of the lens is called a **cataract.**

e. Interior

The interior of the eyeball is divided into two cavities. These are called the anterior cavity and the vitreous chamber and are separated from each other by the lens. The **anterior cavity,** in turn, has two subdivisions known as the anterior chamber and the posterior chamber. The **anterior chamber** lies posterior to the cornea and anterior to the iris. The **posterior chamber** lies posterior to the iris and anterior to the zonular fibers and lens. The anterior cavity is filled with a clear, watery fluid known as the **aqueous** (*aqua* = water) **humor,** which is continually secreted by the ciliary processes and nourishes the lens and cornea. From the posterior chamber, the fluid permeates the posterior cavity and then passes anteriorly between the iris and the lens, through the pupil into the anterior chamber. From the anterior chamber, the aqueous humor is drained off into the scleral venous sinus and passes into the blood. Pressure in the eye, called **intraocular pressure (IOP),** is produced mainly by the aqueous humor. Intraocular pressure maintains the shape of the eyeball and prevents the eyeball from collapsing. Abnormal elevation of intraocular pressure, called **glaucoma** (glaw-KŌ-ma), results in degeneration of the retina and blindness.

The second, larger cavity of the eyeball is the **vitreous chamber.** It is located between the lens and retina and contains a jellylike substance called the **vitreous body.** This substance contributes to intraocular pressure, helps to prevent the eyeball from collapsing, and holds the retina flush against the internal portions of the eyeball. Unlike the aqueous humor, the vitreous body does not undergo constant replacement.

Label the parts of the eyeball in Figure 14.9.

3. Surface Anatomy

Refer to Figure 14.10 on page 312 for a summary of several surface anatomy features of the eyeball and accessory structures of the eye.

4. Dissection of Vertebrate Eye (Cow or Sheep)

▲ **CAUTION!** *Please reread Section D, "Precautions Related to Dissection," at the beginning of the laboratory manual on page xiii before you begin your dissection.*

FIGURE 14.9 **Parts of the eyeball.**

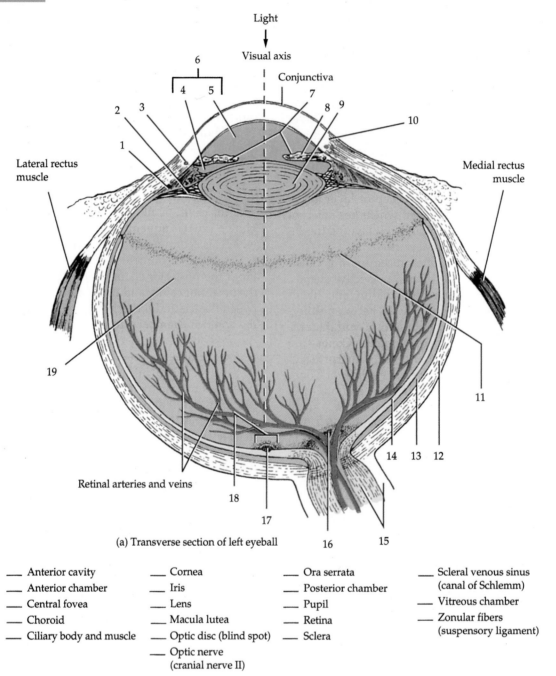

(a) Transverse section of left eyeball

____ Anterior cavity
____ Anterior chamber
____ Central fovea
____ Choroid
____ Ciliary body and muscle

____ Cornea
____ Iris
____ Lens
____ Macula lutea
____ Optic disc (blind spot)
____ Optic nerve
(cranial nerve II)

____ Ora serrata
____ Posterior chamber
____ Pupil
____ Retina
____ Sclera

____ Scleral venous sinus
(canal of Schlemm)
____ Vitreous chamber
____ Zonular fibers
(suspensory ligament)

a. External Examination
PROCEDURE

1. Note any **fat** on the surface of the eyeball that protects the eyeball from shock in the orbit. Remove the fat.

2. Locate the **sclera,** the tough external white coat, and the **conjunctiva,** a delicate membrane that covers the anterior surface of the eyeball and is attached near the edge of the

cornea. The **cornea** is the anterior, transparent portion of the sclera. It is probably opaque in your specimen due to the preservative.

3. Locate the **optic nerve (cranial nerve II),** a solid, white cord of nerve fibers on the posterior surface of the eyeball.

4. If possible, identify six **extrinsic eye muscles** that appear as flat bands near the posterior part of the eyeball.

FIGURE 14.9 **Parts of the eyeball. (Continued)**

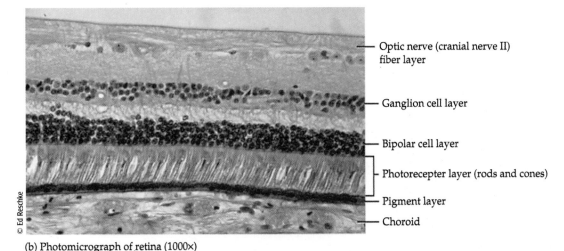

— Optic nerve (cranial nerve II) fiber layer

— Ganglion cell layer

— Bipolar cell layer

— Photorecepter layer (rods and cones)

— Pigment layer

— Choroid

© Ed Reschke

(b) Photomicrograph of retina (1000×)

b. Internal Examination

PROCEDURE

1. With a sharp scalpel, make an incision about 0.6 cm (¼ in.) lateral to the cornea (Figure 14.11 on page 313).

2. Insert scissors into the incision and carefully and slowly cut all the way around the corneal region. The eyeball contains fluid, so take care that it does not squirt out when you make your first incision. Examine the inside of the anterior part of the eyeball.

3. The **lens** is held in position by **zonular fibers (suspensory ligaments),** which are delicate fibers. Around the outer margin of the lens, with a pleated appearance, is the black **ciliary body,** which also functions to hold the lens in place. Free the lens and notice how hard it is.

4. The **iris** can be seen just anterior to the lens and is also heavily pigmented or black.

5. The **pupil** is the circular opening in the center of the iris.

6. Examine the inside of the posterior part of the eyeball, identifying the thick **vitreous humor** that fills the space between the lens and retina.

7. The **retina** is the white inner coat beneath the choroid coat and is easily separated from it.

8. The **choroid coat** is a dark, iridescent tissue that gets its iridescence from a special structure called the **tapetum lucidum.** The tapetum lucidum, which is not present in the human eye, functions to reflect some light back onto the retina.

9. Finally, identify the **blind spot,** the point at which the retina is attached to the back of the eyeball.

5. Testing for Visual Acuity

The acuteness of vision may be tested by means of a **Snellen Chart.** It consists of letters of different sizes which are read at a distance normally designated at 20 ft. If the subject reads to the line that is marked "50," he or she is said to possess 20/50 vision in that eye, meaning that he or she is reading at 20 ft what a person who has normal vision can read at 50 ft. If he or she reads to the line marked "20," he or she has 20/20 vision in that eye. The normal eye can sufficiently refract light rays from an object 20 ft away to focus a clear object on the retina. Therefore, if you have 20/20 vision, your eyes are perfectly normal. The higher the bottom number, the larger the letter must be for you to see if clearly, and of course the worse or weaker are your eyes.

PROCEDURE

1. Have the subject stand 20 ft from the Snellen Chart and cover the right eye with a 3″ × 5″ (3-in.-by-5-in.) card.

2. Instruct the subject to slowly read down the chart until he or she can no longer focus the letters.

3. Record the number of the last line (20/20, 20/30, or whichever) that can be successfully read.

Surface anatomy of eyeball and accessory structures.

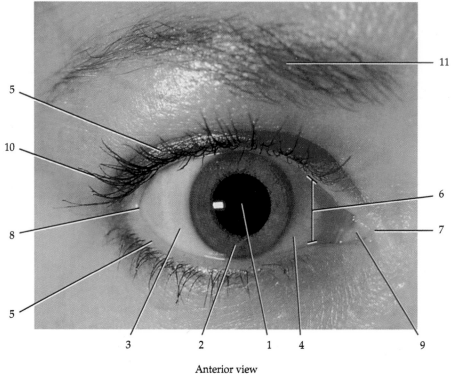

Anterior view

1. **Pupil.** Opening of center of iris of eyeball for light transmission.
2. **Iris.** Circular pigmented muscular membrane behind cornea.
3. **Sclera.** "White" of eye, a coat of fibrous tissue that covers entire eyeball except for cornea.
4. **Conjunctiva.** Membrane that covers exposed surface of eyeball and lines eyelids.
5. **Palpebrae (eyelids).** Folds of skin and muscle lined by conjunctiva.
6. **Palpebral fissure.** Space between eyelids when they are open.
7. **Medial commissure.** Site of union of upper and lower eyelids near nose.
8. **Lateral commissure.** Site of union of upper and lower eyelids away from nose.
9. **Lacrimal caruncle.** Fleshy, yellowish projection of medial commissure that contains modified sweat and sebaceous glands.
10. **Eyelashes.** Hairs on margins of eyelids, usually arranged in two or three rows.
11. **Eyebrows.** Several rows of hair superior to upper eyelids.

4. Repeat this procedure covering the left eye.
5. Now the subject should read the chart using both eyes.
6. Record your results in Section H.5 of the LABORATORY REPORT RESULTS at the end of the exercise and change places.

6. Testing for Astigmatism

The eye, with normal ability to refract light, is referred to as an **emmetropic** (em′-e-TROP-ik) eye. It can sufficiently refract light rays from an object 20 ft away to focus a clear object on the retina. If the lens is normal, objects as far away as the horizon and as close as about 20 ft will form images on the sensitive part of the retina. When objects

are closer than 20 ft, however, the lens has to sharpen its focus by using the ciliary muscles. Many individuals, however, have abnormalities related to improper refraction. Among these are **myopia** (mī-Ō-pē-a) (nearsightedness), **hyperopia** (hī-per-Ō-pē-a) or **hypermetropia** (hī′-per-mē-TRŌ-pē-a) (farsightedness), and **astigmatism** (a-STIG-ma-tizm) (irregular curvature of the lens or cornea).

Why do you think nearsightedness can be corrected with glasses containing biconcave lenses?

_____

_____

_____

FIGURE 14.11 **Procedure for dissecting a vertebrate eye.**

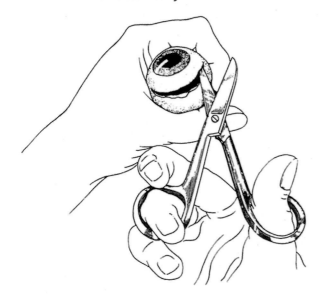

How would you correct farsightedness? _____

_____

_____

PROCEDURE

1. In order to determine the presence of astigmatism, remove any corrective lenses if you are wearing them and look at the center of the astigmatism test chart that follows, first with one eye, then the other:

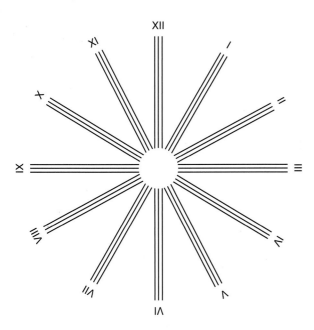

2. If all the radiating lines appear equally sharp and equally black, there is no astigmatism.

3. If some of the lines are blurred or less dark than others, astigmatism is present.

4. If you wear corrective lenses, try the test with them on.

7. Testing for the Blind Spot

PROCEDURE

1. Hold this page about 20 in. from your face with the cross shown below directly in front of your right eye. You should be able to see the cross and the circle when you close your left eye.

2. Now, keeping the left eye closed, slowly bring the page closer to your face while fixing the right eye on the cross.

3. At a certain distance the circle will disappear from your field of vision because its image falls on the blind spot.

8. Image Formation

Formation of an image on the retina requires three basic processes, all concerned with focusing light rays. These are (1) refraction of light rays, (2) accommodation of the lens, and (3) constriction of the pupil.

When light rays traveling through a transparent substance (such as air) pass into a second transparent substance with a different density (such as water), the rays bend at the junction between the two substances. This is called **refraction.** As light rays enter the eye, they are refracted at the anterior and posterior surfaces of the cornea. Both surfaces of the lens of the eye further refract the light rays so that they come into exact focus on the retina.

The lens of the eye has the unique ability to change the focusing power of the eye by becoming moderately curved at one moment and greatly curved the next. When the eye is focusing on a close object, the lens becomes more curved in order to bend the rays toward the central fovea of the eye. This increase in the curvature of the lens is called **accommodation.**

TABLE 14.1	Correlation of age and near-point accommodation	
Age	**Inches**	**Centimeters**
10	2.95	7.5
20	3.54	9.0
30	4.53	11.5
40	6.77	17.2
50	20.67	52.5
60	32.80	83.3

a. Testing for Near-Point Accommodation

The following test determines your **near-point accommodation:**

PROCEDURE

1. Using any card that has a letter printed on it, close one eye and focus on the letter.

2. Measure the distance of the card from the eye using a ruler or a meter stick.

3. Now *slowly* bring the card as close as possible to your open eye, and stop when you no longer see a clear, detailed letter.

4. Measure and record this distance. This value is your near-point accommodation.

5. Repeat this procedure three times and then test your other eye.

6. Check Table 14.1 to see whether the near point for your eyes corresponds with that recorded for your age group. (*Note: Use a letter that is the size of typical newsprint.*)

b. Testing for Constriction of the Pupil
PROCEDURE

1. Place a 3″ × 5″ (3-in.-by-5-in.) card on the side of the nose so that a light shining on one side of the face will not affect the eye on the other side.

2. Shine the light from a lamp or a flashlight on one eye, 6 in. away (approximately 15 cm), for about 5 sec. Note the change in the size of the pupil of this eye.

3. Remove the light, wait about 3 min, and repeat, but this time observe the pupil of the opposite eye.

4. Wait a few minutes, and repeat the test, observing the pupils of both eyes.

9. Testing for Convergence

In humans, both eyes focus on only one set of objects—a characteristic called **binocular vision.** The term **convergence** refers to a medial movement of the two eyeballs so that they are both directed toward the object being viewed. The nearer the object, the greater the degree of convergence necessary to maintain single binocular vision.

PROCEDURE

1. Hold a pencil or pen about 2 ft from your nose and focus on its point. Now slowly bring the pencil toward your nose.

2. At some moment you should suddenly see two pencil points, or a blurring of the point.

3. Observe your partner's eyes when he or she does this test.

Images are actually focused upside down on the retina. They also undergo mirror reversal. That is, light reflected from the right side of an object hits the left side of the retina and vice versa. Reflected light from the top of the object crosses light from the bottom of the object and strikes the retina below the central fovea. Reflected light from the bottom of the object crosses light from the top of the object and strikes the retina above the central fovea.

The reason why we do not see a topsy-turvy world is that the brain learns early in life to coordinate visual images with the exact location of objects. The brain stores memories of reaching and touching objects and automatically turns visual images right-side up and right-side around.

10. Testing for Binocular Vision, Depth Perception, Diplopia, and Dominance

Humans are endowed with binocular vision. Each eye sees a different view, although there is a large degree of overlap. The cerebral cortex uses these discrepancies to produce **stereopsis,** or three-dimensional vision, an orientation of the object in space that allows us to perceive its distance from us **(depth perception).**

Even though there are slight differences in the views from each eye, we see a single image because the cerebral cortex integrates the images from each eye into a single perception. If this integration is not present, **diplopia** (dip-LŌ-pē-a), or double vision, results. We do not perceive an "average" or "mean" of both left and right views; rather the view from one of our eyes is **dominant,** that is, the view we always perceive with both eyes open.

PROCEDURE

1. Place a book at arm's length on a table. Place your left hand at your side and touch the nearest corner of the book with your right index finger. Close the left eye and repeat; close the right eye and repeat. Note the accuracy of your attempts.

   How accurate was your attempt to touch a

   point with one eye closed? _____

   If the right eye was closed, did you have error

   to the left or the right? _____

2. Use a depth perception tester with both eyes open and with each eye closed. Attempt to align the two arrows and measure monocular perception errors using the scale on the base.

   Length of depth perception error _____

3. Focus on a discrete object in the distance and press *very gently* on your left eyelid. Note how

   many objects you see: _____

4. Stab a pencil through a piece of paper and remove the scrap. Hold the paper at arm's length and focus on a small discrete object which just fills the hole. (You may want to use an X on the blackboard.) *Without moving the paper,* close the left eye and note whether the object is still present. Repeat with right eye

   closed. Note your dominant eye: _____

11. Testing for Afterimages

The rods and cones of the retina are photoreceptors; that is, they contain light-sensitive pigments that absorb light of different wavelengths. **Rhodopsin,** the light-sensitive pigment in rods, consists of opsin and retinal. When light strikes a molecule of rhodopsin, the retinal changes from a curved to a straight shape and separates from the opsin. The energy released as a result of the exchange initiates the nerve impulse that causes us to perceive light.

In bright light, all the rhodopsin is decomposed, so that the rods are nonfunctional. Cones function in bright light, in much the same way as rods, but they absorb light of specific wavelengths (red, blue, or green). The particular color perceived depends on which combinations of cones are stimulated. If all three are stimulated, we see white; if none is stimulated, we see black.

If photoreceptors are stimulated by staring at a bright object for a long period of time, they will continue firing briefly after the stimulus has been removed **(positive afterimage).** However, after prolonged firing, these photoreceptors become "bleached," or fatigued, so that they can no longer fire, resulting in a reversal of the image **(negative afterimage).**

PROCEDURE

1. Stare at a light source, *but not a bright light,* for 30 sec. Close the eyes briefly and note the positive afterimage. Open the eyes and focus on a piece of white paper. Note the negative afterimage.

2. Draw a red cross on paper about 1 in. high, with lines ½ in. thick. Stare at the cross without moving the eyes for at least 30 sec. Close the eyes very briefly; open them and stare at a piece of white paper. Note the positive and negative afterimages. Record your observations in Section H.11 of the LABORATORY REPORT RESULTS at the end of the exercise.

12. Testing for Color Blindness

Color blindness is an inherited inability to distinguish between certain colors, resulting from the absence or a deficiency of one of the three cone photo pigments. It is a sex-linked disorder. About 0.5% of all females and about 8% of all males are affected. The most common form is red-green color blindness.

PROCEDURE

1. View Ishihara plates in bright light and compare your responses to those in the plate book or use Holmgren's test for matching colored threads.

2. Note the accuracy of your responses: _____

13. Visual Pathway

From the rods and cones, impulses are transmitted through bipolar cells to ganglion cells. The cell bodies of the ganglion cells lie in the retina, and their axons leave the eye via the **optic nerve (cranial nerve II).** The axons pass through the **optic chiasm** (kī-AZ-em), a crossing point of the optic nerves. Fibers from the medial retina cross to the opposite side. Fibers from the lateral retina remain uncrossed. Upon passing through the optic chiasm, the fibers, now part of the **optic tract,** enter the

Visual pathway.

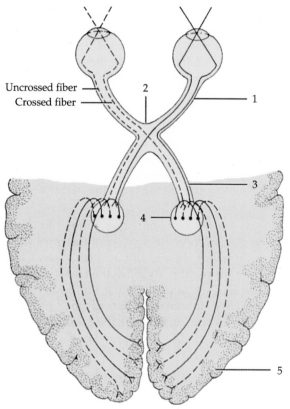

Uncrossed fiber ——
Crossed fiber ——

_____ Occipital lobe of cerebral cortex (primary visual area)
_____ Optic chiasm
_____ Optic nerve (cranial nerve II)
_____ Optic tract
_____ Lateral geniculate nucleus of thalamus

brain and terminate in the lateral geniculate nucleus of the thalamus. Here the fibers synapse with the neurons whose axons pass to the primary visual areas located in the occipital lobes of the cerebral cortex. Label the visual pathway in Figure 14.12.

I. Hearing and Equilibrium

In addition to containing receptors for sound waves, the **ear** also contains receptors for equilibrium. The ear is subdivided into three principal regions: (1) external (outer) ear, (2) middle ear, and (3) internal (inner) ear.

1. Structure of Ear

a. External (Outer) Ear

The **external (outer) ear** collects sound waves and directs them inward. Its structure consists of the auricle, external auditory canal, and eardrum. The **auricle (pinna)** is a flap of elastic cartilage shaped like the flared end of a trumpet covered by thick skin. The rim of the auricle is called the **helix,** and the inferior portion is referred to as the **lobule.** The auricle is attached to the head by ligaments and muscles. The **external auditory canal** is a curved tube about 2.5 cm (1 in.) in length that leads from the auricle to the eardrum. The walls of the canal consist of bone lined with cartilage that is continuous with the cartilage of the auricle. Near the exterior opening, the canal contains a few hairs and specialized sweat glands called **ceruminous (se-RU-mi-nus) glands,** which secrete **cerumen** (earwax). The combination of hairs and cerumen help prevent dust and foreign objects from entering the ear. The **eardrum,** or **tympanic (tim-PAN-ik) membrane,** is a thin, semitransparent partition (of fibrous connective tissue) between the external auditory canal and middle ear.

Examine a model or charts and label the parts of the external ear in Figure 14.13.

b. Middle Ear

The **middle ear** is a small, air-filled cavity in the temporal bone that is lined by epithelium. The area is separated from the external ear by the eardrum and from the internal ear by a very thin bony partition that contains two small membrane-covered openings, called the **oval window** and the **round window.**

The anterior wall of the middle ear contains an opening that leads directly into the **auditory (pharyngotympanic) tube,** commonly known as the **eustachian tube.** The auditory tube connects the middle ear with the nasopharynx (upper portion of the throat). The function of the tube is to equalize air pressure on both sides of the eardrum. Any sudden pressure changes against the eardrum may be equalized by deliberately swallowing.

Extending across the middle ear and attached to it by ligaments are the three smallest bones in the body, called **auditory ossicles** (OS-si-kuls). The bones, named for their shapes, are known as the malleus, incus, and stapes. They are commonly named the hammer, anvil, and stirrup, respectively. The "handle" of the **malleus** is attached to the internal surface of the eardrum. Its head articulates with the body of the **incus,** the middle bone in the series, which articulates with the stapes. The base of the **stapes** fits into a small opening between the middle and inner ear called the **oval window.** Directly below to the oval window is another opening, the **round window.** This opening, which separates the middle and inner

FIGURE 14.13 Principal subdivisions of ear.

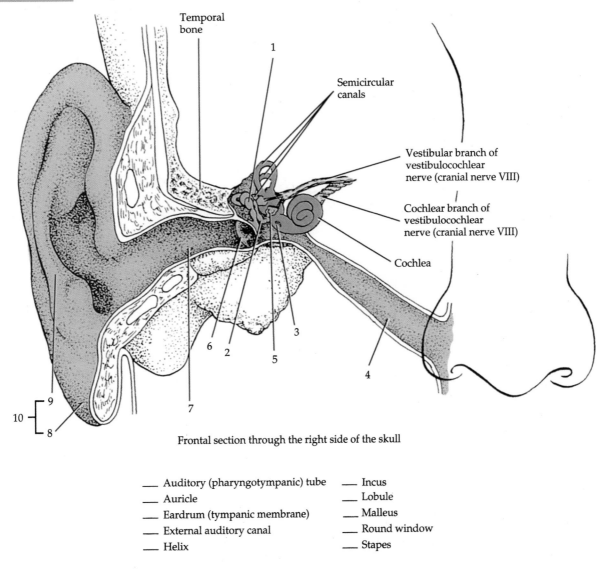

Frontal section through the right side of the skull

___ Auditory (pharyngotympanic) tube
___ Auricle
___ Eardrum (tympanic membrane)
___ External auditory canal
___ Helix

___ Incus
___ Lobule
___ Malleus
___ Round window
___ Stapes

ears, is enclosed by a membrane called the **secondary tympanic membrane.**

Examine a model or charts and label the parts of the middle ear in Figures 14.13 and 14.14 on page 318.

c. Internal (Inner) Ear

The **internal (inner) ear** is also known as the **labyrinth** (LAB-i-rinth) because of its complicated series of canals. Structurally, it consists of two main divisions (1) an outer bony labyrinth and (2) an inner membranous labyrinth that fits within the bony labyrinth. The **bony labyrinth** is a series of cavities in the temporal bone that can be divided into three regions, named on the basis of shape:

vestibule, cochlea, and semicircular canals. The bony labyrinth is lined with periosteum and contains a fluid called the **perilymph.** This fluid surrounds the **membranous labyrinth,** a series of sacs and tubes lying inside and having the same general form as the body labyrinth. Epithelium lines the membranous labyrinth, which is filled with a fluid called the **endolymph.**

The **vestibule** is the oval, central portion of the bony labyrinth (see Figure 14.14). The membranous labyrinth in the vestibule consists of two sacs called the **utricle** (Ū-tri-kl) and **saccule** (SAK-ūl). These sacs are connected to each other by a small duct.

Projecting superiorly and posteriorly from the vestibule are the three bony **semicircular canals**

FIGURE 14.14 Ossicles of middle ear.

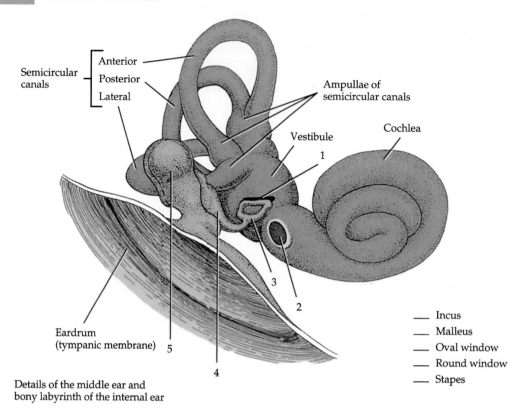

Details of the middle ear and
bony labyrinth of the internal ear

____ Incus

____ Malleus

____ Oval window

____ Round window

____ Stapes

(see Figure 14.14). Each is oriented at approximately right angles to the other two. They are called the anterior, posterior, and lateral canals. At one end of each canal is a swollen enlargement called the **ampulla** (am-PUL-la; = saclike duct). Inside the bony semicircular canals lie portions of the membranous labyrinth, the **semicircular ducts.** These structures communicate with the utricle of the vestibule. Label these structures in Figure 14.15 on page 319.

Anterior to the vestibule is the **cochlea** (KOK-lē-a; = snail-shaped) (label it in Figure 14.15). The cochlea consists of a bony spiral canal that makes about three turns around a central bony core called the **modiolus.** A transverse section through the cochlea shows that the canal is divided by partitions into three separate channels shaped like the letter Y lying on its side. The stem of the Y is a bony shelf that protrudes into the canal. The wings of the Y are composed mainly of the vestibular labyrinth. The channel above the partition is called the **scala vestibuli.** The channel below is known as the **scala tympani.** The cochlea adjoins the wall of the vestibule, into which the scala vestibuli opens. The scala tympani terminates at the round window. The perilymph of the vestibule is continuous with that of the scala

vestibuli. The third channel (between the wings of the Y) is the **cochlear duct (scala media).** This cochlear duct is separated from the scala vestibuli by the **vestibular membrane** and from the scala tympani by the **basilar membrane.** Resting on the basilar membrane is the **spiral organ (organ of Corti),** the organ of hearing. Label these structures in Figure 14.16 on page 320.

The spiral organ (organ of Corti) is a coiled sheet of epithelial cells on the inner surface of the basilar membrane. This structure is composed of supporting cells and about 16,000 **hair cells,** which are receptors for hearing. At the apical tip of each hair cell is a **hair bundle,** consisting of 30–100 **stereocilia** that extend into the endolymph of the cochlear duct. The basal ends of the hair cells synapse with fibers of the cochlear branch of the vestibulocochlear nerve (cranial nerve VIII). Cell bodies of sensory neurons are located in the **spiral ganglion.** Projecting over and in contact with the hair cells of the spiral organ is the **tectorial** (*tector* = cover) **membrane,** a flexible gelatinous membrane. Label the tectorial membrane in Figure 14.16.

Obtain a prepared microscope slide of the spiral organ (organ of Corti) and examine under high power. Now label Figure 14.17 on page 321.

FIGURE 14.15 Details of inner ear.

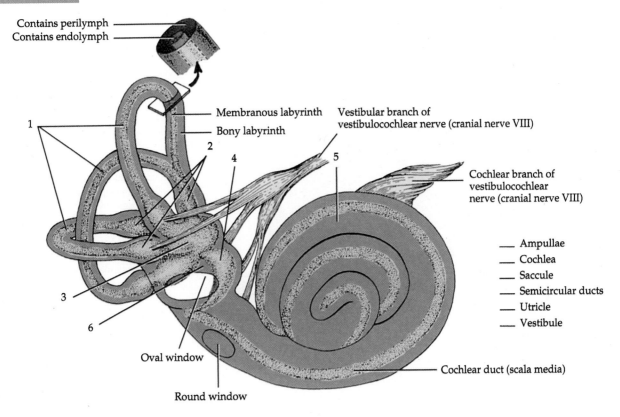

Contains perilymph
Contains endolymph

Membranous labyrinth
Bony labyrinth

Vestibular branch of
vestibulocochlear nerve (cranial nerve VIII)

Cochlear branch of
vestibulocochlear
nerve (cranial nerve VIII)

___ Ampullae
___ Cochlea
___ Saccule
___ Semicircular ducts
___ Utricle
___ Vestibule

Cochlear duct (scala media)

Oval window

Round window

2. Surface Anatomy

Refer to Figure 14.18 on page 322 for a summary of several surface anatomy features of the ear.

3. Tests for Auditory Acuity

Sound waves are a series of alternating high- and low-pressure regions traveling in the same direction through some medium (such as air).

Sound waves result from the alternate compression and decompression of air. The waves have both a frequency and an amplitude. The **frequency** is the distance between crests of a sound wave. It is measured in **Hertz (Hz),** or **cycles per second,** and is perceived as **pitch.** The greater the frequency of a sound, the higher its pitch. Different frequencies displace different areas of the basilar membrane. The **amplitude** of a sound wave is perceived as **intensity (loudness).** Intensity is measured in **decibels (dB).** Population norms have been calculated that measure the intensity of a sound that can just be heard, that is, the **threshold** of sound. These levels are said to have an intensity of 0 dB. Each 10 dB reflects a tenfold in-

crease in intensity; thus a sound is 10 times louder than threshold at 10 dB, 100 times greater at 20 dB, 1 million times greater at 60 dB, and so on.

a. Sound Localization Test

The cerebral cortex localizes the source of a sound by evaluating the time lag between the entry of sound into each ear. Without turning the head, it is impossible to distinguish between the origin of sounds that arise at 30° ahead of the right ear and 30° behind the right ear.

PROCEDURE

1. In a quiet room, have a subject occlude one ear with a cotton plug and close the eyes.

2. Move a ticking watch to various points 6 in. away from the ear (side, front, top, back) and ask the subject to point to the location of the sound.

3. Note in which positions the sound is best localized. In what regions was sound localization most accurate? _____

_____

FIGURE 14.16 **Cochlea.**

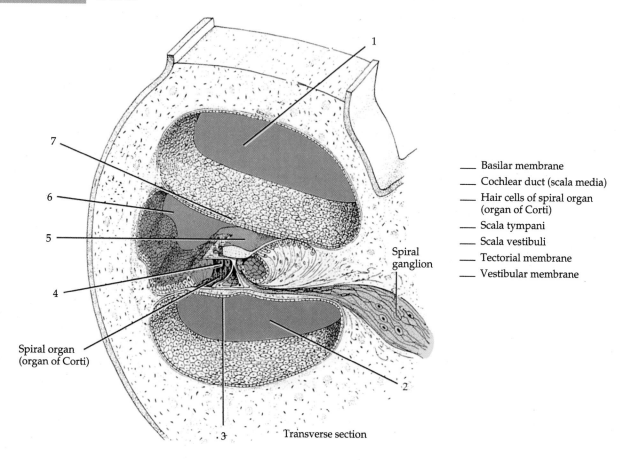

Basilar membrane
Cochlear duct (scala media)
Hair cells of spiral organ (organ of Corti)
Scala tympani
Scala vestibuli
Tectorial membrane
Vestibular membrane

Spiral ganglion

Spiral organ (organ of Corti)

Transverse section

b. Weber's Test

Weber's test is diagnostic for both conduction and sensorineural deafness. **Conduction deafness** results from interference with sound waves reaching the inner ear (plugged external auditory canal, fusion of the ossicles, etc.); **sensorineural deafness** is caused by either impairment of hair cells in the cochlea or damage of the cochlear branch of the vestibulocochlear nerve (cranial nerve VIII). This test should be performed in a quiet room.

PROCEDURE

1. Strike a tuning fork with a mallet and place the tip of the handle on the midline of a subject's forehead.

2. Determine if the tone is equally loud in both ears. _____

3. Have the subject occlude the left ear with cotton and repeat the exercise; occlude the right ear and repeat.

Describe your results: _____

_____

If the sound is equally loud in both unoccluded ears, hearing is normal. If the sound is louder in the occluded ear, hearing is normal. This situation parallels conduction deafness, when the cochlea is not receiving environmental background noise but only the vibrations of the tuning fork. If there is sensorineural deafness, the sound is louder in the normal (or nonoccluded) ear, because sensorineural loss produces a deficit in transmission to the cerebral cortex.

c. Rinne Test for Conduction Deafness
PROCEDURE

1. In a quiet room, strike a tuning fork with a mallet and place the handle of the tuning fork on the right mastoid process of a subject.

FIGURE 14.17 Photomicrographs of the spiral organ (organ of Corti).

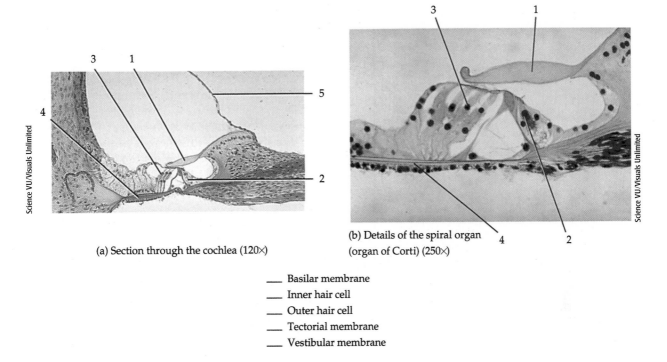

(a) Section through the cochlea (120×)

(b) Details of the spiral organ
(organ of Corti) (250×)

Science VU/Visuals Unlimited

____ Basilar membrane

____ Inner hair cell

____ Outer hair cell

____ Tectorial membrane

____ Vestibular membrane

2. When the sound is no longer audible to the subject, move the vibrating fork close to the external auditory canal.

3. Ask the subject to note whether sound is still heard.

4. If the subject hears the fork again when moved near the external auditory canal (by air conduction), hearing is *not* impaired.

5. Repeat the test by testing air conduction hearing first by striking a tuning fork and holding it near the external auditory canal. When the sound is no longer audible, place the handle of the tuning fork on the mastoid process.

6. If the subject now hears the tuning fork again (by bone conduction), this indicates some degree of conduction deafness.

7. Repeat steps 1 through 6 using the left ear of the subject.

8. Record your results in Section I.3.c of the LABORATORY REPORT RESULTS at the end of the exercise.

With normal hearing ability, air conduction lasts longer than bone conduction, because the sound waves are not absorbed by the tissues. Therefore, the sound is heard outside the external ear. In conduction deafness, bone conduction is better than air conduction because the sound waves can go to the cochlea without interference by damaged structures. Therefore, the sound will not be heard outside the external ear.

4. Equilibrium Apparatus

There are two kinds of **equilibrium** (balance). One kind of equilibrium, called **static equilibrium,** refers to the maintenance of the position of the body (mainly the head) relative to the force of gravity. The second kind of equilibrium, called **dynamic equilibrium,** is the maintenance of the position of the body (mainly the head) in response to sudden movements (rotation, acceleration, and deceleration). Collectively, the receptor organs for equilibrium are called the **vestibular apparatus,** which includes the maculae in the saccule and utricle and the cristae in the semicircular ducts.

The **maculae** (MAK-ū-lē) in the walls of the **utricle** and **saccule** are the receptors concerned mainly with static equilibrium. The maculae are small, thickened regions that resemble the spiral organ (organ of Corti) microscopically. Maculae are located in planes perpendicular to each other and contain two kinds of cells: **hair cells,** which are the sensory receptors, and **supporting cells.**

FIGURE 14.18 Surface anatomy of ear.

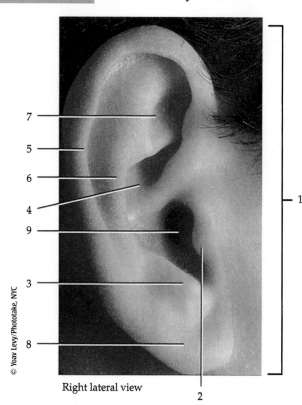

Right lateral view

© Yoav Levy/Phototake, NYC

1. **Auricle.** Portion of external ear not contained in head, also called the pinna.
2. **Tragus.** Cartilaginous projection.
3. **Antitragus.** Cartilaginous projection opposite tragus.
4. **Concha.** Hollow of auricle.
5. **Helix.** Superior and posterior free margin of auricle.
6. **Antihelix.** Semicircular ridge posterior and superior to concha.
7. **Triangular fossa.** Depression in superior portion of antihelix.
8. **Lobule.** Inferior portion of auricle devoid of cartilage.
9. **External auditory canal (meatus).** Canal extending from external ear to eardrum.

The hair cells feature **hair bundles** that consist of 70 or more **stereocilia** (microvilli) plus one **kinocilium** (conventional cilium). The columnar supporting cells are scattered among the hair cells. Resting over the hair cells is a thick, gelatinous glycoprotein layer, the **otolithic membrane.** A layer of dense calcium carbonate crystals, called **otoliths** (*oto* = ear; *liths* = stone), extends over the entire surface of the otolithic membrane. When

the head is tilted, the membrane (and the otoliths as well) slides over the hair cells in the direction determined by the tilt of the head. This sliding causes the membrane to bend the hair bundles, thus initiating a nerve impulse that is conveyed via the vestibular branch of the vestibulocochlear nerve (cranial nerve VIII) to the brain (cerebellum). The cerebellum sends continuous nerve impulses to the motor areas of the cerebral cortex in response to input from the maculae in the utricle and saccule, causing the motor system to increase or decrease its nerve impulses to specific skeletal muscles to maintain static equilibrium.

Label the parts of the macula in Figure 14.19a on page 323.

Now consider the role of the cristae in the semicircular ducts in maintaining dynamic equilibrium. The three semicircular ducts are positioned at right angles to one another in three planes: the two vertical ones are called the **anterior** and **posterior semicircular ducts,** and the horizontal one is called the **lateral semicircular duct.** This positioning permits detection of rotational acceleration or deceleration. In the **ampulla,** the dilated portion of each duct, is a small elevation called the **crista.** Each crista is composed of a group of **hair cells** and **supporting cells** covered by a mass of gelatinous material called the **cupula** (KŪ-pū-la). When the head moves, endolymph in the semicircular ducts flows over the hairs and bends them as water in a stream bends the plant life growing at its bottom. Bending of the hair bundles provides receptor potentials, which lead to nerve impulses that pass over the vestibular branch of the vestibulocochlear nerve (cranial nerve VIII). The nerve impulses follow the same pathway as those involved in static equilibrium and are eventually sent to the muscles that contract to maintain body balance in the new position.

Label the parts of the crista in Figure 14.19b.

5. Tests for Equilibrium

You can test equilibrium by using a few simple procedures.

a. Balance Test
This test is used to evaluate static equilibrium.

PROCEDURE

1. Instruct a subject to stand perfectly still with the arms at the sides and the eyes closed.
2. Observe any swaying movements. These are easier to detect if the subject stands in front of a blackboard with a light in front of the subject.

FIGURE 14.19 Equilibrium apparatus.

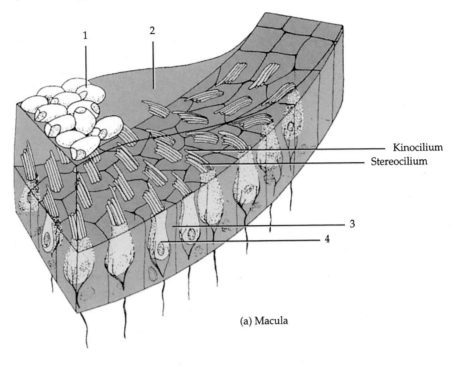

Kinocilium
Stereocilium

(a) Macula

___ Hair cell
___ Otoliths
___ Otolithic membrane
___ Supporting cell

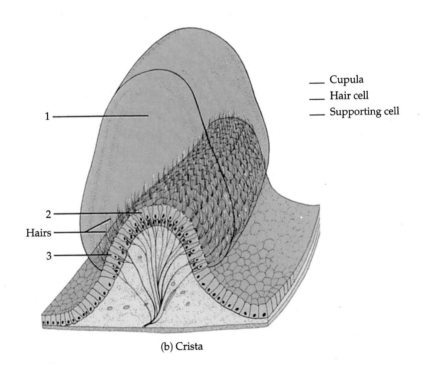

___ Cupula
___ Hair cell
___ Supporting cell

Hairs

(b) Crista

3. A mark is made at the edge of the shadow of the subject's shoulders and movement of the shadow is then observed.

If the static equilibrium system is dysfunctional the subject will sway or fall, although other proprioceptors may compensate for the defect.

Record your observations in Section I.5.a of the LABORATORY REPORT RESULTS at the end of the exercise.

b. Barany Test

This test evaluates the function of each of the semicircular canals. A subject is rotated, and when rotation is stopped, the momentum causes the endolymph in the lateral (horizontal) canal to continue spinning. The spinning bends the cupula in the direction of rotation. Other sensory input informs us that we are no longer rotating. **Nystagmus,** the rapid involuntary movement of the eyeballs, attempts to compensate for the loss of balance by visual fixation on an object. When the head is tipped toward the shoulder, the anterior semicircular canals are stimulated, so nystagmus is vertical; when the head is tipped slightly forward, the lateral canals are stimulated by the rotation, so nystagmus occurs laterally; when the head is bent onto the chest, the posterior canal receptors are stimulated, so nystagmus is rotational.

PROCEDURE

1. Use a person who is not subject to vertigo (dizziness).

2. The subject sits firmly anchored on a stool, legs up on the stool rung, head tilted onto one shoulder. Then the stool is *very carefully* revolved to the right about 10 times (about one turn every 2 sec) and suddenly stopped.

3. Observe the subject's eye movements.

⚠ **CAUTION!** *Watch the subject carefully and be prepared to provide support until the dizziness has passed.*

4. The subject will experience the sensation that the stool is still rotating, which means that the semicircular canals are functioning properly.

5. Repeat on a different subject with the head tipped slightly forward. Use still another subject with the chin touching the chest.

6. Record the direction of nystagmus in all three subjects in Section I.5.b of the LABORATORY REPORT RESULTS at the end of the exercise.

J. Sensory-Motor Integration

Sensory systems provide the input that keeps the CNS informed of changes in the external and internal environment. Responses to this information are conveyed to motor systems, which enable us to move about, alter glandular secretions, and change our relationship to the world around us. As sensory information reaches the CNS, it becomes part of a large pool of sensory input. We do not actively respond to every bit of input the CNS receives. Rather, the incoming information is integrated with other information arriving from all other operating sensory receptors. The integration process occurs not just once, but at many stations along the pathways of the CNS and at both conscious and subconscious levels. It occurs within the spinal cord, brain stem, cerebellum, basal ganglia, and cerebral cortex. As a result, a motor response to make a muscle contract or a gland secrete can be modified at any of these levels. Motor portions of the cerebral motor cortex play the major role in initiating and controlling precise, discrete muscular movements. The basal ganglia largely integrate semivoluntary, automatic movements like walking, swimming, and laughing. The cerebellum assists the motor cortex and basal ganglia by making body movements smooth and coordinated and by contributing significantly to maintaining normal posture and balance.

K. Somatic Motor Pathways

After receiving and interpreting sensory information, the CNS generates nerve impulses to direct responses to that sensory input. The nerve impulses are sent down the spinal cord in two major motor pathways: the direct pathways and the indirect pathways.

1. Direct Pathways

Voluntary motor impulses propagate from the motor cortex to voluntary motor neurons (somatic motor neurons) that innervate skeletal muscles via the **direct** or **pyramidal** (pi-RAM-i-dal) **pathways.** The direct pathways convey impulses from the cortex that result in precise, voluntary movements and include three tracts: lateral corticospinal (control muscles in the distal portions of the limbs), anterior corticospinal (control muscles of the neck and trunk), and corticobulbar (control

FIGURE 14.20 Direct (pyramidal) pathways: lateral and anterior corticospinal tracts.

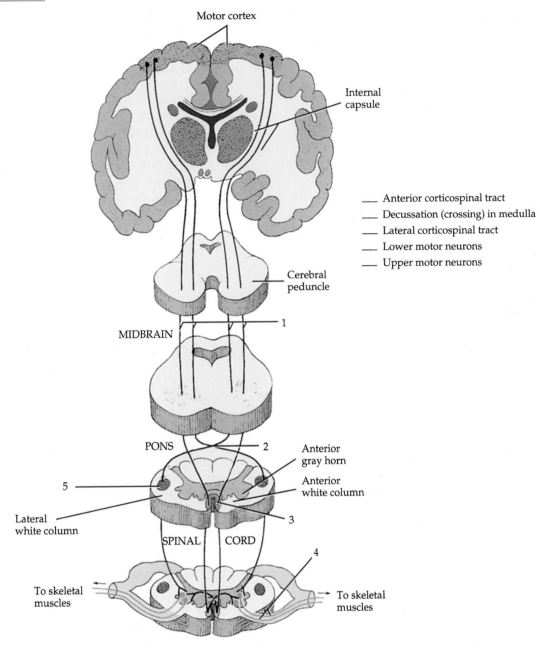

Motor cortex

Internal capsule

___ Anterior corticospinal tract
___ Decussation (crossing) in medulla
___ Lateral corticospinal tract
___ Lower motor neurons
___ Upper motor neurons

Cerebral peduncle

MIDBRAIN

1

PONS

2

Anterior gray horn

Anterior white column

5

Lateral white column

3

SPINAL   CORD

4

To skeletal muscles

To skeletal muscles

muscles of the eyes, tongue, and neck; chewing; facial expression; and speech) (see page 261).

Based on the description provided on page 261, label the components of the lateral corticospinal tracts in Figure 14.20.

2. Indirect Pathways

The **indirect pathways** include all motor tracts other than the corticospinal and corticobulbar tracts (see page 262). These are the rubrospinal, tectospinal, vestibulospinal, and lateral and medial reticulospinal.

ANSWER THE LABORATORY REPORT QUESTIONS AT THE END OF THE EXERCISE.

EXERCISE 14

Sensory Receptors and Sensory and Motor Pathways

Name _____   Date _____

Laboratory Section _____   Score/Grade _____

SECTION D ■ Tests for Somatic Senses

1. Two-Point Discrimination Test

Part of body	Least distance at which two points can be detected
Tip of tongue	1.4 mm
Tip of finger	_____
Side of nose	_____
Back of hand	_____
Back of neck	36.2 mm

3. Identifying Pressure Receptors

Part of body	Distances between points touched by chalk
Palm	_____
Arm	_____
Forearm	_____
Back of neck	_____

6. Identifying Proprioceptors

First trial	_____
Second trial	_____
Third trial	_____

SECTION F ▪ Olfaction

2. Olfactory Adaptation

Substance	Adaptation time

SECTION G ▪ Gustation

2. Identifying Taste Zones

Areas of Tongue in Which Basic Tastes Are Detected

	Sweet	Bitter	Salty	Sour
Tip of tongue				
Back of tongue				
Sides of tongue				

4. Taste and Smell

	Sensations when placed on dry tongue	Sensations while chewing (nostrils closed)	Sensations while chewing with nostrils opened
Carrot			
Onion			
Potato			
Apple			

SECTION H ■ Vision

5. Testing for Visual Acuity

Visual acuity, left eye   _____

Visual acuity, right eye   _____

Visual acuity, both eyes   _____

11. Testing for Afterimages

	Appearance of positive afterimage	Appearance of negative afterimage
Bright light		
Red cross		

SECTION I ■ Hearing and Equilibrium

3. Tests for Auditory Acuity

c. Rinne Test for Conduction Deafness

	Air conduction	Bone conduction
Right ear		
Left ear		

5. Tests for Equilibrium

a. Balance Test

Describe the response of the subject. _____

_____

_____

b. Barany Test

Subject 1 _____

Subject 2 _____

Subject 3 _____

EXERCISE 14

Sensory Receptors and Sensory and Motor Pathways

Name _____ Date _____

Laboratory Section _____ Score/Grade _____

PART 1 ■ Multiple Choice

_____ 1. The process by which the brain refers sensations to their point of stimulation is referred to as (a) modality (b) projection (c) accommodation (d) convergence

_____ 2. An awareness of the activities of muscles, tendons, and joints is known as (a) referred pain (b) adaptation (c) refraction (d) proprioception

_____ 3. The papillae located in an inverted V-shaped row at the posterior portion of the tongue are the (a) circumvallate (b) filiform (c) fungiform (d) gustatory

_____ 4. The technical name for the "white of the eye" is the (a) cornea (b) conjunctiva (c) choroids (d) sclera

_____ 5. Which is *not* a component of the vascular tunic? (a) choroids (b) macula lutea (c) iris (d) ciliary body

_____ 6. The amount of light entering the eyeball is regulated by the (a) lens (b) iris (c) cornea (d) conjunctiva

_____ 7. Which region of the eye is concerned primarily with image formation? (a) retina (b) choroids (c) lens (d) ciliary body

_____ 8. The densest concentration of cones is found at the (a) blind spot (b) macula lutea (c) central fovea (d) optic disc

_____ 9. Which region of the eyeball contains the vitreous body? (a) anterior chamber (b) vitreous chamber (c) posterior chamber (d) conjunctiva

_____ 10. Among the structures found in the middle ear are the (a) vestibule (b) auditory ossicles (c) semicircular canals (d) external auditory canal

_____ 11. The receptors for dynamic equilibrium are the (a) saccules (b) utricles (c) cristae in semi-circular ducts (d) spiral organs (organs of Corti)

_____ 12. The auditory ossicles are attached to the eardrum, to each other, and to the (a) semicircular ducts (b) semicircular canals (c) oval window (d) labyrinth

_____ 13. Another name for the internal ear is the (a) labyrinth (b) fenestra (c) cochlea (d) vestibule

_____ 14. Orientation of the body relative to the force of gravity is termed (a) the postural reflex (b) the tonal reflex (c) dynamic equilibrium (d) static equilibrium

_____ 15. The inability to feel a sensation consciously even though a stimulus is still being applied is called (a) modality (b) projection (c) adaptation (d) afterimage formation

_____ **16.** Which sequence best describes the normal flow of tears from the eyes into the nose? (a) lacrimal canals, lacrimal sacs, nasolacrimal ducts (b) lacrimal sacs, lacrimal canals, nasolacrimal ducts (c) nasolacrimal ducts, lacrimal sacs, lacrimal canals (d) lacrimal sacs, nasolacrimal ducts, lacrimal canals

_____ **17.** A patient whose lens has lost transparency is suffering from (a) glaucoma (b) conjunctivitis (c) cataract (d) trachoma

_____ **18.** The portion of the eyeball that contains aqueous humor is the (a) anterior cavity (b) lens (c) posterior chamber (d) macula lutea

_____ **19.** The organ of hearing located within the inner ear is the (a) vestibule (b) oval window (c) modiolus (d) spiral organ (organ of Corti)

_____ **20.** The sense organs of static equilibrium are the (a) semicircular ducts (b) membranous labyrinths (c) maculae in the utricle and saccule (d) pinnae

_____ **21.** Which receptor does *not* belong with the others? (a) muscle spindle (b) tendon organ (Golgi tendon organ) (c) joint kinesthetic receptor (d) lamellated (pacinian) corpuscle

_____ **22.** The membrane that is reflected from the eyelids onto the eyeball is the (a) retina (b) bulbar conjunctiva (c) sclera (d) choroids

_____ **23.** A characteristic of sensations by which one sensation may be distinguished from another is called (a) modality (b) projection (c) adaptation (d) afterimage formation

_____ **24.** Which are *not* cutaneous receptors? (a) hair root plexuses (b) muscle spindles (c) type I cutaneous mechanoreceptors (tactile or Merkel discs) (d) corpuscles of touch (Meissner's corpuscles)

_____ **25.** Which region of the tongue reacts strongest to bitter tastes? (a) tip (b) center (c) back (d) sides

_____ **26.** Which of the following values indicates the best visual acuity (a) 20/30 (b) 20/40 (c) 20/50 (d) 20/60

_____ **27.** Nearsightedness is referred to as (a) emmetropia (b) hypermetropia (c) eumetropia (d) myopia

_____ **28.** Which is an indirect pathway? (a) lateral corticospinal (b) anterior corticospinal (c) tectospinal (d) corticobulbar

_____ **29.** The presence of a sensation after a stimulus has been removed is called (a) adaptation (b) afterimage (c) translation (d) learned origin

_____ **30.** Receptors that provide information about the external environment are called (a) visceroreceptors (b) proprioceptors (c) baroreceptors (d) exteroceptors

PART 2 ▪ Completion

31. Receptors found in blood vessels and viscera are classified as _____.

32. Structures that collectively produce and drain tears are referred to as the

_____.

33. The three zones of the inner nervous layer of the retina (nervous tunic) are the photoreceptor layer, bipolar cell layer, and _____.

34. The small area of the retina where no images are formed is referred to as the

_____.

35. Abnormal elevation of intraocular pressure (IOP) is called _____.

36. In the visual pathway, nerve impulses pass from the optic chiasm to the _____ before passing to the thalamus.

37. The openings between the middle and inner ears are the oval and _____ windows.

38. The fluid within the bony labyrinth is called _____.

39. Tactile sensations include touch, pressure, itch, tickle, and _____.

40. Receptors for pressure are type II cutaneous mechanoreceptors (Ruffini corpuscles), and _____ corpuscles.

41. Proprioceptive receptors that provide information about the degree and rate of angulations of joints are _____.

42. The neural pathway for olfaction includes olfactory receptors, olfactory bulbs, _____, and the cerebral cortex.

43. At the end of each semicircular canal is a swollen enlargement called the _____.

44. A thin, semitransparent partition of fibrous connective tissue that separates the external auditory canal from the middle ear is the _____.

45. The fluid in the membranous labyrinth is called _____.

46. The cochlear duct is separated from the scala vestibule by the _____.

47. The gelatinous glycoprotein layer over the hair cells in the maculae is called the _____.

48. _____ is blurred vision caused by an irregular curvature of the surface of the cornea or lens.

49. Image formation requires refraction, accommodation, and _____.

50. The _____ tract conveys sensory impulses for pain and temperature.

51. The _____ tract conveys motor impulses for precise contraction of muscles in the distal portions of the limbs.

52. Pain receptors are called _____.

53. The ability to recognize exactly what point of the body is touched is called _____.

54. _____ neurons extend from cranial nerve motor nuclei or spinal cord anterior horns to skeletal muscle fibers.

55. _____ tracts convey impulses that control voluntary movements of the head and neck.

Endocrine System

Objectives

At the completion of this exercise you should understand

A The difference between endocrine and exocrine glands

B The location, histology, and functions of the following endocrine glands: pituitary, thyroid, parathyroid, adrenal, pancreatic islets, testes, ovaries, pineal, and thymus glands

You have learned how the nervous system controls the body through nerve impulses that are delivered over neurons. Another system of the body, the **endocrine system,** is also involved in controlling bodily functions. The endocrine glands affect bodily activities by releasing chemical mediators, called **hormones,** into the bloodstream (the term *hormone* means "to excite or get moving"). The nervous and endocrine systems coordinate their regulatory activities via a complex series of interacting activities. In some instances the nervous and endocrine systems function together as an interlocking "supersystem" termed the **neuroendocrine system.** Certain parts of the nervous system stimulate or inhibit the release of hormones. The hormones, in turn, are quite capable of stimulating or inhibiting the flow of particular nerve impulses.

The body contains two different kinds of glands: exocrine and endocrine. **Exocrine glands** secrete their products into ducts. The ducts then carry the secretions into body cavities, into the lumens of various organs, or to the outer surface of the body. Examples are sudoriferous (sweat), sebaceous (oil), mucous, and digestive glands. **Endocrine glands,** by contrast, secrete their products (hormones) into the interstitial fluid around the secretory cells. The secretion then diffuses into capillaries and blood carries them away. Because they have no ducts, endocrine glands are also called **ductless glands.**

A. Endocrine Glands

The endocrine glands include the pituitary gland, thyroid gland, parathyroid glands, adrenal glands, and pineal gland. In addition, several organs and tissues are not exclusively classified as endocrine glands but contain cells that secrete hormones. They include the hypothalamus, thymus gland, pancreas, testes, ovaries, kidneys, stomach, liver, small intestine, skin, heart, adipose tissue, and placenta. The endocrine glands and hormone secreting tissue constitute the **endocrine system.**

Locate the endocrine glands on a torso and, using your textbook or charts for reference, label Figure 15.1.

All hormones maintain homeostasis by changing the physiological activities of cells. A hormone may stimulate changes in the cells of one or more organs. The cells that respond to the effects of a hormone are called **target cells.**

B. Pituitary Gland (Hypophysis)

The **pituitary** (pi-TOO-i-tār-ē) **gland,** also called the **hypophysis** (hī-POF-i-sis), is nicknamed the "master" endocrine gland because it secretes several hormones that control other endocrine glands. This nickname is a misnomer since we now know that the pituitary gland is controlled by the **hypothalamus.** The pituitary gland lies in the hypophyseal fossa of the sella turcica of the sphenoid bone and

FIGURE 15.1 Location of endocrine glands, organs containing endocrine tissue, and associated structures.

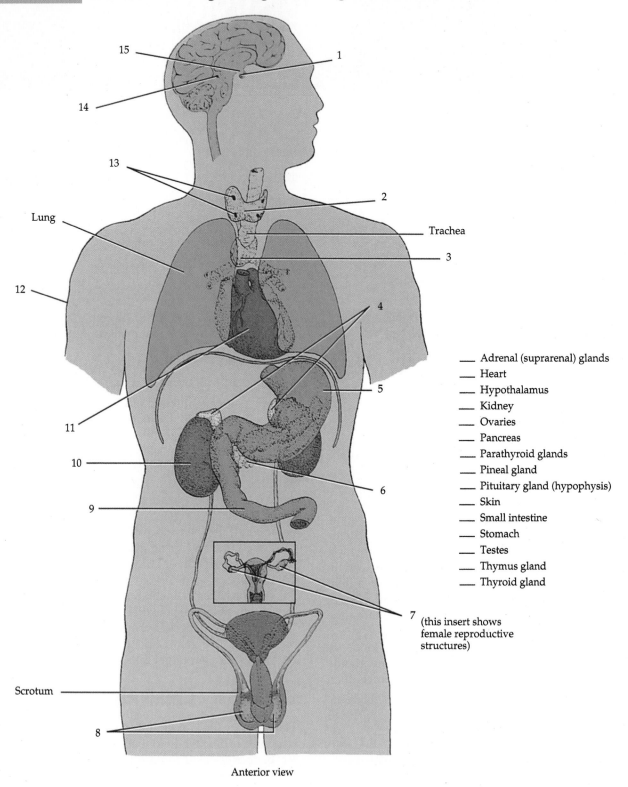

Lung

Trachea

Scrotum

Anterior view

___ Adrenal (suprarenal) glands
___ Heart
___ Hypothalamus
___ Kidney
___ Ovaries
___ Pancreas
___ Parathyroid glands
___ Pineal gland
___ Pituitary gland (hypophysis)
___ Skin
___ Small intestine
___ Stomach
___ Testes
___ Thymus gland
___ Thyroid gland

(this insert shows female reproductive structures)

attaches to the hypothalamus via a stalklike structure termed the **infundibulum,** meaning funnel. Not only is the hypothalamus of the brain an important regulatory center in the nervous system; it is also a crucial endocrine gland. Cells in the hypothalamus synthesize at least nine different hormones, and the pituitary gland secretes seven more. Together, these 16 hormones play important roles in the regulation of virtually all aspects of growth, development, metabolism, and homeostasis.

The pituitary gland is divided structurally and functionally into an **anterior pituitary gland,** also called the **adenohypophysis** (ad'-e-nō-hī-POF-i-sis), and a **posterior pituitary gland,** also called the **neurohypophysis** (noo'-rō-hī-POF-i-sis). The anterior pituitary gland, which accounts for about 75% of the total weight of the gland, contains many glandular epitheloid (epithelial-like) cells and forms the glandular part of the pituitary gland. The hypothalamus is connected to the anterior pituitary gland by a series of blood vessels, the **hypothalamic–hypophyseal portal system.** The posterior pituitary gland contains the axons and axon terminals of more than 10,000 neurons whose cell bodies are in the hypothalamus. These terminals form the neural part of the pituitary gland. Between the anterior and posterior pituitary glands is a small, relatively avascular rudimentary zone, the **pars intermedia,** whose role in humans is unknown.

1. Histology of the Pituitary Gland

The anterior pituitary gland releases hormones that regulate a wide range of body activities, from growth to reproduction. However, the release of these hormones is stimulated by **releasing hormones** or suppressed by **inhibiting hormones** that are produced by neurosecretory cells in the hypothalamus of the brain. The hypothalamic hormones reach the anterior pituitary gland through a network of blood vessels.

There are five types of anterior pituitary cells—somatotrophs, thyrotrophs, gonadotrophs, lactotrophs, and corticotrophs—which secrete seven hormones.

When the anterior pituitary gland receives proper stimulation from the hypothalamus via releasing or inhibiting hormones, its glandular cells increase or decrease the secretion of any one of seven hormones. They are as follows:

1. **Somatotrophs** (*somato* = body; *tropin* = change) secrete **human growth hormone (hGH)** or **somatotropin.** Human growth hormone in turn stimulates several tissues to secrete insulinlike growth factors, hormones that stimulate general body growth and certain aspects of metabolism.

2. **Thyrotrophs** (*thyro* = pertaining to the thyroid glands) secrete **thyroid-stimulating hormone (TSH)** or **thyrotropin,** which controls the secretions and activities of the thyroid gland.

3. **Gonadotrophs** (*gonado* = seed) secrete **follicle-stimulating hormone (FSH)** and **luteinizing hormone (LH).** These hormones stimulate secretion of estrogens and progesterone by the ovaries and maturation of oocytes, and secretion of testosterone and production of sperm in the testes.

4. **Lactotrophs** (*lacto* = milk) secrete **prolactin (PRL),** which initiates milk production in the mammary glands.

5. **Corticotrophs** (*cortex* = rind or bark) secrete **adrenocorticotropic** (ad-rē-nō-kor'-ti-kō-TRŌ-pik) **hormone (ACTH)** or **corticotrophin,** which stimulates the adrenal cortex to secrete glucocorticoids, and **melanocyte-stimulating hormone (MSH),** whose exact role in humans is unknown.

Except for human growth hormone (hGH), melanocyte-stimulating hormone (MSH), and prolactin (PRL), all the secretions are referred to as **tropins** or **tropic hormones,** which means that their target organs are other endocrine glands. Follicle-stimulating hormone (FSH) and luteinizing hormone (LH) are also called **gonadotropic** (gō-nad-ō-TRŌ-pik) **hormones** or **gonadotropins** because they regulate the functions of the gonads (ovaries and testes). The gonads are the endocrine glands that produce sex hormones.

Examine a prepared slide of the anterior pituitary gland and identify as many types of cells as possible with the aid of your textbook or a histology textbook.

The posterior pituitary gland does not synthesize hormones; it does *store* and *release* hormones synthesized by cells of the hypothalamus. The posterior pituitary gland consists of (1) cells called **pituicytes** (pi-TOO-i-sītz), which are similar in appearance to the neuroglia of the nervous system, and (2) axon terminations of hypothalamic **neurosecretory cells.** The cell bodies of these neurons originate in nuclei in the hypothalamus. The fibers project from the hypothalamus, form the **hypothalamohypophyseal** (hī-po-thal-a-mō-hī'-pō-FIZ-ē-al) **tract,** and terminate near blood capillaries in the posterior pituitary gland. The cell bodies of the neurosecretory cells produce the hormones oxytocin (ok′-sē-TŌ-sin), or **OT,** and **antidiuretic hormone (ADH),** also called **vasopressin.** These hormones are transported by fast axonal transport to the axon terminals in the posterior pituitary gland, and are stored in the axon terminals resting on the capillaries. When properly stimulated, the hypothalamus sends impulses over the neurons. The impulses cause release of hormones from the axon terminals into the blood.

Examine a prepared slide of the posterior pituitary gland under high power. Identify the pituicytes and axon terminations of neurosecretory cells with the aid of your textbook.

2. Hormones of the Pituitary Gland

Using your textbook as a reference, give the major functions for the following hormones.

a. Hormones Secreted by the Anterior Pituitary Gland

Human growth hormone (hGH). Also called **somatotropin** _____

_____

_____

_____

Thyroid-stimulating hormone (TSH). Also called **thyrotropin** _____

_____

_____

Adrenocorticotropic hormone (ACTH). Also called **corticotropin** _____

_____

_____

_____

Follicle-stimulating hormone (FSH).

In female: _____

_____

_____

_____

In male: _____

_____

_____

_____

Luteinizing hormone (LH).

In female: _____

_____

_____

_____

In male: _____

_____

_____

_____

Prolactin (PRL). _____

_____

_____

Melanocyte-stimulating hormone (MSH). _____

_____

_____

_____

_____

b. Hormones Stored and Released by the Posterior Pituitary Gland

Oxytocin (OT). _____

_____

_____

Antidiuretic hormone (ADH). Also called **vasopressin** _____

_____

_____

_____

C. Thyroid Gland

The **thyroid gland** is a butterfly-shaped endocrine gland located just inferior to the larynx (voice box). It is composed of right and left **lateral lobes,** one on either side of the trachea, that are connected by a mass of tissue called an **isthmus** (IS-mus) that lies anterior to the trachea. A small **pyramidal lobe,** when present, extends upward from the isthmus.

1. Histology of the Thyroid Gland

Histologically, the thyroid gland consists of microscopic spherical sacs called **thyroid follicles.** The wall of each follicle consists primarily of cells called **follicular cells,** most of which extend to the lumen (internal space) of the follicle. In addition to the follicular cells, the thyroid gland also contains **parafollicular cells** (also termed **C cells**). These cells lie between follicles. Follicular cells synthesize the hormones **thyroxine** (thī-ROK-sēn) (also termed T_4 or **tetraiodothyronine**) and **triiodothyronine** (trī-ī-ō-dō-THĪ-rō-nēn) (T_3). Together these hormones are referred to as the **thyroid hormones.** Approximately 90% of the hormone secreted by the follicular cells is thyroxine, while the remaining 10% is triiodothyronine. The functions of these two hormones are essentially the same, but they differ in rapidity and intensity of action. T_4 has a significantly longer life within the bloodstream but is also significantly weaker than triiodothyronine. The parafollicular cells produce the hormone **calcitonin** (kal-si-TŌ-nin) (**CT**). Each thyroid follicle is filled with a glycoprotein called **thyroglobulin** (**TBG**), which is also called **thyroid colloid.**

Examine a prepared slide of the thyroid gland under high power. Identify the thyroid follicles, epithelial cells forming the follicle, and thyroid colloid. Now label the photomicrograph in Figure 15.2.

2. Hormones of the Thyroid Gland

Using your textbook as a reference, give the major functions of the following hormones.

Thyroxine (T_4) and *triiodothyronine (T_3).* _____

_____

_____

Calcitonin (CT). _____

_____

_____

_____

D. Parathyroid Glands

Partially embedded in the posterior surface of the lateral lobes of the thyroid gland are several small, round masses of tissue called the **parathyroid** (*para* = beside) **glands.** Usually, one superior and one inferior parathyroid gland are attached to each lateral thyroid lobe.

FIGURE 15.2 Histology of thyroid gland (400×).

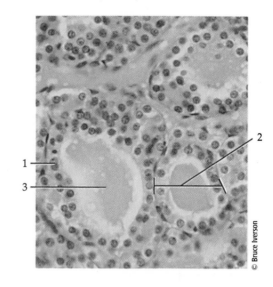

© Bruce Iverson

____ Epithelium of follicle
____ Thyroglobulin (TGB)
____ Thyroid follicle

1. Histology of the Parathyroid Glands

Microscopically, the parathyroid glands contain two kinds of epithelial cells. The more numerous cells called **chief (principal) cells** produce **parathyroid hormone (PTH)** or **parathormone.** The second cell, called an **oxyphil cell,** has an unknown function. The number of oxyphil cells increases with age.

Examine a prepared slide of the parathyroid glands under high power. Identify the principal and oxyphil cells. Now label the photomicrograph in Figure 15.3.

2. Hormone of the Parathyroid Glands

Using your textbook as a reference, give the major functions of the following hormone:

Parathyroid hormone (PTH). Also called **para-**

thormone _____

_____

_____

_____

FIGURE 15.3 Histology of parathyroid glands (320×).

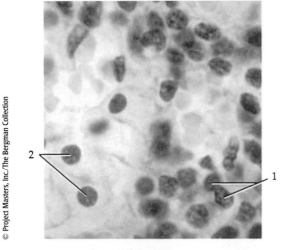

© Project Masters, Inc./The Bergman Collection

____ Oxyphil cells

____ Chief cells

E. Adrenal (Suprarenal) Glands

The paired **adrenal (suprarenal) glands** lie superior to each kidney, and each is structurally and functionally differentiated into two distinct regions: the large, peripherally located **adrenal cortex,** which forms 80% to 90% of the gland in humans, and the small, centrally located **adrenal medulla.** Covering the gland is a connective tissue **capsule.**

1. Histology of the Adrenal Cortex

Histologically, the adrenal cortex is subdivided into three zones, each of which secretes different hormones. The outer zone, just deep to the capsule, is called the **zona glomerulosa** (*glomerul* = little ball). Its cells, arranged in spherical clusters and arched columns, secrete hormones called **mineralocorticoids** (min′-er-al-ō-KOR-ti-koyds). The major mineralocorticoid produced by this region is **aldosterone.**

The middle zone of the adrenal cortex is the **zona fasciculata** (*fascicul* = little bundle). This zone, which is the widest of the three, consists of cells arranged in long, straight columns. The zona fasciculata secretes mainly **glucocorticoids** (gloo′-kō-KOR-ti-koyds). The zona fasciculata secretes three glucocorticoids, 95% of which is **cortisol** (also known as **hydrocortisone**). The re-

maining glucocorticoids synthesized include a small amount of **corticosterone** and a minute amount of **cortisone.**

The cells of the inner zone, the **zona reticularis** (*reticul* = net), are arranged in branching cords. This zone synthesizes minute amounts of male hormones **(androgens),** steroid hormones that have masculinizing effects. The major androgen secreted is dehydroepiandrosterone (DHEA) (dē-hī-drō-ep′-ē-an-DROS-ter-ōn).

Examine a prepared slide of the adrenal cortex under high power. Identify the capsule, zona glomerulosa, zona fasciculata, and zona reticularis. Now label Figure 15.4.

2. Hormones of the Adrenal Cortex

Using your textbook as a reference, give the major functions of the hormones listed on the next page.

FIGURE 15.4 Histology of adrenal (suprarenal) glands (100×).

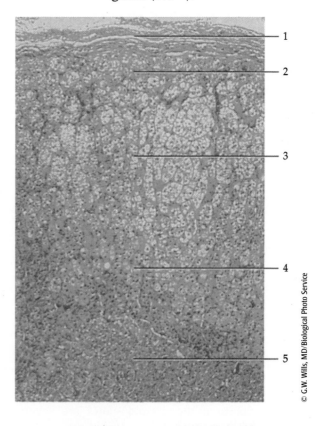

© G.W. Willis, MD/Biological Photo Service

____ Adrenal medulla   ____ Zona glomerulosa

____ Capsule   ____ Zona reticularis

____ Zona fasciculata

Mineralocorticoids (mainly *aldosterone*). _____

_____

_____

_____

Glucocorticoids (mainly *cortisol*). _____

_____

_____

_____

Androgens.

In male: _____

_____

In female: _____

_____

3. Histology of the Adrenal Medulla

The adrenal medulla consists of hormone-producing cells called **chromaffin** (KRŌ-maf-in; *chrom* = color; *affin* = affinity for) **cells.** These cells develop from the same embryonic tissue as all other sympathetic ganglia, but they lack axons and form clusters around large blood vessels. They are directly innervated by preganglionic cells of the sympathetic division of the autonomic nervous system and may be regarded as postganglionic cells that are specialized to secrete hormones. Secretion of hormones from the chromaffin cells is directly controlled by the sympathetic division of the autonomic nervous system, and innervation by the preganglionic fibers allows the gland to respond extremely rapidly to a stimulus. The adrenal medulla secretes the hormones **epinephrine** and **norepinephrine (NE),** also called adrenaline and noradrenaline, respectively.

Examine a prepared slide of the adrenal medulla under high power. Identify the chromaffin cells. Now locate the cells in Figure 15.4.

4. Hormones of the Adrenal Medulla

Using your textbook as a reference, give the major functions of the following hormones.

Epinephrine and *norepinephrine (NE).* _____

_____

_____

F. Pancreatic Islets

The **pancreas** is both an endocrine and an exocrine gland. Thus, it is referred to as a **heterocrine gland.** We shall discuss only its endocrine functions now. The pancreas is a flattened organ located in the curve of the duodenum. The adult pancreas consists of a head, body, and tail.

1. Histology of the Pancreas

The endocrine portion of the pancreas consists of 1 to 2 million tiny clusters of cells called **pancreatic islets (islets of Langerhans).** They contain four types of hormone-secreting cells: (1) **alpha cells** or **A cells,** which secrete the hormone **glucagon;** (2) **beta cells** or **B cells,** which secrete the hormone **insulin;** (3) **delta cells** or **D cells,** which secrete **somatostatin;** and (4) **F cells,** which secrete pancreatic polypeptide. The pancreatic islets are surrounded by blood capillaries and by cells called **acini** that form the exocrine part of the gland.

Examine a prepared slide of the pancreas under high power. Identify the alpha cells, beta cells, and acini (clusters of cells that secrete digestive enzymes) around the pancreatic islets. Now label Figure 15.5.

2. Hormones of the Pancreas

Using your textbook as a reference, give the major functions of the following hormones:

Glucagon. _____

_____

_____

Insulin. _____

_____

_____

Somatostatin. _____

_____

_____

Pancreatic polypeptide. _____

_____

_____

FIGURE 15.5   Histology of pancreas (500×).

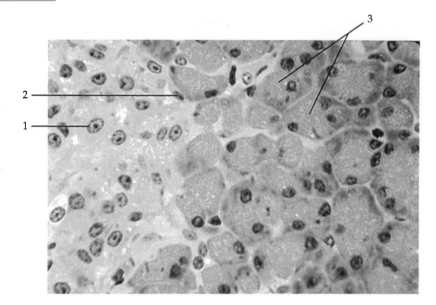

_____ Exocrine acini
_____ Alpha cell
_____ Beta cell

G. Testes

The **testes** (male gonads) are paired oval glands that lie in the scrotum. They are partially covered by a serous membrane called the **tunica** (*tunica* = sheath) **vaginalis,** which is derived from the peritoneum and forms during descent of the testes into the scrotum. Internal to the tunica vaginalis is a capsule of dense irregular connective tissue, the **tunica albuginea** (al'-bū-JIN-ē-a; *albus* = white), which extends inward, forming septa that divide each testis into a series of internal compartments called **lobules.** Each of the 200 to 300 lobules contains one to three tightly coiled tubules, the **seminiferous** (*semin* = seed; *fer* = to carry) **tubules,** where sperm are produced. The process by which the seminiferous tubules of the testes produce sperm is called **spermatogenesis** (sper'-ma-tō-JEN-e-sis).

1. Histology of the Testes

The seminiferous tubules contain two types of cells: spermatogenic cells and Sertoli cells. **Spermatogenic cells** are sperm-forming cells in various stages that undergo mitosis and differentiation to eventually produce sperm. Together with supporting cells, they line the seminiferous tubules. Sperm production begins at the periphery of the seminiferous tubules in stem cells called **spermatogonia** (sper'-ma'-tō-GŌ-nē-a; *gonia* = generation or offspring; the singular is **spermatogonium**). Toward

the lumen of the tubules are layers of progressively more mature cells. In order of advancing maturity, these are **primary spermatocytes** (SPER-ma-tō-sītz), **secondary spermatocytes, spermatids,** and **sperm.** By the time a **sperm cell** or **spermatozoon** (sper'-ma'-tō-ZŌ-on; *zoon* = life; the plural is **sperm** or **spermatozoa**) has nearly reached maturity, it is released into the lumen of the tubule.

Embedded among the spermatogenic cells in the tubules are large **Sertoli** or **sustentacular** (sus'-ten-TAK-ū-lar) **cells,** which extend from the basement membrane to the lumen of the tubule. Sertoli cells support and protect developing spermatogenic cells; nourish spermatocytes, spermatids, and sperm; phagocytize excess spermatid cytoplasm as development proceeds; and mediate the effects of testosterone and follicle-stimulating hormone (FSH). Sertoli cells also control movements of spermatogenic cells and the release of sperm into the lumen of the seminiferous tubule. They produce fluid for sperm transport, secrete androgen-binding protein, and secrete the hormone inhibin, which helps regulate sperm production by inhibiting the secretion of FSH.

In the spaces between adjacent seminiferous tubules are clusters of cells called **Leydig cells (interstitial endocrinocytes).** These cells secrete **testosterone,** the most prevalent androgen (male sex hormone).

Examine a prepared slide of the testes under high power and identify all of the structures listed. Now label Figure 15.6.

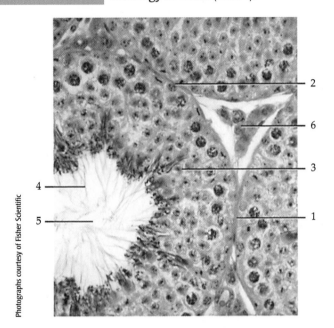

Photographs courtesy of Fisher Scientific

FIGURE 15.6 Histology of testes (100×).

___ Basement membrane
___ Leydig cell (interstitial endocrinocyte)
___ Lumen of seminiferous tubule
___ Spermatid
___ Spermatogonium
___ Sperm cell

2. Hormones of the Testes

Using your textbook as a reference, give the major functions of the following hormones.

Testosterone. _____

_____

_____

Inhibin. _____

_____

_____

H. Ovaries

The **ovaries** (= egg receptacle), or female gonads, are paired glands resembling unshelled almonds in size and shape. They are positioned in the superior pelvic cavity, one on each side of the uterus, and are maintained in position by a series of ligaments. The ovaries are (1) attached to the **broad ligament** of the uterus, which is itself part of the parietal peritoneum, by a double-layered fold of peritoneum called the **mesovarium;** (2) anchored to the uterus by the **ovarian ligament;** and (3) attached to the pelvic wall by the **suspensory ligament.** Each ovary also contains a **hilum,** which is the point of entrance and exit for blood vessels and nerves, and along which the mesovarium is attached.

1. Histology of the Ovaries

Histologically, each ovary consists of the following parts:

1. *Germinal epithelium.* A layer of simple epithelium (low cuboidal or squamous) that covers the free surface of the ovary and is continuous with the mesothelium that covers the mesovarium. The term *germinal epithelium* is a misnomer because it does not give rise to oocytes, although at one time it was believed that it did.

2. *Tunica albuginea.* A whitish capsule of dense, irregular connective tissue immediately deep to the germinal epithelium.

3. *Ovarian cortex.* A region just deep to the tunica albuginea that consists of dense connective tissue and contains ovarian follicles (described shortly).

4. *Ovarian medulla.* A region deep to the ovarian cortex that consists of loose connective tissue and contains blood vessels, lymphatic vessels, and nerves.

5. *Ovarian follicles* (*folliculus* = little bag). Are in the cortex and consist of **oocytes** (immature ova) in various stages of development and their surrounding cells. When the surrounding cells form a single layer, they are called **follicular cells.** Later in development, when they form several layers, they are referred to as **granulosa cells.** The surrounding cells nourish the developing oocyte and begin to secrete estrogens as the follicle grows larger. Ovarian follicles undergo a series of changes prior to ovulation, progressing through several distinct stages. The most numerous and peripherally arranged follicles are termed **primordial follicles.** If a primordial follicle progresses to ovulation (release of a mature ovum), it will sequentially transform into a **primary follicle,** then a **secondary follicle,** and finally a **mature (graafian) follicle.**

FIGURE 15.7   **Histology of ovary.**

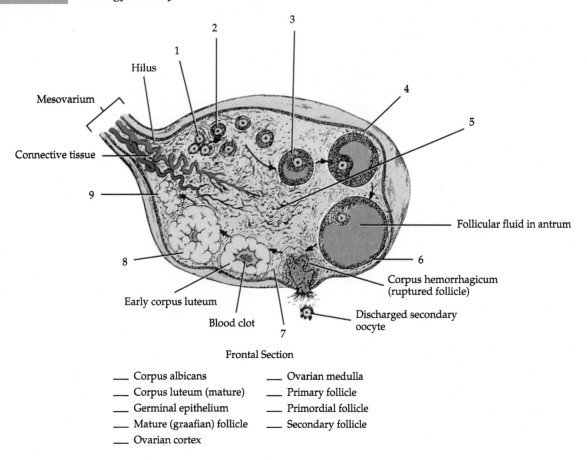

Frontal Section

___ Corpus albicans   ___ Ovarian medulla
___ Corpus luteum (mature)   ___ Primary follicle
___ Germinal epithelium   ___ Primordial follicle
___ Mature (graafian) follicle ___ Secondary follicle
___ Ovarian cortex

6. *Mature (graafian) follicle.* A large, fluid-filled follicle that is preparing to rupture and expel a secondary oocyte, a process called **ovulation.**

7. *Corpus luteum* (= yellow body). Contains the remnants of a mature follicle after ovulation. The corpus luteum produces progesterone, estrogens, relaxin, and inhibin until it degenerates and turns into fibrous tissue called a **corpus albicans** (= white body).

Examine a prepared slide of the ovary under high power. (You may need to examine more than one slide to see all of the structures listed.) Label Figure 15.7.

2. Hormones of the Ovaries

Using your textbook as a reference, give the major functions of the following hormones.

Estrogens. _____

_____

_____

_____

Progesterone. _____

_____

_____

_____

Relaxin. _____

_____

_____

_____

Inhibin. _____

_____

_____

I. Pineal Gland

The small, cone-shaped endocrine gland attached to the roof of the third ventricle of the brain at the midline is known as the **pineal** (PĪN-ē-al; *pinealis* = shaped like a pine cone) **gland.**

1. Histology of the Pineal Gland

The gland is covered by a capsule formed by the pia mater and consists of masses of neuroglial cells and secretory cells called **pinealocytes** (pin-ē-AL-ō-sītz). Sympathatic postganglionic axons from the superior cervical ganglion terminate in the pineal gangli.

The physiological roles of the pineal gland are still unclear. The gland secretes **melatonin.**

2. Hormone of the Pineal Gland

Using your textbook as a reference, give the major functions of the following hormone.

Melatonin. _____

_____

_____

J. Thymus Gland

The **thymus gland** is a bilobed lymphatic gland located in the upper mediastinum posterior to the sternum and between the lungs. The gland is conspicuous in infants and, during puberty, reaches maximum size.

1. Histology of Thymus Gland

After puberty, thymic tissue, which consists primarily of **lymphocytes,** is replaced by fat. By the time a person reaches maturity, the gland has atrophied. Lymphoid tissue of the body consists primarily of lymphocytes that may be distinguished into two kinds: B cells and T cells. Both are derived originally in the embryo from lymphocytic stem cells in red bone marrow. Before migrating to their positions in lymphoid tissue, the descendants of the stem cells follow two distinct pathways. About half of them migrate to the thymus gland, where they are processed to become thymus-dependent lymphocytes, or **T cells.** The thymus gland confers on them the ability to destroy antigens (foreign microbes and substances). The remaining stem cells are processed in some as yet undetermined area of the body,

possibly the fetal liver and spleen, and are known as **B cells.** Hormones produced by the thymus gland are **thymosin, thymic humoral factor (THF), thymic factor (TF),** and **thymopoietin.**

2. Hormones of the Thymus Gland

Using your textbook as a reference, write the major functions of the following hormones:

Thymosin, thymic humoral factor (THF), thymic factor (TF), and thymopoietin. _____

_____

_____

K. Other Endocrine Tissues

Body tissues other than endocrine glands also secrete hormones. The gastrointestinal tract synthesizes several hormones that regulate digestion in the stomach and small intestine. Among these hormones are **gastrin, secretin, cholecystokinin (CCK),** and **glucose-dependent insulinotropic peptide (GIP).**

The placenta produces **human chorionic gonadotropin (hCG), estrogens, progesterone,** and **human chorionic somatomammotropin (hCS),** all of which are related to pregnancy.

When the kidneys (and liver, to a lesser extent) become hypoxic (subject to below-normal levels of oxygen), they release an enzyme called **renal erythropoietic factor.** This is secreted into the blood, where it acts on a plasma protein produced in the liver to form a hormone called **erythropoietin** (ē-rith′-rō-POY-ē-tin), or **EPO,** which stimulates red bone marrow to produce more red blood cells and hemoglobin. This ultimately reverses the original stimulus (hypoxia). The kidneys also secrete **rennin,** which helps regulate blood pressure.

Vitamin D, produced by the skin, liver, and kidneys in the presence of sunlight, is converted to its active hormone, **calcitriol,** in the kidneys and liver.

The atria of the heart secrete a hormone called **atrial natriuretic peptide (ANP),** released in response to increased blood volume.

Adipose tissue secretes a hormone called **leptin,** that influences appetite, and may affect GnRH activity and gonadotropins.

Using your textbook as a reference, write the major functions of the following hormones.

Gastrin. _____

_____

Secretin. _____

_____

Cholecystokinin (CCK). _____

_____

Glucose-dependent insulinotropic peptide (GIP).

_____

Human chorionic gonadotropin (hCG). _____

_____

Human chorionic somatomammotropin (hCS). ___

_____

Renin. _____

_____

Erythropoietin (EPO). _____

_____

Calcitriol. _____

_____

Atrial natriuretic peptide (ANP). _____

_____

Leptin. _____

_____

**ANSWER THE LABORATORY REPORT QUESTIONS
AT THE END OF THE EXERCISE.**

EXERCISE 15

Endocrine System

Name _____   Date _____

Laboratory Section _____   Score/Grade _____

PART 1 ■ Multiple Choice

_____ 1. Somatotrophs, gonadotrophs, and corticotrophs are associated with the (a) thyroid gland (b) anterior pituitary gland (c) parathyroid glands (d) adrenal glands

_____ 2. The posterior pituitary gland is *not* an endocrine gland because it (a) has a rich blood supply (b) is not near the brain (c) does not synthesize hormones (d) contains ducts

_____ 3. Which hormone assumes a role in the development and discharge of a secondary oocyte? (a) hGH (b) TSH (c) LH (d) PRL

_____ 4. The endocrine gland that is probably malfunctioning if a person has a high metabolic rate is the (a) thymus gland (b) posterior pituitary gland (c) anterior pituitary gland (d) thyroid gland

_____ 5. The antagonistic hormones that regulate blood calcium level are (a) hGH-TSH (b) insulin-glucagon (c) aldosterone-cortisone (d) CT-PTH

_____ 6. The endocrine gland that develops from the sympathetic nervous system is the (a) adrenal medulla (b) pancreas (c) thyroid gland (d) anterior pituitary gland

_____ 7. The hormone that aids in sodium conservation and potassium excretion is (a) hydrocortisone (b) CT (c) ADH (d) aldosterone

_____ 8. Which of the following hormones produce effects that mimic those of the sympathetic nervous system? (a) insulin (b) oxytocin (OT) (c) epinephrine (d) testosterone

_____ 9. Which hormone lowers blood glucose level? (a) glucagon (b) melatonin (c) insulin (d) cortisone

_____ 10. The endocrine gland that may contribute to setting the body's biological clock is the (a) pineal gland (b) thymus gland (c) thyroid gland (d) adrenal gland

PART 2 ■ Completion

11. The pituitary gland is attached to the hypothalamus by a stalklike structure called the

_____ .

12. A hormone that acts on another endocrine gland and causes that gland to secrete its own hormones is called a _____ .

13. _____ cells of the anterior pituitary gland synthesize ACTH.

14. The hormone that helps cause contraction of the smooth muscle of the uterus during childbirth is

     _____.

15. Histologically, the spherical sacs that compose the thyroid gland are called thyroid

     _____.

16. The thyroid hormones associated with metabolism are triiodothyronine and _____.

17. Chief and oxyphil cells are associated with the _____ gland.

18. The zona glomerulosa of the adrenal cortex secretes a group of hormones called

     _____.

19. The hormones that promote normal metabolism, provide resistance to stress, function as antiin-

     flammatories and depress immune responses are _____.

20. The pancreatic hormone that raises blood glucose level is _____.

21. In spermatogenesis, the most immature cells near the periphery of the seminiferous tubules are

     called _____.

22. Cells within the testes that secrete testosterone are known as _____.

23. The ovaries are attached to the uterus by means of the _____ ligament.

24. The female hormones that help cause the development of secondary sex characteristics are called

     _____.

25. The endocrine gland that assumes a direct function in the proliferation and maturation of T cells is

     the _____.

26. Any hormone that regulates the functions of the gonads is classified as a(n)

     _____ hormone.

27. The hormone that is stored in the posterior pituitary gland that decreases urine volume is

     _____.

28. The region of the adrenal cortex that synthesizes androgens is the _____.

29. The hormone-producing cells of the adrenal medulla are called _____ cells.

30. Together, alpha cells, beta cells, delta cells, and F cells constitute the _____.

31. Regulating hormones are produced by the _____ and reach the anterior pitu-

     itary gland by a network of blood vessels.

32. FSH and LH are produced by _____ cells of the anterior pituitary gland.

33. The structure in an ovary that produces progesterone, estrogens, relaxin, and inhibin after ovula-

     tion is the _____.

34. A gland that is both an exocrine and an endocrine gland is known as a _____

     gland.

Blood

Objectives

At the completion of this exercise you should understand

A The various components of human blood

B How to conduct various red blood cell tests, including a red blood cell count (utilizing a hemocytometer), hematocrit, sedimentation rate, and various hemoglobin tests

C How to conduct a white blood cell count, utilizing a hemocytometer, and a differential white blood cell count

D The concept of blood groups and how to determine them

The blood, heart, and blood vessels constitute the **cardiovascular** (*cardio* = heart; *vascular* = blood vessels) **system.** In this exercise you will examine the characteristics of **blood,** a connective tissue also known as **vascular tissue.**

⚠ **CAUTION!** *Please reread Section A, "General Safety Precautions and Procedures," and Section B, "Precautions for Working with Blood, Blood Products, or Other Body Fluids," on pages xi–xii at the beginning of the laboratory manual before you begin any of the following experiments. Read the experiments before you perform them, to be sure that you understand all the procedures and safety precautions.*

When working with whole blood, wear tight-fitting surgical gloves and safety goggles. Avoid any kind of contact with an open sore, cut, or wound.

After you have completed your experiments, place all glassware in a fresh solution of household bleach or other comparable disinfectant, wash the laboratory tabletop with a fresh solution of household bleach or comparable disinfectant, and dispose of your gloves in the appropriate biohazard container provided by your instructor.

A. Components and Origin of Blood

Blood is a connective tissue, composed of a liquid, extracellular matrix called **plasma,** that dissolves and suspends **formed elements,** cells and cell fragments. In clinical practice, the most common classification of the formed elements of the blood is the following:

Red blood cells, or **erythrocytes** (e-RITH-rō-sīts)

White blood cells, or **leukocytes** (loo-kō-sīts)

Granular leukocytes

Neutrophils

Eosinophils

Basophils

Agranular leukocytes

Lymphocytes (T cells, B cells, and natural killer cells)

Monocytes

Platelets (thrombocytes)

The process by which the formed elements of blood develop is called **hemopoiesis** (hē-mō-poy-Ē-sis; *hemo* = blood; *poiesis* = making) or **hematopoiesis.** About 0.05% to 0.1% of red bone marrow cells are **pluripotent stem cells,** derived from mesenchyme (Figure 16.1). These cells reproduce themselves and give rise to *all* the formed elements of the blood.

Originating from the pluripotential stem cells are two further types of **multipotential stem cells,**

FIGURE 16.1 Origin, development, and structure of blood cells.

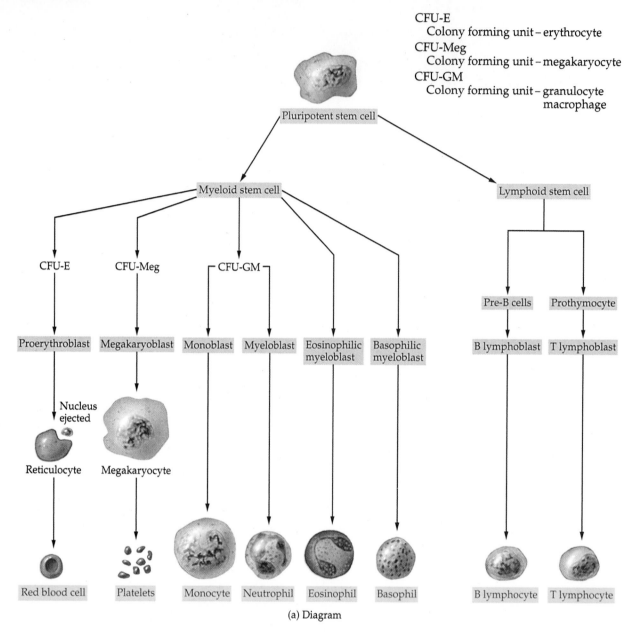

CFU-E
    Colony forming unit – erythrocyte
CFU-Meg
    Colony forming unit – megakaryocyte
CFU-GM
    Colony forming unit – granulocyte
        macrophage

Pluripotent stem cell

Myeloid stem cell   Lymphoid stem cell

CFU-E   CFU-Meg   CFU-GM   Pre-B cells   Prothymocyte

Proerythroblast Megakaryoblast Monoblast Myeloblast Eosinophilic myeloblast Basophilic myeloblast   B lymphoblast T lymphoblast

Nucleus ejected

Reticulocyte Megakaryocyte

Red blood cell Platelets Monocyte Neutrophil Eosinophil Basophil   B lymphocyte T lymphocyte

(a) Diagram

which also have the ability to reproduce themselves but that differentiate to give rise to more specific formed elements of blood. For example, *myeloid stem cells* give rise to red blood cells, platelets, neutrophils, eosinophils, basophils, and monocytes. *Lymphoid stem cells* give rise to lymphocytes. Whereas myeloid stem cells begin and complete their development in red bone marrow, lymphoid stem cells begin their development in red bone marrow but complete it in lymphatic tissue.

As the development of the formed elements of blood continues, myeloid stem cells form **progenitor cells**—cells that no longer are capable of reproducing themselves and can only give rise to more specific formed elements of blood. The indi-

vidual forms of progenitor cells are called *colony-forming units (CFUs)* and are named on the basis of the mature elements in blood that they will ultimately produce. Accordingly, CFU-E ultimately develop into erythrocytes (red blood cells), CFU-GM develop into granulocytes (specifically neutrophils) and monocytes, and CFU-Meg produce megakaryocytes, the source of platelets. Stem cells and progenitor cells cannot be distinguished morphologically (on the basis of their appearance under the microscope) and resemble lymphocytes. In the next generation, the cells are called **precursor cells,** known as **blasts,** which will develop over several cell divisions into the actual formed elements of blood. For example, *proerythroblasts* de-

Origin, development, and structure of blood cells. (Continued)

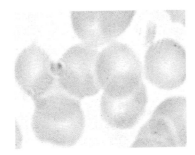

Erythrocytes
(1600×)

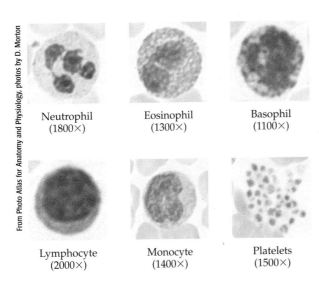

| Neutrophil (1800×) | Eosinophil (1300×) | Basophil (1100×) |
| Lymphocyte (2000×) | Monocyte (1400×) | Platelets (1500×) |

From Photo Atlas for Anatomy and Physiology, photos by D. Morton

velop into erythrocytes, *lymphoblasts* develop into lymphocytes, and so on. Precursor cells exhibit recognizable morphological characteristics.

During embryonic and fetal life, the yolk sac, liver, spleen, thymus, lymph nodes, and red bone marrow all participate at various times in producing the formed elements. After birth, however, hemopoiesis takes place only in red bone marrow, although small numbers of stem cells circulate in the bloodstream. Red bone marrow is found chiefly in bones of the axial skeleton, pectoral and pelvic girdles, and the proximal epiphyses of the humerus and femur. Once they leave the bone marrow, most of the formed elements do not divide, and they survive only a few hours or days. The exceptions are the immature lymphocytes, which may live for years and can be stimulated to undergo further proliferation by mitosis within lymphatic tissues.

Mature blood cells are constantly being replaced as they die; fixed macrophages in the spleen and the liver sinusoids have the responsibility of clearing away the dead, disintegrating cell bodies so that small blood vessels are not clogged. The fixed macrophages of the liver are called **stellate reticuloendothelial cells.**

As you will see later, the shapes of the nuclei, staining characteristics, and color of cytoplasmic granules are all useful in differentiation and identification of the various white blood cells. Red blood cells are biconcave discs without nuclei and can be identified easily.

B. Plasma

When the formed elements are removed from blood, a straw-colored liquid called **blood plasma** (or simply **plasma**) is left. Plasma consists of about 91.5% water and 8.5% solutes. Among the solutes are proteins (albumins, globulins, and fibrinogen), wastes (urea, uric acid, and creatine), nutrients (amino acids, glucose, fatty acids, glycerides, and glycerol), regulatory substances (enzymes and hormones), gases (oxygen, carbon dioxide, nitrogen), and electrolytes (Na^+, K^+, Ca^{2+}, Mg^{2+}, Cl^-, HCO_3^-, SO_4^{2-}, and HPO_4^{3-}).

1. Physical Characteristics

⚠ **CAUTION!** *Please reread Section A, "General Safety Precautions and Procedures," and Section B, "Precautions for Working with Blood, Blood Products, or Other Body Fluids," on pages xi–xii at the beginning of the laboratory manual before you begin any of the following experiments. Read the experiments before you perform them, to be sure that you understand all the procedures and safety precautions.*

When working with whole blood, wear tight-fitting surgical gloves and safety goggles. Avoid any kind of contact with an open sore, cut, or wound.

PROCEDURE

NOTE *Obtain the plasma or whole blood for this procedure from a clinical laboratory where it has been tested and certified as noninfectious[1] or from a mammal (other than a human).*

1. Obtain about 5 ml of plasma as follows.

2. Centrifuge the whole blood obtained from a clinical laboratory where it has been tested and certified as noninfectious or from a mammal (other than a human). After centrifuga-

---

[1]Whole blood that has been tested for syphilis; hepatitis A, B, and C; and HIV is available from Carolina Biological Supply Company.

tion, the bottom of the test tube will contain an anticoagulant (usually EDTA). Right above this will be a layer of erythrocytes, followed by a layer of leukocytes and platelets (buffy coat). The top layer is plasma. Visual inspection of these layers will give an indication of the relative proportions of the various components of blood.

3. Remove 5 ml of plasma using a disposable sterile Pasteur pipette attached to a bulb and place the plasma in a disposable test tube. Note its color. Test the pH of the plasma by dipping the end of a length of pH paper into the plasma and comparing the paper color with the color scale on the pH container. Record the pH in Section B.1 of the LABORATORY REPORT RESULTS at the end of the exercise.

4. Hold the test tube up to a source of natural light and note the color and transparency of the plasma. Record the color and transparency in Section B.1 of the LABORATORY REPORT RESULTS at the end of the exercise.

5. *Dispose of the pipettes and test tubes in the appropriate biohazard container provided by your instructor. Dispose of the pH paper and plasma as per your instructor's directions.*

2. Chemical Constituents

▲ **CAUTION!** *Please reread Section A, "General Safety Precautions and Procedures," and Section B, "Precautions for Working with Blood, Blood Products, or Other Body Fluids," on pages xi–xii at the beginning of the laboratory manual before you begin any of the following experiments. Read the experiments before you perform them, to be sure that you understand all the procedures and safety precautions.*

When working with whole blood, wear tight-fitting surgical gloves and safety goggles. Avoid any kind of contact with an open sore, cut, or wound.

PROCEDURE

NOTE *Obtain the plasma or whole blood for this procedure from a clinical laboratory where it has been tested and certified as noninfectious or from a mammal (other than a human).*

1. Using a medicine dropper, add 10 ml of distilled water to 5 ml of plasma in a Pyrex test tube.

▲ **CAUTION!** *Place the test tube in a test-tube rack because it will become too hot to handle.*

2. With forceps, add one Clinitest tablet.

▲ **CAUTION!** *The concentrated sodium hydroxide in the tablet generates enough heat to make the liquid in the test tube boil. Make sure that the mouth of the test tube is pointed away from you and all other persons in the area.*

3. The color of the solution is graded as follows:

Color	Results
Blue	Negative
Greenish-yellow	1 + (0.5 g/100 ml)
Olive green	2 + (1 g/100 ml)
Orange-yellow	3 + (1.5 g/100 ml)
Brick red (with precipitate)	4 + (more than 2 g/100 ml)

4. Fifteen seconds after boiling has stopped, shake the test tube *gently* and evaluate the color according to the test table above.

▲ **CAUTION!** *Make sure that the mouth of the test tube is pointed away from you and others.*

A green, yellow, orange, or red color indicates the presence of glucose. Record your results in Section B.2 of the LABORATORY REPORT RESULTS at the end of the exercise.

5. Next, test the contents of filter paper for the presence of protein. If protein is present, it will be coagulated by acetic acid and filtered out of the plasma. Place the contents of the filter paper in a test tube and add 3 ml of water and 3 ml of Biuret reagent. A violet or purple color indicates the presence of protein. Record your results in Section B.2 of the LABORATORY REPORT RESULTS at the end of the exercise.

C. Red Blood Cells

Red blood cells (RBCs), or **erythrocytes,** are bioconcave discs, have no nucleus, and can neither reproduce nor carry on extensive metabolic activities (Figure 16.1). The cell contains the oxygen-carrying protein **hemoglobin,** which is a pigment that gives whole blood its red color. The heme portion of hemoglobin combines with oxygen and, to a lesser extent, carbon dioxide and transports them through the blood vessels. An average red blood cell has a lifespan of about 120 days. A healthy adult male has about 5.4 million red blood cells per cubic millimeter (mm^3) or per microliter (μL) of blood; a healthy adult female, about 4.8 million. Erythropoiesis, the production of RBCs, and red blood cell destruction normally proceed at the same pace. A diagnostic test that informs the physician about the rate of erythropoiesis is the **reticulocyte** (re-TIK-ū-lō-sīt) **count.** **Reticulocytes** (Figure 16.1) are precursor cells of mature red blood cells. Normally, a reticulocyte count is 0.5% to 1.5%. The reticulocyte count is an important diagnostic tool because it is a relatively accurate predictor of the status of red blood cell production in red bone marrow. Normally, red bone marrow replaces about 1% of the adult red blood cells each day. A decreased reticulocyte count (reticulocytopenia) is seen in aplastic anemia and in conditions in which the red bone marrow is not producing red blood cells. An increase in reticulocytes (reticulocytosis) is found in acute and chronic blood loss and certain kinds of anemias such as iron-deficiency anemia. It could also point to illegal use of Epoetin alfa by an athlete.

NOTE *Although many blood tests in this exercise are sufficient for laboratory demonstration, they are not necessarily clinically accurate.*

D. Red Blood Cell Tests

▲ **CAUTION!** *Please reread Section A, "General Safety Precautions and Procedures," and Section B, "Precautions for Working with Blood, Blood Products, or Other Body Fluids," on pages xi–xii at the beginning of the laboratory manual before you begin any of the following experiments. Read the experiments before you perform them, to be sure that you understand all the procedures and safety precautions.*

When working with whole blood, wear tight-fitting surgical gloves and safety goggles. Avoid any kind of contact with an open sore, cut, or wound.

1. Source of Blood

Blood should be obtained from a clinical laboratory where it has been tested and certified as noninfectious or from a mammal (other than a human).

Whole blood that has been tested for syphilis; hepatitis A, B, and C; and HIV is available from Carolina Biological Supply Company. This blood can be used for blood grouping and typing, blood glucose, hematocrit, hemoglobin, red blood cell count, white blood cell count, differential white blood cell count, platelet count, sedimentation rate, osmotic fragility studies, and various plasma chemistries. Carolina Biological Supply Company also provides aseptic red blood cells that can be used for blood grouping and typing, blood glucose, hematocrit, hemoglobin, red blood cell count, and osmotic fragility studies. These blood cells are suspended in a modified Alsevere's solution, to which several antibiotics have been added. The cells have been tested for syphilis; hepatitis A, B, and C; and HIV.

2. Filling of Hemocytometer (Counting Chamber)

The procedure for filling the hemocytometer is as follows. Please read it carefully *before* doing 3, "Red Blood Cell Count."

PROCEDURE

NOTE *The blood used for this procedure should be obtained from a clinical laboratory where it has been tested and certified as noninfectious or from a mammal (other than a human).*

1. Obtain a hemocytometer and cover slip (Figure 16.2).

2. Clean the hemocytometer thoroughly and carefully with alcohol.

3. Place the cover slip on the hemocytometer.

4. Using a Unopette Reservoir System® pipette filled with the proper dilution of blood, place the tip of the pipette on the polished surface of the hemocytometer next to the edge of the cover slip. Although we recommend use of this system, your instructor might wish to use an alternate procedure. *If a different system is used for pipetting and diluting, use rubber suction bulbs or pipette pumps. Do not pipette by mouth.*

5. Deposit a small drop of diluted blood by squeezing the sides of the reservoir, but do not leave the tip of the pipette in contact with the hemocytometer for more than an instant because this will cause the chamber to overfill. The diluted blood must not overflow the moat; overfilling the moat results in an inaccurate cell count. A properly filled hemocytometer has a blood specimen only within the space between the cover glass and counting area.

3. Red Blood Cell Count

The purpose of a red blood cell count is to determine the number of circulating red blood cells per cubic millimeter (mm^3), or per microliter (μL), of blood. Red blood cells carry oxygen to all tissues; thus a drastic reduction in the red cell count will cause immediate reduction in available oxygen.

A decrease in red blood cells can result from a variety of conditions, including impaired cell

FIGURE 16.2 Various parts of a hemocytometer (counting chamber). In (b), the areas used for red blood cell counts are indicated with an R and those for white blood cell counts are indicated with a W.

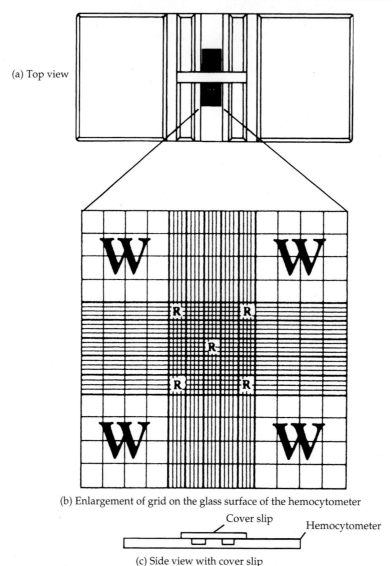

(a) Top view

(b) Enlargement of grid on the glass surface of the hemocytometer

(c) Side view with cover slip

production, increased cell destruction, and acute blood loss. When the red blood cell count is increased above normal limits, the condition is called **polycythemia** (pol'-ē-sī-THĒ-mē-a).

The procedure we will use for determining the number of red blood cells per microliter (μL) of blood is as follows.

PROCEDURE

WARD's Natural Science Establishment, Inc., has made available WARD'S *Simulated Blood*. It is a solution that contains microcomponents that simu-late red blood cells, white blood cells, and platelets. The microcomponents are similar in relative proportion to those found in human blood and can be observed under a microscope without staining.

1. Obtain a Simulated Blood Activity kit.

2. Follow the instructions to perform the red blood cell count.

3. Complete the portion of the Data Table related to red blood cells.

4. Answer the questions related to red blood cells.

ALTERNATE PROCEDURE

NOTE *The blood used for this procedure should be obtained from a clinical laboratory where it has been tested and certified as noninfectious or from a mammal (other than a human).*

1. Obtain a Unopette Reservoir System® for red blood cell determination (the color of the reservoir's bottom surface is red). Identify the reservoir chamber, diluent fluid inside the chamber, pipette, and protective shield for the pipette (Figure 16.3). The diluent fluid contains isotonic saline and sodium azide, an antibacterial agent.

▲ **CAUTION!** *Do not ingest this fluid; it is for in vitro diagnostic use only.*

2. Hold the reservoir on a flat surface in one hand and grasp the pipette assembly in the other hand.

3. Push the tip of the pipette shield firmly through the diaphragm in the neck of the reservoir (Figure 16.4a).

4. Pull out the assembly unit and with a twist remove the protective shield from the pipette assembly.

5. Hold the pipette *almost* horizontally and touch the tip of it to a forming drop of blood that has been placed on a microscope slide (Figure 16.4b). The pipette will fill by capillary action and, when the blood reaches the end of the capillary bore in the neck of the pipette, the filling action will stop.

6. Carefully wipe any excess blood from the tip of the pipette using a Kimwipe or similar wiping tissue, *making certain that no sample is removed from the capillary bore.*

7. Squeeze the reservoir *slightly* to expel a small amount of air.

▲ **CAUTION!** *Do not expel any liquid. If the reservoir is squeezed too hard, the specimen may be expelled through the overflow chamber, resulting in contamination of the fingers. Reagent contains sodium azide, which is extremely toxic and yields explosive products in metal sinks or pipes. Azide compounds should be diluted with running water before being discarded and disposed of per the directions of your instructor.*

While still maintaining pressure on the reservoir, cover the opening of the overflow chamber of the pipette with your index finger and push the pipette securely into the reservoir neck (Figure 16.4c).

FIGURE 16.3 **Photograph of the Unopette Reservoir System®.**

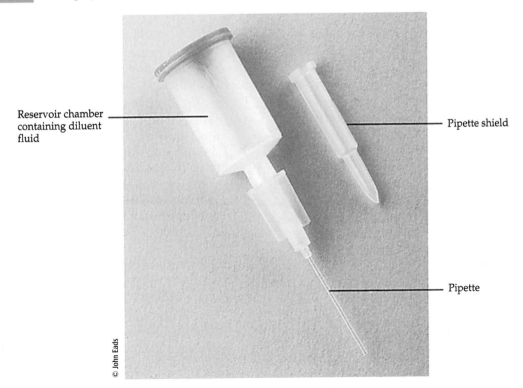

Reservoir chamber containing diluent fluid

Pipette shield

Pipette

© John Eads

FIGURE 16.4   Preparation of the Unopette Reservoir System®. (Modified from *RBC Determination for Manual Methods and WBC Determination for Manual Methods,* Becton-Dickinson, Division of Becton, Dickinson and Company.)

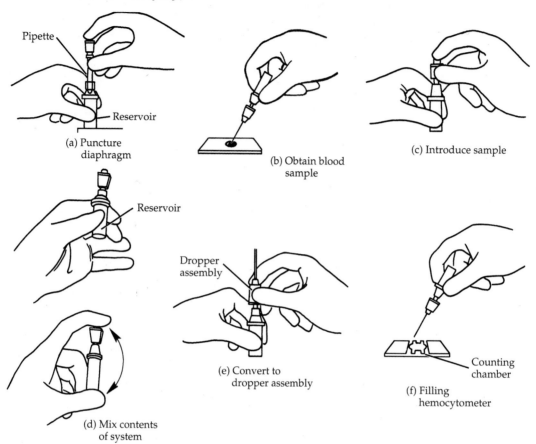

(a) Puncture diaphragm

(b) Obtain blood sample

(c) Introduce sample

(d) Mix contents of system

(e) Convert to dropper assembly

(f) Filling hemocytometer

8. Release the pressure on the reservoir and remove your finger from the pipette opening. This will draw the blood into the diluent fluid.

9. Mix the contents of the reservoir chamber by squeezing the reservoir gently several times. It is important to *squeeze gently* so that the diluent fluid is not forced out of the chamber. In addition to squeezing the reservoir chamber, invert it *gently* several times (Figure 16.4d).

10. Remove the pipette from the reservoir chamber, reverse its position, and replace it on the reservoir chamber (Figure 16.4e). This converts the apparatus into a dropper assembly (Figure 16.4f).

11. Squeeze a few drops out of the reservoir chamber into a disposal container or wipe with Kimwipe or similar wiping tissue. You are now ready to fill the hemocytometer.

12. Follow the procedure outlined in D.2, filling the hemocytometer.

13. Place the hemocytometer on the microscope stage and allow the red blood cells to settle in the hemocytometer (approximately 1 to 2 min).

14. When counting red blood cells with the hemocytometer, the loaded (filled) chamber should be first located using the low-power objective of the microscope. Upon finding that, switch to the high-power objective via parfocal procedure. Focus the field using the fine adjustment of the microscope. Each red blood cell counted is equal to 10,000, so the degree of error in manual red blood cell counting is quite high, generally in the range of 15% to 20%.

15. Using the high-power lens, count the cells in each of the five squares (E, F, G, H, I), as shown in Figure 16.5. It is suggested that you use a hand counter. As you count, move up and down in a systematic manner.

Hemocytometer (counting chamber). Areas E, F, G, H, and I are all counted in the red blood cell count. The large square containing the smaller squares E, F, G, H, and I is seen in a microscopic field under the 10× magnification. To count the smaller squares (E, F, G, H, and I), the 45× lens is used and therefore E, F, G, H, and I each encompasses the entire microscopic field.

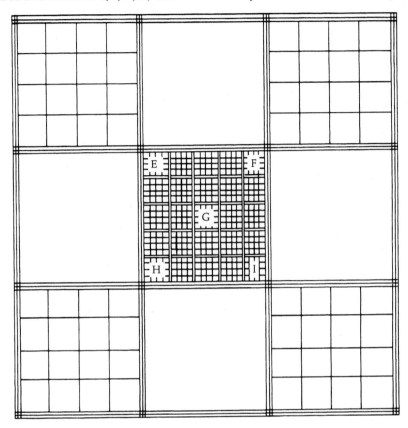

NOTE *To avoid overcounting cells at the boundaries, cells that touch the lines on the left and top sides of the hemocytometer should be counted, but not the ones that touch the boundary lines on the right and bottom sides.*

16. Multiply the total number of red blood cells counted in the five squares by 10,000 to obtain the number of red blood cells in 1 μL of blood. Record your value in Section D.1 of the LABORATORY REPORT RESULTS at the end of the exercise.

17. The hemocytometer and cover slip should be saved, not discarded.

▲ **CAUTION!** *Place the hemocytometer and cover slip in a container of fresh bleach solution and dispose of all other materials as directed by your instructor.*

4. Red Blood Cell Volume (Hematocrit)

The percentage of blood volume occupied by the red blood cells is called the **hematocrit** **(Hct).** It is an important component of a complete blood count and is a standard test for almost all persons having a physical examination or for diagnostic purposes. When a tube of blood is centrifuged, the red blood cells pack into the bottom part of the tube with the plasma on top. The white blood cells and platelets are found in a thin area, the buffy layer, above the red blood cells.

A centrifuge (Figure 16.6) is used for spinning blood to determine hematocrit. The Readacrit® centrifuge incorporates a built-in hematocrit scale and tube-holding compartments which, when used with special precalibrated capillary tubes, permit direct reading of the hematocrit value by measuring the length of the packed red blood cell column. Readacrit® centrifuges present the final hematocrit value without requiring computation by the operator. If the Readacrit® centrifuge or tube reader is not used for direct reading, measure the total length of blood volume, divide this

FIGURE 16.6 Centrifuge.

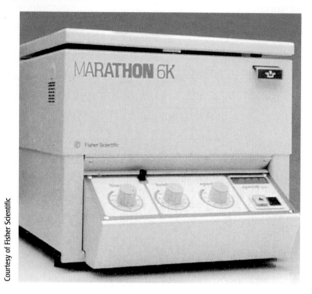

quantity into length of packed red blood cells, and multiply by 100 for percentage.

The calculation is as follows:

$$\frac{\text{length of packed red blood cells}}{\text{total length of blood volume}} \times 100$$

$$= \%\ \text{volume of whole blood occupied by red blood cells (hematocrit)}$$

In males the normal range is between 40% and 54%, with an average of 47%. In females the normal range is between 38% and 46%, with an average of 42%. Anemic blood may have a hematocrit of 15%; polycythemic blood may have a hematocrit of 65%.

The following procedure for testing red blood cell volume is a micromethod requiring only a drop of blood:

PROCEDURE

NOTE *The blood used for this procedure should be obtained from a clinical laboratory where it has been tested and certified as noninfectious or from a mammal (other than a human).*

1. Place a drop of blood on a microscope slide.
2. Place the unmarked (clear) end of a disposable capillary tube into the drop of blood. Hold the tube slightly below the level of the blood and do not move the tip from the blood or an air bubble will enter the column.

▲ **CAUTION!** *If an air bubble is present, immediately dispose of the capillary tube in the appropriate biohazard container provided by your instructor and begin again.*

3. Allow blood to flow about two-thirds of the way into the tube. The tube will fill easily if you hold the open end level with or below the blood source.
4. Place your finger over the marked (red) end of the tube and keep it in that position as you seal the blood end of the tube with Seal-ease® or clay or other similar sealing material.
5. According to the directions of your laboratory instructor, place the tube into the centrifuge, making sure that the sealed end is against the rubber ring at the circumference of the centrifuge rotor. Have your laboratory instructor make sure that the tubes are properly balanced. Write the number that indicates the location of your tube. _____
6. When all the tubes from your laboratory section are in place, secure the inside cover of the centrifuge and close the outside cover. Centrifuge on high speed for 4 min.
7. Remove your tube and determine the hematocrit value by reading the length of packed red blood cells directly on the centrifuge scale, or by placing the tube in the tube reader (Figure 16.7) and following instructions on the reader, or by calculation using

FIGURE 16.7 Microhematocrit tube reader.

the equation mentioned previously. When using the tube reader, place the base of the Seal-ease® or clay on the base line and move the tube until the top of the plasma is at the line marked 100.

▲ **CAUTION!** *Immediately dispose of your capillary tube in the appropriate biohazard container provided by your instructor.*

Record your results in Section D.2 of the LABORATORY REPORT RESULTS at the end of the exercise.

What effect would dehydration have on hematocrit? _____

_____

5. Sedimentation Rate

If blood is allowed to stand vertically in a tube, the red blood cells fall to the bottom of the tube and leave clear plasma above them. The distance that the cells fall in 1 hr can be measured and is called the **sedimentation rate.** It is a function of the amount of fibrinogen and gamma globulin present in plasma and the tendency of the red blood cells to adhere to one another. This rate is greater than normal during menstruation, pregnancy, malignancy, and most infections. Sedimentation rate may be decreased in liver disease. A high rate may indicate tissue destruction in some part of the body; therefore, sedimentation rate is considered a valuable nonspecific diagnostic tool. The normal rate for adults is 0 to 6 mm per hour, for children 0 to 8 mm per hour.

The following method for determining sedimentation rate requires only one drop of blood and is called the Landau micromethod.

PROCEDURE

NOTE *The blood used for this procedure should be obtained from a clinical laboratory where it has been tested and certified as noninfectious or from a mammal (other than a human).*

1. Place a drop of blood on a microscope slide.

2. Using the mechanical suction device, draw the sodium citrate up to the first line that encircles the pipette.

3. Draw the blood up into a disposable sedimentation pipette to the second line. Do not remove the tip of the pipette from the blood or air bubbles will enter the column.

▲ **CAUTION!** *If air bubbles are drawn into the blood, carefully expel the mixture onto a clean microscope slide, dispose of the slide and pipette in the appropriate biohazard container provided by your instructor, and begin again.*

4. Draw the mixture of fluids up into the bulb and mix by expelling it into the lumen of the tube. Draw and mix the fluids six times. If any air bubbles appear, use the procedure described in step 3.

5. Adjust the level of the blood as close to zero as possible. (Exactly zero is very difficult to get.)

6. Remove the suction device by placing the lower end of the pipette on the *gloved* index finger of the left hand before removing the device from the other end. The blood will leave the pipette if the end is not held closed.

7. Place the lower end of the pipette on the base of the pipette rack (Figure 16.8) and the opposite end at the top of the rack. The tube must be exactly perpendicular. Record the time at which it is put in the rack.

Time _____.

8. One hour later, measure the distance from the top of the plasma to the top of red blood cell layer with an accurate millimeter scale, and record your results in Section D.3 of the LABORATORY REPORT RESULTS at the end of the exercise.

▲ **CAUTION!** *Dispose of the pipette in the appropriate biohazard container provided by your instructor. Using a medicine dropper, rinse the base of the pipette rack with fresh bleach solution.*

FIGURE 16.8 **One type of sedimentation tube rack.**

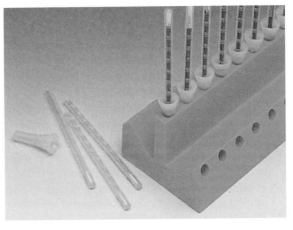

6. Hemoglobin Determination

It is possible for one to have anemia even if one's red blood cell count is normal. For example, the red blood cells present may be deficient in hemoglobin or they may be smaller than normal. Thus, the amount of hemoglobin per unit volume of blood, and not necessarily the number of red blood cells, is the determining factor for anemia. The hemoglobin content of blood is expressed in grams (g) per 100 ml of blood, and the normal ranges vary with the technique used. An average range is 12 to 15 g/100 ml in females and 13 to 16 g/100 ml in males.

One procedure that can be used for determining hemoglobin uses an apparatus called a **hemoglobinometer** (Figure 16.9a). This instrument compares the absorption of light by hemoglobin in a blood sample of known depth to that of a standardized glass plate.[2]

PROCEDURE

NOTE *The blood used for this procedure should be obtained from a clinical laboratory where it has been tested and certified as noninfectious or from a mammal (other than a human).*

1. Obtain a hemoglobinometer and examine its parts (see Figure 16.9a). Note the four scales on the side of the instrument. The top scale gives hemoglobin content in g/100 ml of blood; the other three scales give comparative readings, that is, the amount of hemoglobin expressed as a percent of 15.6, 14.5, or 13.8 g/100 ml of blood.

2. Remove the two pieces of glass from the blood chamber assembly (Figure 16.9b) and clean them with alcohol and wipe them with lens paper. The piece of glass that contains the moat will receive the blood sample; the other piece of glass serves as a cover slip.

3. Replace the glass pieces halfway onto the clip with the moat plate on the bottom.

4. Obtain a sample of blood.

5. Apply the blood to the moat and completely cover the raised surface of the plate with blood.

6. Hemolyze the blood by agitating it with the tip of a hemolysis applicator for about 30 to 45 sec. Hemolysis is complete when the appearance of the blood changes from cloudy to transparent. This change reflects the lysis of the red blood cell membranes and the liberation of hemoglobin.

7. Push the specimen chamber into the clip and push the clip into the hemocytometer.

8. Holding the hemocytometer in your left hand, look through the eyepiece while depressing the light switch button. You will see a split field.

9. Move the slide button back and forth with the right index finger until the two sides of the field match in color and shading.

10. Read the value indicated on the scale marked 15.6. Determine the grams of hemoglobin per 100 ml of the blood sample by reading the number above the index mark on the hemocytometer. Record your results in Section D.4 of the LABORATORY REPORT RESULTS at the end of the exercise.

⚠ **CAUTION!** *Remove the glass pieces from the chamber and wash in fresh bleach solution. Dry the glass pieces with lens paper.*

11. What type of anemia would reveal a normal or slightly below-normal hematocrit, yet a significantly reduced hemoglobin level?

———

ALTERNATE PROCEDURE

The **Tallquist measurement** of hemoglobin is an old and somewhat inaccurate technique, although it is quick and inexpensive. The hemoglobinometer gives more accurate ($\pm 5\%$) results.

NOTE *The blood used for this procedure should be obtained from a clinical laboratory where it has been tested and certified as noninfectious or from a mammal (other than a human).*

1. Obtain a sample of blood.

2. Place one drop of blood on the Tallquist® paper.

3. As soon as the blood no longer appears shiny, match its color with the scale provided. Record your observations in Section D.4 of the LABORATORY REPORT RESULTS at the end of the exercise.

7. Oxyhemoglobin Saturation

The ability of blood to carry oxygen is determined by proper respiratory system function, hematocrit, and the ability of hemoglobin to bind reversibly with oxygen within the alveoli (air sacs) of the

———

[2] If Unopette tests are available, follow the procedure outlined in Unopette test no. 5857, "Cyanmethemoglobin Determination for Manual Methods," for hemoglobin estimation.

FIGURE 16.9 Hemoglobinometer.

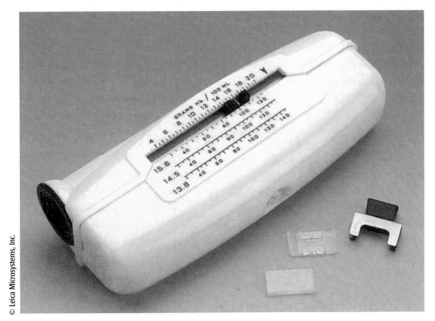

© Leica Microsystems, Inc.

(a) Cambridge Instruments Hemoglobin-Meter

© Leica Microsystems, Inc.

(b) Specimen chamber

lungs and release oxygen at the tissue level. Hemoglobin's interaction with oxygen is measured by an **oxygen-hemoglobin dissociation curve.** Such a curve demonstrates that the number of available oxygen-binding sites on hemoglobin that actually bind to oxygen is proportional to the partial pressure of oxygen (pO_2). Arterial blood usually has a pO_2 of 95, causing 97% of the available hemoglobin to be saturated with oxygen. Venous blood, however, has a pO_2 of 40, so only 75%

of the hemoglobin is saturated with oxygen; 25% is given off to tissues. The determination of percent saturation for hemoglobin is a very sensitive test of pulmonary function. However, such a test needs to be interpreted carefully, as an individual with a normally functioning pulmonary system may exhibit abnormally low oxyhemoglobin saturation due to a variety of possible causes, such as **methemoglobinemia** (transformation of normal oxyhemoglobin due to the reduction of normal

Fe^{2+} to Fe^{3+} within the hemoglobin molecule) or **carbon monoxide (CO) poisoning.** In order to determine the percent hemoglobin saturation in an unknown sample of blood, one would need to compare the absorption spectrum of the blood sample to that obtained from a pure sample of **oxyhemoglobin (HbO_2), carboxyhemoglobin (HbCO),** and **reduced hemoglobin (Hb)** (carboxyhemoglobin is included due to the normal presence of carbon monoxide in today's polluted atmosphere). This is shown in Figure 16.10. The spectrum obtained with the unknown sample of blood would be some combination of these three, since it would contain a certain percentage of each form of hemoglobin. Such a test is possible due to the fact that all three of these types of hemoglobin compounds are different colors and would therefore absorb different portions of the light spectrum. Through a complex determination of the relative contribution of each type of hemoglobin to the resulting spectrum for the unknown sample, the percentage of each hemoglobin form may be determined.

8. Spectrum of Oxyhemoglobin and Reduced Hemoglobin

The following procedure details a simplified, although less accurate method for determining the absorption spectra for fully oxygenated and de-

oxygenated hemoglobin. Exposing blood to air saturates it with oxygen, while sodium dithionate reduces it by removing oxygen.

PROCEDURE

NOTE *The blood used for this procedure should be obtained from a clinical laboratory where it has been tested and certified as noninfectious or from a mammal (other than a human).*

1. Turn on the spectrophotometer and let it warm up for several minutes.

2. After the instrument has warmed up, set the wavelength at 500 nm and adjust the meter needle to 0% transmittance (absorbance at infinity) by using the zero control knob.

3. Place a cuvette containing distilled water into the cuvette holder and adjust the needle to 100% transmittance by using the light control knob.

4. Place 8 ml of distilled water in a clean test tube.

5. Obtain several drops of blood.

6. Mix the distilled water with the blood by placing a stopper in the mouth of the test tube and then inverting the test tube, thereby adding the blood to the distilled water.

7. Transfer half of the test tube contents (4 ml) into a second clean test tube.

FIGURE 16.10 Absorption spectra of carboxyhemoglobin (HbCO), reduced hemoglobin (Hb), and oxyhemoglobin (HbO_2).

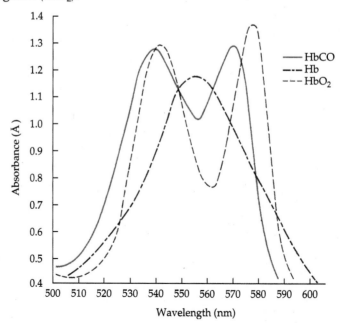

8. Add 0.2 ml of 1% sodium dithionite solution to the second test tube and mix thoroughly.

(**NOTE** *In order for the procedure to work properly, the sodium dithionite solution must be fresh and made up just prior to use, and a spectral determination must be completed within 5 min of addition of sodium dithionite to the second tube.*)

9. Record the absorbances of solutions 1 (oxyhemoglobin) and 2 (reduced hemoglobin) at 500 nm wavelength.

10. Standardize the spectrophotometer at 510 nm utilizing the cuvette containing only distilled water as outlined in steps 2 and 3.

11. Determine the absorbances for solutions 1 and 2 at 510 nm.

12. Repeat steps 2, 3, and 9 at each of the following wavelengths: 520, 530, 540, 550, 560, 570, 580, 590, and 600 nm.

13. Record and graph your results in Section D.5 of the LABORATORY REPORT RESULTS at the end of the exercise. Be sure to compare and give physiological reasons for the observed differences in the two spectra.

E. White Blood Cells

White blood cells (WBCs), or **leukocytes,** are different from red blood cells in that they have nuclei and do not contain hemoglobin (see Figure 16.1). They are less numerous than red blood cells, ranging from 5000 to 10,000 cells per cubic millimeter (mm^3) or per microliter (μL) of blood. The ratio, therefore, of red blood cells to white blood cells is about 700:1.

As noted earlier, the two major groups of WBCs are granular leukocytes and agranular leukocytes.

1. Granular Leukocytes

Granular leukocytes display conspicuous granules in the cytoplasm that can be seen under a light microscope. The three types of granular leukocytes are eosinophils, basophils, and neutrophils. Each class of cells displays a distinctive coloration under a light microscope after staining.

The nucleus of an **eosinophil** (ē-ō-SIN-ō-fil) usually has two lobes connected by a thick strand (see Figure 16.1). Large, uniform-sized granules pack the cytoplasm but usually do not cover or obscure the nucleus. These eosinophilic (= eosin-loving) granules stain red-orange with acidic dyes. The nucleus of a **basophil** (BĀ-sō-fil) has two lobes (see Figure 16.1). The cytoplasmic basophilic granules are round, are variable in size, stain blue-purple with basic dyes, and commonly obscure the nucleus.

The nucleus of a **neutrophil** (NOO-tro-fil) has two to five lobes, connected by very thin strands of chromatin (see Figure 16.1). As the cells age, the extent of nuclear lobulation increases. Because older neutrophils appear to have many differently shaped nuclei, they are often called *polymorphonuclear leukocytes (PMNs),* polymorphs, or "polys." Younger neutrophils are often called *bands* because their nucleus is more rod-shaped. When stained, the cytoplasm of neutrophils includes fine, evenly distributed pale lilac-colored granules, which may be difficult to see.

2. Agranular Leukocytes

Agranular leukocytes do not have cytoplasmic granules that can be seen under a light microscope, owing to their small size and poor staining qualities. The two kinds of agranular leukocytes are **lymphocytes** (LIM-fō-sīts) and **monocytes** (MON-ō-sīts).

Small lymphocytes are 6 to 9 μm in diameter, whereas large lymphocytes are 10 to 14 μm (see Figure 16.1). (Although the functional significance of the size difference between small and large lymphocytes is unclear, the distinction is still useful clinically because an increase in the number of large lymphocytes has diagnostic significance in acute viral infections and some immunodeficiency diseases.) The nuclei of lymphocytes are darkly stained and round, or slightly indented. The cytoplasm stains sky blue and forms a rim around the nucleus. The larger the cell, the more cytoplasm is visible.

Monocytes are 12 to 20 μm in diameter (see Figure 16.1). The nucleus of a monocyte is usually kidney-shaped or horseshoe-shaped, and the cytoplasm is blue-gray and has a foamy appearance. The color and appearance are due to very fine *azurophilic granules,* which are lysosomes. The blood is merely a conduit for monocytes, which migrate out into the tissues, enlarge, and differentiate into **macrophages** (large eaters). Some are **fixed macrophages,** which means they reside in a particular tissue—for example, alveolar macrophages in the lungs, spleen macrophages, or stellate reticuloendothelial (Kupffer) cells in the liver. Others are **wandering macrophages,** which roam the tissues and gather at sites of infection or inflammation.

Leukocytes, as a group, function in phagocytosis, producing antibodies, and combating allergies.

The lifespan of a leukocyte usually ranges from a few hours to a few months. Some lymphocytes, called T and B memory cells, can live throughout one's life once they are formed.

F. White Blood Cell Tests

▲ **CAUTION!** *Please reread Section A, "General Safety Precautions and Procedures," and Section B, "Precautions for Working with Blood, Blood Products, or Other Body Fluids," on pages xi–xii at the beginning of the laboratory manual before you begin any of the following experiments. Read the experiments before you perform them, to be sure that you understand all the procedures and safety precautions.*

When working with whole blood, wear tight-fitting surgical gloves and safety goggles. Avoid any kind of contact with an open sore, cut, or wound.

1. White Blood Cell Count

This procedure determines the number of circulating white blood cells in the body. Because white blood cells are a vital part of the body's immune defense system, any abnormalities in the white blood cell count must be carefully noted.

An increase in number (**leukocytosis**) may result from such conditions as bacterial or viral infection, allergic reactions, leukemia, hypothyroidism, anesthesia, and surgery. A decrease in number (**leukopenia**) may result from radiation, shock, systemic lupus erythematosus (SLE), and certain chemotherapeutic agents.

PROCEDURE

WARD'S Natural Science Establishment, Inc., has made available WARD'S *Simulated Blood.* It is a solution that contains microcomponents that simulate red blood cells, white blood cells, and platelets. The microcomponents are similar in relative proportion to those found in human blood and can be observed under a microscope without staining.

1. Obtain a Simulated Blood Activity kit.

2. Follow the instructions to perform the white blood cell count.

3. Complete the portion of the Data Table related to counted white blood cells.

4. Answer the questions related to white blood cells.

ALTERNATE PROCEDURE

NOTE *The blood used for this procedure should be obtained from a clinical laboratory where it has been tested and certified as noninfectious or from a mammal (other than a human).*

Whole blood that has been tested for syphilis; hepatitis A, B, and C; and HIV is available from Carolina Biological Supply Company. This blood can be used for white blood cell count and differential white blood cell count.

1. Follow steps 1 through 14 of D.3, "Red Blood Cell Count," but use the Unopette Reservoir System® for white blood cell determination (the color of the reservoir's bottom surface is white).

2. Using the low-power objective (10×), count the cells in each of the four corner squares (A, B, C, D in Figure 16.11) of the hemocytometer. The direction to follow when counting white blood cells is indicated in square A in Figure 16.11.

NOTE *To avoid overcounting cells at the boundaries, cells that touch the lines on the left and top sides of the hemocytometer should be counted, but not the ones that touch the boundary lines on the right and bottom sides. It is suggested that you use a hand counter.*

3. Multiply the results by 50 to obtain the amount of circulating white blood cells per μL of blood and record your results in Section F.6 of the LABORATORY REPORT RESULTS at the end of the exercise.

4. The factor of 50 is the dilution factor, or $2.5 \times 20 = 50$. The volume correction factor of 2.5 is arrived at in this manner: each of the corner areas (A, B, C, and D) is exactly 1 mm^3 by 0.1 mm deep (Figure 16.11). Therefore, the volume of each of these corner areas is 0.1 mm^3. Because four of them are counted, the total volume of diluted blood examined is 0.4 mm^3. However, because we want to know the number of cells in 1 mm^3 instead of 0.4 mm^3, we must multiply our count by 2.5 ($0.4 \times 2.5 = 1.0$).

▲ **CAUTION!** *Place the hemocytometer and cover slip in a container of fresh bleach solution and dispose of all other materials as directed by your instructor.*

5. The hemocytometer should be saved, not discarded.

2. Differential White Blood Cell Count

The purpose of a differential white blood cell count is to determine the relative *percentages* of each of the five normally circulating types of white blood cells in a total count of 100 white

Hemocytometer (counting chamber) for white blood cell count (as seen through scanning lens). A, B, C, and D are areas counted in the white blood cell count (when viewed through 10× lens, 1 mm² is seen in the microscopic field). Areas A, B, C, and D each equal 1 mm²; therefore, a total of 4 mm² is counted in the white blood cell count.

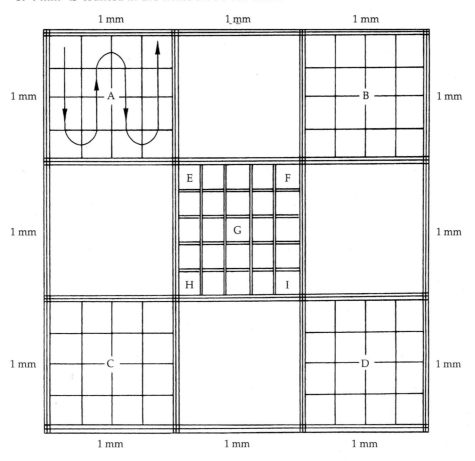

blood cells. A normal differential white blood cell count might appear as follows:

Type of white blood cell	Normal
Neutrophils	60%–70%
Eosinophils	2%–4%
Basophils	0.5%–1%
Lymphocytes	20%–25%
Monocytes	3%–8%

Significant elevations of different types of white blood cells usually indicate specific pathological conditions. For example, a high neutrophil count may indicate acute appendicitis, inflamma-tion, burns, stress, or bacterial infection. Lympho-cytes predominate in antigen-antibody reactions, specific leukemias, and in infectious mononucleo-sis. An increase in eosinophils may be seen in aller-gic reactions, parasitic infections, and autoimmune diseases. An elevated percentage of monocytes may result from chronic infections. An increase in basophils is rare and denotes a specific type of leu-kemia, allergic reactions, and hypothyroidism.

The procedure for making a differential white blood cell count is as follows:

PROCEDURE

NOTE *The blood used for this procedure should be ob-tained from a clinical laboratory where it has been tested and certified as noninfectious or from a mammal (other than a human).*

FIGURE 16.12   **Procedure to prepare a blood smear.**

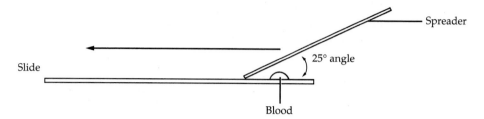

1. Obtain a drop of blood.
2. Use a second slide as a spreader (Figure 16.12).
3. Draw the spreader toward the drop of blood (in this direction →) until it touches the drop. The blood should fan out to the edges of the spreader slide.
4. Keeping the spreader at a 25° angle, press the edge of the spreader firmly against the slide and push the spreader rapidly over the entire length of the slide (in this direction ←). The drop of blood will thin out toward the end of the slide.
5. Let the smear dry.
6. To stain the slide follow this procedure:
   a. Place the slide on a staining rack.
   b. Cover the entire slide with Wright's stain.
   c. Let stain stand for 1 min.
   d. Add 25 drops of buffer solution and mix completely with Wright's stain.
   e. Let the mixture stand for 8 min.
   f. Wash the mixture off completely with distilled water.
7. Let the slide dry completely before counting.

8. To proceed with counting the cells, use an area of slide where blood is thinnest (one cell thick). This is called the *feathering edge*. Count cells under an oil-immersion lens. Count a total of 100 white cells.
9. The actual counting may be done by moving the slide either up and down or from side to side as follows:

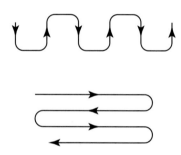

10. Record each white blood cell you observe by making a mark in the following chart until you have recorded 100 cells.

Type of white blood cell	Number observed	Percentage
Neutrophils		
Eosinophils		
Basophils		
Lymphocytes		
Monocytes		

ALTERNATE PROCEDURE

1. Obtain a prepared slide of stained blood.
2. Repeat steps 9 and 10 in the preceding procedure.

G. Platelets

Platelets (thrombocytes) are formed from fragments of the cytoplasm of megakaryocytes (see Figure 16.1). The fragments become enclosed in pieces of plasma membrane from the megakaryocytes and develop into platelets. Platelets are very small, disc-shaped cell fragments and have many granules but no nuclei. Between 150,000 and 400,000 are found in each cubic millimeter (mm^3) or microliter (μL) of blood. They function to prevent fluid loss by starting a chain of reactions that results in blood clotting. They have a short lifespan, normally just 5–9 days, because they are expended in clotting and are just too simple to carry on extensive metabolic activity.

PROCEDURE

WARD'S Natural Science Establishment, Inc., has made available WARD'S *Simulated Blood.* It is a solution that contains microcomponents that simulate red blood cells, white blood cells, and platelets. The microcomponents are similar in relative proportion to those found in human blood and can be observed under a microscope without staining.

1. Obtain a Simulated Blood Activity kit.
2. Follow the instructions to perform the platelet count.
3. Complete the portion of the Data Table related to platelets.
4. Answer the questions related to platelets.

H. Drawings of Blood Cells

Examine a prepared slide of a stained smear of blood cells using the oil-immersion objective. With the aid of your textbook, identify red blood cells, neutrophils, basophils, eosinophils, lymphocytes, monocytes, and platelets. Using colored crayons or pencil crayons, draw and label all of the different blood cells in Section H of the LABORATORY REPORT RESULTS at the end of the exercise. Be very accurate in drawing and coloring the granules and the nuclear shapes, because both of these are used to identify the various types of cells.

I. Blood Grouping (Typing)

The surfaces of erythrocytes contain genetically determined **antigens** called **agglutinogens.** The plasma of blood contains genetically determined **antibodies** called **agglutinins.** The antibodies cause the **agglutination (clumping)** of the red blood cells carrying the corresponding antigen.

These proteins, antigens, and antibodies are responsible for the two major classifications of blood groups: the ABO group and the Rh system. In addition to the ABO group and the Rh system, other human blood groups include MNSs, P, Lutheran, Kell, Lewis, Duffy, Kidd, Diego, and Sutter. Fortunately these different antigenic factors do not exhibit extreme degrees of antigenicity and, therefore, usually cause very weak transfusion reactions or no reaction at all.

1. ABO Group

The ABO blood group is based on two antigens called A and B (Figure 16.13). Individuals whose red blood cells display only antigen *A* have blood **type A.** Individuals who display only antigen *B* have blood **type B.** If both *A* and *B* antigens are displayed, the result is **type AB,** whereas the absence of both *A* and *B* antigens results in the so-called **type O.**

The antibodies in blood plasma are anti-*A* antibody, which reacts with antigen *A,* and anti-*B* antibody, which reacts with antigen *B.* The antigens and antibodies formed by each of the four blood types and the various agglutination reactions that occur when whole blood samples are mixed with serum are shown in Figure 16.13. You do not have antibodies that attack the antigens of your own red blood cells. For example, a type A person has antigen *A* but not antibody anti-*A.*

Antigens and antibodies are of critical importance in blood transfusions. In an incompatible blood transfusion, the donated red blood cells are attacked by the recipient antibodies, causing the blood cells to agglutinate. Agglutinated cells become lodged in small capillaries throughout the body and, over a period of hours, the cells swell, rupture, and release hemoglobin into the blood. Such a reaction as it relates to red blood cells is called **hemolysis.** The degree of agglutination depends on the titer (strength or amount) of antibody in the blood. Agglutinated cells can block blood vessels and may lead to kidney or brain damage and death, and the liberated hemoglobin may also cause kidney damage.

FIGURE 16.13 **ABO blood groupings.**

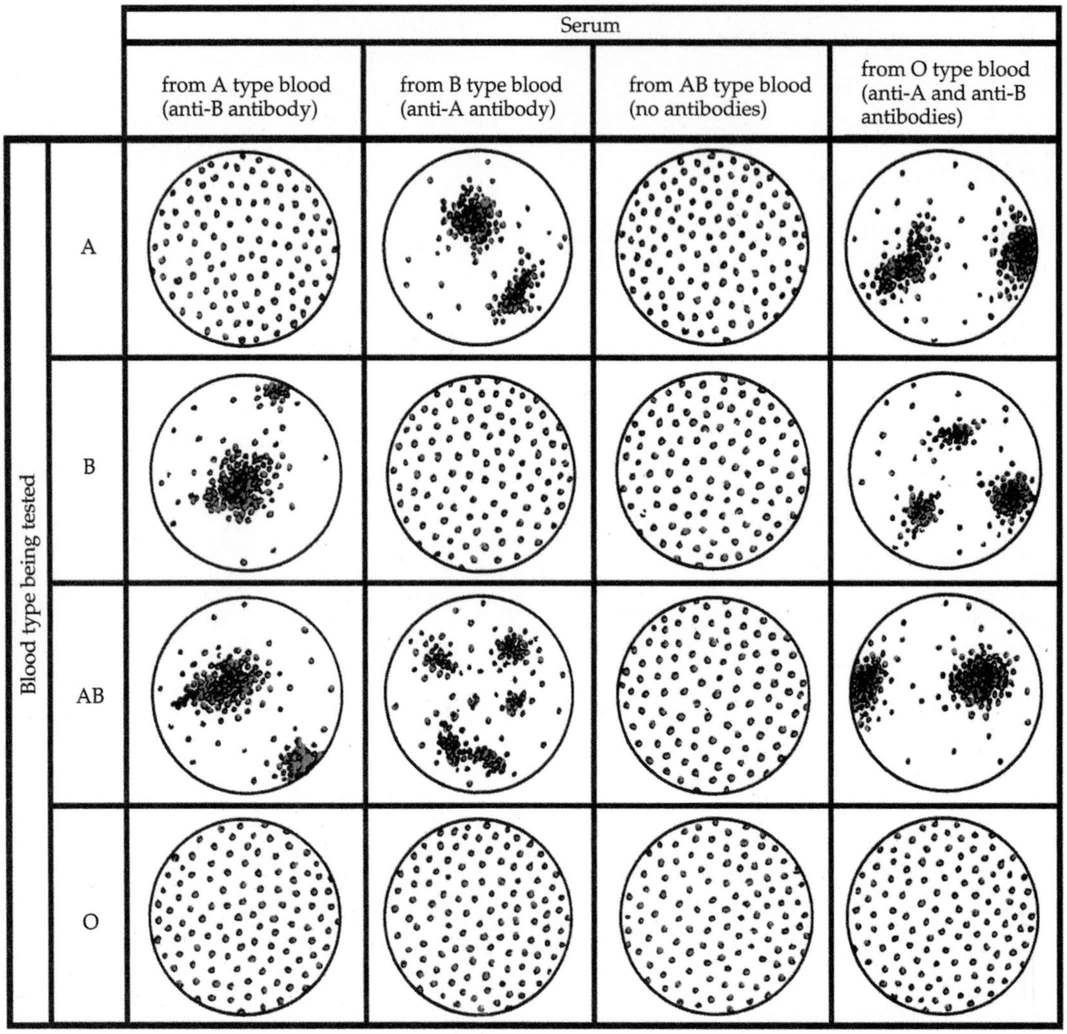

Because cells of type O blood contain neither of the two antigens (*A* or *B*), moderate amounts of this blood can be transfused into a recipient of any ABO blood type without *immediate* agglutination. For this reason, type O blood is referred to as the **universal donor.** However, transfusing large amounts of type O blood into a recipient of type A, B, or AB can cause *delayed* agglutination of the recipient's red blood cells, because the transfused antibodies (type O blood contains anti-*A* and anti-*B* antibodies) are not sufficiently diluted to prevent the reaction.

Persons with type AB blood are sometimes called **universal recipients** because their blood contains no antibodies to agglutinate donor red blood cells. *Small quantities* of blood from all other ABO types can be transfused into type AB blood without adverse reaction. However, again, if *large quantities* of type A, B, or O blood are transfused

into type AB, the antibodies in the donor blood might accumulate in sufficient quantity to clump the recipient's red blood cells.

Table 16.1 summarizes the various interactions of the four blood types. Table 16.2 lists the incidence of these different blood types in the United States, comparing some races.

2. ABO Blood Grouping Test

▲ **CAUTION!** *Please reread Section A, "General Safety Precautions and Procedures," and Section B, "Precautions for Working with Blood, Blood Products, or Other Body Fluids," on pages xi–xii at the beginning of the laboratory manual before you begin any of the following experiments. Read the experiments before you perform them, to be sure that you understand all the procedures and safety precautions.*

TABLE 16.1　Summary of ABO blood group interactions

Blood type	A	B	AB	O
Antigen on RBCs	A	B	Both A and B	Neither A nor B
Antibody in plasma	anti-B	anti-A	Neither anti-A nor anti-B	Both anti-A and B and anti-B
Compatible donor blood types (no hemolysis)	A, O	B, O	A, B, AB, O	O
Incompatible donor blood types (hemolysis)	B, AB	A, AB	—	A, B, AB
Alleles (genotype)	$I^A I^A$ or $I^A i$	$I^B I^B$ or $I^B i$	$I^A I^B$	ii
Expressed blood type (phenotype)	A	B	AB	O

When working with whole blood, wear tight-fitting surgical gloves and safety goggles. Avoid any kind of contact with an open sore, cut, or wound.

Blood should be obtained from a clinical laboratory where it has been tested and certified as noninfectious or from a mammal (other than a human).

Whole blood that has been tested for syphilis; hepatitis A, B, and C; and HIV is available from Carolina Biological Supply Company. This blood can be used for blood grouping and typing. Carolina Biological Supply Company also provides aseptic red blood cells that can be used for blood grouping and typing. These blood cells are suspended in a modified Alsever's solution to which several antibodies have been added. The cells have been tested for syphilis; hepatitis A, B, and C; and HIV.

Also available for this exercise is WARD'S simulated ABO and Rh Blood Typing Activity.

The procedure for ABO sampling is as follows:

PROCEDURE

1. Using a wax pencil, divide a glass slide in half and label the left side A and the right side B.

2. On the left side, place one large drop of anti-*A* serum, and on the right side place one large drop of anti-*B* serum.

3. Next to the drops of antisera, place one drop of blood, being careful not to mix samples on the left and right sides.

4. Using a mixing stick or a toothpick, mix the blood on the left side with the anti-*A* serum,

TABLE 16.2　Incidence of blood types in the United States

Population group	Blood type (percentage)				
	O	A	B	AB	Rh⁺
European-American	45	40	11	4	85
African-American	49	27	20	4	95
Korean-American	32	28	30	10	100
Japanese-American	31	38	21	10	100
Chinese-American	42	27	25	6	100
Native American	79	16	4	1	100

and then, using a *different stick* or *different toothpick*, mix the blood on the right side with the anti-*B*.

▲ **CAUTION!** *Immediately dispose of the stick or toothpick in the appropriate biohazard container provided by your instructor.*

5. *Gently* tilt the slide back and forth and observe it for 1 min.

6. Record your results in Section I.1 of the LABORATORY REPORT RESULTS at the end of the exercise, using "+" for clumping (agglutination) and "−" for no clumping.

▲ **CAUTION!** *Dispose of your slide in the appropriate biohazard container provided by your instructor.*

3. Rh Blood Group

The Rh factor is the other major classification of blood grouping. This group was designated Rh because the blood of the *Rhesus* monkey was used in the first research and development. As is the ABO grouping, this classification is based on antigens that lie on the surfaces of red blood cells. The designation Rh^+ (Rh positive) is given to those that have the antigen, and Rh^- (Rh negative) is for those that lack the antigen. The estimation is that 85% of European-Americans and 95% of African-Americans in the United States are Rh^+, whereas 15% of European-Americans and 5% of African-Americans are Rh^- (see Table 16.2).

The Rh factor is extremely important in pregnancy and childbirth. Under normal circumstances, human plasma does not contain anti-Rh antibodies. If, however, a woman who is Rh^- becomes pregnant with an Rh^+ child, her blood may produce antibodies that will react with the blood of a subsequent child.[3] The first child is unaffected because the mother's body has not yet produced these antibodies. This is a serious reaction and hemolysis may occur in the fetal blood.

The hemolysis produced by this fetal-maternal incompatibility is called **hemolytic disease of newborn,** or **HDN,** and could be fatal for the newborn infant. A drug called anti-Rh gamma globulin (Rho-GAM®), given to Rh^- mothers immediately after delivery or abortion, prevents the production of antibodies by the mother so that the fetus of the next pregnancy is protected.

4. Rh Blood Grouping Test

▲ **CAUTION!** *Please reread Section A, "General Safety Precautions and Procedures," and Section B, "Precautions for Working with Blood, Blood Products, or Other Body Fluids," on pages xi–xii at the beginning of the laboratory manual before you begin any of the following experiments. Read the experiments before you perform them, to be sure that you understand all the procedures and safety precautions.*

When working with whole blood, wear tight-fitting surgical gloves and safety goggles. Avoid any kind of contact with an open sore, cut, or wound.

NOTE *Remember that this procedure is sufficient for laboratory demonstration, but it should not be considered clinically accurate, since the anti-Rh (anti-D) serum deteriorates rapidly at room temperature and the anti-Rh antibodies are far less potent in their agglutinizing capability than the anti-A or anti-B antibodies. This test is less accurate than the ABO determination.*

Blood should be obtained from a clinical laboratory where it has been tested and certified as noninfectious or from a mammal (other than a human).

Whole blood that has been tested for syphilis; hepatitis A, B, and C; and HIV is available from Carolina Biological Supply Company. This blood can be used for blood grouping and typing. Carolina Biological Supply Company also provides aseptic red blood cells that can be used for blood grouping and typing. These blood cells are suspended in a modified Alsever's solution to which several antibodies have been added. The cells have been tested for syphilis; hepatitis A, B, and C; and HIV.

Also available for this exercise is WARD'S simulated ABO and Rh Blood Typing Activity.

The procedure for Rh grouping is as follows:

PROCEDURE

1. Place one large drop of anti-Rh (anti-D) serum on a glass slide.[4]

2. Add one drop of blood and mix using a mixing stick or toothpick.

▲ **CAUTION!** *Immediately dispose of the stick or toothpick in the appropriate biohazard container provided by your instructor.*

3. Place the slide on a preheated warming box, and gently rock the box back and forth for 2 min. (Unlike ABO typing, Rh typing is better done on a heated warming box.)

4. Record your results, using "+" for clumping and "−" for no clumping. Record whether you are Rh^+ or Rh^- in Section I.5 of the LABORATORY REPORT RESULTS at the end of the exercise.

▲ **CAUTION!** *Dispose of your slide in the appropriate biohazard container provided by your instructor.*

ANSWER THE LABORATORY REPORT QUESTIONS AT THE END OF THE EXERCISE.

---

[3]Be sure to note the important difference in the production of antibodies in the ABO and Rh systems. An Rh^- person can *produce antibodies in response to the stimulus of invading Rh antigen;* by contrast, any antibodies of the ABO system that exist in the blood of a person *occur naturally and are present regardless of whether or not ABO antigens are introduced.*

[4]The Rh antigen is more specifically termed the D antigen after the Fisher-Race nomenclature, which is based on genetic concepts or theories of inheritance.

EXERCISE 16 **Blood**

Name _____ Date _____

Laboratory Section _____ Score/Grade _____

SECTION B ■ Plasma

1. *Physical Characteristics*

   pH _____

   Color _____

   Transparency (clear, translucent, opaque) _____

2. *Chemical Constituents*

   Is glucose present? _____

   Is protein present? _____

SECTION D ■ Red Blood Cell Tests

1. Red blood cell count results: _____ RBCs per μL.

2. Red blood cell volume (hematocrit) results: _____ %.

3. Sedimentation rate results: _____ mm per hour.

4. Hemoglobin determination results: hemoglobinometer, _____ g per 100 ml.

   Tallquist® paper, _____ g per 100 ml.

5. Complete the following table:

Wavelength (nm)

	500	510	520	530	540	550	560	570	580	590	600
Solution 1											
Solution 2											

Graph your results here.

SECTION F ▪ White Blood Cell Tests

6. White blood cell count results: _____ WBCs per μL.

SECTION H ▪ Drawings of Blood Cells

SECTION I ▪ Blood Grouping (Typing)

1. In determining the ABO blood grouping, did you observe clumping when the blood was mixed with

   _____ anti-A serum only

   _____ anti-B serum only

   _____ both anti-A and anti-B sera

   _____ neither anti-A nor anti-B serum

2. Based on your observations, what is the ABO grouping?

   _____ A   _____ B   _____ AB   _____ O

3. Based on your observations, briefly explain why you identify your ABO blood grouping as you do.

   _____

   _____

4. Record the results of the ABO blood grouping test done by your class.

Type	Anti-A (present or absent)	Anti-B (present or absent)	Number of individuals	Class percentage
A				
B				
AB				
O				

5. Based on your observations, are you Rh^+ or Rh^-?

    _____ Rh^+ _____ Rh^-

6. Record the results of the Rh test done by your class.

Type	Number of individuals	Class percentage
Rh^+		
Rh^-		

Blood

Name _____   Date _____

Laboratory Section _____   Score/Grade _____

PART 1 ■ Multiple Choice

_____ 1. The process by which all blood cells are formed is called (a) hemocytoblastosis (b) erythropoiesis (c) hemopoiesis (d) leukocytosis

_____ 2. An inability of body cells to receive adequate amounts of oxygen may indicate a malfunction of (a) neutrophils (b) leukocytes (c) lymphocytes (d) erythrocytes

_____ 3. Special cells of the body that have the responsibility of clearing away dead, disintegrating bodies of red and white blood cells are called (a) agranular leukocytes (b) reticuloendothelial cells (c) erythrocytes (d) thrombocytes

_____ 4. The name of the test procedure that informs the physician about the rate of erythropoiesis is called the (a) reticulocyte count (b) sedimentation rate (c) hemoglobin count (d) differential white blood cell count

_____ 5. The normal red blood cell count per cubic millimeter (mm^3) or microliter (μL) in males is about (a) 5.4 million (b) 7 million (c) 4 million (d) more than 9 million

_____ 6. The normal number of leukocytes per cubic millimeter (mm^3) or microliter (μL) is (a) 5000 to 10,000 (b) 8000 to 12,000 (c) 2000 to 4000 (d) over 15,000

_____ 7. Under the microscope, red blood cells appear as (a) circular discs with centrally located nuclei (b) circular discs with lobed nuclei (c) oval discs with many nuclei (d) biconcave discs without nuclei

_____ 8. An increase in the number of white blood cells is called (a) leukopenia (b) hematocrit (c) polycythemia (d) leukocytosis

_____ 9. Platelets are formed from a special large cell that breaks up into small fragments. This cell is called a(n) (a) eosinophil (b) hemocytoblast (c) megakaryocyte (d) platelet

_____ 10. The blood type showing the highest incidence in whites in the United States is (a) A (b) O (c) AB (d) B

PART 2 ■ Completion

11. Another name for red blood cells is _____.

12. Blood gets its red color from the presence of _____.

13. The lifespan of a red blood cell is approximately _____.

14. A good method for routine testing for anemia is _____.

15. The normal sedimentation rate value for adults is _____.

16. The normal ratio of red blood cells to white blood cells is about _____.

17. The granular leukocytes are formed from _____ tissue.

18. The number of platelets per cubic millimeter (mm^3) or microliter (μL) found normally in blood is

    _____.

19. The function of platelets is to prevent blood loss by starting a chain of reactions resulting in

    _____.

20. In the ABO blood grouping system, the genetically determined structures on the surface of red blood cells are called _____.

21. The hemolysis produced by fetal-maternal incompatibility of blood cells is called

    _____.

22. The part of the cell where agglutinogens are located is _____.

23. Cells derived from mesenchyme that give rise to all the formed elements in blood are called

    _____.

PART 3 ▪ Matching

_____ 24. Pluripotent stem cell

_____ 25. Polycythemia

_____ 26. A high neutrophil count

_____ 27. A high monocyte count

_____ 28. Leukopenia

_____ 29. A high lymphocyte count

_____ 30. Plasma

_____ 31. A high eosinophil count

_____ 32. Serum

A. Acute appendicitis or bacterial infection

B. An increase in the normal red blood cell count

C. Leukemia and infectious mononucleosis

D. Immature cells that develop into mature blood cells

E. Chronic infections

F. Liquid portion of blood without the formed elements and clotting substances

G. An allergic reaction

H. A decrease in the normal white blood cell count

I. Liquid portion of blood without the formed elements

Heart

Objectives

A The location and surface projection of the heart

B The external and internal characteristics of the heart

C The blood supply of the heart

D The anatomical features of a sheep heart

As noted in Exercise 16, the cardiovascular system consists of the blood, heart, and blood vessels. Having considered the origin, structural features, and functions of blood, we will now examine the heart, the center of the cardiovascular system. This exercise discusses the location and surface projection of the heart, pericardium, wall, chambers, great vessels, valves, and blood supply of the heart.

A. Location and Surface Projection of Heart

The heart rests on the diaphragm, near the midline of the thoracic cavity in the **mediastinum** (mē′-dē-a-STĪ-num), a mass of tissue that extends from the sternum to the vertebral column and between the lungs (Figure 17.1a). About two-thirds of the mass of the heart lies to the left of the body's midline. The position of the heart in the mediastinum is more readily appreciated by examining its ends, surfaces, and borders. First, visualize the heart as a cone lying on its side. The pointed end of the heart is the **apex,** which is directed anteriorly, inferiorly, and to the left. The end opposite the apex is the **base,** the broad portion of the heart that points posteriorly, superiorly, and to the right. In addition to the apex and base, there are several surfaces and borders (margins) that are useful in describing anatomical relationships and pathological conditions that relate to deviations from normal shape and contour. The **anterior surface** is deep to the sternum and ribs. The **inferior surface** is the portion of the heart that rests mostly on the diaphragm. It is found be-

tween the apex and right border. The **right border** faces the right lung and extends from the inferior surface to the base. Finally, the **left border,** also called the *pulmonary border,* faces the left lung and extends from the base to the apex.

Identify these structures using a specimen, model, or chart of a heart and label Figure 17.1a.

Surface projection refers to outlining the dimensions of an organ with respect to landmarks on the surface of the body. This practice is useful for diagnostic procedures (for example, a lumbar puncture), auscultation (e.g., listening to heart and lung sounds), and anatomical studies. We can project the heart on the anterior surface of the chest by locating the following landmarks. The **superior right point** is located at the superior border of the third right costal cartilage, about 3 cm (1 in.) to the right of the midline (Figure 17.1b). The **superior left point** is located at the inferior border of the second left costal cartilage, about 3 cm to the left of the midline. A line connecting these two points corresponds to the base of the heart.

The **inferior left point** is located at the apex of the heart in the fifth left intercostal space, about 9 cm (3.5 in.) to the left of the midline. A line connecting the superior and inferior left points corresponds to the left border of the heart. The **inferior right point** is located at the superior border of the sixth right costal cartilage, about 3 cm to the right of the midline. A line connecting the inferior left and right points corresponds to the inferior surface of the heart, and a line connecting the inferior and superior right points corresponds to the right border of the heart. When all four points are con-

FIGURE 17.1 Location and surface projection of the heart.

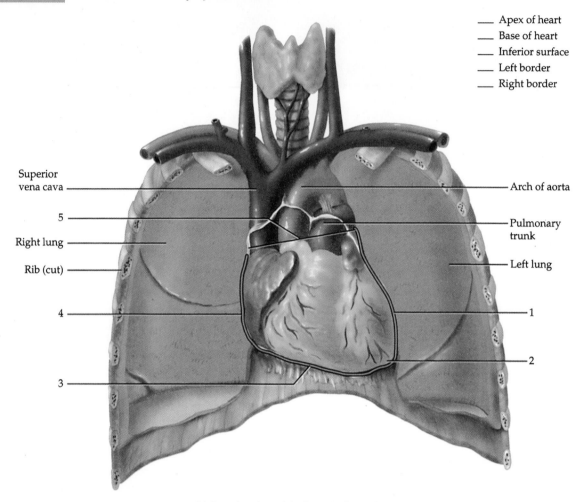

—— Apex of heart
—— Base of heart
—— Inferior surface
—— Left border
—— Right border

Superior vena cava

Arch of aorta

5

Pulmonary trunk

Right lung

Rib (cut)

Left lung

4

1

2

3

(a) Anterior view of the heart in the mediastinum

nected, they form an outline that roughly reveals the size and shape of the heart. Identify the points using a specimen, model, or chart of the heart and label Figure 17.1b.

B. Pericardium

The **pericardium** (*peri* = around; *cardio* = heart) is the membrane that surrounds and protects the heart. It confines the heart to its position in the mediastinum, while allowing sufficient freedom of movement for vigorous and rapid contraction. It consists of two principal portions: the fibrous pericardium and the serous pericardium (Figure 17.2). The superficial **fibrous pericardium** is a

tough, inelastic, dense irregular connective tissue membrane. The fibrous pericardium prevents overstretching of the heart, provides protection, and anchors the heart in the mediastinum.

The deeper **serous pericardium** is a thinner, more delicate membrane that forms a double layer around the heart. The outer **parietal layer** of the serous pericardium is fused to the fibrous pericardium. The inner **visceral layer** of the serous pericardium, also called the **epicardium** (*epi* = on top), adheres tightly to the muscle of the heart. Between the parietal and visceral layers of the serous pericardium is a thin film of serous fluid known as **pericardial fluid.** It is a slippery secretion of the pericardial cells that reduces friction between the membranes as the heart moves. The space that con-

FIGURE 17.1 Location and surface projection of the heart. (Continued)

___ Inferior left point
___ Inferior right point
___ Superior left point
___ Superior right point

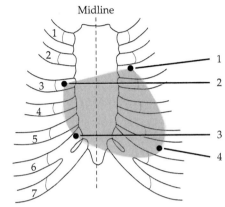

(b) Surface projection of the heart in the mediastinum

tains a few milliliters of pericardial fluid is called the **pericardial cavity.**

Identify these structures using a specimen, model, or chart of a heart and label Figure 17.2.

C. Heart Wall

The wall of the heart consists of three layers: the epicardium (external layer), the myocardium (middle layer), and the endocardium (inner layer). The **epicardium** (*epi* = above), also called the *visceral layer of the serous pericardium,* is the thin, transparent outer layer of the heart wall. The middle **myocardium** (*myo* = muscle), which is composed of cardiac muscle tissue, forms the bulk of the heart and is responsible for its pumping action. The inner most **endocardium** (*endo* = within) is a thin layer of endothelium overlying a thin layer of connective tissue. It provides a smooth lining for the chambers of the heart and covers the heart valves. Label the layers of the heart in Figure 17.2.

D. Chambers and Great Vessels of Heart

The heart contains four **chambers.** The two superior chambers are the **atria** (= entry halls or chambers), and the two inferior chambers are the **ventricles** (= little bellies). The **right atrium** collects deoxygenated blood from systemic circulation, and the **left atrium** collects oxygenated blood

FIGURE 17.2 Structure of pericardium and heart wall.

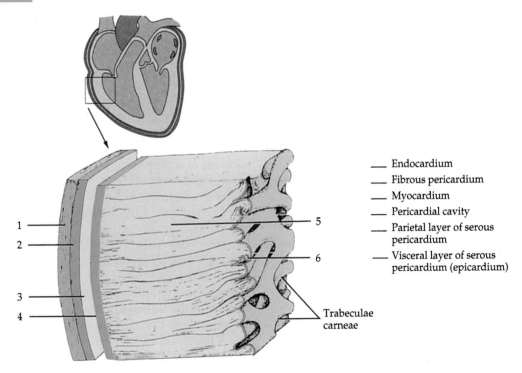

___ Endocardium
___ Fibrous pericardium
___ Myocardium
___ Pericardial cavity
___ Parietal layer of serous pericardium
___ Visceral layer of serous pericardium (epicardium)

Trabeculae carneae

from pulmonary circulation. The **right ventricle** receives deoxygenated blood from the right atrium and pumps it to the lungs to become oxygenated. The **left ventricle** receives oxygenated blood from the left atrium and pumps it to all body cells, except the air sacs of the lungs.

During one phase of a heartbeat, the atria empty of blood and become smaller. This produces the appearance of a wrinkled pouch on the anterior surface of each atrium called an **auricle** (OR-i-kul; *auri* = ear), so named because of its resemblance to a dog's ear. Each auricle slightly increases the capacity of an atrium so that it can hold a greater volume of blood. On the surface of the heart are a series of grooves, called **sulci** (SUL-sē), that contain coronary blood vessels and a variable amount of fat. Each sulcus marks the external boundary between two chambers of the heart. The deep **coronary** (*coron* = crown) **sulcus** encircles most of the heart and marks the boundary between the superior atria and inferior ventricles. The **anterior interventricular sulcus** is a shallow groove on the anterior surface of the heart that marks the boundary between the right and left ventricles. This sulcus continues around to the posterior surface of the heart as the **posterior interventricular sulcus,** which marks the boundary between the ventricles of the posterior aspect of the heart.

Label these structures in Figures 17.3 and 17.4.

The **right atrium** forms the right border of the heart (see Figure 17.1a). It receives deoxygenated blood from systemic circulation through three veins: *superior vena cava, inferior vena cava,* and *coronary sinus.*

The anterior and posterior walls of the right atrium differ considerably. Whereas the posterior wall is smooth, the anterior wall is rough due to the presence of muscular ridges called **pectinate** (*pectin* = comb) **muscles.** Between the right atrium and left atrium is a thin partition called the **interatrial septum.** A prominent feature of this septum is an oval depression called the **fossa ovalis.** It is the remnant of the foramen ovale, an opening in the interatrial septum of the fetal heart that normally closes soon after birth (see Figure 17.4a).

The **right ventricle** forms most of the anterior surface of the heart. Deoxygenated blood passes from the right atrium into the right ventricle through a valve called the *tricuspid valve.* The cusps of the valve are connected to tendonlike cords, the **chordae tendineae** (KOR-dē ten-DIN-ē-ē; *chord* = cord; *tend* = tendon), which, in turn, are connected to cone-shaped muscular columns (trabeculae carneae) in the inner surface of the ventricle called

papillary (*papill* = nipple) **muscles.** (Details about the tricuspid and other valves mentioned in this discussion of the heart chambers will follow shortly.) Another prominent feature of the right ventricle is a series of ridges and folds called **trabeculae carneae** (tra-BEK-ū-lē KAR-nē-ē; *trabecula* = little beam; *carneae* = fleshy). The right ventricle is separated from the left ventricle by a partition called the **interventricular septum.**

Deoxygenated blood passes from the right ventricle through a valve called the **pulmonary valve** into the **pulmonary** (*pulmo* = lung) **trunk** to enter pulmonary circulation. The pulmonary trunk divides into a right and left **pulmonary artery,** each of which carries blood to its respective lung. In the lungs, the deoxygenated blood releases carbon dioxide and takes on oxygen. The blood is then oxygenated blood.

The **left atrium** forms most of the base of the heart (see Figure 17.1a). It receives oxygenated blood from pulmonary circulation in the lungs through four *pulmonary veins.* Like the right atrium, the left atrium is characterized by a smooth posterior wall. Pectinate muscles, however, are found only in the right atrium.

The **left ventricle** forms the apex of the heart (see Figure 17.1a). Oxygenated blood passes from the left atrium into the left ventricle through a valve called the **bicuspid (mitral) valve.** Like the tricuspid valve, the bicuspid valve has chordae tendineae attached to papillary muscles. However, as the name indicates, the bicuspid valve has only two cusps (not three as in the tricuspid valve). The left ventricle also contains trabeculae carneae.

Oxygenated blood passes into systemic circulation from the left ventricle through a valve called the **aortic valve** into the **ascending aorta.** From here some of the blood flows into the left and right **coronary arteries,** which branch from the ascending aorta and carry the blood to the heart wall. The remainder of the blood passes into the **arch of the aorta** and **descending aorta (thoracic aorta** and **abdominal aorta).** Branches of the arch of the aorta and descending aorta carry the blood throughout the systemic circulation.

During fetal life, a temporary blood vessel, called the *ductus arteriosus,* connects the pulmonary trunk with the aorta. It shunts blood so that only a small amount enters the nonfunctioning fetal lungs. The ductus arteriosus normally closes shortly after birth, leaving a remnant known as the **ligamentum arteriosum,** which interconnects the arch of the aorta and pulmonary trunk (see Figure 17.3).

Label these structures in Figures 17.3 and 17.4.

FIGURE 17.3 External surface of the human heart. Red-colored vessels carry oxygenated blood; blue-colored vessels carry deoxygenated blood.

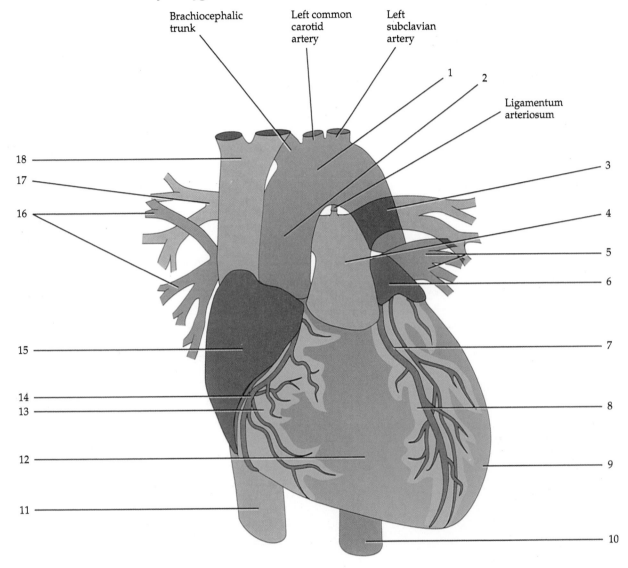

(a) Diagram of anterior external view

___ Anterior interventricular sulcus

___ Arch of aorta

___ Ascending aorta

___ Coronary sulcus

___ Descending aorta

___ Inferior vena cava

___ Left atrium

___ Left coronary artery

___ Left pulmonary artery

___ Left pulmonary veins

___ Left ventricle

___ Pulmonary trunk

___ Right atrium

___ Right coronary artery

___ Right pulmonary artery

___ Right pulmonary veins

___ Right ventricle

___ Superior vena cava

FIGURE 17.3   External surface of the human heart. (Continued)

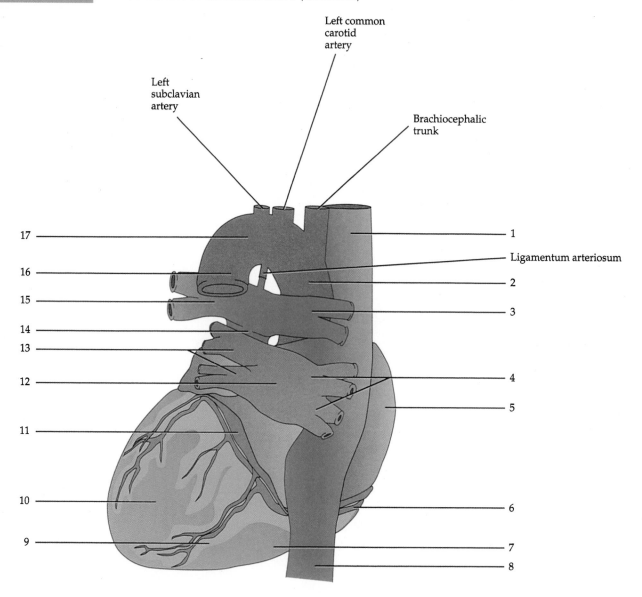

Left common
carotid
artery

Left
subclavian
artery

Brachiocephalic
trunk

17

16

15

14

13

12

11

10

9

1

Ligamentum arteriosum

2

3

4

5

6

7

8

(b) Diagram of posterior external view

____ Arch of aorta
____ Ascending aorta
____ Coronary sinus
____ Descending aorta
____ Inferior vena cava
____ Left atrium
____ Left pulmonary artery
____ Left pulmonary veins
____ Left ventricle

____ Posterior interventricular sulcus
____ Pulmonary trunk
____ Right atrium
____ Right coronary artery
____ Right pulmonary artery
____ Right pulmonary veins
____ Right ventricle
____ Superior vena cava

FIGURE 17.3 External surface of the human heart. (Continued)

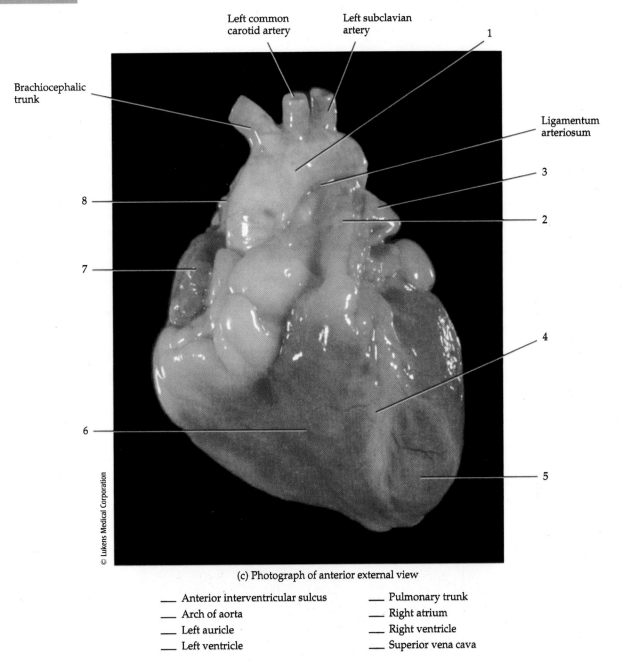

Left common
carotid artery

Left subclavian
artery

1

Brachiocephalic
trunk

Ligamentum
arteriosum

3

8

2

7

4

6

5

© Lukens Medical Corporation

(c) Photograph of anterior external view

___ Anterior interventricular sulcus   ___ Pulmonary trunk
___ Arch of aorta   ___ Right atrium
___ Left auricle   ___ Right ventricle
___ Left ventricle   ___ Superior vena cava

E. Valves of Heart

As each chamber of the heart contracts, it pushes a volume of blood into a ventricle or out of the heart through an artery. To prevent backflow of blood, the heart has **valves.** These structures are composed of dense connective tissue covered by endo-

cardium. Valves open and close in response to pressure changes as the heart contracts and relaxes.

Atrioventricular (AV) valves lie between the atria and ventricles. The right AV valve between the right atrium and right ventricle is also called the **tricuspid** (trī-KUS-pid) **valve** because it consists of three cusps (flaps). The left AV valve be-

FIGURE 17.4 Internal structure of the human heart.

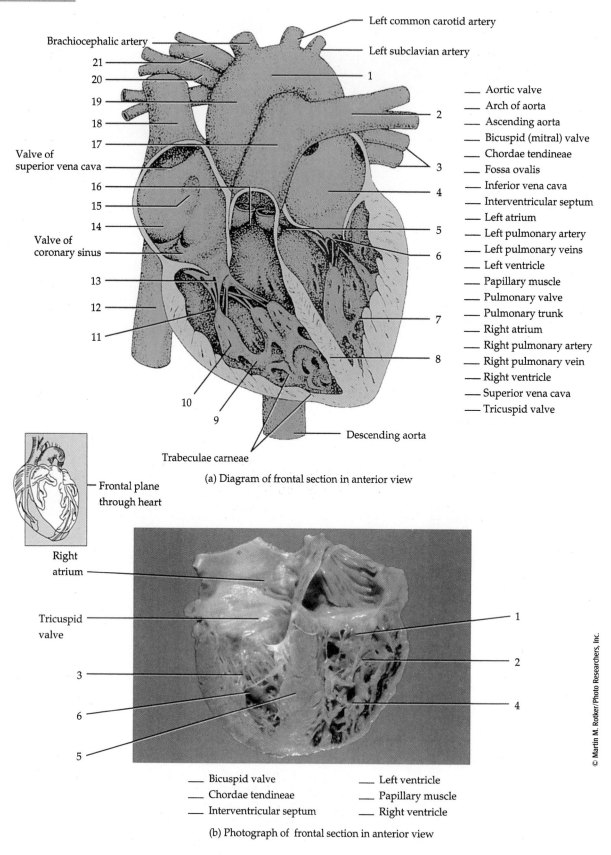

Left common carotid artery

Brachiocephalic artery

Left subclavian artery

21

20

19

18

17

Valve of
superior vena cava

16

15

14

Valve of
coronary sinus

13

12

11

10

9

1

2

3

4

5

6

7

8

_____ Aortic valve
_____ Arch of aorta
_____ Ascending aorta
_____ Bicuspid (mitral) valve
_____ Chordae tendineae
_____ Fossa ovalis
_____ Inferior vena cava
_____ Interventricular septum
_____ Left atrium
_____ Left pulmonary artery
_____ Left pulmonary veins
_____ Left ventricle
_____ Papillary muscle
_____ Pulmonary valve
_____ Pulmonary trunk
_____ Right atrium
_____ Right pulmonary artery
_____ Right pulmonary vein
_____ Right ventricle
_____ Superior vena cava
_____ Tricuspid valve

Trabeculae carneae

Descending aorta

Frontal plane
through heart

(a) Diagram of frontal section in anterior view

Right
atrium

Tricuspid
valve

3

6

5

1

2

4

_____ Bicuspid valve   _____ Left ventricle
_____ Chordae tendineae   _____ Papillary muscle
_____ Interventricular septum _____ Right ventricle

(b) Photograph of frontal section in anterior view

tween the left atrium and left ventricle has two cusps and is called the **bicuspid (mitral) valve.** When an AV valve is open, the pointed ends of the cusps project into a ventricle. Tendonlike cords called **chordae tendineae** connect the pointed ends and undersurfaces to **papillary muscles** (muscular columns) that are located on the inner surface of the ventricles.

Near the origin of both arteries that emerge from the heart, there are heart valves that allow ejection of blood from the heart but prevent blood from flowing backward into the heart. These are the **semilunar (SL) valves.** The **pulmonary valve** lies in the opening where the pulmonary trunk leaves the right ventricle. The **aortic valve** is situated at the opening between the left ventricle and the aorta.

Both valves consist of three semilunar (half-moon, or crescent-shaped) cusps. Each cusp attaches to the artery wall by its convex outer margin. The free borders of the cusps project into the lumen of the blood vessel. As the ventricles relax, blood starts to flow back toward the heart, filling the cusps and tightly closing the semilunar valves.

Label the atrioventricular and semilunar valves in Figure 17.4.

F. Blood Supply of Heart

Nutrients are not able to diffuse quickly enough from blood in the chambers of the heart to supply all the layers of cells that make up the heart tissue. For this reason, the myocardium has its own blood vessels. The flow of blood through the many vessels that pierce the myocardium is called the **coronary (cardiac) circulation.** The arteries of the heart encircle it like a crown encircles the head (*coron* = crown). While it is contracting, the heart receives little flow of oxygenated blood by way of the **coronary arteries** because they are squeezed shut. When the heart relaxes, however, the high pressure of blood in the aorta propels blood through the coronary arteries, into capillaries, and then into **coronary veins.**

Two coronary arteries, the right and left coronary arteries, branch from the ascending aorta and supply oxygenated blood to the myocardium. The **left coronary artery** passes inferior to the left auricle and divides into the anterior interventricular and circumflex branches. The **anterior interventricular branch** or **left anterior descending (LAD) artery** is in the anterior interventricular sulcus and supplies oxygenated blood to the walls of both ventricles. The **circumflex branch** lies in the coronary sulcus and distributes oxygenated blood to the walls of the left ventricle and left atrium.

The **right coronary artery** supplies small branches (arterial branches) to the right atrium. It continues inferior to the right auricle and divides into the posterior interventricular and marginal branches. The **posterior interventricular branch** follows the posterior interventricular sulcus, supplying walls of the two ventricles with oxygenated blood. The **marginal branch** in the coronary sulcus transports oxygenated blood to the myocardium of the right ventricle.

Label the arteries in Figure 17.5a.

After blood passes through the arteries of coronary circulation it flows into capillaries, where it delivers oxygen and nutrients and collects carbon dioxide and wastes, and then moves into coronary veins. The deoxygenated blood then drains into a large vascular sinus on the posterior surface of the heart, called the **coronary sinus,** which empties into the right atrium. A vascular sinus is a vein with a thin wall that has no smooth muscle to alter its diameter. The principal tributaries carrying blood into the coronary sinus are the **great cardiac vein,** which drains the anterior aspect of the heart, the **middle cardiac vein,** which drains the posterior aspect of the heart, the **small cardiac vein,** which drains the right atrium and right ventricle, and the **anterior cardiac veins,** which drain the right ventricle and open directly into the right atrium.

Label the veins in Figure 17.5b.

G. Dissection of Sheep Heart

The anatomy of the sheep heart closely resembles that of the human heart. Use Figures 17.6 and 17.7 as references for this dissection. In addition, models of human hearts can also be used as references.

▲ **CAUTION!** *Please reread Section D, "Precautions Related to Dissection," at the beginning of the laboratory manual on page xiii before you begin your dissection.*

First, examine the **pericardium,** a fibroserous membrane that encloses the heart, which may have already been removed in preparing the sheep heart for preservation. The **myocardium** is the middle layer and constitutes the main muscle portion of the heart. The **endocardium** (the third layer) is the inner lining of the heart. Use the figures to determine which is the ventral surface of the heart and then identify the **pulmonary trunk** emerging from the anterior ventral surface, near the midline, and medial to the **left auricle.** A longitudinal depression on the ventral surface, called the **anterior longitudinal sulcus,** separates the right ventricle from the left ventricle. Locate the **coronary blood vessels** lying in this sulcus.

FIGURE 17.5 Coronary (cardiac) circulation.

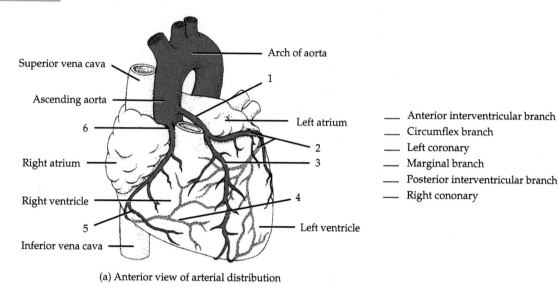

Superior vena cava

Ascending aorta

6

Right atrium

Right ventricle

5

Inferior vena cava

Arch of aorta

1

Left atrium

2

3

4

Left ventricle

___ Anterior interventricular branch
___ Circumflex branch
___ Left coronary
___ Marginal branch
___ Posterior interventricular branch
___ Right cononary

(a) Anterior view of arterial distribution

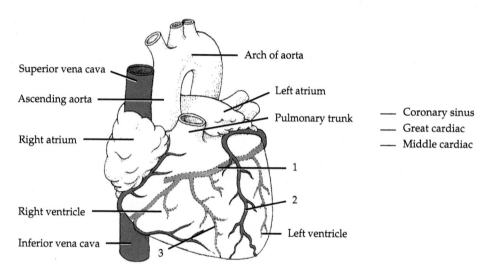

Superior vena cava

Ascending aorta

Right atrium

Right ventricle

Inferior vena cava

Arch of aorta

Left atrium

Pulmonary trunk

1

2

3

Left ventricle

___ Coronary sinus
___ Great cardiac
___ Middle cardiac

(b) Anterior view of venous drainage

PROCEDURE

1. Remove any fat or pulmonary tissue that is present.

2. In cutting the sheep heart open to examine the chambers, valves, and vessels, the anterior longitudinal sulcus is used as a guide.

3. Carefully make a shallow incision through the ventral wall of the pulmonary trunk and the right ventricle, trying not to cut the dorsal surface of either structure.

4. The incision is best made *less than an inch to the right of, and parallel to,* the previously mentioned anterior longitudinal sulcus.

5. If necessary, the incision can be continued to where the pulmonary trunk branches into a **right pulmonary artery,** which goes to the right lung, and a **left pulmonary artery,** which goes to the left lung. The **pulmonary valve** of the pulmonary artery can be clearly seen upon opening it. In any of these internal dissections of the heart, any coagulated blood or latex should be immediately removed so that all important structures can be located and identified.

6. Keeping the cut still parallel to the sulcus, extend the incision around and through the dorsal ventricular wall until you reach the **interventricular septum.**

FIGURE 17.6 External structure of a sheep heart.

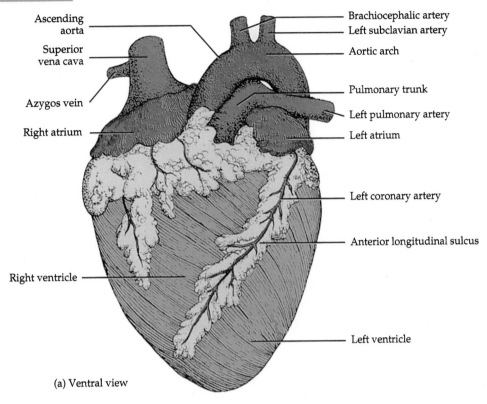

Ascending aorta
Superior vena cava
Azygos vein
Right atrium
Right ventricle

Brachiocephalic artery
Left subclavian artery
Aortic arch
Pulmonary trunk
Left pulmonary artery
Left atrium
Left coronary artery
Anterior longitudinal sulcus
Left ventricle

(a) Ventral view

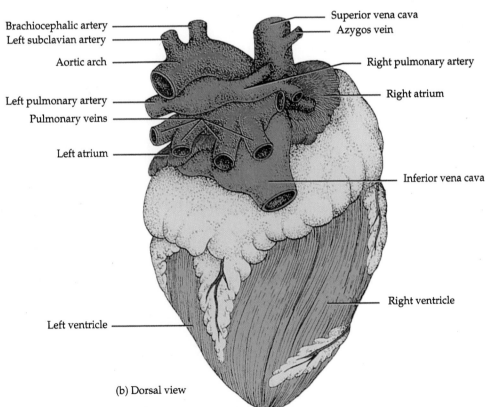

Brachiocephalic artery
Left subclavian artery
Aortic arch
Left pulmonary artery
Pulmonary veins
Left atrium
Left ventricle

Superior vena cava
Azygos vein
Right pulmonary artery
Right atrium
Inferior vena cava
Right ventricle

(b) Dorsal view

FIGURE 17.7 Internal structure of a sheep heart.

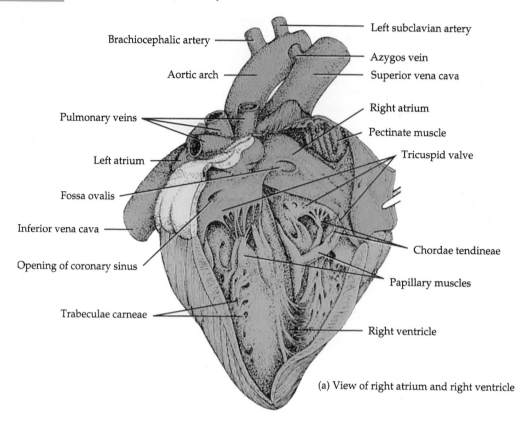

Left subclavian artery

Brachiocephalic artery

Azygos vein

Aortic arch

Superior vena cava

Right atrium

Pectinate muscle

Pulmonary veins

Tricuspid valve

Left atrium

Fossa ovalis

Inferior vena cava

Opening of coronary sinus

Chordae tendineae

Papillary muscles

Trabeculae carneae

Right ventricle

(a) View of right atrium and right ventricle

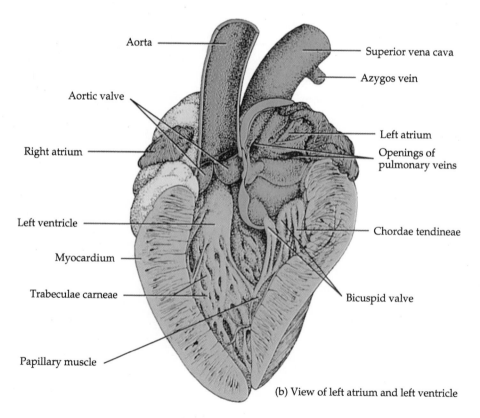

Aorta

Superior vena cava

Azygos vein

Aortic valve

Right atrium

Left atrium

Openings of pulmonary veins

Left ventricle

Chordae tendineae

Myocardium

Trabeculae carneae

Bicuspid valve

Papillary muscle

(b) View of left atrium and left ventricle

7. Now examine the dorsal surface of the heart and locate the thin-walled **superior vena cava** directly above the **right auricle.** This vein proceeds posteriorly straight into the right atrium.

8. Make a second longitudinal cut, this time through the superior vena cava (dorsal wall).

9. Extend the cut posteriorly through the right atrium on the left of the right auricle. Proceed posteriorly to the dorsal right ventricular wall and join your first incision.

10. The entire internal right side of the heart should now be clearly seen when carefully spread apart. The interior of the superior vena cava, right atrium, and right ventricle will now be examined. Start with the right auricle and locate the **pectinate muscle,** the large opening of the **inferior vena cava** on the left side of the right atrium, and the opening of the **coronary sinus** just below the opening of the inferior vena cava. By using a dull probe and gentle pressure, most of the vessels can be traced to the dorsal surface of the heart.

11. Now find the wall that separates the two atria, the **interatrial septum.** Also find the **fossa ovalis,** an oval depression ventral to the entrance of the inferior vena cava.

12. The **tricuspid valve** between the right atrium and the right ventricle should be examined to locate the three cusps, as its name indicates. From the cusps of the valve itself, and tracing posteriorly, the **chordae tendineae,** which hold the valve in place, should be identified. Still tracing posteriorly, the chordae are seen to originate from the **papillary muscles,** which themselves originate from the wall of the right ventricle itself.

13. Look carefully again at the dorsal surface of the left atrium and locate as many **pulmonary veins** (normally, four) as possible.

14. Make your third longitudinal cut through the most lateral of the pulmonary veins that you have located.

15. Continue posteriorly through the left atrial wall and the left ventricle to the **apex** of the heart.

16. Compare the difference in the thickness of the wall between the right and left ventricles. Explain your answer.

17. Examine the **bicuspid (mitral) valve,** again counting the cusps. Determine if the left side of the heart has basically the same structures as studied on the right side.

18. Probe from the left ventricle to the **aorta** as it emerges from the heart, examining the **aortic valve.** Find the openings of the right and left main coronary arteries.

19. Now locate the **brachiocephalic artery,** which is one of the first branches from the arch of the aorta. This artery continues branching and terminates by supplying the arms and head as its name indicates.

20. Connecting the aorta with the pulmonary artery is the remnant of the **ductus arteriosus,** called the **ligamentum arteriosum.** It may not be present in your sheep heart.

ANSWER THE LABORATORY REPORT QUESTIONS AT THE END OF THE EXERCISE.

EXERCISE 17 — Heart

Name _____ Date _____

Laboratory Section _____ Score/Grade _____

PART 1 ■ Multiple Choice

_____ 1. Which of the following veins drain the blood from most of the vessels supplying the heart wall? (a) vasa vasorum (b) superior vena cava (c) coronary sinus (d) inferior vena cava

_____ 2. The atrioventricular valve on the same side of the heart as the origin of the aorta is the (a) aortic (b) tricuspid (c) bicuspid (d) pulmonary

_____ 3. Which valve does the blood go through just before entering the pulmonary trunk on the way to the lungs? (a) tricuspid (b) pulmonary (c) aortic (d) bicuspid

_____ 4. The pointed end of the heart that projects inferiorly and to the left is the (a) costal surface (b) base (c) apex (d) coronary sulcus

_____ 5. Which of these structures is more internal? (a) fibrous pericardium (b) visceral layer of serous pericardium (c) parietal layer of serous pericardium (d) myocardium

_____ 6. The musculature of the heart is referred to as the (a) endocardium (b) myocardium (c) epicardium (d) pericardium

_____ 7. The depression in the interatrial septum corresponding to the foramen ovale of fetal circulation is the (a) interventricular sulcus (b) pectinate muscle (c) chordae tendineae (d) fossa ovalis

PART 2 ■ Completion

8. Malfunction of the _____ valve would interfere with the flow of blood from the right atrium to the right ventricle.

9. Deoxygenated blood is sent to the lungs through the _____.

10. The membrane that encloses the heart is called the _____.

11. The two inferior chambers of the heart are separated by the _____.

12. The earlike flap of tissue on each atrium is called a(n) _____.

13. The large vein that drains blood from most parts of the body superior to the heart and empties into the right atrium is the _____.

14. The cusps of atrioventricular valves are prevented from inverting by the presence of cords called _____, which are attached to papillary muscle.

15. Grooves on the surface of the heart that house blood vessels and variable amounts of fat are called _____.

16. The branch of the left coronary artery that distributes blood to the left atrium and left ventricle is the _____.

17. The _____ border of the heart extends from the base to the apex.

18. The _____ is located at the inferior border of the second costal cartilage, about 3 cm to the left of the midline.

19. The _____ sulcus marks the boundary between the atria and ventricles.

20. The ridges and folds in the ventricles are called _____.

21. The _____ vein drains the anterior aspect of the heart.

PART 3 ▪ Special Exercise

Draw a model of the heart and carefully label the four chambers, the four valves in their proper places, and the major blood vessels entering and exiting from the heart.

Blood Vessels

Objectives

At the completion of this exercise you should understand

A The histological characteristics of arteries, arterioles, capillaries, venules, and veins

B The locations of the systemic arteries, and the organs they supply

C The locations of the systemic veins, and the organs they drain

D The anatomical characteristics of the pulmonary circulation, and how it compares to the systemic circulation

E The anatomical characteristics of the fetal circulation, and how it compares to the circulatory pattern of the adult

Blood vessels are networks of tubes that carry blood throughout the body. They are called arteries, arterioles, capillaries, venules, and veins. In this exercise, you will study the histology of blood vessels and identify the principal arteries and veins of the cardiovascular system.

A. Arteries and Arterioles

Arteries (AR-ter-ēz; *ar* = air; *ter* = to carry) are blood vessels that carry blood *away from the heart* to body tissues. Arterial walls are constructed of three coats or **tunics** and surround a hollow core, called a **lumen,** through which blood flows (Figure 18.1). The innermost coat is called the **tunica interna** or **intima** and consists of a lining of endothelium, a basement membrane, and a layer of elastic tissue called the *internal elastic lamina.* The middle coat, or **tunica media,** is usually the thickest layer and consists of elastic fibers and smooth muscle fibers. This tunic is responsible for two major properties of arteries: **compliance** and **contractility.** The outer coat, or **tunica externa,** is composed principally of elastic and collagen fibers. An *external elastic lamina* may separate the tunica externa from the tunica media.

Obtain a prepared slide of a transverse section of an artery and identify the tunics using Figure 18.1 as a guide.

As arteries approach various tissues of the body, they become smaller and are known as **arterioles.** Arterioles play a key role in regulating blood flow from arteries into capillaries. When arterioles enter a tissue, they branch into countless microscopic blood vessels called capillaries.

B. Capillaries

Capillaries (KAP-i-lar′-ēz; = hairlike) are microscopic blood vessels that connect arterioles and venules. Their prime function is to permit the exchange of nutrients and wastes between blood and tissue cells through the interstitial fluid. This function is related to the fact that capillaries consist of only a single layer of endothelium.

C. Venules and Veins

When several capillaries unite, they form small veins called **venules** (= little vein). They collect blood from capillaries and drain it into veins.

Veins (VĀNZ) are composed of the same three coats as arteries, but there are variations in their relative thicknesses. The tunica interna of veins is thinner than that of arteries. In addition, the tunica media of veins is much thinner than

FIGURE 18.1   **Histology of blood vessels.**

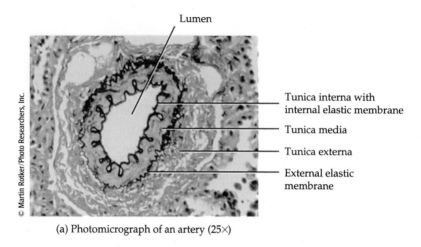

(a) Photomicrograph of an artery (25×)

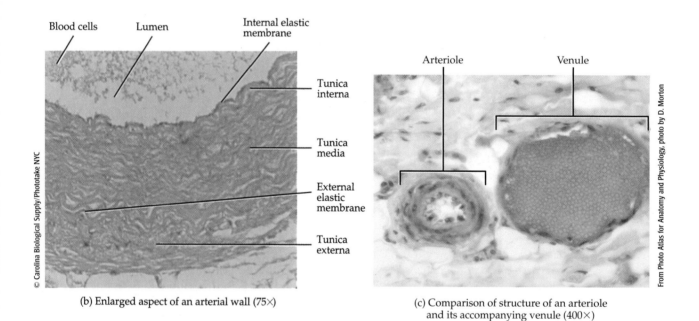

(b) Enlarged aspect of an arterial wall (75×)

(c) Comparison of structure of an arteriole
and its accompanying venule (400×)

that in arteries with relatively little smooth muscle or elastic fibers. The tunica externa is the thickest layer consisting of collagen and elastic fibers (Figure 18.1). Functionally, veins return blood from tissues *to* the heart.

Obtain a prepared slide of a transverse section of an artery and its accompanying vein and compare them, using Figure 18.1 as a guide.

D. Circulatory Routes

The two basic postnatal (after birth) circulatory routes are systemic and pulmonary circulation (Figure 18.2). Some other circulatory routes, which are all subdivisions of systemic circulation, include hepatic portal circulation, coronary (cardiac) circulation, fetal circulation, and the cerebral arterial circle (circle of Willis). The latter is found at the base of the brain (see Table 18.3 on page 399).

1. Systemic Circulation

The largest route is the **systemic circulation** (see Figures 18.3 through 18.11 and Tables 18.1 through 18.12). This route includes the flow of blood from the left ventricle to all parts of the body. The function of the systemic circulation is to carry oxygen

FIGURE 18.2 Circulatory routes. Systemic circulation is indicated by solid arrows; pulmonary circulation by broken arrows; and hepatic portal circulation by dotted arrows.

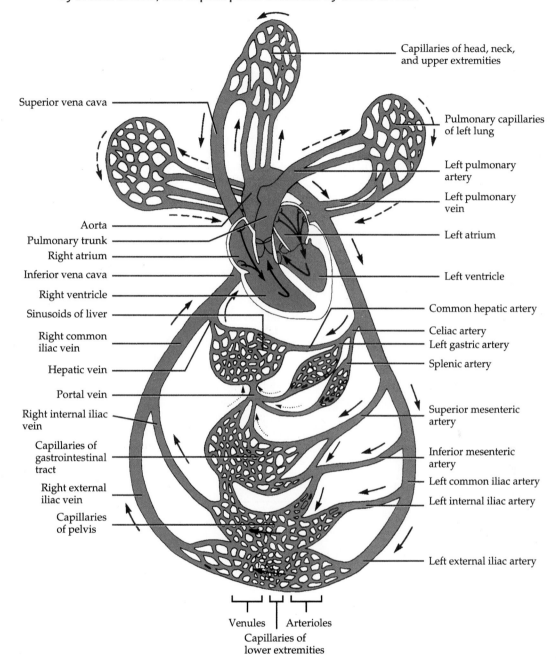

Capillaries of head, neck, and upper extremities

Superior vena cava

Pulmonary capillaries of left lung

Left pulmonary artery

Left pulmonary vein

Aorta
Pulmonary trunk
Right atrium

Left atrium

Inferior vena cava

Left ventricle

Right ventricle

Sinusoids of liver

Common hepatic artery

Right common iliac vein

Celiac artery
Left gastric artery
Splenic artery

Hepatic vein

Portal vein

Right internal iliac vein

Superior mesenteric artery

Capillaries of gastrointestinal tract

Inferior mesenteric artery

Left common iliac artery

Right external iliac vein

Left internal iliac artery

Capillaries of pelvis

Left external iliac artery

Venules   Arterioles
Capillaries of lower extremities

and nutrients to body tissues and to remove carbon dioxide and other wastes from them. All systemic arteries branch from the **aorta.** As the aorta emerges from the left ventricle, it passes superiorly and posteriorly to the pulmonary trunk. At this point, it is called the **ascending aorta.** The ascending aorta gives off two coronary branches (right and left coronary arteries) to the heart muscle. Then it turns to the left, forming the **arch of the aorta** before descending to the level of the intervertebral disc between the fourth and the fifth thoracic vertebrae as

the **descending aorta.** The descending aorta lies close to the vertebral bodies, passes through the diaphragm, and divides at the level of the fourth lumbar vertebra into two **common iliac arteries,** which carry blood to the lower limbs. The section of the descending aorta between the arch of the aorta and the diaphragm is referred to as the **thoracic aorta.** The section between the diaphragm and the common iliac arteries is termed the **abdominal aorta.** Each section of the aorta gives off arteries that continue to branch into distributing arteries leading to organs and finally into the arterioles and capillaries that service the systemic tissues (except the air sacs of the lungs).

Deoxygenated blood is returned to the heart through the systemic veins. All the veins of the systemic circulation flow into either the **superior vena cava,** the **inferior vena cava,** or the **coronary sinus.** They, in turn, empty into the right atrium.

Refer to Tables 18.1 through 18.12 and Figures 18.3 through 18.11.

TABLE 18.1 Aorta and its branches (Figure 18.3)

Overview: The **aorta** (ā-OR-ta) is the largest artery of the body, about 2 to 3 cm (about 1 in.) in diameter. It begins at the left ventricle and contains a valve at its origin, called the aortic valve (see Figure 17.4a), which prevents backflow of blood into the left ventricle during its diastole (relaxation). The four principal divisions of the aorta are the ascending aorta, arch of the aorta, thoracic aorta, and abdominal aorta.

Division of aorta	Arterial branch	Region supplied
Ascending aorta (ā-OR-ta)	Right and left coronary	Heart
Arch of aorta	Brachiocephalic (brā'-kē-ō-se-FAL-ik) trunk → Right common carotid (ka-ROT-id)	Right side of head and neck
	→ Right subclavian (sub-KLĀ-vē-an)	Right upper limb
	Left common carotid	Left side of head and neck
	Left subclavian	Left upper limb
Thoracic (thō-RAS-ik) **aorta**	Intercostals (in'-ter-KOS-tals)	Intercostal and chest muscles, pleurae
	Superior phrenics (FREN-iks)	Posterior and superior surfaces of diaphragm
	Bronchials (BRONG-kē-als)	Bronchi of lungs
	Esophageals (e-sof'-a-JĒ-als)	Esophagus
Abdominal (ab-DOM-i-nal) **aorta**	Inferior phrenics (FREN-iks)	Inferior surface of diaphragm
	Celiac (SĒ-lē-ak) → Common hepatic (he-PAT-ik)	Liver
	→ Left gastric (GAS-trik)	Stomach and esophagus
	→ Splenic (SPLĒN-ik)	Spleen, pancreas, stomach
	Superior mesenteric (MES-en-ter'-ik)	Small intestine, cecum, ascending and transverse colons, and pancreas
	Suprarenals (soo-pra-RĒ-nals)	Suprarenal (adrenal) glands
	Renals (RĒ-nals)	Kidneys
	Gonadals (gō-NAD-als) → Testiculars (tes-TIK-ū-lars) or	Testes
	→ Ovarians (ō-VAR-ē-ans)	Ovaries
	Inferior mesenteric (MES-en-ter'-ik)	Transverse, descending, sigmoid colons and rectum
	Common iliacs (IL-ē-aks) → External iliacs	Lower limbs
	→ Internal iliacs	Uterus (female), prostate (male), muscles of buttocks, and urinary bladder

Label these arteries in Figure 18.3.

FIGURE 18.3 Principal branches of the aorta.

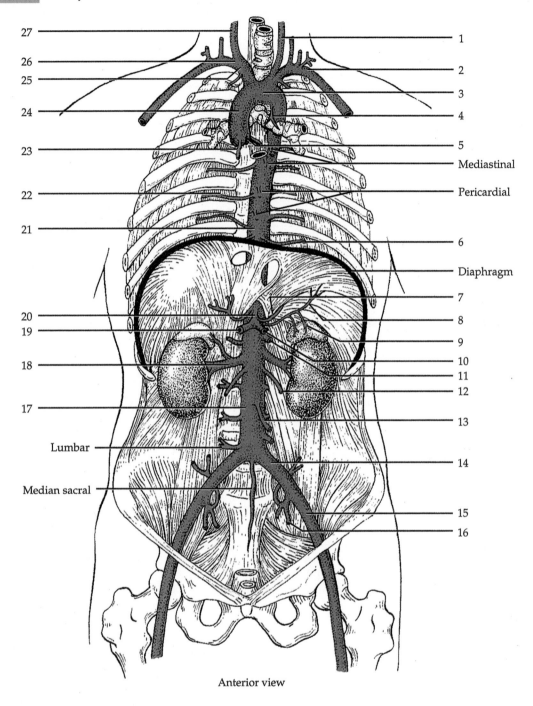

Mediastinal

Pericardial

Diaphragm

Lumbar

Median sacral

Anterior view

___ Abdominal aorta

___ Arch of aorta

___ Ascending aorta

___ Brachiocephalic trunk

___ Bronchial

___ Celiac trunk

___ Common hepatic

___ Inferior mesenteric

___ Left common carotid

___ Left coronary

___ Left common iliac

___ Left external iliac

___ Left gastric

___ Left gonadal
(testicular or ovarian)

___ Left inferior phrenic

___ Left internal iliac

___ Left subclavian

___ Left superior phrenic

___ Left suprarenal

___ Posterior intercostal

___ Right common carotid

___ Right coronary

___ Right renal

___ Right subclavian

___ Splenic

___ Superior mesenteric

___ Thoracic aorta

TABLE 18.2	Ascending aorta (see Figure 17.5a)

Overview: The **ascending aorta** is the first division of the aorta, about 5 cm (2 in.) in length. It begins at the aortic valve in the left ventricle and is directed superiorly, slightly anteriorly, and to the right and ends at the level of the sternal angle, where it becomes the arch of the aorta. The beginning of the ascending aorta is posterior to the pulmonary trunk and right auricle; the right pulmonary artery is posterior to it. At its origin, the ascending aorta contains three dilations, called aortic sinuses. Two of these, the right and left sinuses, give rise to the right and left coronary arteries, respectively. The ascending aorta is more susceptible to aneurysms than succeeding parts of the aorta because of the higher pressure of blood from the left ventricle during ventricular systole (contraction).

Branch	Description and region supplied
Coronary (KOR-ō-nar-ē; *coron* = crown) **arteries**	The right and left **coronary arteries** arise from the ascending aorta just superior to the aortic valve. They form a crownlike ring around the heart, giving off branches to the atrial and ventricular myocardium. The **posterior interventricular** (in′-ter-ven-TRIK-ū-lar; *inter* = between) **branch** of the right coronary artery supplies both ventricles, and the **marginal branch** supplies the right ventricle. The **anterior interventicular branch (left anterior descending)** of the left coronary artery supplies both ventricles, and the **circumflex** (SER-kum-flex; *circum* = around; *flex* = to bend) **branch** supplies the left atrium and left ventricle.

Write the names of the missing arteries in the following scheme of circulation. Be sure to indicate left or right where applicable.

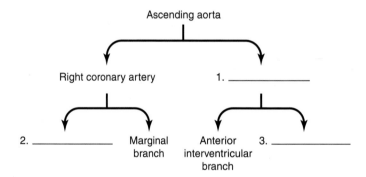

TABLE 18.3 Arch of aorta (Figure 18.4)

Overview: The **arch of the aorta** is about 4–5 cm (almost 2 in.) in length and is the continuation of the ascending aorta. It emerges from the pericardium posterior to the sternum at the level of the sternal angle. The arch is directed superiorly, posteriorly, and to the left, and then inferiorly; it ends at the level of the intervertebral disc between the fourth and fifth thoracic vertebrae, where it become the thoracic aorta. The thymus gland lies anterior to the arch of the aorta, whereas the trachea lies posterior to it.

Interconnecting the arch of the aorta and the pulmonary trunk is a structure called the **ligamentum arteriosum.** It is a remnant of a fetal blood vessel called the ductus arteriosus (see Figure 18.14). Within the arch of the aorta are receptors (nerve cells) called baroreceptors that are sensitive to changes in blood pressure. Situated near the arch of the aorta are other receptors, referred to as **aortic bodies,** that are sensitive to changes in blood concentrations of chemicals such as O_2, CO_2, and H^+.

Three major arteries branch from the superior aspect of the arch of the aorta. In order of their origination, they are the **brachiocephalic trunk, left common carotid artery,** and **left subclavian artery.**

Branch	Description and region supplied
Brachiocephalic (brā′-kē-ō-se-FAL-ik; *brachio* = arm; *cephalic* = head) **trunk**	The **brachiocephalic trunk,** which is found only on the right side, is the first and largest branch off the arch of the aorta. There is no left brachiocephalic artery. It bifurcates (divides) at the right sternoclavicular joint to form the right subclavian artery and right common carotid artery.

The **right subclavian** (sub-CLĀ-vē-an) **artery** extends from the brachiocephalic trunk to the first rib and then passes into the armpit (axilla). The general distribution of the artery is to the brain and spinal cord, neck, shoulder, thoracic viscera and wall, and scapular muscles.

Continuation of the right subclavian artery into the axilla is called the **axillary** (AK-si-ler-ē) **artery.** (Note that the right subclavian artery, which passes deep to the clavicle, is a good example of the practice of giving the same vessel different names as it passes through different regions.) Its general distribution is the shoulder, thoracic and scapular muscles, and humerus.

The **brachial** (BRĀ-kē-al) **artery** is the continuation of the axillary artery into the arm. The brachial artery provides the main blood supply to the arm and is superficial and palpable along its course. It begins at the tendon of the teres major muscle and ends just distal to the bend of the elbow. At first, the brachial artery is medial to the humerus, but as it descends it gradually curves laterally and passes through the cubital fossa, a triangular depression anterior to the elbow. It is here that you can easily detect the pulse of the brachial artery and listen to the various sounds when taking a person's blood pressure. Just distal to the bend in the elbow, the brachial artery divides into the radial artery and ulnar artery.

The **radial** (RĀ-dē-al = radius) **artery** is the smaller branch and is a direct continuation of the brachial artery. It passes along the lateral (radial) aspect of the forearm and then through the wrist and hand, supplying these structures with blood. At the wrist, the radial artery comes into contact with the distal end of the radius, where it is covered only by fascia and skin. Because of its superficial location at this point, it is a common site for measuring radial pulse.

The **ulnar** (UL-nar = ulna) **artery,** the larger branch of the brachial artery, passes along the medial (ulnar) aspect of the forearm and then into the wrist and hand, also supplying these structures with blood. In the palm, branches of the radial and ulnar arteries anastomose to form the superficial palmar arch and the deep palmar arch.

The **superficial palmar** (*palma* = palm) **arch** is formed mainly by the ulnar artery, with a contribution from a branch of the radial artery. The arch is superficial to the long flexor tendons of the fingers and extends across the palm at the bases of the metacarpals. It gives rise to **common palmar digital arteries,** which supply the palm. Each divides into a pair of **proper palmar digital arteries,** which supply the fingers.

The **deep palmar arch** is formed mainly by the radial artery, with a contribution from a branch of the ulnar artery. The arch is deep to the long flexor tendons of the fingers and extends across the palm, just distal to the bases of the metacarpals. Arising from the deep palmar arch are **palmar metacarpal arteries,** which supply the palm and anastomose with the common palmar digital arteries of the superficial palmar arch.

Label these arteries in Figure 18.4a.

(Continued)

TABLE 18.3   Arch of aorta (Figure 18.4) (Continued)

Before passing into the axilla, the right subclavian artery gives off a major branch to the brain called the **right vertebral** (VER-te-bral) **artery.** The right vertebral artery passes through the foramina of the transverse processes of the sixth through first cervical vertebrae and enters the skull through the foramen magnum to reach the inferior surface of the brain. Here it unites with the left vertebral artery to form the **basilar** (BAS-i-lar) **artery.** The vertebral artery supplies the posterior portion of the brain with blood. The basilar artery passes along the midline of the anterior aspect of the brain stem and supplies the cerebellum and pons of the brain and the inner ear.

The **right common carotid** (ka-ROT-id) **artery** begins at the bifurcation of the brachiocephalic trunk, posterior to the right sternoclavicular joint, and passes superiorly in the neck to supply structures in the head. At the superior border of the larynx (voice box), it divides into the right external and right internal carotid arteries.

The **external carotid artery** begins at the superior border of the larynx and terminates near the temporomandibular joint in the substance of the parotid gland, where it divides into two branches: the superficial temporal and maxillary arteries. The carotid pulse can be detected in the external carotid artery just anterior to the sternocleidomastoid muscle at the superior border of the larynx. The general distribution of the external carotid artery is to structures *external* to the skull.

The **internal carotid artery** has no branches in the neck and supplies structures *internal* to the skull. It enters the cranial cavity through the carotid foramen in the temporal bone. At the proximal portion of the internal carotid artery there is a slight dilation called the **carotid sinus,** which contains receptors (nerve cells) called **baroreceptors** that monitor changes in blood pressure. Near the baroreceptors is a small mass of tissue at the bifurcation of the common carotid artery called the **carotid body.** It contains receptors that monitor changes in concentrations of chemicals in blood such as O_2, CO_2, and H^+. The internal carotid artery supplies blood to the eyeball and other orbital structures, ear, most of the cerebrum of the brain, pituitary gland, and external nose. The terminal branches of the internal carotid artery are the **anterior cerebral artery,** which supplies most of the medial surface of the cerebrum, and the **middle cerebral artery,** which supplies most of the lateral surface of the cerebrum.

Label the vertebral, basilar, common carotid, external carotid, and internal carotid arteries in Figure 18.4b.

Inside the cranium, anastomoses of the left and right internal carotid arteries along with the basilar artery form an arrangement of blood vessels at the base of the brain near the sella turcica called the **cerebral** (se-RĒ-bral) **arterial circle (circle of Willis).** From this circle arise arteries supplying most of the brain. Essentially the cerebral arterial circle is formed by the union of the **anterior cerebral arteries** (branches of internal carotids) and **posterior cerebral arteries** (branches of basilar artery). The posterior cerebral arteries are connected with the internal carotid arteries by the **posterior communicating arteries.** The anterior cerebral arteries are connected by the **anterior communicating arteries.** The **internal carotid arteries** are also considered part of the cerebral arterial circle. The functions of the cerebral arterial circle are to equalize blood pressure to the brain and provide alternate routes for the blood to the brain, should the arteries become damaged.

Label the arteries of the cerebral arterial circle in Figure 18.4c.

Left common carotid (ka-ROT-id) **artery**	The **left common carotid artery** is the second branch off the arch of the aorta (see Figure 17.3a). Corresponding to the right common carotid, it divides into basically the same branches with the same names, except that the arteries are now labeled "left" instead of "right."
Left subclavian (sub-KLĀ-vē-an) **artery**	The **left subclavian artery** is the third branch off the arch of the aorta (see Figure 17.3a). It distributes blood to the left vertebral artery and vessels of the left upper limb. Arteries branching from the left subclavian artery are named like those of the right subclavian artery.

FIGURE 18.4 Branches of the arch of the aorta.

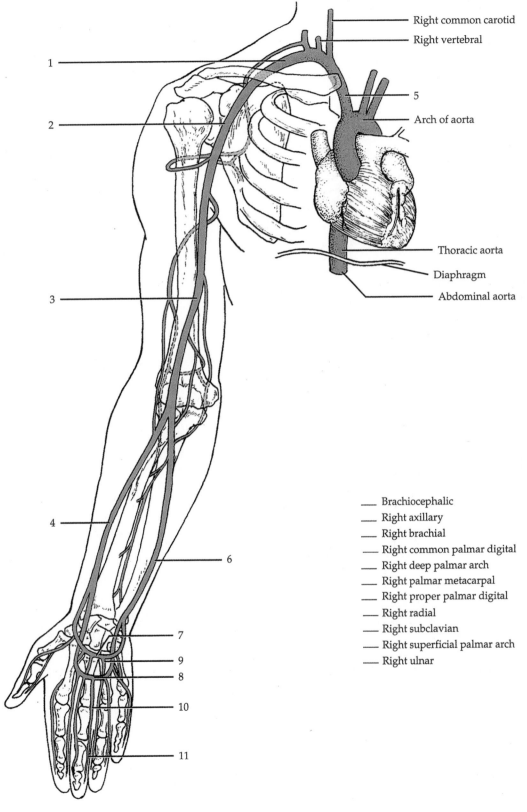

Right common carotid

Right vertebral

5

Arch of aorta

Thoracic aorta

Diaphragm

Abdominal aorta

___ Brachiocephalic
___ Right axillary
___ Right brachial
___ Right common palmar digital
___ Right deep palmar arch
___ Right palmar metacarpal
___ Right proper palmar digital
___ Right radial
___ Right subclavian
___ Right superficial palmar arch
___ Right ulnar

(a) Anterior view of arteries of right upper limb

FIGURE 18.4 Branches of the arch of the aorta. (Continued)

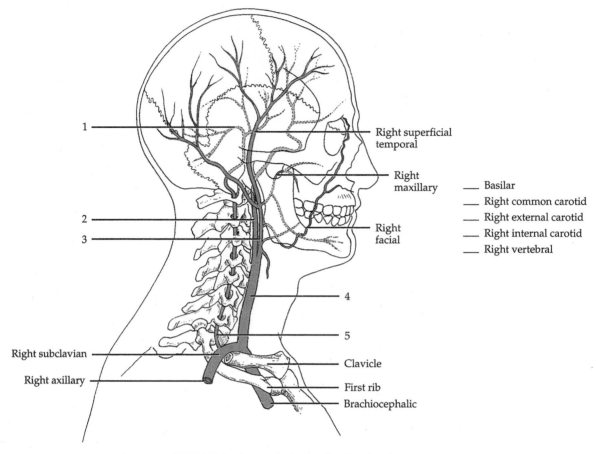

1 — Right superficial temporal

Right maxillary

Right facial

__ Basilar
__ Right common carotid
__ Right external carotid
__ Right internal carotid
__ Right vertebral

2

3

4

5

Right subclavian

Right axillary

Clavicle

First rib

Brachiocephalic

(b) Right lateral view of arteries of head and neck

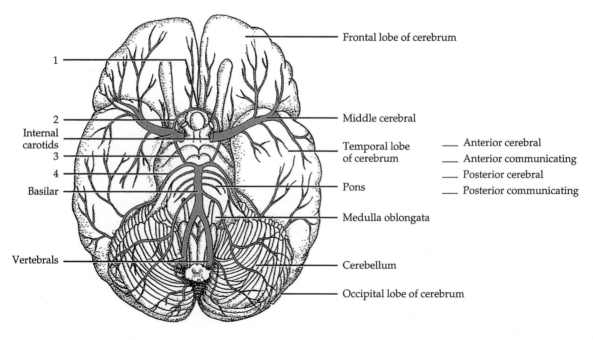

1

2

Internal carotids

3

4

Basilar

Vertebrals

Frontal lobe of cerebrum

Middle cerebral

Temporal lobe of cerebrum

Pons

Medulla oblongata

Cerebellum

Occipital lobe of cerebrum

__ Anterior cerebral
__ Anterior communicating
__ Posterior cerebral
__ Posterior communicating

(c) Inferior view of arteries of base of brain

Write the names of the missing arteries in the following scheme of circulation. Be sure to indicate left or right where applicable.

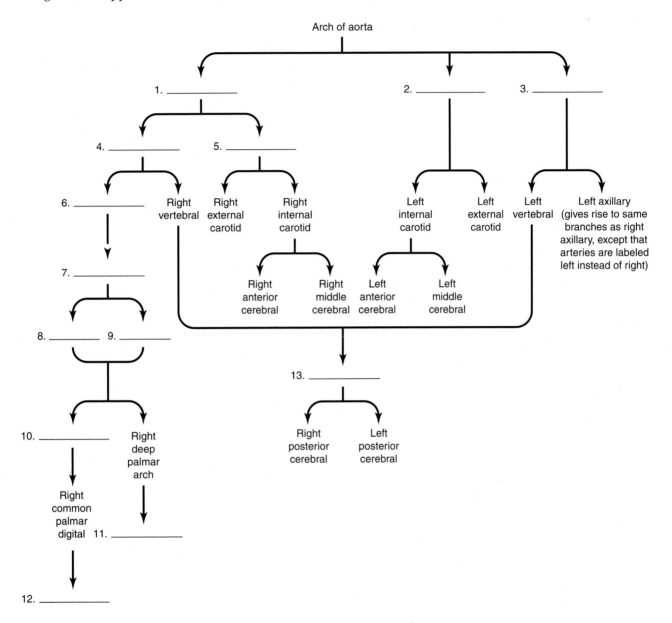

Arch of aorta

1. _____

2. _____

3. _____

4. _____

5. _____

6. _____

Right vertebral

Right external carotid

Right internal carotid

Left internal carotid

Left external carotid

Left vertebral

Left axillary (gives rise to same branches as right axillary, except that arteries are labeled left instead of right)

Right anterior cerebral

Right middle cerebral

Left anterior cerebral

Left middle cerebral

7. _____

8. _____ 9. _____

10. _____

Right deep palmar arch

Right common palmar digital   11. _____

12. _____

13. _____

Right posterior cerebral

Left posterior cerebral

TABLE 18.4	Thoracic aorta (see Figure 18.3)

Overview: The **thoracic aorta** is about 20 cm (8 in.) long and is a continuation of the arch of the aorta. It begins at the level of the intervertebral disc between the fourth and fifth thoracic vertebrae, where it lies to the left of the vertebral column. As it descends, it moves closer to the midline and ends at an opening in the diaphragm (aortic hiatus) anterior to the vertebral column at the level of the intervertebral disc between the twelfth thoracic and first lumbar vertebrae.

Along its course, the thoracic aorta sends off numerous small arteries to the viscera **(visceral branches)** and body wall structures **(parietal branches).**

Branch	Description and region supplied
Visceral	
Pericardial (per'-i-KAR-dē-al; *peri* = around; *cardia* = heart) **arteries**	Two or three minute **pericardial arteries** supply blood to the pericardium.
Bronchial (BRONG-kē-al; *bronchus* = windpipe) **arteries**	One right and two left **bronchial arteries** supply the bronchial tubes, pleurae, bronchial lymph nodes, and esophagus. (Whereas the right bronchial artery arises from the third posterior intercostal artery, the two left bronchial arteries arise from the thoracic aorta.)
Esophageal (e-sof'-a-JĒ-al; *eso* = to carry; *phage* = food) **arteries**	Four or five **esophageal arteries** supply the esophagus.
Mediastinal (mē'-dē-as-TĪ-nal) **arteries**	Numerous small **mediastinal arteries** supply blood to structures in the mediastinum.
Parietal	
Posterior intercostal (in'-ter-KOS-tal; *inter* = between; *costa* = rib) **arteries**	Nine pairs of **posterior intercostals arteries** supply the intercostals, pectoralis major and minor, and serratus anterior muscles; overlying subcutaneous tissue and skin; mammary glands; and vertebrae, meninges, and spinal cord.
Subcostal (sub-KOS-tal; *sub* = under) **arteries**	The left and right **subcostal arteries** have a distribution similar to that of the posterior intercostals.
Superior phrenic (FREN-ik; *phren* = diaphragm) **arteries**	Small **superior phrenic arteries** supply the superior and posterior surfaces of the diaphragm.

Write the names of the missing arteries in the following scheme of circulation. Be sure to indicate left or right where applicable.

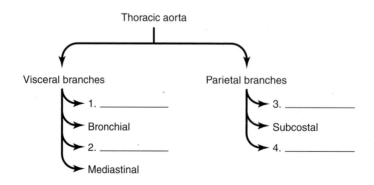

TABLE 18.5   Abdominal aorta (Figure 18.5)

Overview: The **abdominal** (ab-DOM-i-nal) **aorta** is the continuation of the thoracic aorta. It begins at the aortic hiatus in the diaphragm and ends at about the level of the fourth lumbar vertebra, where it divides into right and left common iliac arteries. The abdominal aorta lies anterior to the vertebral column.

As with the thoracic aorta, the abdominal aorta gives off visceral and parietal branches. The unpaired visceral branches arise from the anterior surface of the aorta and include the **celiac, superior mesenteric,** and **inferior mesenteric arteries.** The paired visceral branches arise from the lateral surfaces of the aorta and include the **suprarenal, renal,** and **gonadal arteries.** The paired parietal branches arise from the posterolateral surfaces of the aorta and include the **inferior phrenic** and **lumbar arteries.** The unpaired parietal artery is the **median sacral.**

Branch	Description and region supplied
Visceral **Celiac** (SĒ-lē-ak = abdominal cavity) **trunk**	The **celiac trunk (artery)** is the first visceral branch off the aorta, inferior to the diaphragm at about the level of the 12th thoracic vertebra. Its general distribution is to the esophagus, stomach, pancreas, liver, gallbladder, part of the small intestine, and greater omentum (a portion of the peritoneum that drapes over the intestines). Almost immediately following its origination, the celiac trunk divides into three branches: left gastric, splenic, and common hepatic arteries. The **left gastric** (GAS-trik = stomach) **artery** is the smallest of the three branches. It passes superiorly to the left toward the esophagus and then turns to follow the lesser curvature of the stomach. It supplies the stomach and esophagus. The **splenic** (SPLĒN-ik) **artery** is the largest branch of the celiac trunk. It arises from the left side of the celiac trunk distal to the left gastric artery. It passes horizontally to the left along the pancreas to reach the spleen. Its branches are the (1) **pancreatic** (pan'-krē-AT-ik) **artery,** which supplies the pancreas; (2) **left gastroepiploic** (gas'-trō-ep'-i-PLŌ-ik; *epiplo* = omentum) **artery,** which supplies the stomach and greater omentum; and (3) **short gastric artery,** which supplies the stomach. The **common hepatic** (he-PAT-ik = liver) **artery** is intermediate in size between the left gastric and splenic arteries. Unlike the other two branches of the celiac trunk, the common hepatic artery arises from the right side. Its branches are the (1) **hepatic artery proper,** which supplies the liver, gallbladder, and stomach; (2) **right gastric artery,** which supplies the stomach; and (3) **gastroduodenal** (gas'-trō-doo-ō-DĒ-nal) **artery,** which supplies the stomach, duodenum of the small intestine, pancreas, and greater omentum. Label these arteries in Figure 18.5a and b.
Superior mesenteric (MES-en-ter'-ik; *meso* = middle; *enteric* = intestine) **artery**	The **superior mesenteric artery** arises from the anterior surface of the abdominal aorta about 1 cm inferior to the celiac trunk at the level of the first lumbar vertebra. It runs inferiorly and anteriorly and between the layers of mesentery, a portion of the peritoneum that attaches the small intestine to the posterior abdominal wall. The general distribution of the superior mesenteric artery is to the pancreas, most of the small intestine, and part of the large intestine. It anastomoses extensively and has the following branches: (1) **inferior pancreaticoduodenal** (pan'-krē-at'-i-kō-doo-ō-DĒ-nal) **artery,** which supplies the pancreas and duodenum; (2) **jejunal** (je-JOO-nal) and **ileal** (IL-ē-al) **arteries,** which supply the jejunum and ileum of the small intestine, respectively; (3) **ileocolic** (il'-ē-ō-KŌL-ik) **artery,** which supplies the ileum and ascending colon of the large intestine; (4) **right colic** (KŌL-ik) **artery,** which supplies the ascending colon; and (5) **middle colic artery,** which supplies the transverse colon of the large intestine. Label these arteries in Figure 18.5c.

(Continued)

TABLE 18.5   Abdominal aorta (Figure 18.5) (Continued)

Inferior mesenteric (MES-en-ter'-ik) **artery**	The **inferior mesenteric artery** arises from the anterior aspect of the abdominal aorta at the level of the third lumbar vertebra. It passes inferiorly to the left of the aorta and anastomoses extensively. The principal branches of the inferior mesenteric artery are the (1) **left colic** (KŌL-ik) **artery,** which supplies the transverse and descending colons of the large intestine; (2) **sigmoid** (SIG-moyd) **arteries,** which supply the descending and sigmoid colons of the large intestine; and (3) **superior rectal** (REK-tal) **artery,** which supplies the rectum of the large intestine. Label these arteries in Figure 18.5a and d.
Suprarenal (soo-pra-RĒ-nal; *supra* = above; *renal* = kidney) **arteries**	Although there are three pairs of **suprarenal (adrenal) arteries** that supply the adrenal (suprarenal) glands (superior, middle, and inferior), only the middle pair originates directly from the abdominal aorta. The middle suprarenal arteries arise at the level of the first lumbar vertebra at or superior to the renal arteries. The superior suprarenal arteries arise from the inferior phrenic artery, and the inferior suprarenal arteries originate from the renal arteries.
Renal (RĒ-nal; *renal* = pertaining to the kidney) **arteries**	The right and left **renal arteries** usually arise from the lateral aspects of the abdominal aorta at the superior border of the second lumbar vertebra, about 1 cm inferior to the superior mesenteric artery. The right renal vein, which is longer than the left, arises slightly lower than the left and passes posterior to the right renal vein and inferior vena cava. The left renal artery is posterior to the left renal vein and is crossed by the inferior mesenteric vein. The renal arteries carry blood to the kidneys, adrenal (suprarenal) glands, and ureters.
Gonadal (gō-NAD-al; *gonos* = seed) [**testicular** (tes-TIK-ū-lar) or **ovarian** (ō-VAR-ē-an)] **arteries**	The **gonadal arteries** arise from the abdominal aorta at the level of the second lumbar vertebra just inferior to the renal arteries. In the male, the gonadal arteries are specifically referred to as the **testicular arteries.** They pass through the inguinal canal and supply the testes, epididymis, and ureters. In the female, the gonadal arteries are called the **ovarian arteries.** They are much shorter than the testicular arteries and supply the ovaries, uterine (fallopian) tubes, and ureters.
Parietal **Median sacral** (SĀ-kral = pertaining to the sacrum) **artery**	The **median sacral artery** arises from the posterior surface of the abdominal aorta about 1 cm superior to the bifurcation of the aorta into the right and left common iliac arteries. The median sacral artery supplies the sacrum and coccyx. Label these arteries in Figure 18.5a.
Inferior phrenic (FREN-ik; *phrenic* = diaphragm) **arteries**	The **inferior phrenic arteries** are the first paired branches of the abdominal aorta, immediately superior to the origin of the celiac trunk. (They may also arise from the renal arteries.) The inferior phrenic arteries are distributed to the inferior surface of the diaphragm and adrenal (suprarenal) glands. These arteries are not illustrated.
Lumbar (LUM-bar = pertaining to the loin) **arteries**	The four pairs of **lumbar arteries** arise from the posterolateral surface of the abdominal aorta. They supply the lumbar vertebrae, spinal cord and its meninges, and the muscles and skin of the lumbar region of the back.

FIGURE 18.5 **Abdominal arteries.**

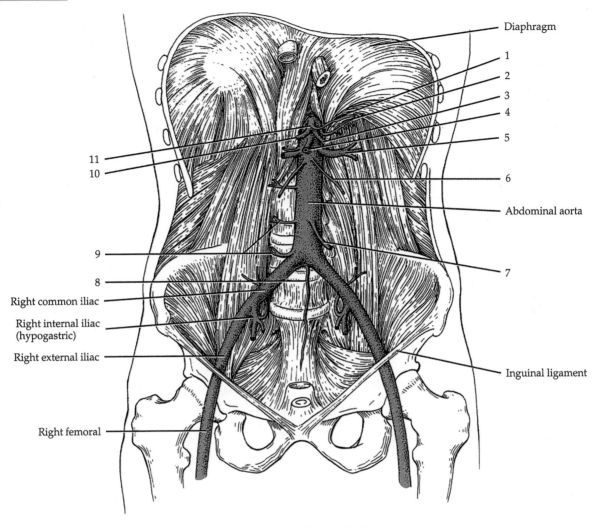

Diaphragm

1
2
3
4
5

6

Abdominal aorta

7

11
10

9

8

Right common iliac

Right internal iliac
(hypogastric)

Right external iliac

Inguinal ligament

Right femoral

(a) Anterior view of abdominal aorta and its principal branches

___ Celiac

___ Common hepatic

___ Inferior mesenteric

___ Left gastric

___ Left gonadal
      (testicular or ovarian)

___ Left renal

___ Left suprarenal

___ Median sacral

___ Right lumbars

___ Splenic

___ Superior mesenteric

FIGURE 18.5   Abdominal arteries. (Continued)

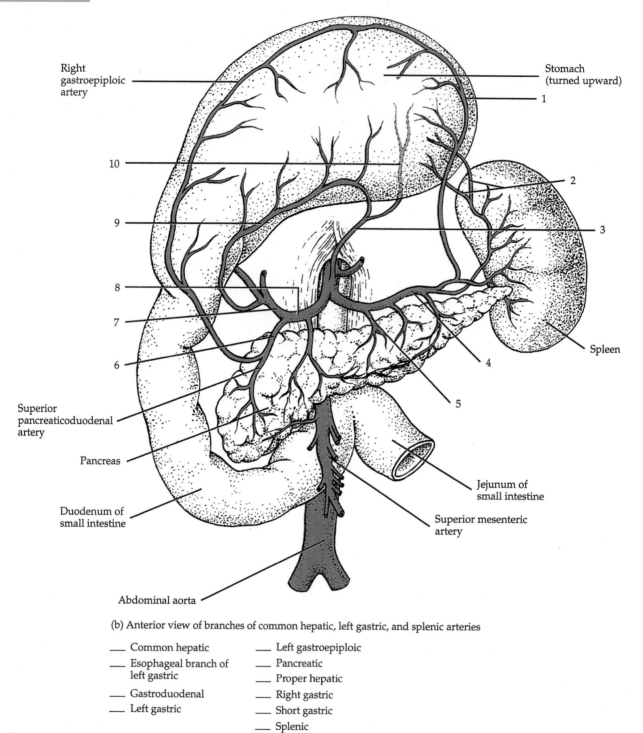

Right gastroepiploic artery

Stomach (turned upward)

1

10

2

9

3

8

7

6

Spleen

5

4

Superior pancreaticoduodenal artery

Pancreas

Duodenum of small intestine

Jejunum of small intestine

Superior mesenteric artery

Abdominal aorta

(b) Anterior view of branches of common hepatic, left gastric, and splenic arteries

___ Common hepatic
___ Esophageal branch of left gastric
___ Gastroduodenal
___ Left gastric

___ Left gastroepiploic
___ Pancreatic
___ Proper hepatic
___ Right gastric
___ Short gastric
___ Splenic

FIGURE 18.5   **Abdominal arteries. (Continued)**

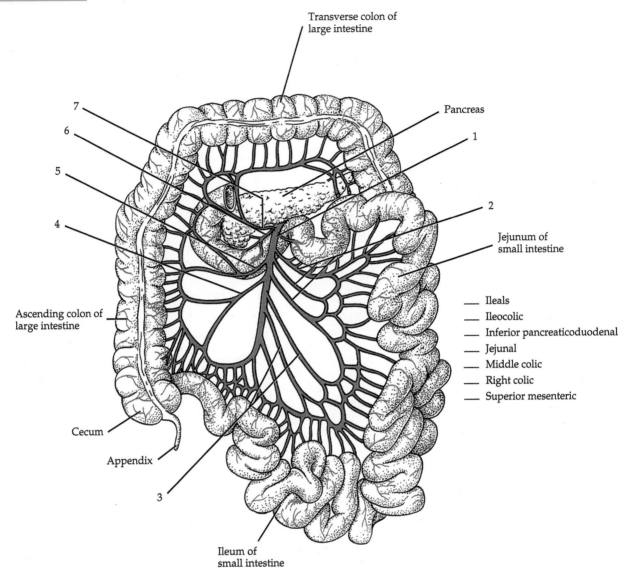

Transverse colon of
large intestine

Pancreas

7

6

1

5

2

4

Jejunum of
small intestine

___ Ileals
___ Ileocolic
___ Inferior pancreaticoduodenal
___ Jejunal
___ Middle colic
___ Right colic
___ Superior mesenteric

Ascending colon of
large intestine

Cecum

Appendix

3

Ileum of
small intestine

(c) Anterior view of branches of superior mesenteric artery

FIGURE 18.5 **Abdominal arteries. (Continued)**

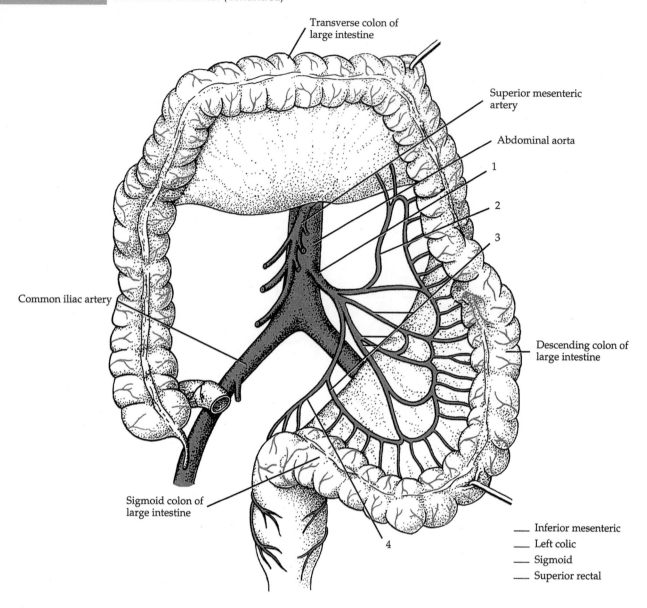

Transverse colon of
large intestine

Superior mesenteric
artery

Abdominal aorta

1

2

3

Common iliac artery

Descending colon of
large intestine

Sigmoid colon of
large intestine

4

—— Inferior mesenteric
—— Left colic
—— Sigmoid
—— Superior rectal

(d) Anterior view of branches of inferior mesenteric artery

Write the names of the missing arteries in the following scheme of circulation. Be sure to indicate left or right where applicable.

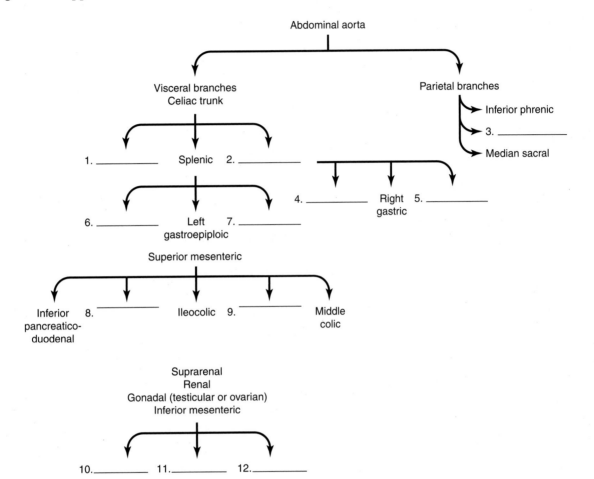

TABLE 18.6   Arteries of the pelvis and lower limbs (Figure 18.6)

Overview: The abdominal aorta ends by dividing into the right and left **common iliac arteries**. These, in turn, divide into the **internal** and **external iliac arteries**. Upon entering the thighs, the external iliac arteries become the **femoral arteries**, then the **popliteal arteries** posterior to the knee, and then the **anterior** and **posterior tibial arteries** at the knees.

Branch	Description and region supplied
Common iliac (IL-ē-ak; *iliac* = pertaining to the ilium) **arteries**	At about the level of the fourth lumbar vertebra, the abdominal aorta divides into the right and left **common iliac arteries,** the terminal branches of the abdominal aorta. Each passes inferiorly about 5 cm (2 in.) and gives rise to two branches: internal iliac and external iliac arteries. The general distribution of the common iliac arteries is to the pelvis, external genitals, and lower limbs.
Internal iliac arteries	The **internal iliac arteries** are the primary arteries of the pelvis. They begin at the bifurcation of the common iliac arteries anterior to the sacroiliac joint at the level of the lumbosacral intervertebral disc. They pass posteromedially as they descend in the pelvis and divide into anterior and posterior divisions. The general distribution of the internal iliac arteries is to the pelvis, buttocks, external genitals, and thighs.
External iliac arteries	The **external iliac arteries** are larger than the internal iliac arteries. Like the internal iliac arteries, they begin at the bifurcation of the common iliac arteries. They descend along the medial border of the psoas major muscles following the pelvic brim, pass posterior to the midportion of the inguinal ligaments, and become the femoral arteries. The general distribution of the external iliac arteries is to the lower limbs. Specifically, branches of the external iliac arteries supply the muscles of the anterior abdominal wall, the cremaster muscle in males and the round ligament of the uterus in females, and the lower limbs. The **femoral** (FEM-ō-ral = pertaining to the thigh) **arteries** descend along the anteriomedial aspects of the thighs to the junction of the middle and lower third of the thighs. Here they pass through an opening in the tendon of the adductor magnus muscle, where they emerge posterior to the femurs as the popliteal arteries. A pulse may be felt in the femoral artery just inferior to the inguinal ligament. The general distribution of the femoral arteries is to the lower abdominal wall, groin, external genitals, and muscles of the thigh. The **popliteal** (pop'-li-TĒ-al = posterior surface of knee) **arteries** are the continuation of the femoral arteries through the popliteal fossa (space behind the knee). They descend to the inferior border of the popliteus muscles, where they divide into the anterior and posterior tibial arteries. A pulse may be detected in the popliteal arteries. In addition to supplying the adductor magnus and hamstring muscles and the skin on the posterior aspect of the legs, branches of the popliteal arteries also supply the gastrocnemius, soleus, and plantaris muscles of the calf, knee joint, femur, patella, and fibula. The **anterior tibial** (TIB-ē-al = pertaining to shin bone) **arteries** descend from the bifurcation of the popliteal arteries. They are smaller than the posterior tibial arteries. The anterior tibial arteries descend through the anterior muscular compartment of the leg. They pass through the interosseous membrane that connects the tibia and fibula, lateral to the tibia. The anterior tibial arteries supply the knee joints, anterior compartment muscles of the legs, skin over the anterior aspects of the legs, and ankle joints. At the ankles, the anterior tibial arteries become the **dorsal arteries of the foot (dorsalis pedis arteries),** also arteries from which a pulse may be detected. The dorsalis pedis arteries supply the muscles, skin, and joints on the dorsal aspects of the feet. On the dorsum of the feet, the dorsalis pedi arteries give off a transverse branch at the first (medial) cuneiform bone called the **arcuate** (*arcuat* = bowed) **arteries** that run laterally over the bases of the metatarsals. From the arcuate arteries branch **dorsal metatarsal arteries,** which supply the feet. The dorsal metatarsal arteries terminate by dividing into the **dorsal digital arteries,** which supply the toes.

Branch	Description and region supplied
	The **posterior tibial arteries,** the direct continuations of the popliteal arteries, descend from the bifurcation of the popliteal arteries. They pass down the posterior muscular compartment of the legs posterior to the medial malleolus of the tibia. They terminate by dividing into the medial and lateral plantar arteries. Their general distribution is to the muscles, bones, and joints of the leg and foot. Major branches of the posterior tibial arteries are the **fibular (peroneal) arteries.** They supply the fibularis, soleus, tibialis posterior, and flexor hallucis muscles, fibula, tarsus, and lateral aspect of the heel.
	The bifurcation of the posterior tibial arteries into the medial and lateral plantar arteries occurs deep to the flexor retinaculum on the medial side of the feet. The **medial plantar** (PLAN-tar = sole of foot) **arteries** supply the abductor hallucis and flexor digitorum brevis muscles and the toes. The **lateral plantar arteries** unite with a branch of the dorsalis pedis arteries to form the **plantar arch.** The arch begins at the base of the fifth metatarsal and extends medially across the metacarpals. As the arch crosses the foot, it gives off **plantar metatarsal arteries,** which supply the feet. These terminate by dividing into **plantar digital arteries,** which supply the toes.
	Label these arteries in Figure 18.6.

FIGURE 18.6   Arteries of pelvis and right lower limb.

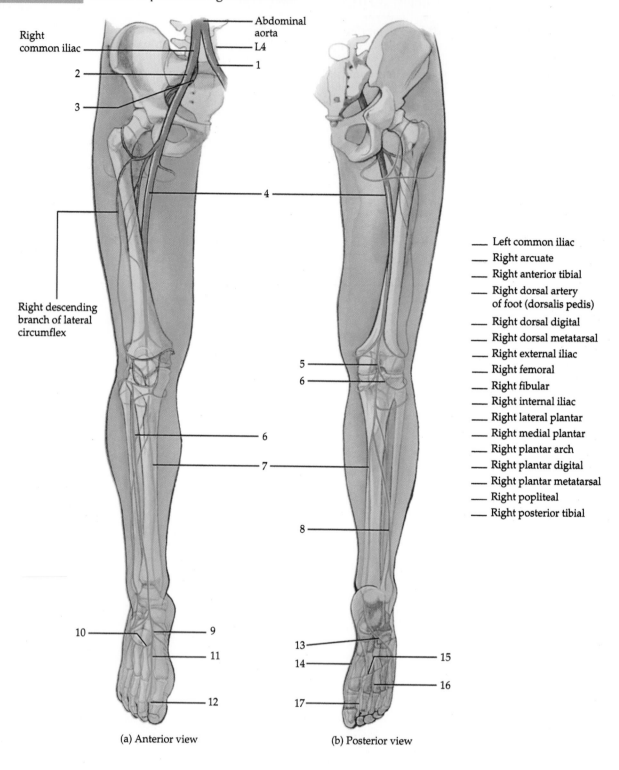

Abdominal aorta

Right common iliac

L4

1

2

3

4

Right descending branch of lateral circumflex

5

6

6

7

8

10   9

11

12

(a) Anterior view

13

14   15

17   16

(b) Posterior view

___ Left common iliac
___ Right arcuate
___ Right anterior tibial
___ Right dorsal artery of foot (dorsalis pedis)
___ Right dorsal digital
___ Right dorsal metatarsal
___ Right external iliac
___ Right femoral
___ Right fibular
___ Right internal iliac
___ Right lateral plantar
___ Right medial plantar
___ Right plantar arch
___ Right plantar digital
___ Right plantar metatarsal
___ Right popliteal
___ Right posterior tibial

Write the names of the missing arteries in the following scheme of circulation. Be sure to indicate left or right where applicable.

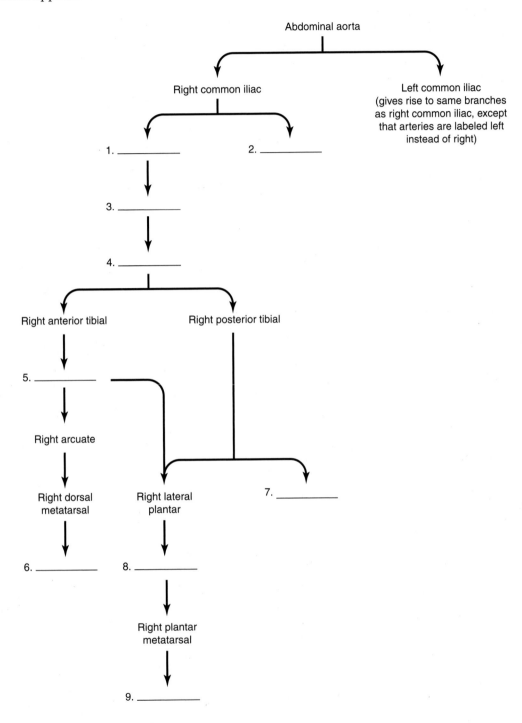

TABLE 18.7   Veins of systemic circulation (see Figure 18.8)

Overview: Whereas arteries distribute blood to various parts of the body, veins drain blood away from them. For the most part, arteries are deep, whereas veins may be superficial or deep. Superficial veins are located just beneath the skin and can easily be seen. Because there are no large superficial veins, the names of superficial veins do not correspond to those of arteries. Superficial veins are clinically important as sites for withdrawing blood or giving injections. Deep veins generally travel alongside arteries and usually bear the same name. Arteries usually follow definite pathways; veins are more difficult to follow because they connect in irregular networks in which many tributaries merge to form a large vein. Although only one systemic artery, the aorta, takes oxygenated blood away from the heart (left ventricle), three systemic veins, the **coronary sinus, superior vena cava,** and **inferior vena cava,** deliver deoxygenated blood to the heart (right atrium). The coronary sinus receives blood from the cardiac veins; the superior vena cava receives blood from other veins superior to the diaphragm, except the air sacs (alveoli) of the lungs. The inferior vena cava receives blood from veins inferior to the diaphragm.

Vein	Description and region supplied
Coronary (KOR-ō-nar-ē; *corona* = crown) **sinus**	The **coronary sinus** is the main vein of the heart; it receives almost all venous blood from the myocardium. It is located in the coronary sulcus (see Figure 17.5b) and opens into the right atrium between the orifice of the inferior vena cava and the tricuspid valve. It is a wide venous channel into which three veins drain. It receives the **great cardiac vein** (in the anterior interventricular sulcus) into its left end and the **middle cardiac vein** (in the posterior interventricular sulcus) and the **small cardiac vein** into its right end. Several **anterior cardiac veins** drain directly into the right atrium.
Superior vena cava (VĒ-na CĀ-va; *vena* = vein; *cava* = cavelike) **(SVC)**	The **SVC** is about 7.5 cm (3 in.) long and 2 cm (1 in.) in diameter and empties its blood into the superior part of the right atrium. It begins posterior to the right first costal cartilage by the union of the right and left brachiocephalic veins and ends at the level of the right third costal cartilage, where it enters the right atrium. The SVC drains the head, neck, chest, and upper limbs.
Inferior vena cava (IVC)	The **IVC** is the largest vein in the body, about 3.5 cm (1.4 in.) in diameter. It begins anterior to the fifth lumbar vertebra by the union of the common iliac veins, ascends behind the peritoneum to the right of the midline, pierces the costal tendon of the diaphragm at the level of the eighth thoracic vertebra, and enters the inferior part of the right atrium. The IVC drains the abdomen, pelvis, and lower limbs. The inferior vena cava is commonly compressed during the later stages of pregnancy by the enlarging uterus. This produces edema of the ankles and feet and temporary varicose veins.

TABLE 18.8	Veins of the head and neck (Figure 18.7)

Overview: The majority of blood draining from the head passes into three pairs of veins: **internal jugular, external jugular,** and **vertebral veins.** Within the brain, all veins drain into dural venous sinuses and then into the internal jugular veins. **Dural venous sinuses** are endothelial-lined venous channels between layers of the cranial dura mater.

Vein	Description and region drained
Internal jugular (JUG-ū-lar; *jugular* = throat) **veins**	The flow of blood from the dural venous sinuses into the internal jugular veins is as follows (Figure 18.7): The **superior sagittal** (SAJ-i-tal = straight) **sinus** begins at the frontal bone, where it receives a vein from the nasal cavity, and passes posteriorly to the occipital bone. Along its course, it receives blood from the superior, medial, and lateral aspects of the cerebral hemispheres, meninges, and cranial bones. The superior sagittal sinus usually turns to the right and drains into the right transverse sinus.

The **inferior sagittal sinus** is much smaller than the superior sagittal sinus and begins posterior to the attachment of the flax cerebri and receives the great cerebral vein to become the straight sinus. The great cerebral vein drains the deeper parts of the brain. Along its course, the inferior sagittal sinus also receives tributaries from the superior and medial aspects of the cerebral hemispheres.

The **straight sinus** runs in the tentorium cerebelli and is formed by the union of the inferior sagittal sinus and the great cerebral vein. The straight sinus also receives blood from the cerebellum and usually drains into the left transverse sinus.

The **transverse sinuses** begin near the occipital bone, pass laterally and anteriorly, and become the sigmoid sinuses near the temporal bone. The transverse sinuses receive blood from the cerebrum, cerebellum, and cranial bones.

The **sigmoid** (SIG-moyd = S-shaped) **sinuses** are located along the temporal bone. They pass through the jugular foramina, where they terminate in the internal jugular veins. The sigmoid sinuses drain the transverse sinuses.

The **cavernous** (KAV-er-nus = cavelike) **sinuses** are located on either side of the sphenoid bone. They receive blood from the ophthalmic veins from the orbits and cerebral veins from the cerebral hemispheres. They ultimately empty into the transverse sinuses and internal jugular veins. The cavernous sinuses are quite unique because they have nerves and a major blood vessel passing through them on their way to the orbit and face. Through the sinuses pass the oculomotor (III), trochlear (IV), and ophthalmic and maxillary divisions of the trigeminal (V) nerve and the internal carotid arteries.

The right and left **internal jugular veins** pass inferiorly on either side of the neck lateral to the internal carotid and common carotid arteries. They then unite with the subclavian veins posterior to the clavicles at the sternoclavicular joints to form the right and left **brachiocephalic** (brā-kē-ō-se-FAL-ik; *brachio* = arm; *cephal* = head) **veins.** From here blood flows into the superior vena cava. The general structures drained by the internal jugular veins are the brain (through the dural venous sinuses), face, and neck. |
| **External jugular veins** | The right and left **external jugular veins** begin in the parotid glands near the angle of the mandible. They are superficial veins that descend through the neck across the sternocleidomastoid muscles. They terminate at a point opposite the middle of the clavicle, where they empty into the subclavian veins. The general structures drained by the external jugular veins are external to the cranium, such as the scalp and superficial and deep regions of the face.

In cases of heart failure, the venous pressure in the right atrium may rise. In such patients the pressure in the column of blood in the external jugular vein rises so that, even with the patient at rest and sitting in a chair, the external jugular vein will be visibly distended. Temporary distention of the vein is often seen in healthy adults when the intrathoracic pressure is raised during coughing and physical exertion. |
| **Vertebral** (VER-te-bral; *vertebra* = vertebrae) **veins** | The right and left **vertebral veins** originate inferior to the occipital condyles. They descend through successive transverse foramina of the first six cervical vertebrae and emerge from the foramina of the sixth cervical vertebrae to enter the brachiocephalic veins in the root of the neck. The vertebral veins drain deep structures in the neck such as the cervical vertebrae, cervical spinal cord, and some neck muscles.

Label these veins in Figure 18.7. |

FIGURE 18.7 **Veins of head and neck.**

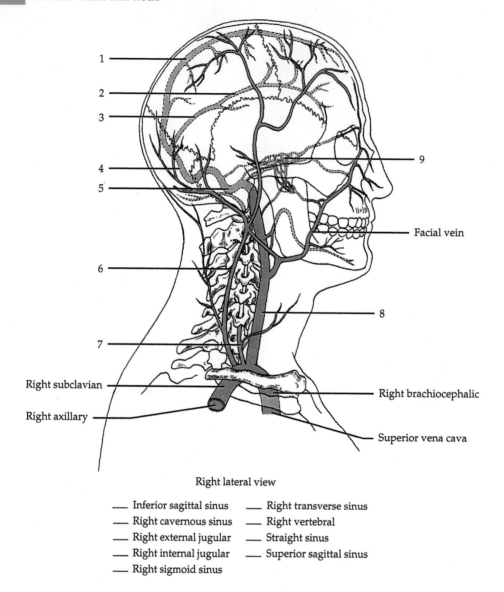

1

2

3

9

4

5

Facial vein

6

8

7

Right subclavian

Right brachiocephalic

Right axillary

Superior vena cava

Right lateral view

___ Inferior sagittal sinus   ___ Right transverse sinus
___ Right cavernous sinus   ___ Right vertebral
___ Right external jugular   ___ Straight sinus
___ Right internal jugular   ___ Superior sagittal sinus
___ Right sigmoid sinus

Write the names of the missing veins in the following scheme of circulation. Be sure to indicate left or right where applicable.

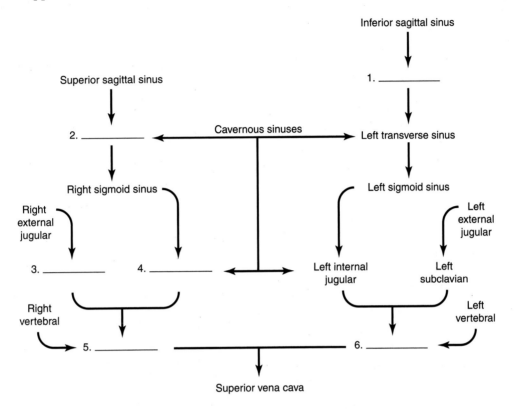

TABLE 18.9 Veins of the upper limbs (Figure 18.8)

Overview: Blood from the upper limbs is returned to the heart by both superficial and deep veins. Both sets of veins have valves, which are more numerous in the deep veins. **Superficial veins** are located just deep to the skin and are often visible. They anastomose extensively with one another and with deep veins, and they do not accompany arteries. Superficial veins are larger than deep veins and return most of the blood from the upper limbs. **Deep veins** are located deep in the body. They usually accompany arteries and have the same names as the corresponding arteries.

Vein	Description and region drained
Superficial **Cephalic** (se-FAL-ik = head) **veins**	The principal superficial veins that drain the upper limbs are the cephalic and basilica veins. They originate in the hand and convey blood from the smaller superficial veins into the axillary veins. The **cephalic veins** begin on the lateral aspect of the **dorsal venous networks of the hands (dorsal venous arches)**, networks of veins on the dorsum of the hands formed by the **dorsal metacarpal veins.** These veins, in turn, drain the **dorsal digital veins,** which pass along the sides of the fingers. Following their formation from the dorsal venous networks of the hands, the cephalic veins arch around the radial side of the forearms to the anterior surface and ascend through the entire limbs along the anterolateral surface. The cephalic veins end where they join the axillary veins, just inferior to the clavicles. **Accessory cephalic veins** originate either from a venous plexus on the dorsum of the forearms or from the medial aspects of the dorsal venous arches and unite with the cephalic veins just inferior to the elbow. The cephalic veins drain blood from the lateral aspect of the upper limbs. Label these veins in Figure 18.8a.
Basilic (ba-SIL-ik; *basilikos* = royal) **veins**	The **basilic veins** begin on the medial aspects of the dorsal venous networks of the hand and ascend along the posteromedial surface of the forearm and anteromedial surface of the arm. They drain blood from the medial aspects of the upper limbs. Anterior to the elbow, the basilic veins are connected to the cephalic veins by the **median cubital** (*cubital* = pertaining to the elbow) **veins,** which drain the forearm. If veins must be punctured for an injection, transfusion, or removal of a blood sample, the medial cubital veins are preferred. After receiving the median cubital veins, the basilic veins continue ascending until they reach the middle of the arm. There they penetrate the tissues deeply and run alongside the brachial arteries until they join the brachial veins. As the basilic and brachial veins merge in the axillary area, they form the axillary veins. Label these veins in Figure 18.8d.
Median antebrachial (an'-tē-BRĀ-kē-al; *ante* = before, in front of; *brachio* = arm) **veins**	The **median antebrachial veins (median veins of the forearm)** begin in the **palmar venous plexuses,** networks of veins on the palms. The plexuses drain the **palmar digital veins** in the fingers. The median antebrachial veins ascend anteriorly in the forearms to join the basilica or median cubital veins, sometimes both. They drain the palms and forearms. Label these veins in Figure 18.8b.
Deep **Radial** (RĀ-dē-al = pertaining to the radius) **veins**	The paired **radial veins** begin at the **deep venous palmar arches.** These arches drain the **palmar metacarpal veins** in the palms. The radial veins drain the lateral aspects of the forearms and pass alongside the radial arteries. Just inferior to the elbow joint, the radial veins unite with the ulnar veins to form the brachial veins. Label these veins in Figure 18.8c.

Vein	Description and region drained
Ulnar (UL-nar = pertaining to the ulna) **veins**	The paired **ulnar veins,** which are larger than the radial veins, begin at the **superficial palmar venous arches.** These arches drain the **common palmar digital veins** and the **proper palmar digital veins** in the fingers. The ulnar veins drain the medial aspect of the forearms, pass alongside the ulnar arteries, and join with the radial veins to form the brachial veins. Label these veins in Figure 18.8c.
Brachial (BRĀ-kē-al; *brachio* = arm) **veins**	The paired **brachial veins** accompany the brachial arteries. They drain the forearms, elbow joints, arms, and humerus. They pass superiorly and join with the basilic veins to form the axillary veins.
Axillary (AK-sil-ār'-ē; *axilla* = armpit) **veins**	The **axillary veins** ascend to the outer borders of the first ribs, where they become the subclavian veins. The axillary veins receive tributaries that correspond to the branches of the axillary arteries. The axillary veins drain the arms, axillas, and superolateral chest wall.
Subclavian (sub-KLĀ-vē-an; *sub* = under; *clavian* = clavicle) **veins**	The **subclavian veins** are continuations of the axillary veins that terminate at the sternal end of the clavicles, where they unite with the internal jugular veins to form the brachiocephalic veins. The subclavian veins drain the arms, neck, and thoracic wall. The thoracic duct of the lymphatic system delivers lymph into the left subclavian vein at the junction with the internal jugular. The right lymphatic duct delivers lymph into the right subclavian vein at the corresponding junction (see Figure 20.1a). In a procedure called *central line placement,* the right subclavian vein is frequently used to administer nutrients and medication. Label these veins in Figure 18.8d.

FIGURE 18.8 **Veins of the right upper limb.**

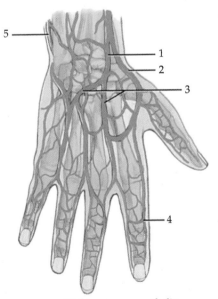

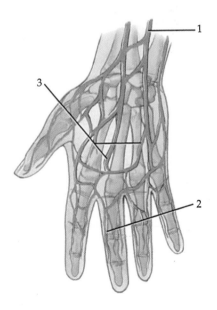

___ Right accessory cephalic

___ Right basilic

___ Right cephalic

___ Right dorsal digital

___ Right dorsal venous networks of the hand (dorsal venous arch)

___ Right median antebrachial

___ Right palmar digital

___ Right palmar venous plexus

(a) Posterior view of superficial veins of hand

(b) Anterior view of superficial veins of hand

FIGURE 18.8   Veins of the right upper limb. (Continued)

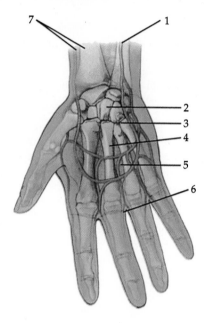

_____ Right common palmar digital
_____ Right deep venous palmar arch
_____ Right palmar metacarpal
_____ Right proper palmar digital
_____ Right radial
_____ Right superficial venous palmar arch
_____ Right ulnar

(c) Anterior view of deep veins of hand

FIGURE 18.8 **Veins of the right upper limb. (Continued)**

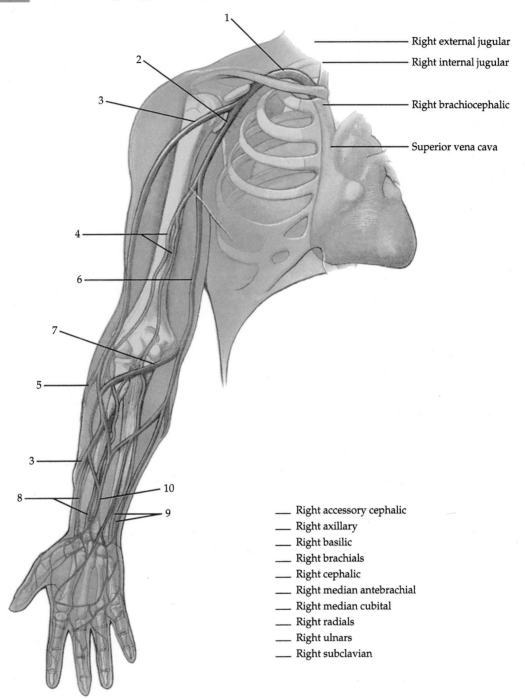

Right external jugular
Right internal jugular
Right brachiocephalic
Superior vena cava

___ Right accessory cephalic
___ Right axillary
___ Right basilic
___ Right brachials
___ Right cephalic
___ Right median antebrachial
___ Right median cubital
___ Right radials
___ Right ulnars
___ Right subclavian

(d) Anterior view of superficial and deep veins of upper limb

Write the names of the missing veins in the following scheme of circulation. Be sure to indicate left or right where applicable.

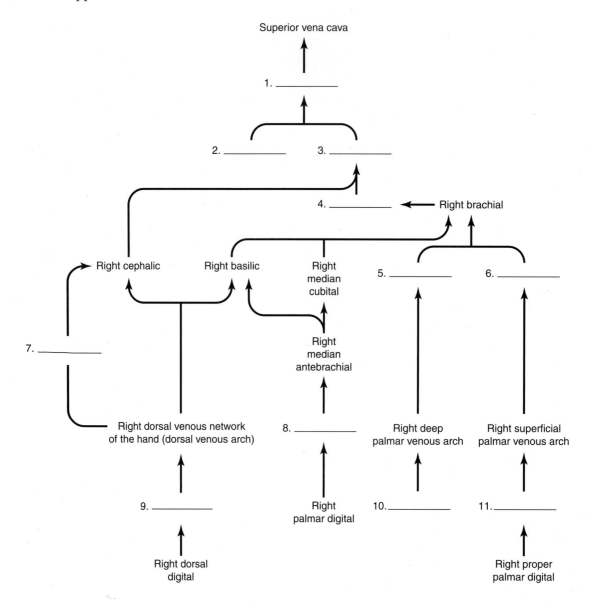

TABLE 18.10 Veins of the thorax (Figure 18.9)

Overview: Although the brachiocephalic veins drain some portions of the thorax, most thoracic structures are drained by a network of veins called the **azygos system.** This is a network of veins on each side of the vertebral column: **azygos, hemiazygos,** and **accessory hemiazygos.** They show considerable variation in origin, course, tributaries, anastomoses, and termination. Ultimately, they empty into the superior vena cava.

Vein	Description and region drained
Brachiocephalic (brā′-kē-ō-se-FAL-ik; *brachio* = arm; *caphalic* = pertaining to the head) **vein**	The right and left **brachiocephalic veins,** formed by the union of the subclavian veins and internal jugular veins, drain blood from the head, neck, upper limbs, mammary glands, and superior thorax. The brachiocephalic veins unite to form the superior vena cava. Because the superior vena cava is to the right, the left brachiocephalic vein is longer than the right. The right brachiocephalic vein is anterior and to the right of the brachiocephalic trunk. The left brachiocephalic vein is anterior to the brachiocephalic trunk, the left common carotid and left subclavian arteries, the trachea, and the left vagus (X) and phrenic nerves.
Azygos (az-Ī-gos = unpaired) **system**	The **azygos system,** besides collecting blood from the thorax and abdominal wall, may serve as a bypass for the inferior vena cava that drains blood from the lower body. Several small veins directly link the azygos system with the inferior vena cava. Large veins that drain the lower limbs and abdomen pass blood into the azygos system. If the inferior vena cava or hepatic portal vein becomes obstructed, the azygos system can return blood from the lower body to the superior vena cava.
Azygos vein	The **azygos vein** is anterior to the vertebral column, slightly to the right of the midline. It begins at the junction of the right ascending lumbar and right subcostal veins near the diaphragm. At the level of the fourth thoracic vertebra, it arches over the root of the right lung to end in the superior vena cava. Generally, the azygos vein drains the right side of the thoracic wall, thoracic viscera, and abdominal wall.
Hemiazygos (HEM-ē-az-ī-gos; *hemi* = half) **vein**	The **hemiazygos vein** is anterior to the vertebral column and slightly to the left of the midline. It begins at the junction of the left ascending lumbar and left subcostal veins. It terminates by joining the azygos vein at about the level of the ninth thoracic vertebra. Generally, the hemiazygos vein drains the left side of the thoracic wall, thoracic viscera, and abdominal wall. Specifically, the hemiazygos vein receives blood from the ninth through eleventh **left posterior intercostal veins, esophageal, mediastinal,** and sometimes the **accessory hemiazygos veins.**
Accessory hemiazygos vein	The **accessory hemiazygos vein** is also anterior to the vertebral column and to the left of the midline. It begins at the fourth or fifth intercostal space and descends from the fifth to the eighth thoracic vertebra. It terminates by joining the azygos vein at about the level of the eighth thoracic vertebra. The accessory hemiazygos vein drains the left side of the thoracic wall. It receives blood from the fourth through eighth **left posterior intercostal veins** (the first through third left intercostal nerves open into the left brachiocephalic vein), **left bronchial,** and **mediastinal veins.** Label the veins of the azygos system in Figure 18.9.

FIGURE 18.9   Veins of thorax, abdomen, and pelvis.

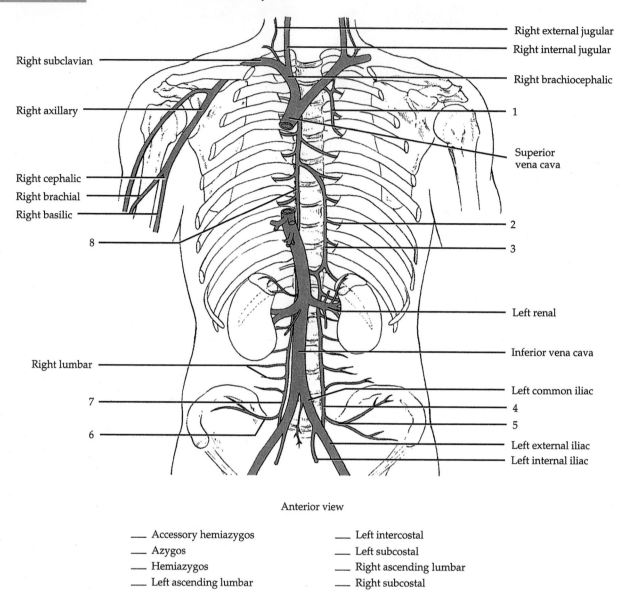

Anterior view

___ Accessory hemiazygos   ___ Left intercostal
___ Azygos   ___ Left subcostal
___ Hemiazygos   ___ Right ascending lumbar
___ Left ascending lumbar   ___ Right subcostal

Write the names of the missing veins in the following scheme of circulation. Be sure to indicate left or right where applicable.

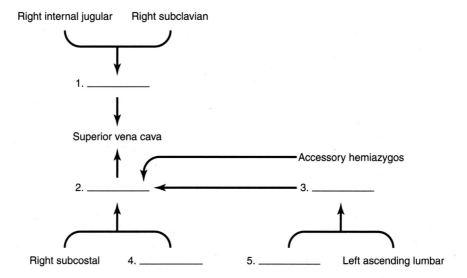

TABLE 18.11 Veins of the abdomen and pelvis (Figure 18.10)

Overview: Blood from the abdominal and pelvic viscera and abdominal wall returns to the heart via the **inferior vena cava.** Many small veins enter the inferior vena cava. Most carry return flow from parietal branches of the abdominal aorta, and their names correspond to the names of the arteries.

The inferior vena cava does not receive veins directly from the gastrointestinal tract, spleen, pancreas, and gallbladder. These organs pass their blood into a common vein, the **hepatic portal vein,** which delivers the blood to the liver. The hepatic portal vein is formed by the union of the superior mesenteric and splenic veins (see Figure 18.12). This special flow of venous blood is called **hepatic portal circulation,** which is described shortly. After passing through the liver for processing, blood drains into the hepatic veins, which empty into the inferior vena cava.

Vein	Description and region drained
Inferior vena cava (VĒ-na CĀ-va; *vena* = vein; *cava* = cavelike)	The **inferior vena cava** is formed by the union of two common iliac veins that drain the lower limbs, pelvis, and abdomen. The inferior vena cava extends superiorly through the abdomen and thorax to the right atrium.
Common iliac (IL-ē-ak; *iliac* = pertaining to the ilium) **veins**	The **common iliac veins** are formed by the union of the internal and external iliac veins anterior to the sacroiliac joint and represent the distal continuation of the inferior vena cava at their bifurcation. The right common iliac vein is much shorter than the left and is also more vertical. Generally, the common iliac veins drain the pelvis, external genitals, and lower limbs.
Internal iliac veins	The **internal iliac veins** begin near the superior portion of the greater sciatic notch and run medial to their corresponding arteries. Generally, the veins drain the thigh, buttocks, external genitals, and pelvis.
External iliac veins	The **external iliac veins** are companions of the internal iliac arteries and begin at the inguinal ligaments as continuations of the femoral veins. They end anterior to the sacroiliac joint where they join with the internal iliac veins to form the common iliac veins. The external iliac veins drain the lower limbs, cremaster muscle in males, and the abdominal wall.
Lumbar (LUM-bar = pertaining to the loin) **veins**	A series of parallel **lumbar veins,** usually four on each side, drain blood from both sides of the posterior abdominal wall, vertebral canal, spinal cord, and meninges. The lumbar veins run horizontally with the lumbar arteries. The lumbar veins connect at right angles with the right and left **ascending lumbar veins,** which form the origin of the corresponding azygos or hemiazygos vein. The lumbar veins drain blood into the ascending lumbars and then run to the inferior vena cava, where they release the remainder of the flow.
Gonadal (gō-NAD-al; *gono* = seed) [**testicular** (tes-TIK-ū-lar) or **ovarian** (ō-VAR-ē-an)] **veins**	The **gonadal veins** ascend with the gonadal arteries along the posterior abdominal wall. In the male, the gonadal veins are called the testicular veins. The **testicular veins** drain the testes (the left testicular vein empties into the left renal vein, and the right testicular vein drains into the inferior vena cava). In the female, the gonadal veins are called the ovarian veins. The **ovarian veins** drain the ovaries (the left ovarian vein empties into the left renal vein, and the right ovarian vein drains into the inferior vena cava).
Renal (RĒ-nal = kidney) **veins**	The **renal veins** are large and pass anterior to the renal arteries. The left renal vein is longer than the right renal vein and passes anterior to the abdominal aorta. It receives the left testicular (or ovarian), left inferior phrenic, and usually left suprarenal veins. The right renal vein passes into the inferior vena cava posterior to the duodenum. The renal veins drain the kidneys.

Vein	Description and region drained
Suprarenal (soo-pra-RĒ-nal; *supra* = above) **veins**	The **suprarenal veins** drain the adrenal (suprarenal) glands (the left suprarenal vein empties into the left renal vein, and the right suprarenal vein empties into the inferior vena cava).
Inferior phrenic (FREN-ik; *phrenic* = pertaining to the diaphragm) **veins**	The **inferior phrenic veins** drain the diaphragm (the left inferior phrenic vein sends a tributary to the left renal vein, and the right inferior phrenic vein empties into the inferior vena cava).
Hepatic (he-PAT-ik = liver) **veins**	The **hepatic veins** drain the liver. Label these veins in Figure 18.10.

FIGURE 18.10 Veins of the abdomen and pelvis.

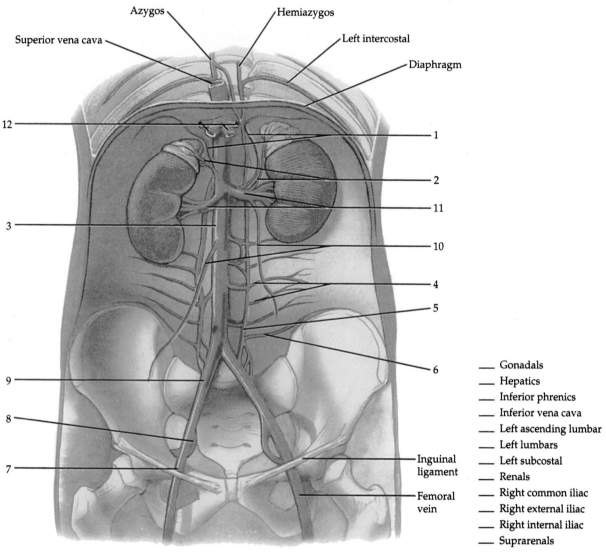

Azygos

Hemiazygos

Superior vena cava

Left intercostal

Diaphragm

12

1

2

11

3

10

4

5

6

9

8

7

Inguinal ligament

Femoral vein

___ Gonadals
___ Hepatics
___ Inferior phrenics
___ Inferior vena cava
___ Left ascending lumbar
___ Left lumbars
___ Left subcostal
___ Renals
___ Right common iliac
___ Right external iliac
___ Right internal iliac
___ Suprarenals

Anterior view

Write the names of the missing veins in the following scheme of circulation. Be sure to indicate left or right where applicable.

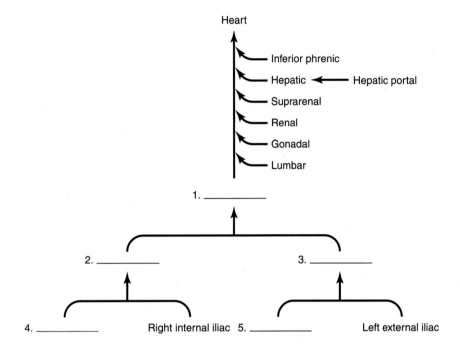

TABLE 18.12   Veins of the lower limbs (Figure 18.11)

Overview: As with the upper limbs, blood from the lower limbs is drained by both **superficial** and **deep veins.** The superficial veins often anastomose with each other and with deep veins along their length. Deep veins, for the most part, have the same names as corresponding arteries. All veins of the lower limbs have valves, which are more numerous than in veins of the upper limbs.

Vein	Description and region drained
Superficial veins **Great saphenous** (sa-FĒ-nus; *saphen* = clearly visible) **veins**	The **great (long) saphenous veins,** the longest veins in the body, ascend from the foot to the groin in the subcutaneous layer. They begin at the medial end of the dorsal venous arches of the foot. The **dorsal venous** (VĒ-nus) **arches** are networks of veins on the dorsum of the foot formed by the **dorsal digital veins,** which collect blood from the toes, and then unite in pairs to form the **dorsal metatarsal veins,** which parallel the metatarsals. As the dorsal metatarsal veins approach the foot, they combine to form the dorsal venous arches. The great saphenous veins pass anterior to the medial malleolus of the tibia and then superiorly along the medial aspect of the leg and thigh just deep to the skin. They receive tributaries from superficial tissues and connect with the deep veins as well. They empty into the femoral veins at the groin. Generally, the great saphenous veins drain mainly the medial side of the leg and thigh, the groin, external genitals, and abdominal wall. Along their length, the great saphenous veins have from 10 to 20 valves, with more located in the leg than the thigh. These veins are more likely to be subject to varicosities than other veins in the lower limbs because they must support a long column of blood and are not well supported by skeletal muscles. The great saphenous veins are often used for prolonged administration of intravenous fluids. This is particularly important in very young children and in patients of any age who are in shock and whose veins are collapsed. The great saphenous veins are also often used as a source of vascular grafts, especially for coronary bypass surgery. In the procedure, the vein is removed and then reversed so that the valves do not obstruct the flow of blood.
Small saphenous veins	The **small (short) saphenous veins** begin at the lateral aspect of the dorsal venous arches of the foot. They pass posterior to the lateral malleolus of the fibula and ascend deep to the skin along the posterior aspect of the leg. They empty into the popliteal veins in the popliteal fossa, posterior to the knee. Along their length, the small saphenous veins have from 9 to 12 valves. The small saphenous veins drain the foot and posterior aspect of the leg. They may communicate with the great saphenous veins in the proximal thigh.

Vein	Description and region drained
Deep veins **Posterior tibial** (TIB-ē-al) **veins**	The **plantar digital veins** on the plantar surfaces of the toes unite to form the **plantar metatarsal veins,** which parallel the metatarsals. They unite to form the **deep plantar venous arches.** From each arch emerges the **medial** and **lateral plantar veins.** The paired **posterior tibial veins,** which sometimes merge to form a single vessel, are formed by the medial and lateral plantar veins, posterior to the medial malleolus of the tibia. They accompany the posterior tibial artery through the leg. They ascend deep to the muscles in the posterior aspect of the leg and drain the foot and posterior compartment muscles. About two-thirds the way up the leg, the posterior tibial veins drain blood from the **fibular (peroneal) veins,** which drain the lateral and posterior leg muscles. The posterior tibial veins unite with the anterior tibial veins just inferior to the popliteal fossa to form the popliteal veins.
Anterior tibial veins	The **paired anterior tibial veins** arise in the dorsal venous arch and accompany the anterior tibial artery. They ascend in the interosseous membrane between the tibia and fibula and unite with the posterior tibial veins to form the popliteal vein. The anterior tibial veins drain the ankle joint, knee joint, tibiofibular joint, and anterior portion of leg.
Popliteal (pop'-li-TĒ-al = pertaining to the hollow behind knee) **veins**	The **popliteal veins** are formed by the union of the anterior and posterior tibial veins. The popliteal veins also receive blood from the small saphenous veins and tributaries that correspond to branches of the popliteal artery. The popliteal veins drain the knee joint and the skin, muscles, and bones of portions of the calf and thigh around the knee joint.
Femoral (FEM-o-ral) **veins**	The **femoral veins** accompany the femoral arteries and are the continuations of the popliteal veins just superior to the knee. The femoral veins extend up the posterior surface of the thighs and drain the muscles of the thighs, femurs, external genitals, and superficial lymph nodes. The largest tributaries of the femoral veins are the **deep femoral veins.** Just before penetrating the abdominal wall, the femoral veins receive the deep femoral veins and the great saphenous veins. The veins formed from their union penetrate the body wall and enter the pelvic cavity. Here they are known as the **external iliac veins.**

FIGURE 18.11 Veins of the lower limb.

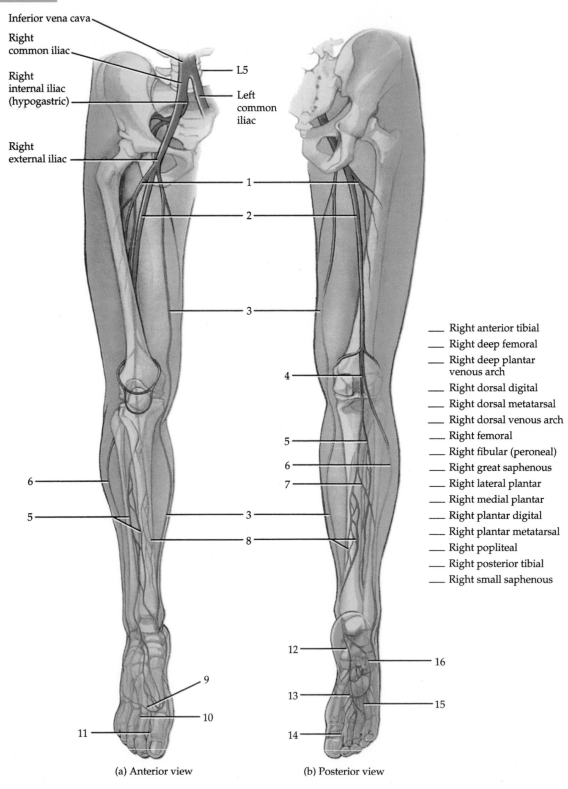

Inferior vena cava

Right common iliac

Right internal iliac (hypogastric)

Right external iliac

L5

Left common iliac

_____ Right anterior tibial
_____ Right deep femoral
_____ Right deep plantar venous arch
_____ Right dorsal digital
_____ Right dorsal metatarsal
_____ Right dorsal venous arch
_____ Right femoral
_____ Right fibular (peroneal)
_____ Right great saphenous
_____ Right lateral plantar
_____ Right medial plantar
_____ Right plantar digital
_____ Right plantar metatarsal
_____ Right popliteal
_____ Right posterior tibial
_____ Right small saphenous

(a) Anterior view

(b) Posterior view

Write the names of the missing veins in the following scheme of circulation. Be sure to indicate left or right where applicable.

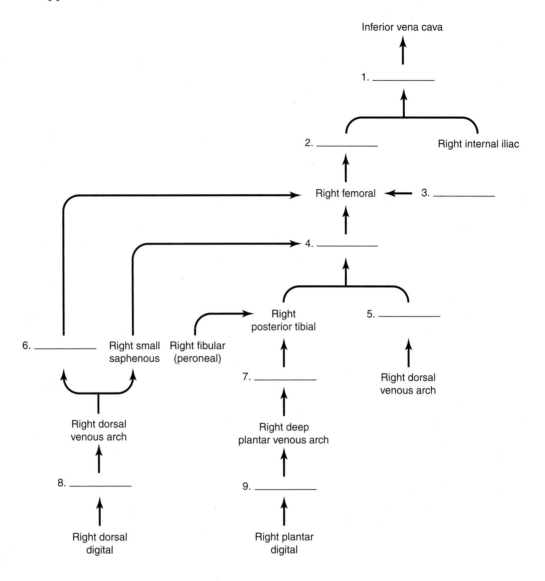

2. Hepatic Portal Circulation

The **hepatic** (*hepato* = liver) **portal circulation** carries venous blood from the gastrointestinal organs and spleen to the liver before it returns to the heart (Figure 18.12). A *portal system* carries blood between two capillary networks, from one location in the body to another without passing through the heart, in this case from capillaries of the gastrointestinal tract to sinusoids of the liver. After a meal, hepatic portal blood is rich with absorbed substances. The liver stores some and modifies others before they pass into the general circulation. For example, the liver converts glucose into glycogen for storage. It also modifies other digested substances so they may be used by cells, detoxifies harmful substances that have been absorbed by the gastrointestinal tract, and destroys bacteria by phagocytosis.

The **hepatic portal vein** is formed by the union of the (1) superior mesenteric and (2) splenic veins.

FIGURE 18.12 **Heptic portal circulation.**

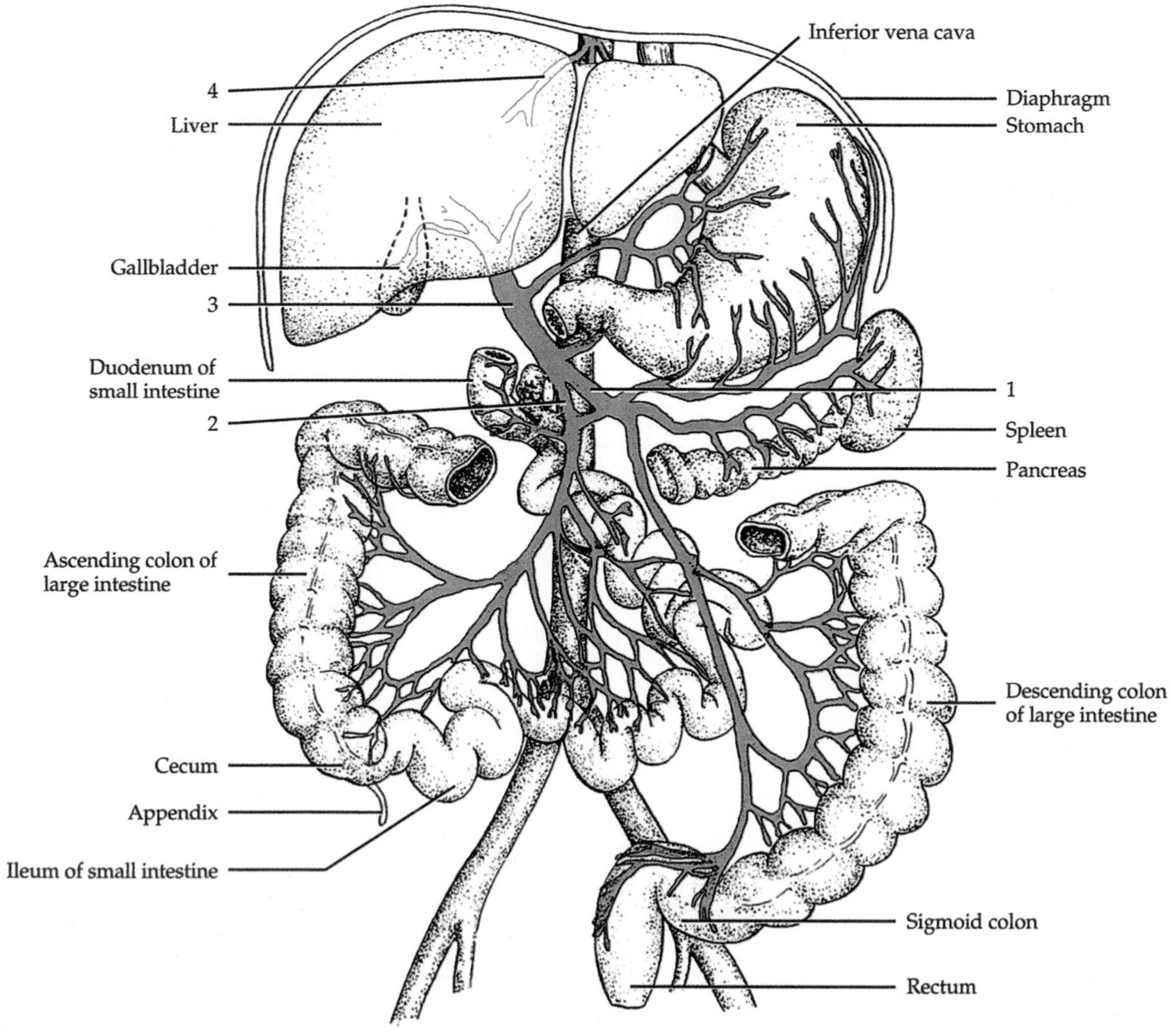

(a) Anterior view

___ Hepatic
___ Hepatic portal
___ Splenic
___ Superior mesenteric

FIGURE 18.12 **Heptic portal circulation. (Continued)**

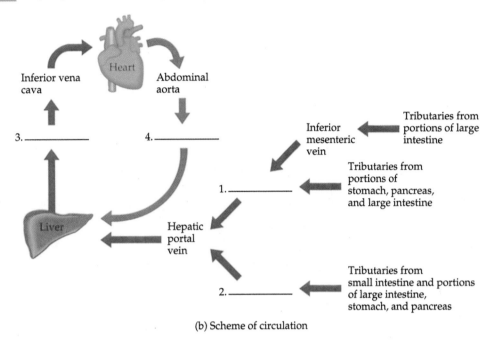

(b) Scheme of circulation

The **superior mesenteric vein** drains blood from the small intestine and portions of the large intestine, stomach, and pancreas through the *jejunal, ileal, ileocolic, right colic, middle colic, pancreaticoduodenal,* and *right gastroepiploic veins.* The **splenic vein** drains blood from the stomach, pancreas, and portions of the large intestine through the *short gastric, left gastroepiploic, pancreatic,* and *inferior mesenteric veins.* The inferior mesenteric vein, which passes into the splenic vein, drains portions of the large intestine through the superior *rectal, sigmoidal,* and *left colic veins.* The *right* and *left gastric veins,* which open directly into the hepatic portal vein, drain the stomach. The *cystic vein,* which also opens into the hepatic portal vein, drains the gallbladder.

At the same time the liver receives deoxygenated blood via the hepatic portal system, it also receives oxygenated blood from the systemic circulation via the hepatic artery. Ultimately, all blood leaves the liver through the **hepatic veins,** which drain into the inferior vena cava.

Using your textbook, charts, or models for reference, label Figure 18.12a.

Using your textbook, chart, or models for reference, label Figure 18.12b by filling in the blanks.

3. Pulmonary Circulation

The **pulmonary** (*pulmo* = lung) **circulation** carries deoxygenated blood from the right ventricle to the air sacs (alveoli) within the lungs and returns oxygenated blood from the air sacs within the lungs to the left atrium (Figure 18.13). The **pulmonary trunk** emerges from the right ventricle and passes superiorly, posteriorly, and to the left. It then divides into two branches: The **right pulmonary artery** extends to the right lung; the **left pulmonary artery** goes to the left lung. The pulmonary arteries are the only postnatal (after birth) arteries that carry deoxygenated blood. On entering the lungs, the branches divide and subdivide until finally they form capillaries around the air sacs within the lungs. CO_2 passes from the blood into these air sacs and is exhaled. Inhaled O_2 passes from the air sacs into the blood. The pulmonary capillaries unite, form venules and veins, and eventually two **pulmonary veins** exit from each lung and transport the oxygenated blood to the left atrium. The pulmonary veins are the only postnatal veins that carry oxygenated blood. Contractions of the left ventricle then send the blood into the systemic circulation.

Study a chart or model of the pulmonary circulation, trace the path of blood through it, and label Figure 18.13a.

Using your textbook, charts, or models for reference, label Figure 18.13b by filling in the blanks.

4. Fetal Circulation

The circulatory system of a fetus, called **fetal circulation,** differs from the postnatal circulation because the lungs, kidneys, and gastrointestinal

FIGURE 18.13   Pulmonary circulation.

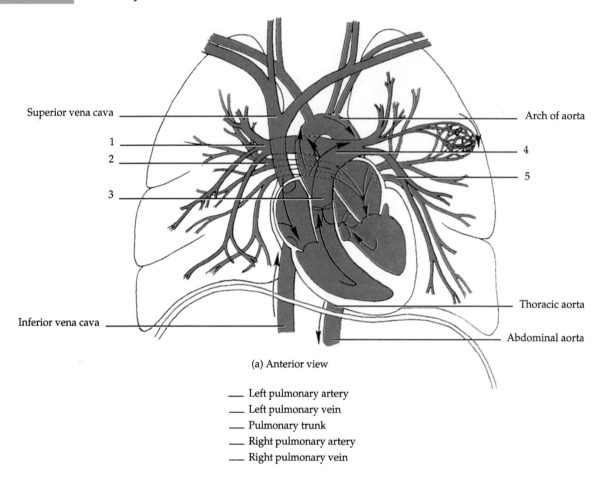

Superior vena cava

Arch of aorta

1

2

3

4

5

Inferior vena cava

Thoracic aorta

Abdominal aorta

(a) Anterior view

___ Left pulmonary artery
___ Left pulmonary vein
___ Pulmonary trunk
___ Right pulmonary artery
___ Right pulmonary vein

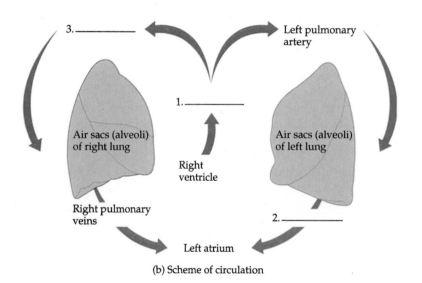

3. _____

Left pulmonary artery

1. _____

Air sacs (alveoli) of right lung

Air sacs (alveoli) of left lung

Right ventricle

Right pulmonary veins

2. _____

Left atrium

(b) Scheme of circulation

tract begin to function at birth. The fetus obtains its O_2 and nutrients by diffusion from the maternal blood and eliminates its CO_2 and wastes by diffusion into the maternal blood (Figure 18.14).

The exchange of materials between fetal and maternal circulation occurs through a structure called the **placenta** (pla-SEN-ta). It is attached to the umbilicus (navel) of the fetus by the **umbilical** (um-BIL-i-kal) **cord,** and it communicates with the mother through countless small blood vessels that emerge from the uterine wall. The umbilical cord contains blood vessels that branch into capillaries in the placenta. Wastes from the fetal blood diffuse out of the capillaries, into spaces containing maternal blood (intervillous spaces) in the placenta, and finally into the mother's uterine blood vessels. Nutrients travel the opposite route—from the maternal blood vessels to the intervillous spaces to the fetal capillaries. Normally, there is no direct mixing of maternal and fetal blood since all exchanges occur by diffusion through capillary walls.

Blood passes from the fetus to the placenta via two **umbilical arteries.** These branches of the internal iliac arteries are within the umbilical cord. At the placenta, fetal blood picks up O_2 and nutrients and eliminates CO_2 and wastes. The oxygenated blood returns from the placenta via a single **umbilical vein.** This vein ascends to the liver of the fetus, where it divides into two branches. Whereas some blood flows through the branch that joins the hepatic portal vein and enters the liver, most of the blood flows into the second branch, the **ductus venosus** (DUK-tus ve-NŌ-sus), which drains into the inferior vena cava. Thus a good portion of the O_2 and nutrients in the blood bypasses the fetal liver and is delivered to the developing fetal brain.

Circulation through other portions of the fetus is similar to postnatal circulation. Deoxygenated blood returning from the lower regions mingles with oxygenated blood from the ductus venosus in the inferior vena cava. This mixed blood then enters the right atrium. Deoxygenated blood returning from the upper regions of the fetus enters the superior vena cava and passes into the right atrium.

Most of the fetal blood does not pass from the right ventricle to the lungs, as it does in postnatal circulation, since the fetal lungs do not operate. In the fetus, an opening called the **foramen ovale** (fō-RĀ-men ō-VAL-ē) exists in the septum between the right and left atria. About one-third of the blood passes through the foramen ovale directly into the systemic circulation. The blood that does pass into the right ventricle is pumped into the pulmonary trunk, but little of this blood reaches the nonfunctioning fetal lungs. Instead most is sent through the **ductus arteriosus** (ar-tē-rē-Ō-sus). This vessel connects the pulmonary trunk with the aorta and allows most blood to bypass the fetal lungs. The blood in the aorta is carried to all parts of the fetus through the systemic circulation. When the common iliac arteries branch into the external and internal iliacs, part of the blood flows into the internal iliacs. It then goes to the umbilical arteries and back to the placenta for another exchange of materials. The only fetal vessel that carries fully oxygenated blood is the umbilical vein.

Using your textbook, charts, or models for reference, label Figure 18.14a.

Using your textbook, charts, or models for reference, label Figure 18.14b by filling in the blanks.

E. Blood Vessel Exercise

For each vessel listed, indicate the region supplied (if an artery) or the region drained (if a vein):

1. *Coronary artery.* _____
   _____

2. *Internal iliac veins.* _____
   _____

3. *Lumbar arteries.* _____
   _____

4. *Renal artery.* _____
   _____

5. *Left gastric artery.* _____
   _____

6. *External jugular vein.* _____
   _____

7. *Left subclavian artery.* _____
   _____

8. *Axillary vein.* _____
   _____

9. *Brachiocephalic veins.* _____
   _____

10. *Transverse sinuses.* _____
    _____

(text continues on page 441)

FIGURE 18.14 Fetal circulation.

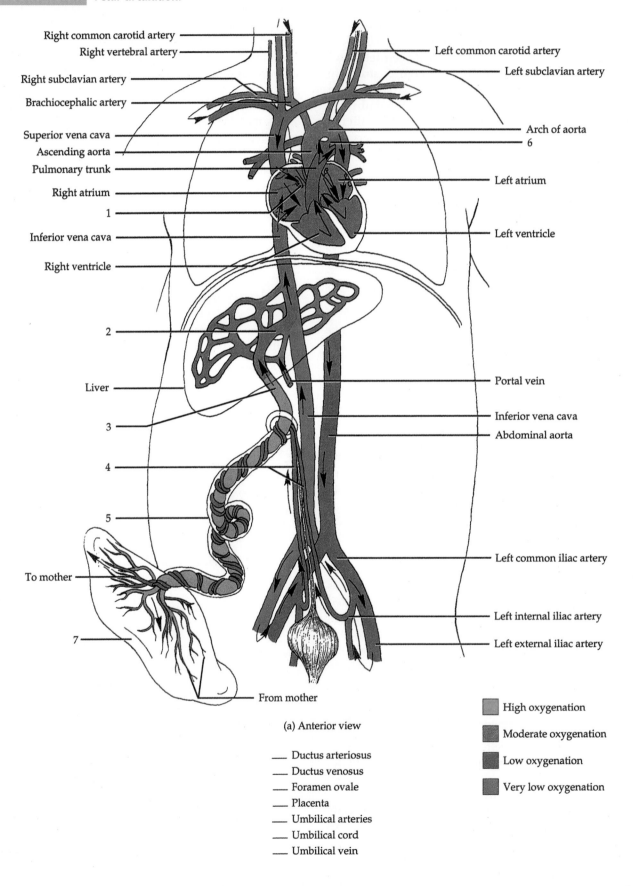

Right common carotid artery

Right vertebral artery

Right subclavian artery

Brachiocephalic artery

Superior vena cava

Ascending aorta

Pulmonary trunk

Right atrium

1

Inferior vena cava

Right ventricle

2

Liver

3

4

5

To mother

7

Left common carotid artery

Left subclavian artery

Arch of aorta

6

Left atrium

Left ventricle

Portal vein

Inferior vena cava

Abdominal aorta

Left common iliac artery

Left internal iliac artery

Left external iliac artery

From mother

(a) Anterior view

___ Ductus arteriosus
___ Ductus venosus
___ Foramen ovale
___ Placenta
___ Umbilical arteries
___ Umbilical cord
___ Umbilical vein

High oxygenation

Moderate oxygenation

Low oxygenation

Very low oxygenation

FIGURE 18.14 Fetal circulation. (Continued)

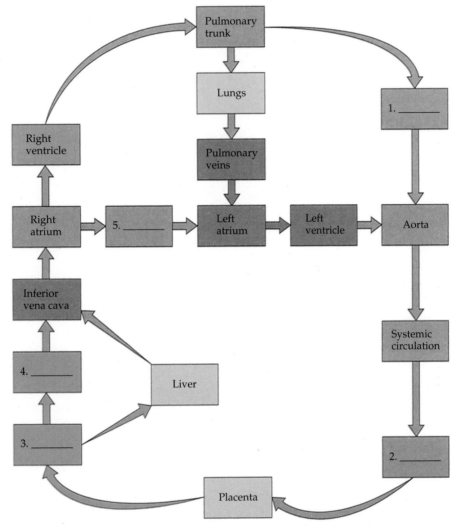

(b) Scheme of fetal circulation

11. *Hepatic artery.* _____

12. *Inferior mesenteric artery.* _____

13. *Suprarenal artery.* _____

14. *Inferior phrenic artery.* _____

15. *Great saphenous vein.* _____

16. *Popliteal vein.* _____

17. *Azygos vein.* _____

18. *Internal iliac artery.* _____

19. *Internal carotid artery.* _____

20. *Cephalic vein.* _____

**ANSWER THE LABORATORY REPORT QUESTIONS
AT THE END OF THE EXERCISE.**

EXERCISE 18	Blood Vessels

Name _____ Date _____

Laboratory Section _____ Score/Grade _____

PART 1 ■ Multiple Choice

_____ 1. The largest of the circulatory routes is (a) systemic (b) pulmonary (c) coronary (d) hepatic portal

_____ 2. All arteries of systemic circulation branch from the (a) superior vena cava (b) aorta (c) pulmonary artery (d) coronary artery

_____ 3. The arterial system that supplies the brain with blood is the (a) hepatic portal system (b) pulmonary system (c) cerebral arterial circle (circle of Willis) (d) carotid system

_____ 4. An obstruction in the inferior vena cava would hamper the return of blood from the (a) head and neck (b) upper limbs (c) thorax (d) abdomen and pelvis

_____ 5. Which statement best describes arteries? (a) all carry oxygenated blood to the heart (b) all contain valves to prevent the backflow of blood (c) all carry blood away from the heart (d) only large arteries are lined with endothelium

_____ 6. Which statement is *not* true of veins? (a) they have less elastic tissue and smooth muscle than arteries (b) their tunica externa is the thickest coat (c) most veins in the limbs have valves (d) they always carry deoxygenated blood

_____ 7. A thrombus in the first branch of the arch of the aorta would affect the flow of blood to the (a) left side of the head and neck (b) myocardium of the heart (c) right side of the head and neck and right upper limb (d) left upper limb

_____ 8. If a vein must be punctured for an injection, transfusion, or removal of a blood sample, the likely site would be the (a) median cubital (b) subclavian (c) hemiazygos (d) anterior tibial

_____ 9. In hepatic portal circulation, blood is eventually returned to the inferior vena cava through the (a) superior mesenteric vein (b) hepatic portal vein (c) hepatic artery (d) hepatic veins

_____ 10. Which of the following are involved in pulmonary circulation? (a) superior vena cava, right atrium, and left ventricle (b) inferior vena cava, right atrium, and left ventricle (c) right ventricle, pulmonary artery, and left atrium (d) left ventricle, aorta, and inferior vena cava

_____ 11. If a thrombus in the left common iliac vein dislodged, into which arteriole system would it first find its way? (a) brain (b) kidneys (c) lungs (d) left arm

_____ 12. In fetal circulation, the blood containing the highest amount of oxygen is found in the (a) umbilical arteries (b) ductus venosus (c) aorta (d) umbilical vein

_____ 13. The greatest amount of elastic tissue found in the arteries is located in which coat? (a) tunica interna (b) tunica media (c) tunica externa (d) tunica adventitia

_____ **14.** Which coat of an artery contains endothelium? (a) tunica interna (b) tunica media (c) tunica externa (d) tunica adventitia

_____ **15.** Permitting the exchange of nutrients and gases between the blood and tissue cells is the primary function of (a) capillaries (b) arteries (c) veins (d) arterioles

_____ **16.** The circulatory route that runs from the gastrointestinal tract to the liver is called (a) coronary circulation (b) pulmonary circulation (c) hepatic portal circulation (d) cerebral circulation

_____ **17.** Which of the following statements about systemic circulation is *not* correct? (a) its purpose is to carry oxygen and nutrients to body tissues and to remove carbon dioxide (b) all systemic arteries branch from the aorta (c) it involves the flow of blood from the left ventricle to all parts of the body except the lungs (d) it involves the flow of blood from the body to the left atrium

_____ **18.** The opening in the septum between the right and left atria of a fetus is called the (a) foramen ovale (b) ductus venosus (c) foramen rotundum (d) foramen spinosum

_____ **19.** The branch of the umbilical vein in the fetus that connects with the inferior vena cava, bypassing the liver, is the (a) foramen ovale (b) ductus venosus (c) ductus arteriosus (d) patent ductus

_____ **20.** Which of the vessels does *not* belong with the others? (a) brachiocephalic artery (b) left common carotid artery (c) celiac artery (d) left subclavian artery

PART 2 ▪ Matching

_____ **21.** Aortic branch that supplies the head and associated structures

_____ **22.** Artery that distributes blood to the small intestine and part of the large intestine

_____ **23.** Vessel into which veins of the head and neck, upper limbs, and thorax enter

_____ **24.** Vessel into which veins of the abdomen, pelvis, and lower limbs enter

_____ **25.** Vein that drains the head and associated structures

_____ **26.** Longest vein in the body

_____ **27.** Vein just behind the knee

_____ **28.** Artery that supplies a major part of the large intestine and rectum

_____ **29.** First branch off of the arch of the aorta

_____ **30.** Arteries supplying the heart

A. Inferior vena cava

B. Superior vena cava

C. Superior mesenteric

D. Common carotid

E. Jugular

F. Brachiocephalic

G. Coronary

H. Inferior mesenteric

I. Popliteal

J. Great saphenous

PART 3 ▪ Completion

31. The arteries that branch from the arch of the aorta, in order of their origination, are the brachio-cephalic trunk, _____, and left subclavian artery.

32. Just distal to the elbow, the _____ artery divides into the radial and ulnar arteries.

33. Whereas the superficial palmar arch is formed mainly by the ulnar artery, the deep palmar arch is formed mainly by the _____ artery.

34. In the skull, the vertebral arteries unite to form the _____ artery.

35. The carotid sinus is associated with the _____ artery.

36. The parietal branches of the thoracic aorta are the posterior intercostal, subcostal, and _____ arteries.

37. The left gastric, splenic, and common hepatic arteries are branches of the _____.

38. The _____ artery supplies the stomach and greater omentum.

39. Blood is supplied to the pancreas, most of the small intestine, and part of the large intestine by the _____ artery.

40. The testicular and ovarian arteries are together known as the _____ arteries.

41. The _____ arteries supply the diaphragm with blood.

42. The terminal branches of the aorta are the _____ arteries.

43. At the ankle, the anterior tibial artery becomes the _____ artery.

44. The plantar metatarsal arteries are branches of the _____.

45. Blood is delivered to the right atrium by the superior vena cava, inferior vena cava, and _____.

46. Blood draining the head passes into the internal jugular, external jugular, and _____ veins.

47. The _____ sinuses drain blood from the cerebrum, cerebellum, and cranial bones.

48. The cephalic veins begin on the lateral aspect of the _____.

49. The medial aspects of the upper limbs are drained by the _____ veins.

50. The _____ veins drain the palms and forearm.

51. The union of the brachial and basilic veins forms the _____ veins.

52. The arms, neck, and thoracic wall are drained by the _____ veins.

53. The azygos system of veins consists of the azygos, hemiazygos, and _____ veins.

54. The inferior vena cava is formed by the union of the _____ veins.

55. The _____ vein is the longest vein in the body.

56. The lateral aspect of the dorsal venous arch is the point of origin for the _____ veins.

57. The anterior and posterior tibial veins unite to form the _____ veins.

Cardiovascular Physiology

Objectives

At the completion of this exercise you should understand

A The components of the conduction system of the heart

B The meaning of an electrocardiogram and the procedure for recording an electrocardiogram

C The definition and phases of a cardiac cycle

D The heart sounds and their significance

E The meaning of pulse and blood pressure and how to measure each

A. Cardiac Conduction System and Electrocardiogram (ECG or EKG)

The heart is innervated by the autonomic nervous system (ANS), which modifies, but does not initiate, the cardiac cycle (heartbeat). The heart can continue to contract if separate from the ANS. This is possible because the heart has an *intrinsic pacemaker* termed the **sinoatrial (SA) node.** The SA node is called an intrinsic pacemaker because of its ability to rhythmically, spontaneously depolarize, thereby reaching threshold value and initiating an action potential (nerve impulse). The spontaneous depolarization is a **pacemaker potential.** This ability to spontaneously depolarize is due to the SA node's permeability characteristics to sodium, potassium, and calcium ions. The SA node is connected to a series of specialized muscle cells that comprise the **(cardiac) conduction system.** This system distributes the action potentials that stimulate the cardiac muscle fibers to contract.

The SA node is located in the right atrial wall just inferior to the opening of the superior vena cava. Once the SA node spontaneously depolarizes and reaches threshold value, the action potential is conducted through the remainder of the heart via the cardiac conduction system. From the SA node the action potential propagates through both atria via gap junctions in the intercalated discs of the atrial fibers. The velocity for this is about 1 m/sec.

By conducting along atrial muscle fibers, the action potential reaches the **atrioventricular (AV) node,** which is located in the septum between the two atria, just anterior to the opening of the coronary sinus. As the action potential spreads to the atrial musculature, it initiates atrial contraction. When the action potential reaches the AV node, it too depolarizes, but in a manner that significantly slows the conduction velocity of the cardiac action potential. From the AV node, the action potential enters the **atrioventricular (AV) bundle (bundle of His).** This bundle is the only site where action potentials can conduct from the atria to the ventricles. After propagating along the AV bundle, the action potential enters both the **right** and **left bundle branches.** The bundle branches, in turn, divide into a complex network of large-diameter fibers called **Purkinje fibers** that emerge from the bundle branches and pass into the cells of the ventricular myocardium. As the action potential passes from the AV node into the AV bundle, the conduction velocity steadily and rapidly increases. Because of the large diameter of the Purkinje fibers, the conduction velocity of the action potential reaches its highest speed within these fibers, thereby significantly increasing the speed with which the action potential is spread throughout the ventricles. As the action potentials reach the ventricular myocardial fibers, they are depolarized, thereby initiating ventricular contraction.

FIGURE 19.1 Conduction system of heart.

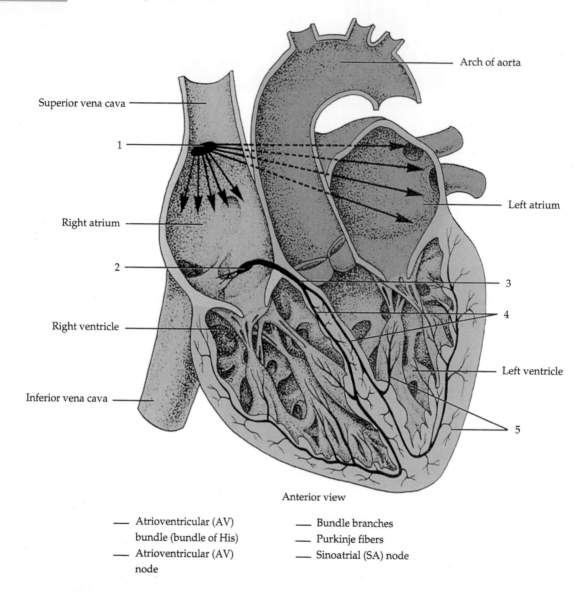

Anterior view

—— Atrioventricular (AV)
    bundle (bundle of His)
—— Atrioventricular (AV)
    node

—— Bundle branches
—— Purkinje fibers
—— Sinoatrial (SA) node

Label the components of the cardiac conduction system shown in Figure 19.1.

The electrocardiogram (ECG) provides a record of the electrical events happening within the heart. The ECG is obtained by placing recording electrodes on the surface of the body. These recording electrodes detect the changing potential differences of the heart on the body's surfaces, and a time-dependence plot of these differences is called an **electrocardiogram.** An **electrocardiograph** is an instrument used to record these changes. As the electrical impulses are transmitted throughout the cardiac conduction system and myocardial cells, a different electrical impulse is generated. These impulses are transmitted from the electrodes to a recording needle that graphs the impulses as a series of up-and-down (vertical) waves called **deflection waves** (Figure 19.2).

1. Electrocardiographic Recordings

Electrocardiograms are usually recorded by *indirect leads* (recording electrodes placed some distance from the heart on the skin of the subject rather than directly on the heart). The scalar electrocardiogram represents a *two-dimensional* tracing of the moment-to-moment *three-dimensional* electrical changes that occur within the heart during the electrical processes of the cardiac cycle. Because of the differences in the size and shape of any one individual's heart and body dimensions, the ECG varies from individual to individual. In addition, within any one individual the ECG pattern varies with the placement of the leads.

The ECG recordings that will be obtained in the laboratory utilize the *three standard limb leads*. A **lead** is defined as two electrodes working in pairs. **Electrodes** are sensing devices of sensitive metal plates or small rods that can detect electrophysiological phenomena such as changes in electrical potential of skin or nerves (see Exercise 9). For convenience, the three standard limb leads are typically attached to the right and left wrists and the left leg. Lead I records the potential difference between the left and right arms (LA and RA, respectively). Lead II records the potential difference between the right arm (RA) and left leg (LL). Lead III, therefore, would record the potential difference between the left arm (LA) and the left leg (LL).

A typical lead II ECG record (Figure 19.2) is composed of an orderly series of deflections and intervals. The first deflection, or wave, is a small upward deflection termed the **P wave,** and represents *atrial depolarization.* After the P wave the ECG returns to baseline level because the electrodes are no longer recording any potential differences. This time period is termed the **P-Q interval.** During this time period, however, the electrical activity of the heart is not quiescent, as the action potential is being propagated through the atria, AV node, and remaining fibers of the conduction system.

The **QRS complex** of the ECG begins as a downward deflection, continues upward as a large, upright, triangular wave, and ends as a downward wave. It represents rapid *ventricular depolarization.* The QRS complex also represents *atrial repolarization,* since the atria repolarize at the same time the ventricles depolarize. No separate deflection for atrial repolarization is found because of the significantly smaller amount of atrial muscle mass as compared to the ventricular muscle mass.

In the time period between the QRS complex and the repolarization of the ventricles (during the T wave described below), the ECG again returns to, or nearly returns to, its baseline level. This time period is termed the **S-T segment.** During the S-T segment of the ECG all of the ventricular muscle is depolarized. This segment of the ECG represents the plateau phase of the ventricular action potential.

The third wave is a dome-shaped upward deflection termed the **T wave,** and represents *ventricular repolarization.* The T wave is smaller and wider than the QRS complex because the rate of repolarization is slower than the rate of depolarization within the ventricles.

FIGURE 19.2   Recordings of a normal electrocardiogram (ECG), lead II.

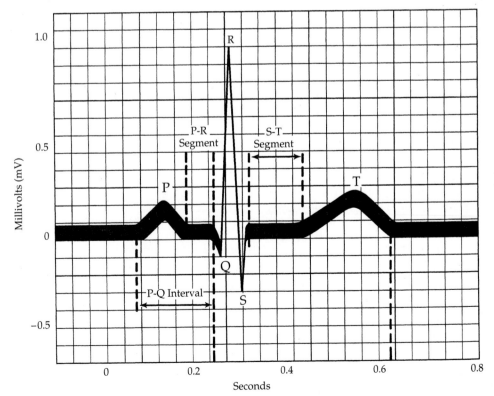

In reading and interpreting an ECG, you must note the *lead* type of configuration from which the ECG was recorded, the size of the deflection waves, and the lengths of the P-Q interval and S-T segment. An attempt will not be made to interpret an ECG, but only to become familiar with a normal ECG. The ECG is valuable in diagnosing abnormal cardiac rhythms and conducting patterns, detecting the presence of fetal life, determining the presence of more than one fetus, and following the course of recovery from a heart attack.

2. Recording an ECG

NOTE *Depending on the type of laboratory equipment available, select from the following procedures related to electrocardiography.*

PROCEDURE USING BIOPAC

Biopac Student Lab (BSL) system is an interactive hardware and software system that permits students to generate and record data from their bodies and animal or tissue preparations. It permits students to perform experiments, analyze data, draw conclusions, and form hypotheses on the basis of the collected data. The BSL system is available from BIOPAC Systems, Inc., 42 Aero Camino, Goleta, CA 93117. To order call: 1-805-685-0066; email: info@biopac.com; web page: www.biopac.com.

For activities related to electrocardiography, select the appropriate experiments from Biopac.

BSL5—Components of the ECG (lead II)

BSL7—ECG & pulse

H01—12-lead ECG

PROCEDURE USING CARDIOCOMP™

CARDIOCOMP™ is a computer hardware and software system designed to acquire data related to electrocardiography by permitting students to analyze their cardiac biopotentials. The program is available from INTELITOOL® (1-800-955-7621).

PROCEDURE USING A PHYSIOGRAPH-TYPE POLYGRAPH, ELECTROCARDIOGRAPH, OR OSCILLOSCOPE

In the following procedure a physiograph-type polygraph is used, but an electrocardiograph or oscilloscope may also be used (Figure 19.3a).

In recording the ECG, we will utilize only the standard three limb leads. Usually only two electrodes will be in actual use at any one time. The third electrode will be automatically switched off by the **lead-selector switch** of the polygraph.

For most general purposes three leads are usually used. However, electrocardiologists use additional leads, including several chest wall electrodes (Figure 19.3b). These leads are positioned around the chest, and encircle the heart so that there are six intersecting lines on a horizontal plane through the atrioventricular node.

The procedure for recording electrocardiograms using the physiograph is as follows:

PROCEDURE

1. Either you or your laboratory partner should lie on a table or cot, roll down long stockings or socks to the ankles, and *remove all metal jewelry.*

2. Swab the skin with alcohol where electrodes will be applied. Electrode cream, jelly, or saline paste is applied to the skin only where the electrode will make contact, such as to the inner forearm just above the wrists and to the inside of the legs just above the ankles.

3. The cream is then applied on the concave surface of the electrode plates, which are then fastened securely to the limb area surfaces with rubber straps or adhesive tape (rings). Adjust the straps to fit snugly, but *not so tight as to impede blood flow.*

4. Firmly connect the proper end of the electrode cables to the electrode plates, using the standard limb leads previously described, and connect the other end to a cardiac preamplifier or to a self-contained electrocardiograph.

5. After you apply the electrodes and connect the instrument to the subject, ascertain the following before actually recording:

   a. Recording power switch is on.

   b. Paper is sufficient for the entire recording, and the pen is centered properly on the paper with ink flowing freely.

   c. Preamplifier sensitivity has been set at 10 mm per millivolt.

   d. Subject is lying quietly and is relaxed.

NOTE *Usually five or six ECG complex recordings from each lead are sufficient. All electrocardiographers have chosen a standard paper speed of 25 mm per second.*

6. Turn the instrument on and record for 30 sec from each lead. Obtain directions from your laboratory instructor on the procedure to be followed when switching from lead to lead. When the recording leads are switched, note the change by marking it on the electrocardiogram. The subject must remain very still and relax completely during the recording period.

FIGURE 19.3   **Electrocardiograph.**

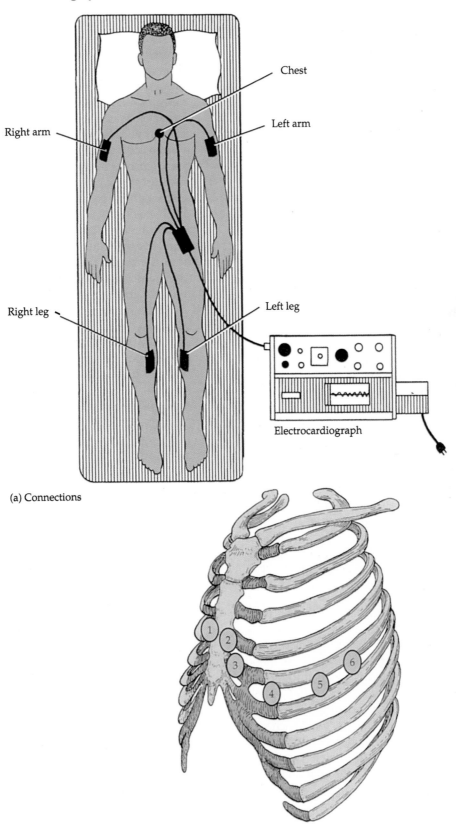

Right arm

Chest

Left arm

Right leg

Left leg

Electrocardiograph

(a) Connections

(b) Normal chest wall leads

7. When the recording is finished, turn off the reading instrument and disconnect the leads from the subject and remove the electrodes, cleaning off the excess cream or paste. The skin can be washed with water to remove electrode cream.

8. Identify and letter the P, QRS, and T waves, and compare the waves with those shown in Figure 19.2.

9. Calculate the duration of the waves and the P-Q interval. Attach this recording to Section A of the LABORATORY REPORT RESULTS at the end of the exercise.

B. Cardiac Cycle

In a normal **cardiac cycle** (heartbeat), the atria and ventricles alternately contract and relax. The term **systole** (SIS-tō-lē; *systole* = contraction) refers to the phase of contraction; the phase of relaxation is **diastole** (dī-AS-tō-lē; *diastole* = expansion). A cardiac cycle consists of a systole and diastole of the atria plus a systole and diastole of the ventricles.

As we discuss the cardiac cycle, for clarity its events will be correlated with the ECG. The electrical events recorded in the ECG always precede the mechanical events of the cardiac cycle. Therefore the P wave, which represents atrial depolarization, precedes atrial systole. Similarly, the QRS complex, which represents ventricular depolarization and atrial repolarization, precedes ventricular systole and atrial diastole. And the T wave, which represents ventricular repolarization, precedes ventricular diastole.

For purposes of our discussion, we will divide the cardiac cycle into three main phases (Figure 19.4).

1. *Atrial systole.* Deplorization of the SA node leads to atrial depolarization, which causes atrial systole (P wave in the ECG). The contraction of the atria exerts pressure on the blood within the chamber forces blood through the AV valves into the ventricles. Atrial systole contributes 25 mL of blood to the 105 mL of blood already in each ventricle. Thus, each ventricle contains approximately 130 mL at the end of its relaxation period (diastole). This blood volume is the **end-diastolic volume (EDV).**

2. *Ventricular systole.* Ventricular depolarization causes ventricular systole (QRS complex in the ECG). With ventricular systole comes a rise in the pressure within the ventricles which forces the blood up against the AV valves, forcing them shut. For about 0.05 seconds, both the SL (semilunar) and AV valves are closed. This period of the cardiac cycle, when all valves are closed and pressure continues to climb is termed the **period of isovolumetric contraction.** During this period, cardiac muscle fibers are contracting and exerting force but are not yet shortening. Thus, the muscle contraction is isometric (same length). In addition, the ventricular volume remains the same (isovolumic) due to all four valves being closed. When the ventricular pressure exceeds the aortic and pulmonary trunk pressure, both semilunar valves open. At this point, ejection of blood from the heart begins. The period when the semilunar valves are open is **ventricular ejection.** The left ventricle ejects approximately 70 mL of blood into the aorta and the right ventricle ejects the same amount of blood into the pulmonary trunk. The volume of blood remaining in each ventricle at the end of systole, about 60 mL, is the **end-systolic volume (ESV).** Stroke volume, the volume of blood ejected per beat from each ventricle, equals the end-diastolic volume minus end-systolic volume: SV = EDV − ESV. At rest the stroke volume is about 130 mL − 60 mL = 70 mL.

3. *Relaxation period.* The T wave in the ECG marks the onset of ventricular repolarization. During the relaxation period, the atria and the ventricles are both relaxed. As the heart beats faster, the relaxation period becomes shorter and shorter, whereas the durations of the atrial systole and ventricular systole shorten only slightly. Ventricular repolarization causes ventricular diastole. As the ventricles relax, pressure within the chambers falls, and blood in the aorta and pulmonary trunk begins to flow back toward the ventricles. As this blood gets caught in the cusp of the aortic valve, the valve closes. Rebound of blood off the closed cusps produce the **dicrotic wave** on the aortic pressure curve. After the semilunar valves close, there is a brief interval when ventricular blood volume does not change because all four valves are closed. This period is called **isovolumetric relaxation.** The pressure falls quickly, as the ventricles relax. When the pressure in the ventricles drop below the atrial pressure, the AV valves open and **ventricular filling** begins. The major part of ventricular filling occurs just after the AV valves open. Blood that has been flowing into and building up in the atria during ventricular systole then rushes rapidly into the ventricles. At the end of relaxation period, the ventricles

FIGURE 19.4 Cardiac cycle. (a) ECG related to the cardiac cycle; (b) left atrial, left ventricular, and aortic pressure changes along with the opening and closing of the valves during cardiac cycle; (c) left ventricular volume during the cardiac cycle; (d) heart sounds related to the cardiac cycle; (e) phases of the cardiac cycle.

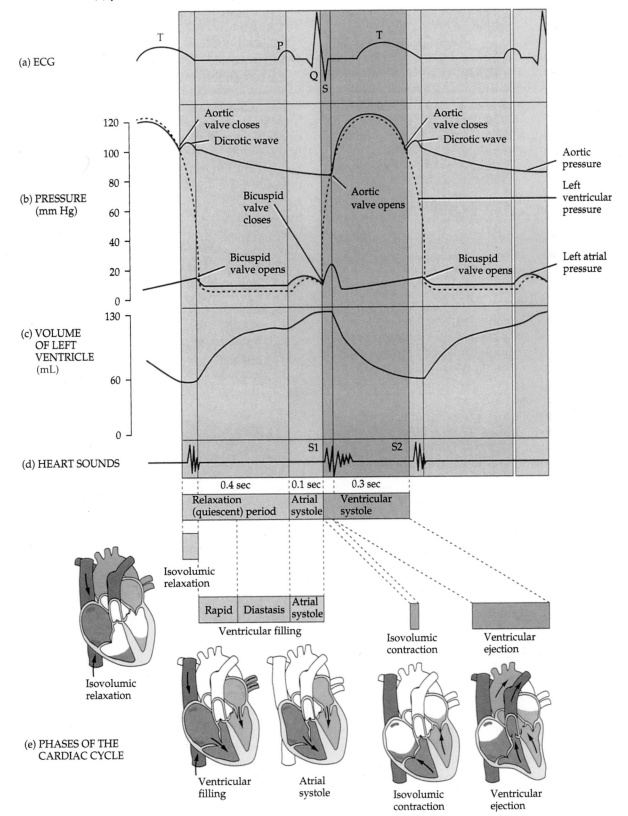

are about three-quarters full. The P wave appears in the ECG, signaling the start of another cardiac cycle.

Since resting heart rate is approximately 75 beats per minute, each cardiac cycle lasts about 0.8 sec. The duration of the various phases of the cardiac cycle, as represented in Figure 19.4, are as follows: the total time for ventricular diastole = 0.4 sec, with isovolumic contraction accounting for approximately 40 milliseconds (msec) and atrial contraction accounting for 0.1 sec of that time period; total time for ventricular systole = 0.3 sec, with rapid ventricular ejection accounting for between 50 and 100 msec of that period.

C. Cardiac Cycle Experiments

NOTE *Depending on the availability of live animals and the type of laboratory equipment, you may select from the following procedures related to heart experiments.*

PROCEDURE USING BIOPAC

Biopac Student Lab (BSL) system is an interactive hardware and software system that permits students to generate and record data from their bodies and animal or tissue preparations. It permits students to perform experiments, analyze data, draw conclusions, and form hypotheses on the basis of the collected data. The BSL system is available from BIOPAC Systems, Inc., 42 Aero Camino, Goleta, CA 93117. To order call: 1-805-685-0066; email: info@biopac.com; web page: www.biopac.com.

For activities related to the cardiac cycle, select the appropriate experiments from Biopac.

H04—Blood Pressure (Isometric or Straining Exercise)

H32—Heart Rate Variability

A04—Frog Heart

D. Heart Sounds

The beating of a human heart usually produces four sounds, but only two may be detected with a **stethoscope.** (The remaining two sounds may be heard if adequately amplified electronically.) The first sound is created by blood turbulence caused by the closure of the atrioventricular valves soon after ventricular

systole begins (see Figure 19.4d). This sound is the louder and longer of the two sounds, and may be best heard over the apex of the heart. The sound produced by the tricuspid valve is best heard in the fifth intercostal space just lateral to the left border of the sternum, while the bicuspid valve sound is best heard in the fifth intercostal space at the apex of the heart (Figure 19.5).

The second heart sound is created by blood turbulence caused by closure of the semilunar valves at the beginning of ventricular diastole. This sound is of shorter duration and lower intensity, and has a more snapping sound as compared to the quality of the first heart sound. The second heart sound produced by the aortic valve closing is best heard in the second intercostal space to the right of the sternum. The second sound produced by the pulmonary valve is best heart in the second intercostal space just to the left of the sternum (Figure 19.5).

Heart sounds provide valuable information about the valves. Abnormal sounds made by the valves are termed **murmurs.** Some murmurs are caused by the noise made by blood flowing back into a chamber of the heart because of improper closure (incompetence) of a valve. Another murmur may be produced by the improper opening (incompetence) of an AV or semilunar valve, which restricts the flow of blood out of a cardiac chamber.

Murmurs do not always indicate that the valves are not functioning properly. Many individuals possess a **functional murmur** that has no clinical significance at all.

Listening to sounds within the body is called **auscultation** (aws-kul-TĀ-shun).

1. Use of Stethoscope

PROCEDURE

1. The stethoscope should be used in a quiet room.

2. The earpieces of the stethoscope *should be cleaned with alcohol* just before using and should also be pointed slightly forward when placed in the ears. They will be more comfortable in this position, and it will be easier to hear through them.

3. Listen to the heart sounds of your laboratory partner by placing the diaphragm of the stethoscope next to the skin at several positions on the chest wall illustrated in Figure 19.5.

4. The first sound is best heard at the apex of the heart.

5. The second sound is best heard in the second intercostal space.

▲ **CAUTION!** *Assuming that your partner has no known or apparent cardiac or other health problems and is capable of such an activity, ask her/him to run in place about 25 steps. Listen to the heart sounds again.*

6. Answer all questions pertaining to Section D of the LABORATORY REPORT RESULTS at the end of the exercise.

E. Pulse Rate

A wave of pressure, called **pulse,** is produced in the arteries due to ventricular contraction. Pulse rate and heart rate are essentially the same. Pulse can be felt readily where an artery is near the surface of the skin and over the surface of a bone. Pulse rates vary considerably in individuals because of time of day, temperature, emotions, stress, and other factors. The normal adult pulse rate of the heart at rest is about 75 beats per minute. With practice you can learn to take accurate pulse rates by counting the beats per 15 sec and multiplying by 4 to obtain beats per minute. The term **tachycardia** (tak′-e-KAR-dē-a) is applied to a rapid heart rate or pulse rate (over 100 beats/minute). **Bradycardia** (brād-e-KAR-dē-a) indicates a slow heart rate or pulse rate (below 50 beats/minute).

Although the pulse may be detected in most surface arteries, the pulse rate is usually determined on the **radial artery** of the wrist.

1. Radial Pulse

PROCEDURE

1. Using your index and middle finger palpate your laboratory partner's radial artery.

2. The thumb should never be used because it has its own prominent pulse.

3. Palpate the area behind your partner's thumb just inside the bony prominence on the lateral aspect of the wrist.

▲ **CAUTION!** *Do not apply too much pressure.*

FIGURE 19.5 Surface areas where heart valve sounds are best heard.

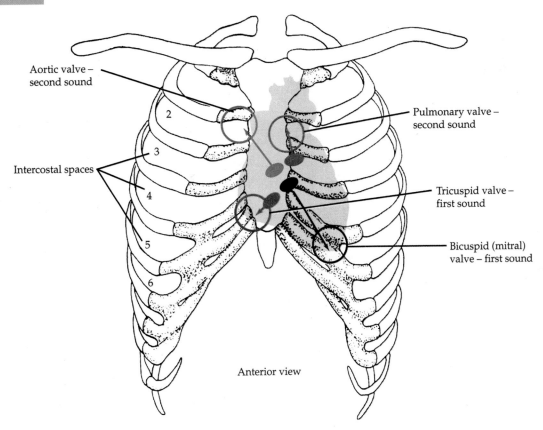

Aortic valve – second sound

Pulmonary valve – second sound

Intercostal spaces

Tricuspid valve – first sound

Bicuspid (mitral) valve – first sound

Anterior view

4. Count the pulse, change positions, and record your results in Section E of the LABORATORY REPORT RESULTS at the end of the exercise.

5. Calculate the **median pulse rate** of the entire class.

2. Carotid Pulse

PROCEDURE

1. Using the same fingers that you used for the radial pulse, place them on either side of your partner's larynx (voice box).

2. *Gently* press downward and toward the back until you feel the pulse. You must feel the pulse clearly with at least two fingers, so adjust your hand accordingly.

3. The radial and carotid pulse can be compared under the following conditions: (a) sitting quietly, (b) standing quietly, (c) right after walking 60 steps, (d) right after running in place 60 steps, *assuming that your partner has no known or apparent cardiac or other health problems and is capable of such an activity.* Notice how long it takes the pulse to return to normal after the walking and running exercises.

4. Record your results in Section E of the LABORATORY REPORT RESULTS at the end of the exercise.

5. Compare your radial and carotid pulse in the table provided.

F. Blood Pressure (Auscultation Method)

In physiological terms, the term **blood pressure** actually refers to the interaction of several different pressures (the pressure only within the arteries, or arterial pressure; the pressure only within the veins, or venous pressure; the pressure within the pulmonary system, or pulmonary pressure; and the pressure within all of the other vascular beds, termed systemic pressure). Clinically, however, the term **blood pressure** only refers to the pressure within the large arteries.

Arterial pressure can be measured either directly, by the insertion of a needle or catheter directly into the artery in such a way that the needle is pointing "upstream," or against the flow of blood within the vessel, or indirectly, by an instrument called a **sphygmomanometer** (sfig'-mō-

ma-NOM-e-ter; *sphygmo* = pulse). Regardless of which method is utilized, blood pressure is recorded in millimeters of mercury (mm Hg) and is normally taken in the brachial artery.

A commonly used sphygmomanometer consists of an inflatable rubber cuff attached by a rubber tube to a compressible hand pump or bulb. Another tube attaches to a cuff and to a mercury column marked off in millimeters or an anaeroid gauge that measures the pressure in mm Hg. The highest pressure in the artery, occurring during ventricular systole, is termed **systolic blood pressure.** The lowest pressure, occurring during ventricular diastole, is termed **diastolic blood pressure.** Blood pressures are usually expressed as a ratio of systolic to diastolic pressure. The normal blood pressure of an adult male is about 120 mm Hg systolic and 80 mm Hg diastolic, abbreviated to 120/80. The difference between systolic and diastolic pressure is called **pulse pressure.** Pulse pressure may be used clinically to indicate several physiological and pathological parameters, and usually averages 40 mm Hg in a healthy individual. The ratio of systolic pressure to diastolic pressure to pulse pressure may also be utilized clinically as a diagnostic tool, and is usually 3:2:1.

This exercise employs the indirect **auscultatory method,** in which the sounds of blood flow are heard with a stethoscope. Blood flow in an artery is impeded by increasing pressure within a sphygmomanometer. When the cuff of the sphygmomanometer applies sufficient pressure to completely occlude blood flow, no sounds can be heard distal to the cuff because no blood can flow through the artery. When cuff pressure drops below the maximal (systolic) pressure in the artery, blood is heard passing through the vessel. When cuff pressure drops below the lowest (diastolic) pressure in the vessel, the sound becomes muffled and usually disappears. The sounds heard through the stethoscope via this procedure are termed **Korotkoff** (kō-ROT-kof) **sounds.**

Indirect blood pressure can be taken in any artery that can be occluded easily. The brachial artery has the advantage of being at approximately the same level as the heart, so brachial pressure closely reflects aortic pressure.

The procedure for determining blood pressure using the sphygmomanometer is as follows:

PROCEDURE

1. Either you or your laboratory partner should be comfortably seated, at ease, with your arm bared, slightly flexed, abducted, and perfectly

relaxed. You may, for convenience, rest the forearm on a table in the supinated position.

2. Wrap the deflated cuff of the sphygmomanometer around the arm with the lower edge about 2.54 cm (1 in.) above the antecubital space. Close the valve on the neck of the rubber bulb.

3. Clean the earpieces of the stethoscope with alcohol before using it. Using the diaphragm of the stethoscope, find the pulse in the brachial artery just above the bend of the elbow, on the inner margin of the biceps brachii muscle.

4. Inflate the cuff by squeezing the bulb until the air pressure within it just exceeds 170 mm Hg. At this point the wall of the brachial artery is compressed tightly, and no blood should be able to flow through.

5. Place the diaphragm of the stethoscope firmly over the brachial artery and, while watching the pressure gauge, slowly turn the valve, releasing air from the cuff. Listen carefully for Korotkoff sounds as you watch the pressure fall. The first loud, rapping sound you hear will be the **systolic pressure.**

6. Continue listening as the pressure falls. The pressure recorded on the mercury column when the sounds become faint or disappear is the **diastolic pressure** reading. It measures the force of blood in the arteries during ventricular relaxation and specifically reflects the peripheral resistance of the arteries.

7. Repeat this procedure for both readings two or three times to see if you get consistent results. *Allow a few minutes between readings.* Record all results in Section F of the LABORATORY REPORT RESULTS at the end of the exercise.

8. Have a partner stand and record his or her blood pressure several times for each arm. Record all results in the table provided in Section F of the LABORATORY REPORT RESULTS at the end of the exercise.

9. *Now, assuming that your partner has no known or apparent cardiac or other health problems, and is capable of such an activity,* have your partner do some exercise, such as running in place 40 to 50 steps, and measure the blood pressure again *immediately after the completion of the exercise.* Record the pulse pressure in the table provided in Section F of the LABORATORY REPORT RESULTS at the end of the exercise.

ANSWER THE LABORATORY REPORT QUESTIONS AT THE END OF THE EXERCISE.

Cardiovascular Physiology

Name _____   Date _____

Laboratory Section _____   Score/Grade _____

SECTION A ■ Cardiac Conduction System and Electrocardiogram (ECG or EKG)

Attach examples of the electrocardiogram strips you obtained.

Lead I
_____

Electrocardiogram Strip

Lead II
_____

Electrocardiogram Strip

Lead III
_____

Electrocardiogram Strip

SECTION D ▪ **Heart Sounds**

1. Which heart sound is the loudest? _____

2. Did you hear a third sound? _____

3. Where does the first sound originate? _____

4. Where does the second sound originate? _____

5. After you exercised, how did the heart sounds differ from before? _____

6. Did they differ in rate and intensity? _____

7. Did the first or second increase in loudness? _____

SECTION E ▪ **Pulse Rate**

1. Radial pulse rate count results: _____ pulses per minute.

2. Have all radial pulse rates put on the blackboard, arranging them from the highest to the lowest.
    The median pulse rate is found exactly halfway down from the top.

    What is the median radial pulse rate of the class? _____

    What was the highest rate? _____

    What was the lowest rate? _____

3. Compare your radial and carotid pulse rates by filling in the following table.

	Radial pulse rate	Carotid pulse rate
Sitting quietly		
Standing quietly		
After walking		
After running in place		

SECTION F ▪ Blood Pressure (Auscultation Method)

Record your systolic and diastolic blood pressures in the following table.

	Systolic pressure		Diastolic pressure		Pulse pressure
	Left arm	Right arm	Left arm	Right arm	
Sitting					
Standing					
After running					

EXERCISE 19 | Cardiovascular Physiology

Name _____   Date _____

Laboratory Section _____   Score/Grade _____

PART 1 ■ Multiple Choice

_____ 1. When the semilunar valves are open during a cardiac cycle, which of the following occur?
I—atrioventricular valves are closed;
II—ventricles are in systole;
III—ventricles are in diastole;
IV—blood enters the aorta;
V—blood enters the pulmonary trunk;
VI—atrial contraction.
(a) I, II, IV, and V (b) I, II, and VI (c) II, IV, and V (d) I, III, IV, and VI

_____ 2. When ventricular pressure drops below atrial pressure, which phase of the cardiac cycle occurs? (a) atrial systole (b) ventricular systole (c) relaxation period

_____ 3. During which phase of the cardiac cycle are all four chambers in diastole? (a) atrial systole (b) relaxation period (c) ventricular systole

_____ 4. Which of the following statements is *not true?* (a) Pulse pressure is the difference between systolic and diastolic blood pressures. (b) Both systolic and diastolic blood pressures can be obtained via the pulse method. (c) Systolic blood pressure is obtained at the first loud, rapping sound you hear when you are measuring blood pressure via the ausculatory method. (d) Diastolic blood pressure is obtained when the sound heard through the stethoscope as you are measuring blood pressure via the ausculatory method becomes muffled and usually disappears.

_____ 5. The two distinct heart sounds, described phonetically as lubb and dupp, represent (a) contraction of the ventricles and relaxation of the atria (b) contraction of the atria and relaxation of the ventricles (c) blood turbulence associated with closing of the atrioventricular and semilunar valves (d) surging of blood into the pulmonary artery and aorta.

PART 2 ■ Completion

6. Systole and diastole of both atria pulse systole and diastole of both ventricles is called

    _____.

7. The period of the cardiac cycle when all valves are closed and pressure continues to climb within the left ventricle is called the period of _____.

8. Heart sounds provide valuable information about the _____.

9. Abnormal or peculiar heart sounds are called _____.

10. Although the pulse may be detected in most surface arteries, pulse rate is usually determined on

    the _____.

11. The heart has an intrinsic regulating system called the cardiac _____ system.

12. Electrical impulses accompanying the cardiac cycle are recorded by the _____.

13. The typical ECG produces three clearly recognizable waves. The first wave, which indicates depo-

    larization of the atria, is called the _____.

14. Various up-and-down impulses produced by an ECG are called _____.

15. The instrument normally used to measure blood pressure is called a(n) _____.

16. The artery that is normally used to evaluate blood pressure is the _____.

17. Rapping or thumping sounds heard clinically when blood pressure is being taken are called

    _____ sounds.

18. The difference between systolic and diastolic pressure is called _____.

19. An average blood pressure value for an adult is _____.

20. An average pulse pressure is _____.

Lymphatic and Immune System

Objectives

At the completion of this exercise you should understand

A The functions of the lymphatic system

B The anatomical characteristics and functions of lymphatic vessels

C The anatomical characteristics and functions of the thymus gland, lymph nodes, spleen, and lymphatic nodules

D How lymph flows through the body

The **lymphatic** (lim-FAT-ik) **system** consists of a fluid called lymph flowing within lymphatic vessels, a number of structures and organs containing lymphatic tissue, and red bone marrow, where stem cells develop into various types of blood cells including lymphocytes (Figure 20.1). Interstitial fluid and lymph are basically the same. The major difference between the two is location. After fluid passes from interstitial spaces into lymphatic vessels, it is called **lymph** (LIMF = clear water). Lymphatic tissue is a specialized form of reticular connective tissue that contains large numbers of lymphocytes.

The lymphatic system has three primary functions:

1. *Draining excess interstitial fluid.* Lymphatic vessels drain excess interstitial fluid from tissue spaces and return it to the blood.

2. *Transporting dietary lipids.* Lymphatic vessels transport lipids and lipid-soluble vitamins (A, D, E, and K) absorbed by the gastrointestinal tract to the blood.

3. *Carry out immune responses.* Lymphatic tissue initiates highly specific responses targeted to particular microbes or abnormal cells. Lymphocytes, aided by macrophages, recognize foreign cells, microbes, and cancer cells and respond to them in two basic ways. In cell-mediated immune responses, T cells destroy the intruders by causing them to rupture or by

releasing cytotoxic (cell-killing) substances. In antibody-mediated immune responses, B cells differentiate into plasma cells that secrete antibodies, proteins that combine with and cause destruction of specific foreign substances. In carrying out immune responses, the lymphatic system concentrates foreign substances in certain lymphatic organs, circulates lymphocytes through the organs to make contact with the foreign substances, destroys the foreign substances, and eliminates them from the body.

A. Lymphatic Vessels

Lymphatic vessels begin as microscopic **lymph capillaries** in spaces between cells. They are found in most parts of the body; they are absent in avascular tissue, the central nervous system, portions of the spleen, and red bone marrow. They are slightly larger in diameter than blood capillaries and have a unique structure that permits interstitial fluid to flow into them but not out. Lymphatic capillaries also differ from blood capillaries in that they are closed at one end; blood capillaries have an arterial and a venous end. In addition, lymphatic capillaries are structurally adapted to ensure the return of proteins to the cardiovascular system when they leak out of blood capillaries.

Just as blood capillaries converge to form venules and veins, lymph capillaries unite to

FIGURE 20.1 Lymphatic system.

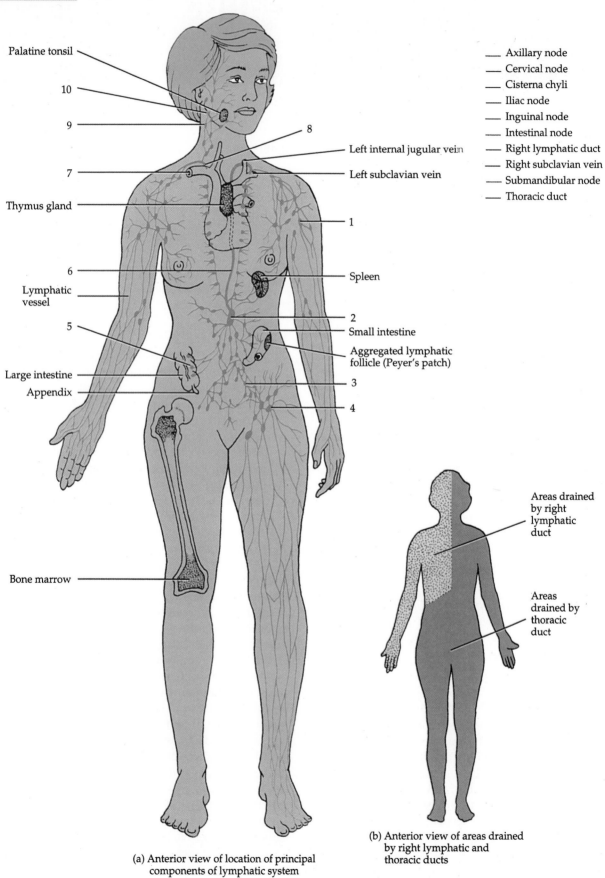

Palatine tonsil

10

9

Thymus gland

7

6

Lymphatic
vessel

5

Large intestine

Appendix

Bone marrow

8

Left internal jugular vein

Left subclavian vein

1

Spleen

2

Small intestine

Aggregated lymphatic
follicle (Peyer's patch)

3

4

___ Axillary node
___ Cervical node
___ Cisterna chyli
___ Iliac node
___ Inguinal node
___ Intestinal node
___ Right lymphatic duct
___ Right subclavian vein
___ Submandibular node
___ Thoracic duct

Areas drained
by right
lymphatic
duct

Areas
drained by
thoracic
duct

(a) Anterior view of location of principal
components of lymphatic system

(b) Anterior view of areas drained
by right lymphatic and
thoracic ducts

FIGURE 20.2 Relationship of lymphatic system to cardiovascular system.

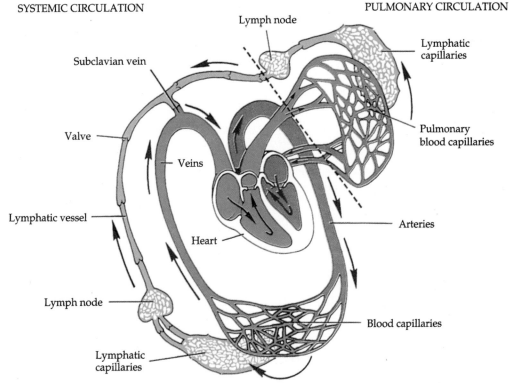

Arrows show direction of flow of lymph and blood

form larger and larger lymph vessels called **lymphatic vessels** (Figure 20.2). Lymphatic vessels resemble veins in structure, but have thinner walls and more valves, and contain lymph nodes at various intervals along their length (Figure 20.2). Ultimately, lymphatic vessels deliver lymph into two main channels—the thoracic duct and the right lymphatic duct. These will be described shortly.

Lymphangiography (lim-fan'-jē-OG-ra-fē) is the x-ray examination of lymphatic vessels and lymph organs after they are filled with a radiopaque substance. Such an x-ray is called a **lymphangiogram** (lim-FAN-jē-ō-gram). Lymphangiograms are useful in detecting edema and carcinomas, and in locating lymph nodes for surgical and radiotherapeutic treatment.

B. Lymphatic Tissues

The **primary lymphatic organs** of the body are the **red bone marrow** (in flat bones and the epiphyses of long bones) and the **thymus.** They are termed primary lymphatic organs because they provide the appropriate environment for stem cells to divide and mature into B and T cells, the lymphocytes that carry out immune responses. Pluripotent stem cells in red bone marrow give rise to immunocompetent B cells and pre-T cells. The pre-T cells then migrate and become immunocompetent in the thymus gland. The major **secondary lymphatic organs,** which are the sites where most immune responses occur, are **lymph nodes,** the **spleen,** and **lymphatic nodules.** Lymphatic nodules are egg-shaped concentrations of lymphatic tissues that are not surrounded by a capsule. They stand guard in all mucous membranes, where invaders might try to enter the body. Mucous membranes line the gastrointestinal tract, respiratory passageways, and urinary and reproductive tracts. Most immune responses occur in secondary lymphatic organs.

1. Thymus Gland

A bilobed lymphatic organ, the **thymus gland** is located in the mediastinum, between the sternum and the aorta (see Figure 20.1). An enveloping layer of connective tissue holds the two **thymic lobes** closely together, but a connective tissue **capsule** encloses each lobe separately. The capsule gives off extensions into the lobes called **trabeculae** (tra-BEK-ū-lē; *trabecula* = little beams), which divide the lobes into **lobules.** Each lobule consists of a deeply staining outer **cortex** and a lighter-staining central **medulla.** The cortex is composed of a large number of T cells and scattered dendritic cells, epithelial cells, and macrophages. Immature T cells (pre-T cells) migrate from red bone marrow to the thymus, where they proliferate and develop into mature T cells. The medulla consists of widely scattered, more mature T cells, epithelial cells, dendritic cells, and macrophages. Some of the epithelial cells cluster together in concentric layers to form **thymic (Hassall's) corpuscles,** whose role is uncertain.

The thymus gland is large in the infant, having a mass of about 70 g (about 2.3 oz.). After puberty, adipose and areolar connective tissue begin to replace the thymic tissue. By the time a person reaches maturity, the gland has atrophied considerably. Although most T cells arise before puberty, some continue to mature throughout life.

Obtain a prepared slide of thymic lobules and, with the aid of your textbook, label the capsule, trabecula, cortex, medulla, and thymic (Hassall's) corpuscle in Figure 20.3.

2. Lymph Nodes

The oval or bean-shaped structures located along the length of lymphatic vessels are called **lymph nodes.** They range from 1 to 25 mm (0.04 to 1 in.) in length. Lymph nodes are scattered throughout the body, usually in groups (see Figure 20.1). They are heavily concentrated in areas such as the mammary glands, axillae, and groin.

Each node is covered by a **capsule** of dense connective tissue that extends strands into the node. The capsular extensions, called **trabeculae,** divide the node into compartments, provide support, and convey blood vessels into the interior of a node. Internal to the capsule is a supporting network of reticular fibers and fibroblasts. The capsule, trabeculae, reticular fibers, and fibroblasts constitute the stroma (framework) of a lymph node. The **parenchyma** (functioning part) of a lymph node is divided into an outer cortex and an inner medulla. The outer **cortex** contains oval-shaped aggregates of B cells called **lymphatic nodules** (follicles). The follicles contain lighter-staining central areas, the **germinal centers,** where **B cells (B lymphocytes)** proliferate into antibody-secreting plasma cells. The inner region of the cortex, also called the **paracortex,** contains **T cells** and **dendritic cells,** which par-

FIGURE 20.3   **Histology of the thymus gland.**

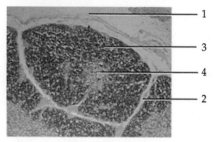

(a) Several thymus lobules (65×)

© Fred Hossler/Visuals Unlimited

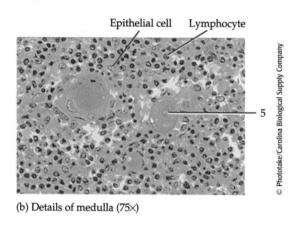

(b) Details of medulla (75×)

© Phototake/Carolina Biological Supply Company

____ Capsule
____ Cortex
____ Medulla
____ Thymic (Hassall's) corpuscle
____ Trabecula

ticipate in activation of T cells. The inner region of a lymph node is called the **medulla.** The medulla contains B cells and antibody-producing plasma cells, plus macrophages.

Lymph flows through a node in one direction. It enters through **afferent** (= to carry toward) **lymphatic vessels,** which penetrate the convex surface of the node at several points. They contain valves that open toward the node so that the lymph is directed *inward.* Inside the node, lymph enters the **sinuses,** which are a series of irregular channels that contain branching reticular fibers, lymphocytes, and macrophages. Lymph flows into the **subcapsular sinus,** then **trabecular sinuses,** and then in the **medullary sinuses** and exits the lymph node via one or two **efferent** (= to carry away) **lymphatic vessels.** Efferent lymphatic vessels are wider and fewer in number than the afferent vessels. They contain valves that open away from the node to convey lymph *out* of the node. Efferent lymphatic vessels emerge from one side of the lymph node at a slight depression called a **hilum** (HĪ-lum). Blood vessels also enter and leave the node at the hilus.

Among lymphatic tissues, only lymph nodes filter lymph by having it enter at one end and exit at another. The lymph nodes filter foreign substances from lymph as it passes back toward the bloodstream. These substances are trapped by the reticular fibers within the node. Then macrophages destroy some foreign substances by phagocytosis, and lymphocytes bring about destruction of others by immune responses. Plasma cells and T cells that have proliferated within a lymph node also can leave and circulate to other parts of the body.

Label the parts of a lymph node in Figure 20.4 and the various groups of lymph nodes in Figure 20.1.

3. Spleen

The oval **spleen** is the largest single mass of lymphatic tissue in the body, measuring about 12 cm

(5 in.) in length (see Figure 20.1). It is located in the left hypochondriac region between the stomach and diaphragm. The superior surface of the spleen is smooth and convex and conforms to the concave surface of the diaphragm. Neighboring organs make indentations in the spleen—the gastric impression (stomach), renal impression (left kidney), and colic impression (left flexure of colon). Like lymph nodes, the spleen has a **hilum**. The splenic artery and vein and the efferent lymphatic vessels pass through the hilum.

A **capsule** of dense connective tissue surrounds the spleen. The capsule, in turn, is covered by a serous membrane, the visceral peritoneum. **Trabeculae** extend inward from the capsule. The capsule plus trabeculae, reticular fibers, and fibroblasts constitute the stroma of the spleen.

The parenchyma of the spleen consists of two different kinds of tissue called white pulp and red pulp. **White pulp** is lymphatic tissue, consisting mostly of lymphocytes and macrophages, arranged around central arteries, branches of the splenic artery. The **red pulp** consists of blood-filled **venous**

FIGURE 20.4 **Structure of a lymph node.**

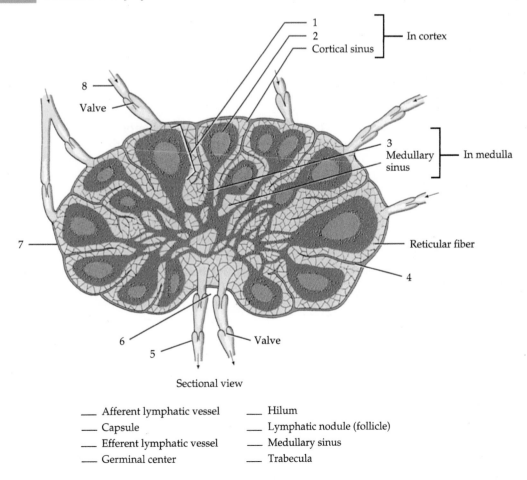

Sectional view

___ Afferent lymphatic vessel   ___ Hilum

___ Capsule   ___ Lymphatic nodule (follicle)

___ Efferent lymphatic vessel   ___ Medullary sinus

___ Germinal center   ___ Trabecula

sinuses and cords of splenic tissue called **splenic (Billroth's) cords.** Splenic cords consist of red blood cells, macrophages, lymphocytes, plasma cells, and granulocytes. Veins are closely associated with the red pulp.

Blood flowing into the spleen through the splenic artery enters the central arteries of the white pulp. Within the white pulp, B cells and T cells carry out immune functions, similar to lymph nodes, while spleen macrophages destroy blood-borne pathogens by phagocytosis. Within the red pulp, the spleen carries out several functions related to blood cells: (1) removal by macrophages of worn-out or defective red blood cells, white blood cells, and platelets; (2) storage of blood platelets, perhaps up to one-third of the body's supply; and (3) production of blood cells (hemopoiesis) during the second trimester of pregnancy (after birth, blood cell formation usually occurs only in the red bone marrow, but the spleen can also participate if necessary).

Obtain a prepared slide of the spleen and, with the aid of your textbook, label the capsule, trabecula, white pulp, central artery, and red pulp in Figure 20.5.

4. Lymphatic Nodules

Lymphatic nodules are egg-shaped masses of lymphatic tissue that are not surrounded by a capsule. They are scattered throughout the lamina propria (connective tissue) of mucous membranes lining the gastrointestinal tract, respiratory airways, and urinary and reproductive tracts. This lymphatic tissue is referred to as **mucosa-associated lymphoid tissue (MALT).**

Although many lymphatic nodules are small and solitary, some lymphatic nodules occur in multiple, large aggregations in specific parts of the body. Among these are the tonsils in the pharyngeal region and aggregated lymphatic follicles (Peyer's patches) in the ileum of the small intestine. Aggregations of lymphatic nodules also occur in the appendix (see Figure 20.1).

Usually there are five **tonsils,** which form a ring at the junction of the oral cavity and oropharynx and at the junction of the nasal cavity and nasopharynx (see Figure 21.2). Thus they are strategically positioned to participate in immune responses against foreign substances that are inhaled or ingested. The single **pharyngeal** (fa-RIN-jē-al) **tonsil** or **adenoid** is embedded in the posterior wall of the nasopharynx. The two **palatine** (PAL-a-tīn) **tonsils** lie at the posterior region of the oral cavity, one on each side. These are the

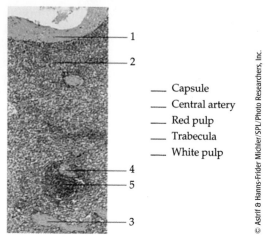

FIGURE 20.5 Histology of the spleen (approximately 23×).

— 1
— 2

___ Capsule
___ Central artery
___ Red pulp
___ Trabecula
___ White pulp

— 4
— 5
— 3

ones commonly removed by a tonsillectomy. The paired **lingual** (LIN-gwal) **tonsils** are located at the base of the tongue and may also have to be removed by a tonsillectomy.

C. Lymph Circulation

When plasma is filtered by blood capillaries, it passes into the interstitial spaces; it is then known as interstitial fluid. When this fluid passes from interstitial spaces into lymph capillaries, it is called **lymph** (*lympha* = clear water). Lymph from lymph capillaries flows into lymphatic vessels that run toward lymph nodes. At the nodes, afferent vessels penetrate the capsules at numerous points, and the lymph passes through the sinuses of the nodes. Efferent vessels from the nodes unite to form **lymph trunks.**

The principal trunks pass their lymph into two main channels, the thoracic duct and the right lymphatic duct. The **thoracic (left lymphatic) duct** begins as a dilation anterior to the second lumbar vertebra called the **cisterna chyli** (sis-TER-na KĪ-lē) (see Figure 20.1a). The thoracic duct is the main collecting duct of the lymphatic system and receives lymph from the left side of the head, neck, and chest, the left upper limb, and the entire body inferior to the ribs (see Figure 20.1b).

The **right lymphatic duct** drains lymph from the upper right side of the body (see Figure 20.1b). Ultimately, the thoracic duct empties all of its lymph into the junction of the left internal jugular vein and left subclavian vein, and the right lymphatic duct empties all of its lymph into the junction of the right internal jugular vein and right subclavian vein (see Figure 20.1a). Thus, lymph is drained back into the blood, and the cycle repeats itself continuously.

The flow of lymph from tissue spaces to the large lymphatic ducts to the subclavian veins is maintained primarily by the contraction of skeletal muscles (milking action). These contractions compress lymphatic vessels and force lymph toward the subclavian veins. One-way valves (similar to those found in veins) within the lymphatic vessels prevent backflow of lymph.

Another factor that maintains lymph flow is a respiratory pump. Breathing movements create a pressure gradient between the two ends of the lymphatic system. With each inhalation, lymph flows from the abdominal region, where the pressure is higher, toward the thoracic region, where

it is lower. In addition, when a lymphatic vessel distends, the smooth muscle in its wall contracts. This helps move lymph from one segment of a lymphatic vessel to another.

Edema, an excessive accumulation of interstitial fluid in tissue spaces, may be caused by an obstruction, such as an infected node or a blockage of vessels, in the pathway between the lymphatic capillaries and the subclavian veins. Another cause is excessive lymph formation and increased permeability of blood capillary walls. A rise in capillary blood pressure, in which interstitial fluid is formed faster than it is passed into lymphatic vessels, also may result in edema.

ANSWER THE LABORATORY REPORT QUESTIONS AT THE END OF THE EXERCISE.

Lymphatic and Immune System

Name _____   Date _____

Laboratory Section _____   Score/Grade _____

PART 1 ■ Multiple Choice

_____ 1. Put the following items in the correct order for the pathway of lymph from the stomach to blood. (1) thoracic duct (2) lymphatic vessels and lymph nodes (3) lymph capillaries (4) interstitial fluid (5) left subclavian vein (a) 2, 3, 4, 5, 1 (b) 3, 2, 4, 1, 5 (c) 4, 2, 1, 5, 3 (d) 4, 3, 2, 1, 5

_____ 2. Which of the following is *not* a major site of lymphatic tissue? (a) tonsils (b) thymus gland (c) kidneys (d) spleen

_____ 3. Which of the following is an important function of the lymphatic system? (a) controlling body temperature by evaporation of sweat (b) manufacturing all white blood cells (c) transporting fluids out to, and back from, body tissues (d) returning fluid and proteins to the cardiovascular system

_____ 4. The spleen (a) serves as a storage site for blood platelets (b) is an organ in which phagocytosis of aged red blood cells occurs (c) is a site of blood formation in the fetus (d) all of the above

_____ 5. Both the thoracic duct and the right lymphatic duct empty directly into the (a) axillary lymph nodes (b) superior vena cava (c) subclavian arteries (d) junction of the internal jugular and subclavian veins

_____ 6. Which of the following statements is true about the flow of lymph? (1) It is aided in its movement toward the thoracic region by a pressure gradient. (2) It is maintained primarily by the milking action of muscles. (3) It is possible because of valves in lymphatic vessels. (a) 1 only (b) 2 only (c) 3 only (d) all of the above

_____ 7. A major difference between the spleen and lymph nodes is that (a) lymph nodes have afferent and efferent lymphatic vessels; the spleen has only efferent lymphatic vessels (b) lymph nodes are not enclosed in a capsule, whereas the spleen is (c) the spleen is located in the abdomen; lymph nodes are not (d) lymph nodes have afferent and efferent lymphatic vessels; the spleen has neither

PART 2 ■ Completion

8. Small masses of lymphatic tissue located along the length of the lymphatic vessels are called

   _____.

9. Lymphatic vessels have thinner walls than veins, but resemble veins in that they also have

   _____.

10. All lymphatic vessels converge, get larger, and eventually merge into two main channels, the thoracic duct and the _____.

11. Lymph is conveyed out of lymph nodes in _____ vessels.

12. B cells in lymph nodes produce certain cells that are responsible for the production of antibodies. These cells are called _____.

13. The x-ray examination of lymphatic vessels and lymph organs after they are filled with a radiopaque substance is called _____.

14. This x-ray examination is useful in detecting edema and _____.

15. The largest mass of lymphatic tissue is the _____.

16. The main collecting duct of the lymphatic system is the _____ duct.

17. _____ are the areas within a lymph node that contain T and B cells.

18. The _____ tonsils are commonly removed by a tonsillectomy.

19. The red bone marrow and the _____ are the primary lymphatic organs because they produce the lymphocytes for the immune system.

20. The thoracic duct starts in the lumbar region as a dilation known as the _____.

21. Lymphatic nodules in the lamina propria of mucous membranes are collectively referred to as _____.

22. The _____ of the spleen consists of two types of tissue, red pulp and white pulp.

Respiratory System

Objectives

At the completion of this exercise you should understand

A The anatomical characteristics and functions of the organs of the respiratory system

B The mechanics of inspiration and expiration

C The physiological significance of respiratory volumes and capacities

C ells continually use oxygen (O_2) for the metabolic reactions that release energy from nutrient molecules and produce ATP. At the same time, these reactions release carbon dioxide (CO_2). Since an excessive amount of CO_2 produces acidity that is toxic to cells, the excess CO_2 must be eliminated quickly and efficiently. The two systems that cooperate to supply O_2 and eliminate CO_2 are the cardiovascular system and the respiratory system. The respiratory system provides for gas exchange, intake of O_2, and elimination of CO_2, whereas the cardiovascular system transports the gases in the blood between the lungs and the cells. Failure of either system has the same effect on the body: disruption of homeostasis and rapid death of cells from oxygen starvation and buildup of waste products. In addition to functioning in gas exchange, the respiratory system also participates in regulating blood pH, contains receptors for the sense of smell, filters inhaled air, produces sounds, and rids the body of some water and heat in exhaled air.

Respiration is the exchange of gases between the atmosphere, blood, and cells. It takes place in three basic steps:

1. *Pulmonary ventilation.* The first process, **pulmonary** (*pulmon* = lung) **ventilation,** or breathing, is the inhalation (inflow) and exhalation (outflow) of air between the atmosphere and the alveoli of the lungs.

2. *External respiration.* This is the exchange of gases between the alveoli of the lungs and blood in pulmonary capillaries. The blood gains O_2 and loses CO_2.

3. *Internal respiration.* The exchange of gases between blood in systemic capillaries and tissue cells is known as internal respiration. The blood loses O_2 and gains CO_2. Within cells, the metabolic reactions that consume O_2 and produce CO_2 during production of ATP are termed **cellular respiration.**

The **respiratory system** consists of the nose, pharynx (throat), larynx (voice box), trachea (windpipe), bronchi, and lungs (Figure 21.1). Structurally, the respiratory system consists of two portions. (1) The **upper respiratory system** includes the nose, pharynx, and associated structures. (2) The **lower respiratory system** includes the larynx, trachea, bronchi, and lungs. Functionally, the respiratory system also consists of two portions. (1) The **conducting portion** consists of a series of interconnecting cavities and tubes both outside and within the lungs—nose, pharynx, larynx, trachea, bronchi, and terminal bronchioles—that filter, warm, and moisten air and conduct air into the lungs. (2) The **respiratory portion** consists of tissues within the lung where the exchange of gases occurs—respiratory bronchioles, alveolar ducts, alveolar sacs, and alveoli, the main sites of gas exchange between air and blood.

Using your textbook, charts, or models for reference, label Figure 21.1.

FIGURE 21.1   **Organs of respiratory system.**

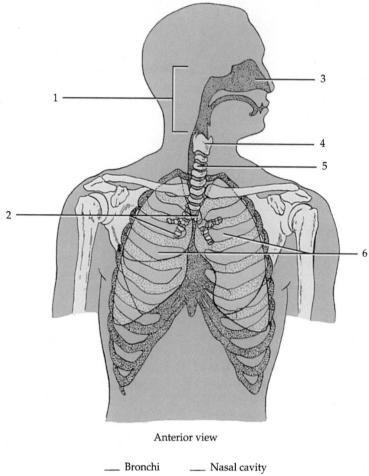

Anterior view

___ Bronchi   ___ Nasal cavity
___ Larynx   ___ Pharynx
___ Lungs   ___ Trachea

A. Organs of the Respiratory System

1. Nose

The **nose** has an external portion and an internal portion inside the skull. The external portion consists of a supporting framework of bone and hyaline cartilage covered with muscle and skin and lined by mucous membrane. The frontal bone, nasal bones, and maxillae form the bony framework of the external nose. Because it has a framework of pliable hyaline cartilage, the rest of the external nose is somewhat flexible. On the undersurface of the external nose are two openings called the **external nares** (NA-rēz; singular is **naris**), or **nostrils**. The interior structures of the nose are specialized for three functions: (1) warming, moistening, and filtering incoming air; (2) detecting olfactory stimuli; and (3) modifying speech vibrations as

they pass through the large, hollow resonating chambers.

The internal portion of the nose is a large cavity in the anterior aspect of the skull that lies inferior to the nasal bone and superior to the mouth. Anteriorly, the internal nose merges with the external nose, and posteriorly it communicates with the pharynx through two openings called the **internal nares (choanae).** Ducts from the paranasal sinuses (frontal, sphenoidal, maxillary, and ethmoidal) and the nasolacrimal ducts also open into the internal nose. The lateral walls of the internal nose are formed by the ethmoid, maxillae, lacrimal, palatine, and inferior nasal conchae bones. The ethmoid also forms the roof of the internal nose. The floor of the internal nose is formed by the palatine bones and palatine processes of the maxillae, which together constitute the hard palate.

The space inside of the internal nose is called the **nasal cavity.** It is divided into right and left

FIGURE 21.2 Upper respiratory system.

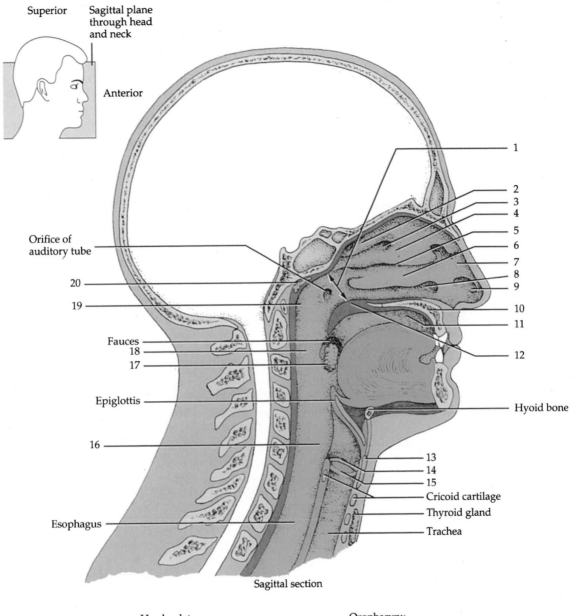

Superior · Sagittal plane through head and neck

Anterior

Orifice of auditory tube

20

19

Fauces
18
17

Epiglottis

16

Esophagus

1
2
3
4
5
6
7
8
9
10
11
12

Hyoid bone

13
14
15

Cricoid cartilage
Thyroid gland
Trachea

Sagittal section

___ Hard palate
___ Inferior meatus
___ Inferior nasal concha (turbinate)
___ Internal naris
___ Laryngopharynx
___ Middle meatus
___ Middle nasal concha (turbinate)
___ Nasal cavity
___ Nasopharynx
___ Oral cavity

___ Oropharynx
___ Palatine tonsil
___ Pharyngeal tonsil
___ Soft palate
___ Superior meatus
___ Superior nasal concha (turbinate)
___ Thyroid cartilage (Adam's apple)
___ Ventricular fold (false vocal cord)
___ Vestibule
___ Vocal fold (true vocal cord)

sides by a vertical partition called the **nasal septum.** The anterior portion of the septum consists primarily of hyaline cartilage. The remainder is formed by the vomer, perpendicular plate of the ethmoid, maxillae, and palatine bones (see Figure 21.2). The anterior portion of the nasal cavity, just inside the nostrils, is called the **vestibule** and is surrounded by cartilage. The superior nasal cavity is surrounded by bone.

When air enters the nostrils, it passes first through the vestibule. The vestibule is lined by skin containing coarse hairs that filter out large dust particles. Three shelves formed by projections of the superior, middle, and inferior nasal conchae extend out of each lateral wall of the cavity. The conchae, almost reaching the septum, subdivide each side of the nasal cavity into a series of groovelike passageways—the **superior, middle,** and **inferior meatuses** (mē-Ā-tus-ez; *meatus* = passage; singular is **meatus**). Mucous membrane lines the cavity and its shelves.

The olfactory receptors lie in the membrane lining the superior nasal conchae and adjacent septum. This region is called the **olfactory epithelium.** Inferior to the olfactory epithelium, the mucous membrane contains capillaries and pseudostratified ciliated columnar epithelium with many goblet cells. As inhaled air whirls around the conchae and meatuses, it is warmed by blood in the capillaries. Mucus secreted by the goblet cells moistens the air and traps dust particles. Drainage from the nasolacrimal ducts and perhaps secretions from the paranasal sinuses also help moisten the air. The cilia move the mucus and trapped dust particles toward the pharynx at which point they can be swallowed or spit out, thus removing particles from the respiratory tract. Substances in cigarette smoke inhibit movement of cilia. When this happens, only coughing can remove mucus–dust particles from the airways. This is one reason that smokers cough often.

Label the hard palate, inferior meatus, inferior nasal concha, internal naris, middle meatus, middle nasal concha, nasal cavity, oral cavity, soft palate, superior meatus, superior nasal concha, and vestibule in Figure 21.2.

The surface anatomy of the nose is shown in Figure 21.3.

2. Pharynx

The **pharynx** (FAIR-inks) (throat) is a funnel-shaped tube about 13 cm (5 in.) long that starts at the internal nares and extends to the level of the cricoid cartilage, the most inferior cartilage of the larynx (voice box). The pharynx lies just posterior to the nasal and oral cavities, superior to the larynx, and just anterior to the cervical vertebrae. The pharynx functions as a passageway for air and food, a resonating chamber for speech sounds, and a housing for tonsils.

The pharynx can be divided into three anatomical regions: a superior portion, called the **nasopharynx;** an intermediate portion, the **oropharynx;** and an inferior portion, the **laryngopharynx** (la-rin′-gō-FAIR-inks) or **hypopharynx.** The na-

---

FIGURE 21.3 Surface anatomy of nose.

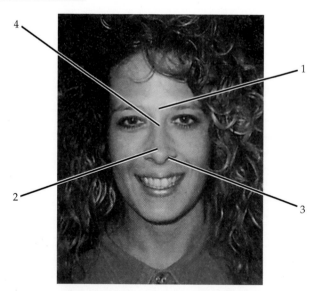

Anterior view

1. **Root.** Superior attachment of nose to the frontal bone.
2. **Apex.** Tip of nose.
3. **External nares.** External openings into nasal cavity (nostrils).
4. **Bridge.** Bony framework of nose formed by nasal bones.

sopharynx consists of **pseudostratified ciliated columnar epithelium** and has five openings in its wall: two **internal nares** plus two openings into the **auditory (pharyngotympanic) tubes** (commonly known as the eustachian tubes), and the opening to the oropharynx. The nasopharynx also contains the **pharyngeal tonsil.** The oropharynx is lined by **nonkeratinized stratified squamous epithelium** and receives one opening: the **fauces (FAW-sēz).** The oropharynx contains the **palatine** and **lingual tonsils.** The laryngopharynx is also lined by nonkeratinized stratified squamous epithelium and becomes continuous with the esophagus posteriorly and the larynx anteriorly.

Label the laryngopharynx, nasopharynx, oropharynx, palatine tonsil, and pharyngeal tonsil in Figure 21.2.

3. Larynx

The **larynx** (LAIR-inks), or voice box, is a short passageway connecting the laryngopharynx with the trachea. Its wall is composed of nine pieces of cartilage.

a. *Thyroid cartilage (Adam's apple).* Large anterior piece that gives the larynx its triangular shape.

b. *Epiglottis* (*epi* = over; *glottis* = tongue). Leaf-shaped piece of elastic cartilage on top of larynx. The "stem" of the epiglottis is attached to the anterior rim of the thyroid cartilage, but the "leaf" portion is unattached and free to move up and down like a trap door. During swallowing, the larynx rises. This causes the free edge of the epiglottis to move down and form a lid over the glottis, closing it off. The **glottis** consists of a pair of folds of mucous membrane, the vocal folds (true vocal cords) in the larynx, and the space between them called the **rima glottidis (RĪ-ma GLOT-ti-dis).** In this way, the larynx is closed off, and liquids and foods are routed into the esophagus and kept out of the larynx and airways inferior to it. When small particles such as dust, smoke, food, or liquids pass into the larynx, a cough reflex occurs to expel the material.

c. *Cricoid* (KRĪ-koyd = ringlike) *cartilage.* Ring of hyaline cartilage forming the inferior wall of the larynx. It is attached to the first ring of tracheal cartilage. The cricoid cartilage is the landmark for making an emergency airway, a procedure called a **tracheotomy** (trā-kē-O-tō-mē).

d. *Arytenoid* (ar-i-TĒ-noyd = ladlelike) *cartilages.* Paired, triangular pieces of hyaline cartilages at the posterior, superior border of

cricoid cartilage that attach vocal folds to the intrinsic pharyngeal muscles.

e. *Corniculate* (kor-NIK-ū-lāt = shaped like a small horn) *cartilages.* Paired, horn-shaped elastic cartilages at apex of each arytenoid cartilage.

f. *Cuneiform* (KŪ-nē-i-form = wedge shaped) *cartilages.* Paired, club-shaped elastic cartilages anterior to the corniculate cartilages that support the vocal folds and epiglottis.

With the aid of your textbook, label the laryngeal cartilages shown in Figure 21.4. Also label the thyroid cartilage in Figure 21.2.

The mucous membrane of the larynx forms two pairs of folds, a superior pair called the **ventricular folds (false vocal cords)** and an inferior pair called the **vocal folds (true vocal cords).** Movement of the vocal folds produces sounds; variations in pitch result from (1) varying degrees of tension and (2) varying lengths in males and females.

With the aid of your textbook, label the ventricular folds, vocal folds, and rima glottidis in Figure 21.5 on page 482. Also label the ventricular and vocal folds in Figures 21.2 and 21.4c.

4. Trachea

The **trachea** (TRĀ-kē-a) (windpipe) is a tubular passageway for air that is about 12 cm (5 in.) in length and 2.5 cm (1 in.) in diameter. It lies anterior to the esophagus and extends from the larynx to the superior border of the fifth thoracic vertebra, where it divides into right and left primary bronchi (Figure 21.6 on page 482). The epithelium of the trachea consists of **pseudostratified ciliated columnar epithelium.** This epithelium contains ciliated columnar cells, goblet cells, and basal cells. The epithelium offers the same protection against dust as the membrane lining the nasal cavity and larynx.

Obtain a prepared slide of pseudostratified ciliated columnar epithelium from the trachea and, with the aid of your textbook, label the ciliated columnar cell, cilia, goblet cell, and basal cell in Figure 21.7 on page 483.

The trachea consists of smooth muscle, elastic connective tissue, and incomplete **rings of cartilage** (hyaline) shaped like a series of letter Cs and stacked one on top of another. The open ends of the Cs are held together by the **trachealis muscle.** The cartilage provides a rigid support so that the tracheal wall does not collapse inward and obstruct the air passageway, and, because the open parts of the Cs face the esophagus, an arrangement that accommodates slight expansion of the

FIGURE 21.4 Larynx.

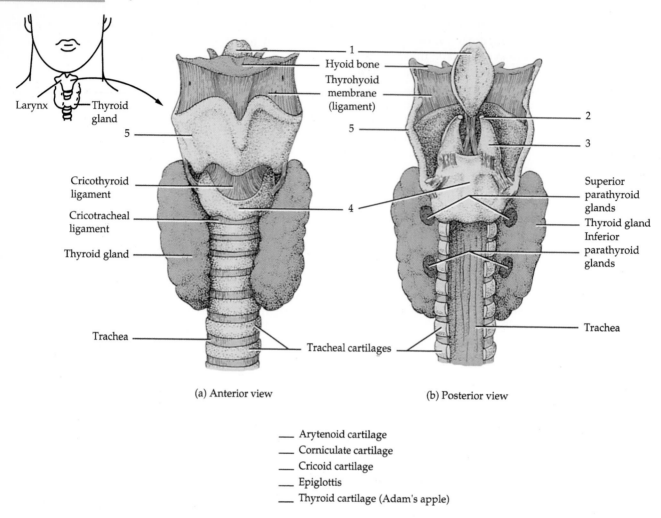

(a) Anterior view (b) Posterior view

____ Arytenoid cartilage
____ Corniculate cartilage
____ Cricoid cartilage
____ Epiglottis
____ Thyroid cartilage (Adam's apple)

esophagus into the trachea during swallowing. If the trachea should become obstructed, a tracheostomy may be performed. Another method of opening the air passageway is called **intubation,** in which a tube is passed into the mouth or nose and down through the larynx and the trachea.

5. Bronchi

The trachea terminates by dividing into a **right primary bronchus** (BRON-kus), which goes into the right lung, and a **left primary bronchus,** which goes into the left lung. They continue dividing in the lungs into smaller bronchi, the **secondary (lobar) bronchi** (BRON-kē), one for each lobe of the lung. These bronchi, in turn, continue dividing into still smaller bronchi, called **tertiary (segmental) bronchi,** that divide into **bronchioles.** The next division is into even smaller tubes called **terminal bronchioles.** This extensive branching of the trachea is commonly referred to as the **bronchial tree.**

Label Figure 21.6.

Bronchography (bron-KOG-ra-fē) is a technique for examining the bronchial tree. With this procedure, an intratracheal catheter is passed transorally or transnasally through the rima glottidis into the trachea. Then an opaque contrast medium is introduced into the trachea and distributed through the bronchial branches. Radiographs of the chest in various positions are taken and the developed film, called a **bronchogram** (BRON-kō-gram), provides a picture of the bronchial tree.

6. Lungs

The **lungs** (= lightweights, because they float) are paired, cone-shaped organs in the thoracic cavity (see Figure 21.1). The **pleural** (*pleur* = side) **membrane** encloses and protects each lung. Whereas the **superficial parietal pleura** lines the wall of the thoracic cavity, the **deep visceral pleura** covers the lungs; the small space between parietal

FIGURE 21.4 **Larynx. (Continued)**

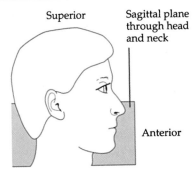

Superior

Sagittal plane through head and neck

Anterior

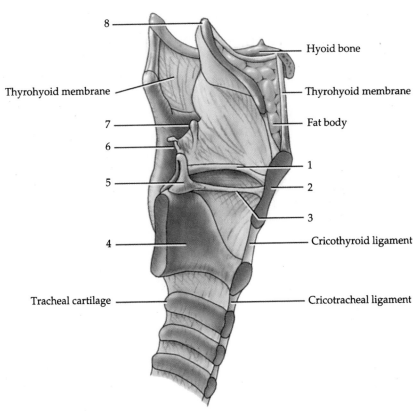

Hyoid bone

Thyrohyoid membrane

Thyrohyoid membrane

Fat body

8

7

6

5

4

1

2

3

Cricothyroid ligament

Tracheal cartilage

Cricotracheal ligament

(c) Sagittal section

___ Arytenoid cartilage
___ Corniculate cartilage
___ Cricoid cartilage
___ Cuneiform cartilage
___ Epiglottis
___ Thyroid cartilage
___ Ventricular fold (false vocal cord)
___ Vocal fold (true vocal fold)

and visceral pleurae, the **pleural cavity,** contains a lubricating fluid to reduce friction as the lungs expand and recoil.

Major surface features of the lungs include:

a. *Base.* Broad inferior portion that fits over the convex area of the diaphragm.

b. *Apex.* Narrow superior portion just superior to the clavicles.

c. *Costal surface.* Surface lying against ribs.

d. *Mediastinal (medial) surface.* Medial surface.

e. *Hilum.* Region in mediastinal surface through which bronchi, pulmonary blood vessels, lymphatic vessels, and nerves enter and exit the lung.

f. *Cardiac notch.* Medial concavity in left lung in which heart lies.

Each lung is divided into **lobes** by one or more **fissures.** The right lung has three lobes, **superior, middle,** and **inferior;** the left lung has two lobes, **superior** and **inferior.** The **horizontal**

FIGURE 21.5 Photograph of larynx.

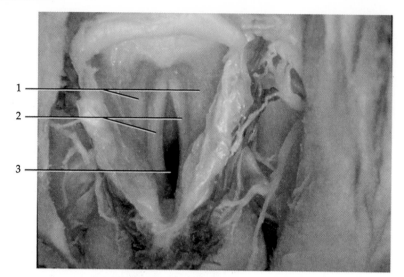

1

2

3

Superior view

___ Rima glottidis
___ Ventricular folds
       (false vocal cords)
___ Vocal folds
       (true vocal cords)

FIGURE 21.6 Air passageways to the lungs. Shown is the bronchial tree in relationship to lungs.

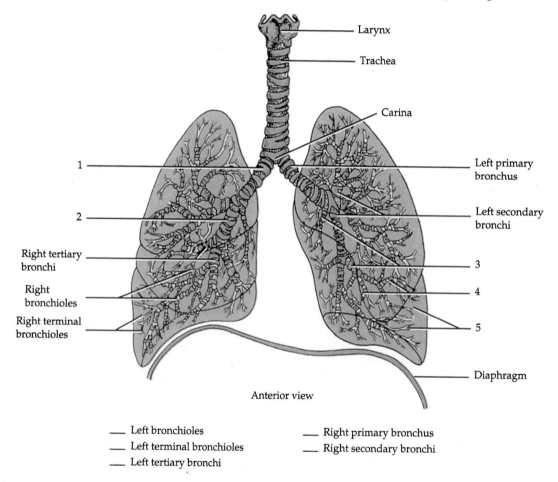

Larynx

Trachea

Carina

Left primary
bronchus

Left secondary
bronchi

1

2

Right tertiary
bronchi

Right
bronchioles

Right terminal
bronchioles

3

4

5

Diaphragm

Anterior view

___ Left bronchioles   ___ Right primary bronchus
___ Left terminal bronchioles   ___ Right secondary bronchi
___ Left tertiary bronchi

FIGURE 21.7 Histology of trachea (250×).

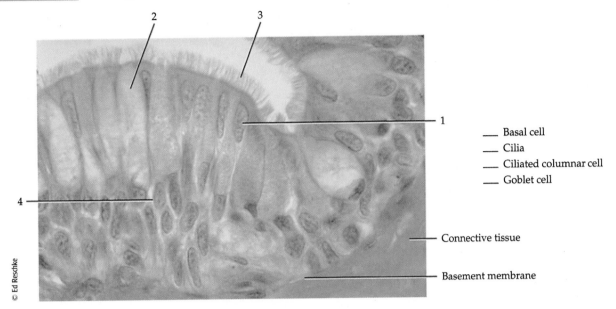

___ Basal cell
___ Cilia
___ Ciliated columnar cell
___ Goblet cell

— Connective tissue

— Basement membrane

© Ed Reschke

fissure separates the superior lobe from the middle lobe in the right lung; an **oblique fissure** separates the middle lobe from the inferior lobe in the right lung and the superior lobe from the inferior lobe in the left lung. Each lobe receives its own secondary bronchus.

Using your textbook as a reference, label Figure 21.8a.

Each lobe of a lung is divided into regions called **bronchopulmonary segments,** each supplied by a tertiary bronchus. Each bronchopulmonary segment is composed of many small compartments called **lobules.** Each lobule is wrapped in elastic connective tissue and contains a lymphatic vessel, arteriole, venule, and branch from a terminal bronchiole. Terminal bronchioles subdivide into **respiratory bronchioles,** which, in turn, subdivide into several **alveolar** (al-VĒ-ō-lar) **ducts.** Around the circumference of alveolar ducts are numerous alveoli and alveolar sacs. **Alveoli** (al-VĒ-ō-lī) are cup-shaped outpouchings lined by simple squamous epithelium and supported by a thin elastic basement membrane. The singular is **alveolus. Alveolar sacs** are two or more alveoli that share a common opening. Over the alveoli, an arteriole and venule disperse into a network of capillaries. Gas is exchanged between the lungs and blood by diffusion across the alveolar and the capillary walls.

Using your textbook as a reference, label Figure 21.8.

Each alveolus consists of:

a. *Type I alveolar cells.* Simple squamous epithelial cells that form a mostly continuous lining of the alveolar wall, except for occasional type II alveolar (septal) cells.

b. *Type II alveolar (septal) cells.* Cuboidal or round epithelial cells dispersed among type I alveolar cells that secrete alveolar fluid, which contains a phospholipid, lipoprotein substance called **surfactant** (sur-FAK-tant), a surface tension–lowering agent.

c. *Alveolar macrophages (dust cells).* Wandering phagocytic cells that remove fine dust particles and other debris from the alveolar spaces.

Obtain a slide of normal lung tissue and examine it under high power. Using your textbook as a reference, see if you can identify a terminal bronchiole, respiratory bronchiole, alveolar duct, alveolar sac, and alveoli.

If available, examine several pathological slides of lung tissue, such as slides that show emphysema and lung cancer. Compare your observations to the normal lung tissue.

The exchange of respiratory gases between the lungs and blood takes place by diffusion across the alveolar and capillary walls. This membrane, through which the respiratory gases move, is collectively known as the **respiratory membrane** (Figure 21.9). It consists of:

1. A layer of type I alveolar cells with type II alveolar cells and alveolar macrophages that constitute the **alveolar wall.**

2. An **epithelial basement membrane** underlying the alveolar wall.

FIGURE 21.8 Lungs.

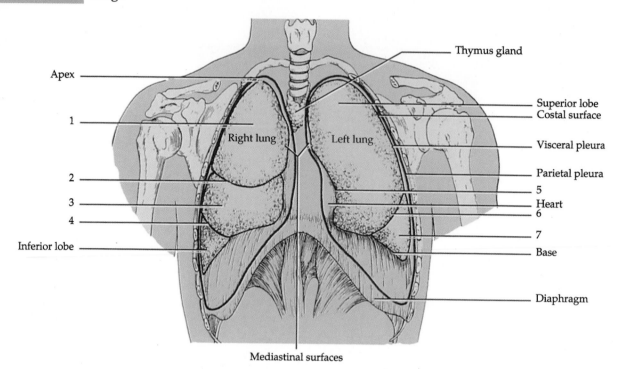

Thymus gland

Apex

Superior lobe
Costal surface

1

Right lung   Left lung

Visceral pleura

Parietal pleura

2   5

3   Heart

4   6

Inferior lobe   7

                                                               Base

                                                               Diaphragm

Mediastinal surfaces

(a) Coverings and external anatomy in anterior view

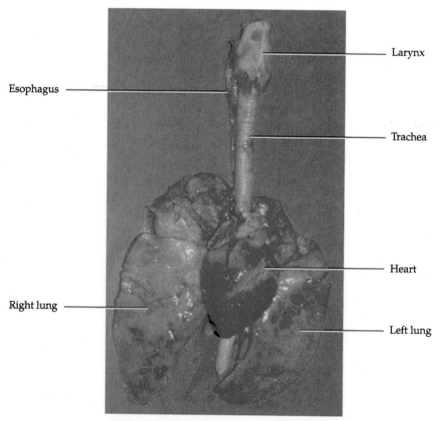

Larynx

Esophagus

Trachea

Heart

Right lung

Left lung

(b) Photograph of the lungs and associated structures

FIGURE 21.8 **Lungs. (Continued)**

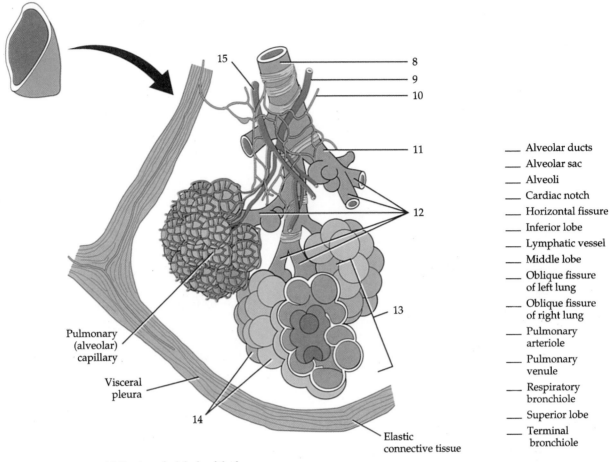

Pulmonary
(alveolar)
capillary

Visceral
pleura

____ Alveolar ducts
____ Alveolar sac
____ Alveoli
____ Cardiac notch
____ Horizontal fissure
____ Inferior lobe
____ Lymphatic vessel
____ Middle lobe
____ Oblique fissure
     of left lung
____ Oblique fissure
     of right lung
____ Pulmonary
     arteriole
____ Pulmonary
     venule
____ Respiratory
     bronchiole
____ Superior lobe
____ Terminal
     bronchiole

Elastic
connective tissue

(c) Portion of a lobule of the lung

3. A **capillary basement membrane** that is often fused to the epithelial basement membrane.

4. The **endothelial cells** of the capillary.

Label the components of the alveolar-capillary (respiratory) membrane in Figure 21.9.

Examine the scanning electron micrograph of the cells of the alveolar wall in Figure 21.10.

B. Dissection of Sheep Pluck

▲ **CAUTION!** *Please reread Section D, "Precautions Related to Dissection," at the beginning of the laboratory manual on page xiii before you begin your dissection.*

PROCEDURE

1. Preserved sheep pluck may or may not be available for dissection.

2. Pluck consists mainly of a sheep **trachea, bronchi, lungs, heart,** and **great vessels,** and a

small portion of the **diaphragm.** It is a good demonstration because it is large and shows the close anatomical correlation between these structures and the systems to which they belong, namely, the respiratory and the cardiovascular systems.

3. The heart and its great blood vessels have been described in detail in Exercise 17.

4. Pluck can also be used to examine in great detail the trachea and its relationship to the development of the bronchi until they branch into each lung. In addition, this specimen is sufficiently large that the bronchial tree can be exposed by careful dissection.

5. This dissection is done by starting at the primary and secondary bronchi and slowly and carefully removing lung tissue as the trees form even smaller branches into lungs.

6. These specimens should be used primarily by the instructor for demonstration, but you can help expose the bronchial tree.

FIGURE 21.9 Respiratory membrane.

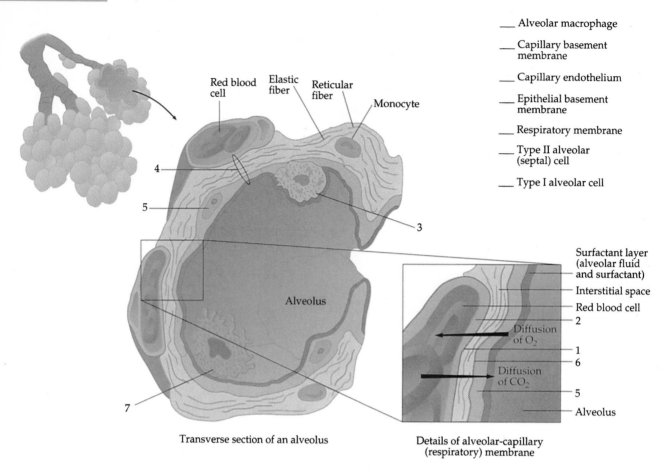

___ Alveolar macrophage

___ Capillary basement membrane

___ Capillary endothelium

___ Epithelial basement membrane

___ Respiratory membrane

___ Type II alveolar (septal) cell

___ Type I alveolar cell

Red blood cell

Elastic fiber

Reticular fiber

Monocyte

Alveolus

Surfactant layer (alveolar fluid and surfactant)

Interstitial space

Red blood cell

Diffusion of O_2

Diffusion of CO_2

Alveolus

Transverse section of an alveolus

Details of alveolar-capillary (respiratory) membrane

FIGURE 21.10 Histology of lung tissue.

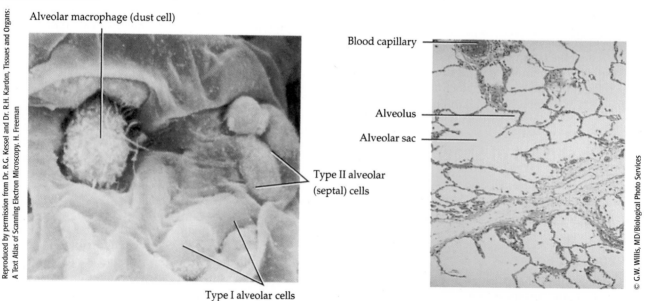

Alveolar macrophage (dust cell)

Type II alveolar (septal) cells

Type I alveolar cells

Blood capillary

Alveolus

Alveolar sac

(a) Scanning electron micrograph of alveolar wall (3420×)

(b) Photomicrograph of lung tissue (26×)

© G.W. Willis, MD/Biological Photo Services

C. Laboratory Tests on Respiration

1. Mechanics of Pulmonary Ventilation (Breathing)

Pulmonary ventilation (breathing) is the process by which gases are exchanged between the atmosphere and the alveoli. Air moves throughout the respiratory system as a result of pressure gradients (differences) between the external environment and the respiratory system. Likewise, oxygen and carbon dioxide then move between the respiratory alveoli and the pulmonary capillaries of the cardiovascular system, and the peripheral capillaries of the cardiovascular system and the tissues of the body as a result of diffusion gradients. We breathe in (inhale) when the pressure inside the lungs is less than the air pressure in the atmosphere; similarly, we breathe out (exhale) when the pressure inside the lungs is greater than the pressure in the atmosphere.

a. Inhalation

Breathing in is called **inhalation (inspiration).** Just before each inhalation, the air pressure inside the lungs equals the pressure of the atmosphere, which is approximately 760 mm Hg, or 1 atmosphere (atm), at sea level. For air to flow into the lungs, the pressure inside the alveoli must become lower than atmospheric pressure. This condition is achieved by increasing the volume of the lungs.

In order for inhalation to occur, the thoracic cavity must be expanded. This increases lung volume and thus decreases pressure in the lungs. The first step toward increasing lung volume (size) involves contraction of the principal inspiratory muscles—the diaphragm and external intercostals.

Contraction of the diaphragm causes it to flatten. This increases the vertical length of the thoracic cavity and accounts for about 75% of the air that enters the lungs during inspiration. At the same time the diaphragm contracts, the external intercostals contract. As a result, the ribs are elevated, increasing the anteroposterior and lateral diameters of the thoracic cavity. This movement of the ribs by the external intercostals is much like the movement of a bucket handle when a bucket is placed on its side and the handle is moved from a vertical to a more horizontal position. Because the thoracic cavity is a closed system, this increase in the vertical length and anteroposterior and lateral diameters of the thoracic cavity causes the volume of the thoracic cavity to increase, and the pressure within the cavity to decrease. As a result of this decreased pressure within the thoracic cavity, pressure within the lungs decreases, thereby causing air to move into the lungs.

During quiet inhalation, the pressure between two pleural layers, called **intrapleural (intrathoracic) pressure,** is always subatmospheric. (It may become temporarily positive, but only during modified respiratory movements such as coughing or straining during childbirth or defecation.) Just before inhalation, the intrapleural pressure is approximately 4 mm Hg lower than atmospheric pressure, or approximately 756 mm Hg. The overall increase in the size of the thoracic cavity causes intrapleural pressure to fall to approximately 754 mm Hg. The parietal and visceral pleural membranes are normally strongly attached to each other due to surface tension created by their moist adjoining surfaces. Therefore, as the walls of the thoracic cavity expand, the parietal pleural lining the thoracic cavity is pulled outward in all directions, and the visceral pleura is pulled along with it. Consequently, the size of the lungs increases, thereby decreasing the pressure within the lungs (called **alveolar [intrapulmonic] pressure**). This decrease in alveolar pressure causes a pressure gradient between the alveoli of the lungs and the external environment. Because the respiratory tree is open to the external environment via the oral and nasal cavities, air moves down the pressure gradient from the atmosphere into the pulmonary alveoli. Air continues to move into the lungs until alveolar pressure equals atmospheric pressure.

b. Exhalation

Breathing out, called **exhalation (expiration),** is also achieved by a pressure gradient, but in this case the gradient is reversed so that the pressure in the lungs is greater than the pressure of the atmosphere. Normal quiet expiration, unlike inspiration, is a *passive process,* in that it does not involve the active contraction of muscles. Instead, exhalation results from **elastic recoil** of the chest wall and lungs. Exhalation starts when the diaphragm and external intercostal muscles relax. As the external intercostals relax, the ribs are depressed, and as the diaphragm relaxes, it returns to its domelike position. These movements decrease the vertical, lateral, and anteroposterior dimensions of the thoracic cavity, returning to its original dimension. These movements return intrapleural pressure to its normal resting value of 756 mm Hg, or 4 mm Hg below atmospheric pressure.

As intrapleural pressure returns to its preinspiration level, the walls of the lungs are no longer pulled outward by the parietal pleura. The elastic recoil of the connective tissue within the lungs and respiratory tree allows the lungs to return to their resting shape and volume, thereby causing alveolar pressure to become slightly greater than atmospheric pressure (or approximately 762 mm Hg). Now air again moves down its pressure gradient, moving from the alveoli through the respiratory tree and out the oral and nasal openings.

In order to demonstrate pulmonary ventilation, a model lung (a bell-jar demonstrator) will be used. This apparatus is basically an artificial thoracic cavity that mimics the organs of the respiratory system and allows the pressure/volume ratios to be manipulated. A diagram of such a device is shown in Figure 21.11. Some of the models will not have a stopcock (rima glottidis); some will not have a tube opening into the pleural space.

PROCEDURE

1. Label the diagram in Figure 21.11 by writing the name that corresponds to the following

parts of the apparatus: rima glottidis, trachea, primary bronchus, lung, intrapleural pressure, alveolar pressure, chest wall, and diaphragm.

2. Pull on the rubber membrane (diaphragm) to simulate inspiration. What changes occur in

intrapleural pressure? _____

_____

_____

In alveolar pressure? _____

_____

3. Push the rubber membrane (diaphragm) upward to simulate expiration. What changes occur in intrapleural pressure? _____

_____

In alveolar pressure? _____

_____

FIGURE 21.11 Model lung.

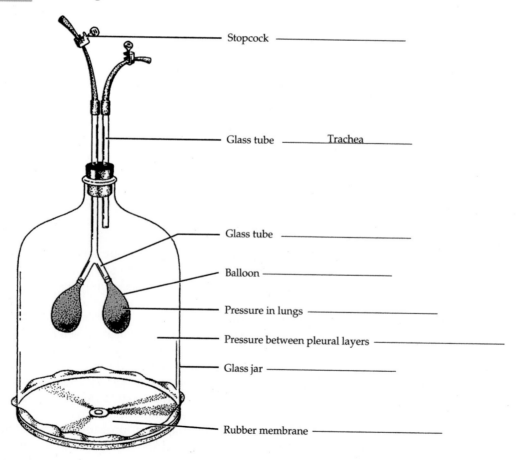

Stopcock _____

Glass tube _____ Trachea _____

Glass tube _____

Balloon _____

Pressure in lungs _____

Pressure between pleural layers _____

Glass jar _____

Rubber membrane _____

Under what normal processes would such changes in intrapleural and alveolar pressures be observed? _____

_____

4. Open the clamp on the tube leading to the intrapleural space. This simulates pneumothorax (air in the pleural cavity). Try to cause inspiration and expiration.

    Explain your results. _____

    _____

5. If present, close the stopcock (rima glottidis) and simulate expiration. This procedure mimics the Valsalva maneuver (forced expiration against a closed rima glottidis as during periods of straining). What changes occur in intrapleural and alveolar pressure? _____

    _____

2. Measurement of Chest and Abdomen in Respiration

When the diaphragm contracts, the dome shape of this muscle flattens; the flattening pushes against the abdominal viscera and decreases thoracic volume, which, in turn, decreases alveolar pressure. The volume changes in the abdomen and chest can easily be measured.

PROCEDURE

1. Place a tape measure at the level of the fifth rib (about armpit level) and determine the size of the chest immediately after a

    a. Normal inspiration
    b. Normal expiration
    c. Maximal inspiration
    d. Maximal expiration

2. Repeat step 1 with the tape measure at the level of the waist to determine abdominal size.

3. Record your results in Section C.1 of the LABORATORY REPORT RESULTS at the end of the exercise.

    Which measurement increased during inspiration? _____

    _____

    Which increased during expiration? _____

    _____

3. Respiratory Sounds

Air flowing through the respiratory tree creates characteristic sounds that can be detected through the use of a stethoscope. Normal breathing sounds include a soft, breezy sound caused by air filling the lungs during inspiration. As the air exits the lungs during expiration, a short, lower-pitched sound may be heard. Perform the following exercises.

PROCEDURE

1. *Clean the earplugs of a stethoscope with alcohol.*
2. Place the diaphragm of the stethoscope just below the larynx and listen for bronchial sounds during both inspiration and expiration.
3. Move the stethoscope slowly downward toward the bronchial tubes until the sounds are no longer heard.
4. Place the stethoscope under the scapula, under the clavicle, and over different intercostal spaces (the spaces between the ribs) on the chest, and listen for any sound during inspiration and expiration.

4. Use of a Pneumograph

The **pneumograph** is an instrument that measures variations in breathing patterns caused by various physical or chemical factors. The chest pneumograph, which is attached to a polygraph recorder via electrical leads, consists of a rubber bellows that fits around the chest just below the rib cage. As the subject breathes, chest movements cause changes in the air pressure within the pneumograph that are transmitted to the recorder. Normal inspiration and expiration can thus be recorded, and the effects of a wide range of physical and chemical factors on these movements can be studied.

PROCEDURE

Perform the following exercises and record your values in Section C.2 of the LABORATORY REPORT RESULTS at the end of the exercise. Label and save all recordings and attach them to Section C.2 of the LABORATORY REPORT RESULTS. In addition, label the inspiratory and expiratory phases of each recording, determine their duration, and then calculate the respiratory rate.

1. Place a respiratory pneumograph around the chest at the level of the sixth rib and attach it at the back. Connect the electrical lead to the recorder and adjust the instrument's centering and sensitivity so that the needle deflects as the subject breathes. If it does not, adjust

the pneumograph bellows up or down on the chest, or loosen the degree of the tightness around the subject's chest.

2. Seat the subject so that the pneumograph recording is not visible to the subject.

3. Set the polygraph at a slow speed (approximately 1 cm/sec) and record normal, quiet breathing (eupnea) for approximately 30 sec.

4. Have the subject inhale deeply and hold his or her breath for as long as possible. Record the breathing pattern during the breath holding and 30 to 60 sec after the resumption of breathing at the end of breath holding.

5. Record the respiratory movements of a subject in Section C.2 of the LABORATORY REPORT RESULTS at the end of the exercise during the following activities:

   a. Reading
   b. Swallowing water
   c. Laughing
   d. Yawning
   e. Coughing
   f. Sniffing
   g. Doing a rather difficult long-division calculation

5. Measurement of Respiratory Volumes

In clinical practice, **respiration** refers to one complete respiratory cycle, that is, one inhalation and one exhalation. While at rest, a healthy adult averages 12 breaths a minute, with each inhalation and exhalation, during which the lungs exchange specific volumes of air with the atmosphere.

As the following respiratory volumes and capacities are discussed, keep in mind that the values given vary with age, height, sex, and physiological state. The volume of air exhaled under normal, quiet inhalation is approximately 500 mL. The volume of one breath is called **tidal volume** (V_T) (Figure 21.12). Of this volume of air, approximately 350 mL reaches the alveoli. The other 150 mL of air does not reach the alveoli because it remains in the spaces not designed for air exchange known as **anatomical (respiratory) dead space** (nose, pharynx, larynx, trachea, bronchi, bronchioles, and terminal bronchioles).

By taking a very deep breath, you can inspire more than the 500 mL tidal volume taken in during normal, quiet respiration. The additional inhaled air, called the **inspiratory reserve volume,** averages 3100 mL in males and 1900 mL in females above the tidal volume of quiet respiration. If we inhale *nor-*

FIGURE 21.12 Spirogram of lung volumes and capacities.

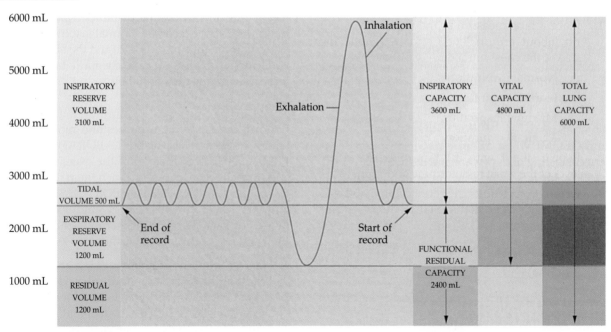

LUNG VOLUMES   LUNG CAPACITIES

mally and then exhale *as forcibly as possible,* a normal individual would be able to exhale the 500 mL of air taken in during a normal, quiet tidal volume, as well as an additional 1200 mL of air in males and 700 mL in females, which is termed **expiratory reserve volume.** Even after a forceful expiration there is still some air remaining in the lungs, which is termed **residual volume** (1200 mL in males and 1100 mL in females). This volume of air ensures gas exchange between the lungs and the cardiovascular system during brief time periods between respiratory cycles, or during extended periods of no respiratory activity (apnea).

Opening the thoracic cavity allows the intrapleural pressure to equal the atmospheric pressure, forcing out some of the residual volume. The air remaining is called the **minimal volume.** Minimal volume provides a medical and legal tool for determining whether a baby was born dead or died after birth. The presence of minimal volume can be demonstrated by placing a piece of lung in water and watching it float. Fetal lungs contain no air, and so the lung of a stillborn will not float in water.

Lung capacities are combinations of specific lung volumes. **Inspiratory capacity,** the total inspiratory ability of the lungs, is the sum of tidal volume plus inspiratory reserve volume (3600 mL in males and 2400 mL in females). **Functional residual capacity** is the sum of residual volume plus expiratory reserve volume (2400 mL in males and 1800 mL in females). **Vital capacity** is the sum of the inspiratory reserve volume, tidal volume, and expiratory reserve volume (4800 mL in males and 3100 mL in females). Finally, **total lung capacity** is the sum of all volumes (6000 mL in males and 4200 mL in females).

In a normal person the volumes and capacities of air in the lungs depend upon body size and build, as well as body position and the physiological state of the pulmonary system. Pulmonary volumes and capacities may change when a person lies down due to the movement of abdominal contents upon assuming a supine position. When the supine position is assumed, the abdominal contents tend to press superiorly upon the diaphragm, thereby decreasing the volume of the thoracic cavity. In addition, an individual assuming the supine position increases the volume of blood within the pulmonary circulation, thereby decreasing the space for pulmonary air.

NOTE *Depending on the type of laboratory equipment available, select from the following procedures related to measurement of respiratory volumes.*

PROCEDURE USING BIOPAC

Biopac Student Lab (BSL) system is an interactive hardware and software system that permits students to generate and record data from their bodies and animal or tissue preparations. It permits students to perform experiments, analyze data, draw conclusions, and form hypotheses on the basis of the collected data. The BSL system is available from BIOPAC Systems, Inc., 42 Aero Camino, Goleta, CA 93117. To order call: 1-805-685-0066; email: info@biopac.com; web page: www.biopac.com.

For activities related to respiratory volumes, select the appropriate experiments from BIOPAC.

BSL 8—Respiratory Cycle

BSL 12—Pulmonary Function: Vol. and Capacities

BSL 13—Pulmonary Flow Rates: FEV and MVV

BSL 15—Aerobic Exercise Physiology

PROCEDURE USING SPIROCOMP™

SPIROCOMP™ is a computerized spirometry system consisting of hardware and software designed to allow for quick and easy measurement of standard lung volumes. A stacked bar graph display of the data clearly depicts the relationship between the various volumes. The built-in group analysis features also allow you to view gender-specific data. The program is available from INTELITOOL® (1-800-955-7621).

PROCEDURE USING HANDHELD RESPIROMETER

A **respirometer (spirometer)** is an instrument used to measure volumes of air exchanged in breathing. Of the several different respirometers available, we will make use of two: a handheld respirometer and the Collins respirometer. One type of handheld respirometer is the Pulmometer (Figure 21.13). It measures and provides a direct digital display of certain respiratory volumes and capacities.

PROCEDURE

1. Set the needle or the digital indicator to zero.

2. Place a *clean* mouthpiece in the spirometer tube and hold the tube in your hand while you breathe normally for a few respirations.

(**NOTE** *Inhale through the nose and exhale through the mouth.*)

FIGURE 21.13 **Handheld respirometer.**

3. Place the mouthpiece in your mouth and inhale; exhale three normal breaths through the mouthpiece.

4. Divide the total volume of air expired by three to determine your average tidal volume. Record in Section C.3 of the LABORATORY REPORT RESULTS at the end of the exercise.

5. Return the scale indicator to zero.

6. Breathe normally for a few respirations. Following a normal expiration of tidal volume, forcibly exhale as much air as possible into the mouthpiece.

7. Repeat step 6 twice. Divide the total amount of air expired by three to determine your average expiratory reserve volume. Record this value in Section C.3 of the LABORATORY REPORT RESULTS at the end of the exercise.

8. To determine vital capacity, breathe deeply a few times and inhale as much air as possible. Exhale as fully and as steadily as possible into the spirometer tube.

9. Repeat step 8 twice. Divide the total volume of expired air by three to determine your average vital capacity. Record this value in Section C.3 of the LABORATORY REPORT RESULTS at the end of the exercise.

10. Calculate your inspiratory reserve volume by substituting the known volumes in this equation: Vital capacity = Inspiratory reserve volume + tidal volume + expiratory reserve volume. Record this volume in Section C.3 of the LABORATORY REPORT RESULTS at the end of the exercise.

11. Compare your vital capacity with the normal values shown in Tables 21.1 and 21.2.

12. Sit quietly and count your respirations per minute. Calculate your minute volume of respirations (MVR) by multiplying the breaths per minute by your tidal volume. Record this value in Section C.3 of the LABORATORY REPORT RESULTS at the end of the exercise.

$$\frac{\text{\# breaths}}{\text{minutes}} \times \frac{\text{\# mL}}{\text{breath}} = \frac{\text{\# mL}}{\text{minute}}$$

13. Repeat the preceding procedures while lying in the supine position. Compare the various lung volumes and capacities obtained while in the supine position with those obtained in the upright position.

PROCEDURE USING COLLINS RESPIROMETER

The Collins respirometer, shown in Figure 21.14, is a closed system that allows the measurement of volumes of air inhaled and exhaled. The instrument consists of a weighted drum, containing air, inverted over a chamber of water. The air-filled chamber is connected to the subject's mouth by a tube. When the subject inspires, air is removed from the chamber, causing the drum to sink and producing an upward deflection. This deflection is recorded by the stylus on the graph paper on the kymograph (rotating drum). When the subject expires, air is added, causing the drum to rise and producing a downward deflection. These deflections are recorded as a **spirogram** (see Figure 21.12). The horizontal (*x*) axis of a spirogram is graduated in millimeters (mm) and records elapsed time; the vertical (*y*) axis is graduated in milliliters (mL) and records air volumes. Spirometric studies measure lung capacities and rates and depths of ventilation for diagnostic purposes. Spirometry is indicated for individuals with labored breathing, and is used to diagnose respiratory disorders such as bronchial asthma and emphysema.

The air in the body is at a different temperature than the air contained in the respirometer; body air is also saturated with water vapor. To make the correction for these differences, one

TABLE 21.1 Predicted vital capacities for females

										Height in centimeters and inches										
	cm	152	154	156	158	160	162	164	166	168	170	172	174	176	178	180	182	184	186	188
Age	in.	59.8	60.6	61.4	62.2	63.0	63.7	64.6	65.4	66.1	66.9	67.7	68.5	69.3	70.1	70.9	71.7	72.4	73.2	74.0
16		3070	3110	3150	3190	3230	3270	3310	3350	3390	3430	3470	3510	3550	3590	3630	3670	3715	3755	3800
17		3055	3095	3135	3175	3215	3255	3295	3335	3375	3415	3455	3495	3535	3575	3615	3655	3695	3740	3780
18		3040	3080	3120	3160	3200	3240	3280	3320	3360	3400	3440	3480	3520	3560	3600	3640	3680	3720	3760
20		3010	3050	3090	3130	3170	3210	3250	3290	3330	3370	3410	3450	3490	3525	3565	3605	3645	3695	3720
22		2980	3020	3060	3095	3135	3175	3215	3255	3290	3330	3370	3410	3450	3490	3530	3570	3610	3650	3685
24		2950	2985	3025	3065	3100	3140	3180	3200	3260	3300	3335	3375	3415	3455	3490	3530	3570	3610	3650
26		2920	2960	3000	3035	3070	3110	3150	3190	3230	3265	3300	3340	3380	3420	3455	3495	3530	3570	3610
28		2890	2930	2965	3000	3040	3070	3115	3155	3190	3230	3270	3305	3345	3380	3420	3460	3495	3535	3570
30		2860	2895	2935	2970	3010	3045	3085	3120	3160	3195	3235	3270	3310	3345	3385	3420	3460	3495	3535
32		2825	2865	2900	2940	2975	3015	3050	3090	3125	3160	3200	3235	3275	3310	3350	3385	3425	3460	3495
34		2795	2835	2870	2910	2945	2980	3020	3055	3090	3130	3165	3200	3240	3275	3310	3350	3385	3425	3460
36		2765	2805	2840	2875	2910	2950	2985	3020	3060	3095	3130	3165	3205	3240	3275	3310	3350	3385	3420
38		2735	2770	2810	2845	2880	2915	2950	2990	3025	3060	3095	3130	3170	3205	3240	3275	3310	3350	3385
40		2705	2740	2775	2810	2850	2885	2920	2955	2990	3025	3060	3095	3135	3170	3205	3240	3275	3310	3345
42		2675	2710	2745	2780	2815	2850	2885	2920	2955	2990	3025	3060	3100	3135	3170	3205	3240	3275	3310
44		2645	2680	2715	2750	2785	2820	2855	2890	2925	2960	2995	3030	3060	3095	3130	3165	3200	3235	3270
46		2615	2650	2685	2715	2750	2785	2820	2855	2890	2925	2960	2995	3030	3060	3095	3130	3165	3200	3235
48		2585	2620	2650	2685	2715	2750	2785	2820	2855	2890	2925	2960	2995	3030	3060	3095	3130	3160	3195
50		2555	2590	2625	2655	2690	2720	2755	2785	2820	2855	2890	2925	2955	2990	3025	3060	3090	3125	3155
52		2525	2555	2590	2625	2655	2690	2720	2755	2790	2820	2855	2890	2925	2955	2990	3020	3055	3090	3125
54		2495	2530	2560	2590	2625	2655	2690	2720	2755	2790	2820	2855	2885	2920	2950	2985	3020	3050	3085
56		2460	2495	2525	2560	2590	2625	2655	2690	2720	2755	2790	2820	2855	2885	2920	2950	2980	3015	3045
58		2430	2460	2495	2525	2560	2590	2625	2655	2690	2720	2750	2785	2815	2850	2880	2920	2945	2975	3010
60		2400	2430	2460	2495	2525	2560	2590	2625	2655	2685	2720	2750	2780	2810	2845	2875	2915	2940	2970
62		2370	2405	2435	2465	2495	2525	2560	2590	2620	2655	2685	2715	2745	2775	2810	2840	2870	2900	2935
64		2340	2370	2400	2430	2465	2495	2525	2555	2585	2620	2650	2680	2710	2740	2770	2805	2835	2865	2895
66		2310	2340	2370	2400	2430	2460	2495	2525	2555	2585	2615	2645	2675	2705	2735	2765	2800	2825	2860
68		2280	2310	2340	2370	2400	2430	2460	2490	2520	2550	2580	2610	2640	2670	2700	2730	2760	2795	2820
70		2250	2280	2310	2340	2370	2400	2425	2455	2485	2515	2545	2575	2605	2635	2665	2695	2725	2755	2780
72		2220	2250	2280	2310	2335	2365	2395	2425	2455	2480	2510	2540	2570	2600	2630	2660	2685	2715	2745
74		2190	2220	2245	2275	2305	2335	2360	2390	2420	2450	2475	2505	2535	2565	2590	2620	2650	2680	2710

Source: E. A. Gaensler and G. W. Wright, *Archives of Environmental Health,* 12 (Feb.):146–189 (1966).

measures the temperature of the respirometer and uses Table 21.3 to obtain a conversion factor: **body temperature, pressure (atmospheric), saturated (with water vapor) (BTPS).** All measured volumes are then multiplied by this factor.

PROCEDURE

1. *When using the Collins respirometer, the disposable mouthpiece is discarded after use by each subject, and the hose is then detached and rinsed with 70% alcohol.*

▲ **CAUTION!** *This procedure must be repeated with every student using the equipment.*

2. Students should work in pairs, with one student operating the respirometer while the other student is tested.

3. Before starting the recording, a little practice may be necessary to learn to inhale and exhale only through your mouth and into the hose. A nose clip may be used to prevent leakage from the nose.

4. Make sure that the respirometer is functioning. Your instructor may have already prepared the instrument for use.

   a. The leveling screws must be raised or lowered so that the bell does not rub on the metal body.

TABLE 21.2   Predicted vital capacities for males

Height in centimeters and inches

Age	cm 152 / in. 59.8	154 60.6	156 61.4	158 62.2	160 63.0	162 63.7	164 64.6	166 65.4	168 66.1	170 66.9	172 67.7	174 68.5	176 69.3	178 70.1	180 70.9	182 71.7	184 72.4	186 73.2	188 74.0
16	3920	3975	4025	4075	4130	4180	4230	4285	4335	4385	4440	4490	4540	4590	4645	4695	4745	4800	4850
18	3890	3940	3995	4045	4095	4145	4200	4250	4300	4350	4405	4455	4505	4555	4610	4660	4710	4760	4815
20	3860	3910	3960	4015	4065	4115	4165	4215	4265	4320	4370	4420	4470	4520	4570	4625	4675	4725	4775
22	3830	3880	3930	3980	4030	4080	4135	4185	4235	4285	4335	4385	4435	4485	4535	4585	4635	4685	4735
24	3785	3835	3885	3935	3985	4035	4085	4135	4185	4235	4285	4330	4380	4430	4480	4530	4580	4630	4680
26	3755	3805	3855	3905	3955	4000	4050	4100	4150	4200	4250	4300	4350	4395	4445	4495	4545	4595	4645
28	3725	3775	3820	3870	3920	3970	4020	4070	4115	4165	4215	4265	4310	4360	4410	4460	4510	4555	4605
30	3695	3740	3790	3840	3890	3935	3985	4035	4080	4130	4180	4230	4275	4325	4375	4425	4470	4520	4570
32	3665	3710	3760	3810	3855	3905	3950	4000	4050	4095	4145	4195	4240	4290	4340	4385	4435	4485	4530
34	3620	3665	3715	3760	3810	3855	3905	3950	4000	4045	4095	4140	4190	4225	4285	4330	4380	4425	4475
36	3585	3635	3680	3730	3775	3825	3870	3920	3965	4010	4060	4105	4155	4200	4250	4295	4340	4390	4435
38	3555	3605	3650	3695	3745	3790	3840	3885	3930	3980	4025	4070	4120	4165	4210	4260	4305	4350	4400
40	3525	3575	3620	3665	3710	3760	3805	3850	3900	3945	3990	4035	4085	4130	4175	4220	4270	4315	4360
42	3495	3540	3590	3635	3680	3725	3770	3820	3865	3910	3955	4000	4050	4095	4140	4185	4230	4280	4325
44	3450	3495	3540	3585	3630	3675	3725	3770	3815	3860	3905	3950	3995	4040	4085	4130	4175	4220	4270
46	3420	3465	3510	3555	3600	3645	3690	3735	3780	3825	3870	3915	3960	4005	4050	4095	4140	4185	4230
48	3390	3435	3480	3525	3570	3615	3655	3700	3745	3790	3835	3880	3925	3970	4015	4060	4105	4150	4190
50	3345	3390	3430	3475	3520	3565	3610	3650	3695	3740	3785	3830	3870	3915	3960	4005	4050	4090	4135
52	3315	3353	3400	3445	3490	3530	3575	3620	3660	3705	3750	3795	3835	3880	3925	3970	4010	4055	4100
54	3285	3325	3370	3415	3455	3500	3540	3585	3630	3670	3715	3760	3800	3845	3890	3930	3975	4020	4060
56	3255	3295	3340	3380	3425	3465	3510	3550	3595	3640	3680	3725	3765	3810	3850	3895	3940	3980	4025
58	3210	3250	3290	3335	3375	3420	3460	3500	3545	3585	3630	3670	3715	3755	3800	3840	3880	3925	3965
60	3175	3220	3260	3300	3345	3385	3430	3470	3500	3555	3595	3635	3680	3720	3760	3805	3845	3885	3930
62	3150	3190	3230	3270	3310	3350	3390	3440	3480	3520	3560	3600	3640	3680	3730	3770	3810	3850	3890
64	3120	3160	3200	3240	3280	3320	3360	3400	3440	3490	3530	3570	3610	3650	3690	3730	3770	3810	3850
66	3070	3110	3150	3190	3230	3270	3310	3350	3390	3430	3470	3510	3550	3600	3640	3680	3720	3760	3800
68	3040	3080	3120	3160	3200	3240	3280	3320	3360	3400	3440	3480	3520	3560	3600	3640	3680	3720	3760
70	3010	3050	3090	3130	3170	3210	3250	3290	3330	3370	3410	3450	3480	3520	3560	3600	3640	3680	3720
72	2980	3020	3060	3100	3140	3180	3210	3250	3290	3330	3370	3410	3450	3490	3530	3570	3610	3650	3680
74	2930	2970	3010	3050	3090	3130	3170	3200	3240	3280	3320	3360	3400	3440	3470	3510	3550	3590	3630

Source: E. A. Gaensler and G. W. Wright, *Archives of Environmental Health,* 12 (Feb.):146–189 (1966).

b. Paper must be fed from a roll onto the kymograph *or* taped onto the kymograph in single sheets.

c. The soda-lime that absorbs carbon dioxide must be pink (fresh) and not purple (saturated with CO_2).

d. Pen reservoirs must be full and primed.

5. Raise and lower the drum several times to flush the respirometer of stale air and position the ventilometer pen (the upper pen, usually with black ink) so that it will begin writing in the center of the spirogram paper.

6. Set the free-breathing valve so that the opening is seen through the side of the valve.

7. Place a *sterile or disposable mouthpiece in your mouth* and close your nose with a clamp or with the fingers.

8. Breathe several times to become accustomed to the use of this instrument.

9. Have your lab partner close the free-breathing valve (turn it completely to the opposite direction so that no opening is seen) and turn on the respirometer to the slow speed. At this setting, the kymograph moves 32 mm (the distance between two vertical lines) each minute.

10. Perform the following exercises and record your values in Section C.4 of the LABORA-

FIGURE 21.14 Collins respirometer. This type is commonly used in college biology laboratories.

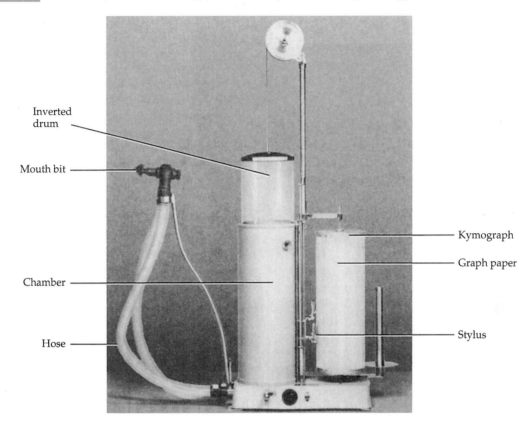

TORY REPORT RESULTS at the end of the exercise. In each case, * indicates the placement of the mouth on the mouthpiece.

11. Inspire normally, then * exhale normally three times. This volume is your tidal volume. Repeat two more times and average and record the values.

12. Expire normally, then * exhale as much air as possible three times, recording this volume. This value is your expiratory reserve volume. Repeat two more times and record the values.

13. After taking a deep breath, * exhale as much air as possible three times. This volume is your vital capacity. Repeat two more times and record the values. Compare your vital capacity to the normal value shown in Tables 21.1 and 21.2.

14. Because your vital capacity consists of tidal volume, inspiratory reserve volume, and expiratory reserve volume, and because you have already measured tidal volume in step 11 and expiratory reserve volume in step 12, you can calculate your inspiratory reserve volume by subtracting tidal volume and expiratory reserve volume from vital capacity. Inspiratory reserve volume = (vital capacity) − (tidal volume + expiratory reserve volume). Record this value.

In some cases, an individual with a pulmonary disorder has a nearly normal vital capacity. If the rate of expiration is timed, however, the extent of the pulmonary disorder becomes apparent. In order to do this, an individual expels air into a Collins respirometer as fast as possible, and the expired volume is measured per unit of time. Such a test is called **forced expiratory volume (FEV$_T$).** The T indicates that the volume of air is timed. FEV$_1$ is the volume of air forcefully expired in 1 sec, FEV$_2$ is the volume expired in 2 sec, and so on. A normal individual should be able to expel 83% of the total capacity during the first second, 94% in 2 sec, and 97% in 3 sec. For individuals with disorders such as emphysema and asthma, the percentage can be considerably lower, depending on the extent of the problem.

FEV$_1$ is determined according to the following procedure.

TABLE 21.3	Temperature variation conversion factors

Temperature °C (°F)	Conversion factor
20 (68.0)	1.102
21 (69.8)	1.096
22 (71.6)	1.091
23 (73.4)	1.085
24 (75.2)	1.080
25 (77.0)	1.075
26 (78.8)	1.068
27 (80.6)	1.063
28 (82.4)	1.057
29 (84.2)	1.051
30 (86.0)	1.045
31 (87.8)	1.039
32 (89.6)	1.032
33 (91.4)	1.026
34 (93.2)	1.020
35 (95.0)	1.014
36 (96.8)	1.007
37 (98.6)	1.000

PROCEDURE

1. Apply a noseclip to prevent leakage of air through the nose.
2. Turn on the kymograph.
3. Before placing the mouthpiece in your mouth, inhale as deeply as possible.
4. Expel all the air you can into the mouthpiece.
5. Turn off the kymograph.
6. Draw a vertical line on the spirogram at the starting point of exhalation. Mark this A.
7. Using the Collins VC timed interval ruler, draw a vertical line to the left of A and label it line B. The time between lines A and B is 1 sec.
8. The FEV$_1$ is the point where the spirogram tracing crosses line B.
9. Record your value here _____
10. Now read your vital capacity from the spirogram and record the value here

_____

11. In order to adjust for differences in temperature in the respirometer, use Table 21.3 as a guide. Determine the temperature in the respirometer, find the appropriate conver-

sion factor, and multiply the conversion factor by your vital capacity.

12. If, for example, the temperature of the respirometer is 75.2°F (24°C), the conversion factor is 1.080. And, if your vital capacity is 5600 mL, then

$$1.080 \times 5600 \text{ mL} = 6048 \text{ mL}$$

13. Use the same conversion factor and multiply it by your FEV$_1$. If your FEV$_1$ is 4000 mL, then

$$1.080 \times 4000 \text{ mL} = 4320 \text{ mL}$$

14. To calculate FEV$_1$, divide 4320 by 6048.

$$\text{FEV}_1 = \frac{4320}{6048} = 71\%$$

15. Repeat the procedure three times and record your FEV$_1$ in Section C.4 of the LABORATORY REPORT RESULTS at the end of the exercise.

Compare your results using the Collins respirometer with those using the handheld respirometer.

D. Laboratory Tests Combining Respiratory and Cardiovascular Interactions

1. **Experimental setup.** Review earlier explanations on recording the following:
   a. Blood pressure with sphygmomanometer and stethoscope (Section F in Exercise 19).
   b. Radial pulse (Section E.1 in Exercise 19).

⚠ **CAUTION!** *Assuming that your partner has no known or apparent cardiac or other health problems and is capable of such an activity, ask her/him to perform the following activities after attaching the various pieces of apparatus in order to record respiratory movements and blood pressure.*

2. **Experimental procedure.** Some form of regulated exercise is necessary for this experiment. Choose one of the following forms of exercise to have your subject participate in:
   a. *Riding exercise cycle.* If this form is chosen, set the resistance to be felt while riding the cycle, but *not so high* that the subject cannot complete the 4-min exercise period without difficulty.
   b. *Harvard Step Test.* In this form of exercise the subject is to step up onto a 20-in.

platform (a chair will substitute quite well) with one foot at a time. The subject is to bring *both* feet up onto the platform before stepping back down to the floor. The subject is also to remain erect at all times, and to do 30 complete cycles (up onto the platform and back down) per minute.

3. *Experimental protocol.* Record respiratory movements, blood pressure, and pulse during each of the procedures listed. Record your data in the table provided in Section D of the LABORATORY REPORT RESULTS at the end of the exercise. All data should be recorded *simultaneously*, thereby requiring participation of all members of the experimental group.

   a. *Basal readings.* Have the subject sit erect and quiet for 3 min. Obtain readings for:

      Respiratory rate and depth

      Systolic and diastolic blood pressure

      Pulse rate

      When the basal readings have been recorded, obtain additional readings after (1) sitting quietly on the exercise cycle for 3 min, if this form of exercise is to be utilized, or (2) standing quietly for 3 min in front of the platform that is to be utilized for the Harvard Step Test.

   b. *Readings after 1 min of exercise.* After the subject has exercised for 1 min, obtain additional readings.

   c. *Readings after 2 and 3 min of exercise.* Again, obtain additional readings after the subject has completed 2 and 3 min of exercise.

   d. *Readings upon completion of exercise.* When the subject has completed 4 min of exercise, obtain readings for respiratory rate, respiratory depth, heart rate, and systolic and diastolic blood pressure immediately upon completion *while the subject remains seated on the exercise cycle or stands erect on the floor,* depending upon the type of exercise utilized.

   e. *Readings 1, 2, 3, and 5 min after completion of exercise.* With the subject *still sitting on the exercise cycle or still standing erect on the floor,* obtain additional readings at the above time intervals after completion of the exercise period.

ANSWER THE LABORATORY REPORT QUESTIONS AT THE END OF THE EXERCISE.

EXERCISE 21

Respiratory System

Name _____ Date _____

Laboratory Section _____ Score/Grade _____

SECTION C ■ Laboratory Tests on Respiration

1. Measurement of Chest and Abdomen in Respiration

Record the results of your chest measurements, in inches, in the following table.

Size after a	Chest	Abdomen
Normal inspiration		
Normal expiration		
Maximal inspiration		
Maximal expiration		

2. Use of Pneumograph

Attach a sample of any one of the following activities.

Reading
Swallowing water
Laughing
Yawning
Coughing
Sniffing
Doing a rather difficult long-division calculation

3. Use of Handheld Respirometer

Tidal volume _____

Expiratory reserve volume _____

Vital capacity _____

Inspiratory reserve volume _____

Minute volume of respiration (MVR) _____

4. Use of Collins Respirometer

Record the results of your exercises using the Collins respirometer in the following table.

	Tidal volume (1)	Expiratory reserve (2)	Vital capacity (3)	Inspiratory reserve (4)	FEV$_1$ (5)
First time	mL	mL	mL	mL	mL
Second time	mL	mL	mL	mL	mL
Third time	mL	mL	mL	mL	mL
Your average	mL	mL	mL	mL	mL
Normal value	mL	mL	mL	mL	mL

SECTION D ▪ Laboratory Tests Combining Respiratory and Cardiovascular Interactions

Record the results of your exercises in the following table.

	Respiratory rate and depth	Systolic and diastolic pressure	Pulse rate
Basal readings while sitting erect and quiet for 3 min			
Basal readings while sitting on cycle or standing in front of platform for 3 min			
Readings after 1 min of exercise			
Readings after 2 min of exercise			
Readings after 3 min of exercise			
Readings after 4 min of exercise			
Readings after 1 min following completion of exercise			
Readings after 2 min following completion of exercise			
Readings after 3 min following completion of exercise			
Readings after 5 min following completion of exercise			

EXERCISE 21 | # Respiratory System

Name _____ Date _____

Laboratory Section _____ Score/Grade _____

PART 1 ■ Multiple Choice

_____ 1. The overall exchange of gases between the atmosphere, blood, and cells is called
(a) inspiration (b) respiration (c) expiration (d) none of these

_____ 2. The lateral walls of the internal nose are formed by the ethmoid bone, maxillae,
lacrimal, inferior conchae, and the (a) hyoid bone (b) nasal bone (c) palatine bone
(d) occipital bone

_____ 3. The portion of the pharynx that contains the pharyngeal tonsils is the (a) oropharynx
(b) laryngopharynx (c) nasopharynx (d) pharyngeal orifice

_____ 4. The Adam's apple is a common term for the (a) thyroid cartilage (b) cricoid cartilage
(c) epiglottis (d) none of these

_____ 5. The C-shaped rings of cartilage of the trachea not only prevent the trachea from collapsing but also aid in the process of (a) lubrication (b) removing foreign particles (c) gas exchange (d) swallowing

_____ 6. Of the following structures, the smallest in diameter is the (a) left primary bronchus
(b) bronchioles (c) secondary bronchi (d) alveolar ducts

_____ 7. The structures of the lung that actually contain the alveoli are the (a) respiratory bronchioles (b) fissures (c) lobules (d) terminal bronchioles

_____ 8. From superficial to deep, the structure(s) that you would encounter first among the following is/are the (a) bronchi (b) parietal pleura (c) pleural cavity (d) secondary bronchi

_____ 9. Which of the following statements is/are true regarding the pharynx? (a) It serves as a
common passageway for both the respiratory and digestive systems. (b) Pharyngeal tonsils are located in the inferior part, the laryngopharynx. (3) The nasal cavity connects
with it, through two internal nares, into the nasopharynx. (a) 1 only (b) 2 only (c) 3 only
(d) 1 and 3

_____ 10. The auditory (pharyngotympanic) tube opens into the lateral wall of the (a) inner ear
(b) nasopharynx (c) oropharynx (d) laryngopharynx

PART 2 ■ Completion

11. An advantage of nasal breathing is that the air is warmed, moistened, and _____.

12. Improper fusion of the palatine and maxillary bones results in a condition called

_____.

13. The protective lid of cartilage that prevents food from entering the trachea is the

    _____.

14. After removal of the _____ an individual would be unable to speak.

15. The upper respiratory tract is able to trap and remove dust because of its lining of

    _____.

16. The passage of a tube into the mouth or nose and down through the larynx and trachea to bypass an obstruction is called _____.

17. A radiograph of the bronchial tree after administration of an iodinated medium is called a

    _____.

18. The sequence of respiratory tubes from largest to smallest is trachea, primary bronchi, secondary bronchi, bronchioles, _____, respiratory bronchioles, and alveolar ducts.

19. Both the external and internal nose are divided internally by a vertical partition called the

    _____.

20. The undersurface of the external nose contains two openings called the nostrils or

    _____.

21. The functions of the pharynx are to serve as a passageway for air and food and to provide a resonating chamber for _____.

22. An inflammation of the membrane that encloses and protects the lungs is called

    _____.

23. The anterior portion of the nasal cavity just inside the nostrils is called the _____.

24. Groovelike passageways in the nasal cavity formed by the conchae are called

    _____.

25. The portion of the pharynx that contains the palatine and lingual tonsils is the

    _____.

26. Each bronchopulmonary segment of a lung is subdivided into many small compartments called

    _____.

27. A(n) _____ is an outpouching lined by epithelium and supported by a thin elastic membrane.

28. The _____ cartilage attaches the larynx to the trachea.

29. The broad, inferior portion of a lung that fits over the convex area of the diaphragm is the

    _____.

30. Phagocytic cells in the alveolar wall are called _____.

31. The surface of a lung lying against the ribs is called the _____ surface.

32. The _____ is a structure in the medial surface of a lung through which bronchi, pulmonary blood vessels, lymphatic vessels, and nerves pass.

33. The nose, pharynx, and associated structures comprise the _____ respiratory system.

34. The _____ is the superior attachment of the nose to the frontal bone.

35. The _____ cartilages of the larynx attach the vocal folds to the intrinsic pharyngeal muscles.

36. Each _____ segment of a lung is supplied by a tertiary bronchus.

PART 3 ■ Matching

_____ 37. Tidal volume

_____ 38. Inspiratory reserve volume

_____ 39. Inspiratory capacity

_____ 40. Expiratory reserve volume

_____ 41. Vital capacity

_____ 42. Total lung capacity

A. 1200 mL of air in males

B. 3600 mL of air in males

C. 4800 mL of air in males

D. 500 mL of air in males

E. 3100 mL of air in males

F. 6000 mL of air in males

PART 4 ■ Matching

_____ 43. Thicker, broader, and shorter

_____ 44. Has a cardiac notch

_____ 45. Has only two secondary bronchi

_____ 46. Has a horizontal fissure

_____ 47. Has a primary bronchus that is shorter, wider, and more vertical

A. Right lung

B. Left lung

PART 5 ▪ **Matching**

_____ 48. Nasal conchae and meatuses are located here

_____ 49. Contains pharyngeal tonsil

_____ 50. Palatine tonsils are located here

_____ 51. This structure leads directly into the esophagus

_____ 52. Cricoid, epiglottis, and thyroid cartilages are here

_____ 53. Vocal cords here enable voice production

_____ 54. Both a respiratory and a digestive pathway (two answers)

_____ 55. Fauces opens into this structure

A. Nasal cavity

B. Oropharynx

C. Laryngopharynx

D. Larynx

E. Nasopharynx

Digestive System

Objectives

At the completion of this exercise you should understand

A The general arrangement of the walls of the gastrointestinal tract

B The anatomical characteristics and functions of the organs of the digestive system

C The chemistry of digestion and how it applies to absorption within the digestive system

The food we eat contains a variety of nutrients, which are used for building new body tissues, repairing damaged tissues, and sustaining needed chemical reactions. Food also is vital for life because it is the source of chemical energy. Energy is needed for muscle contraction, conduction of nerve impulses, and secretory and absorptive activities of many cells. As consumed, however, most food cannot be used to either build tissues or to power the energy-requiring chemical reactions of the body. First, it must be broken down into molecules small enough to enter the body cells, a process known as **digestion.** The passage of these smaller molecules through the plasma membranes of cells lining the stomach and intestines and then into the blood and lymph is termed **absorption.** The organs that collectively perform these functions compose the **digestive system.**

Digestion occurs basically as two events—mechanical digestion and chemical digestion. **Mechanical digestion** consists of various movements of the gastrointestinal tract that help chemical digestion. These movements include the physical breakdown of food by the teeth and complete churning and mixing of this food with enzymes by the smooth muscles of the stomach and small intestine. **Chemical digestion** consists of a series of catabolic (hydrolysis) reactions that break down the large nutrient molecules that we eat, such as carbohydrates, lipids, and proteins, into much smaller molecules that can be absorbed and used by body cells.

A. General Organization of Digestive System

Digestive organs are usually divided into two main groups. The first is the **gastrointestinal (GI) tract,** or **alimentary** (*alimentary* = nourishment) **canal,** a continuous tube running from the mouth to the anus, and measuring about 9 m (30 ft) in length in a cadaver. This tract is composed of the mouth, pharynx, esophagus, stomach, small intestine, and large intestine. The small intestine has three regions: duodenum, jejunum, and ileum. The large intestine has four regions: cecum, colon, rectum, and anal canal. The colon is divided into ascending colon, transverse colon, descending colon, and sigmoid colon.

The second group of organs composing the digestive system consists of the **accessory digestive organs** such as the teeth, tongue, salivary glands, liver, gallbladder, and pancreas (see Figure 22.1).

The medical specialty that deals with the structure, function, diagnosis, and treatment of diseases of the stomach and intestines is called

FIGURE 22.1   Organs of the digestive system and related structures.

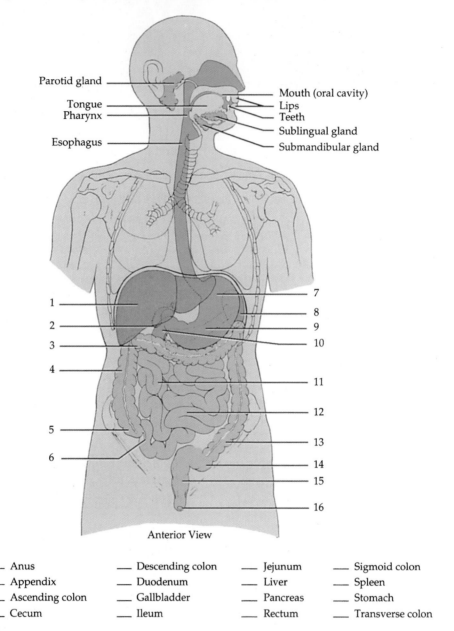

Parotid gland

Tongue
Pharynx

Esophagus

Mouth (oral cavity)
Lips
Teeth
Sublingual gland
Submandibular gland

1
2
3
4

5
6

7
8
9
10

11

12

13

14
15

16

Anterior View

___ Anus	___ Descending colon	___ Jejunum	___ Sigmoid colon
___ Appendix	___ Duodenum	___ Liver	___ Spleen
___ Ascending colon	___ Gallbladder	___ Pancreas	___ Stomach
___ Cecum	___ Ileum	___ Rectum	___ Transverse colon

gastroenterology (gas′-trō-en′ter-OL-ō-jē; *gastro* = stomach; *entero* = intestines; *logy* = study of). The medical specialty that deals with the diagnosis and treatment of disorders of the rectum and anus is called **proctology** (prok-TOL-ō-jē; *proct* = rectum).

Using your textbook, charts, or models for reference, label Figure 22.1.

The wall of the gastrointestinal tract, from the lower esophagus to the anal canal, has the same basic, four-layered arrangement of tissues. The four layers of the tract, from deep to superficial, are the **mucosa, submucosa, muscularis,** and **serosa** (see Figure 22.9).

Inferior to the diaphragm, the serosa is also called the **peritoneum** (per′-i-tō-NĒ-um; *peri* = around; *tonos* = tension). The peritoneum is composed of a layer of simple squamous epithelium (called mesothelium) with an underlying layer of areolar connective tissue. The **parietal peritoneum** lines the wall of the abdominopelvic cavity, and the **visceral peritoneum** covers some of the organs in the cavity and is their serosa. The slim space between the parietal and visceral portions of the peritoneum is called the **peritoneal cavity.** Unlike the two other serous membranes of the body, the pericardium and the pleurae, which smoothly cover the heart and lungs, the peritoneum contains large folds that weave in between the viscera. The important extensions of the peritoneum are the **mesentery** (MEZ-en-ter'-ē; *meso* = middle; *enteron* = intestine), **mesocolon, falciform** (FAL-si-form) **ligament, lesser omentum** (ō-MENT-um), and **greater omentum.**

Acute inflammation of the peritoneum, called **peritonitis,** is a serious condition because the peritoneal membranes are continuous with one another, enabling the infection to spread to all the organs in the cavity.

B. Organs of Digestive System

1. Mouth (Oral Cavity)

The **mouth,** also called the **oral,** or **buccal** (BUK-al; *bucca* = cheeks), **cavity,** is formed by the cheeks, hard and soft palates, and tongue. The **hard palate** forms the anterior portion of the roof of the mouth, and the **soft palate** forms the posterior portion. The **tongue** forms the floor of the oral cavity and is composed of skeletal muscle covered by mucous membrane. Partial digestion of carbohydrates and triglycerides occurs in the mouth.

a. *Cheeks.* Lateral walls of oral cavity. Muscular structures covered externally by skin and internally by nonkeratinized stratified squamous epithelium; anterior portions terminate in the **superior** and **inferior labia** (lips).

b. *Vermilion* (ver-MIL-yon). Transition zone of lips where outer skin and inner mucous membranes meet.

c. *Labial frenulum* (LĀ-bē-al FREN-ū-lum; *labium* = fleshy border; *frenulum* = small bridle). Midline fold of mucous membrane that attaches the inner surface of each lip to its corresponding gum.

d. *Vestibule* (= entrance to a canal). Space bounded externally by cheeks and lips and internally by gums and teeth.

e. *Oral cavity proper.* Space extending from the gums and teeth to the **fauces** (FAW-sēs = passages), opening between the oral cavity and the pharynx. Area is enclosed by the dental arches.

f. *Hard palate.* Formed by maxillae and palatine bones and covered by mucous membrane.

g. *Soft palate.* Arch-shaped muscular partition between oropharynx and nasopharynx lined by mucous membrane. Hanging from free border of soft palate is a conical muscular projection, the **uvula** (Ū-vū-la = little grape).

h. *Palatoglossal arch.* Muscular fold that extends inferiorly, laterally, and anteriorly to the side of the base of tongue.

i. *Palatopharyngeal* (PAL-a-tō-fa-rin'-jē-al) *arch.* Muscular fold that extends inferiorly, laterally, and posteriorly to the side of pharynx. **Palatine tonsils** are between arches, and **lingual tonsil** is at base of tongue.

j. *Tongue.* Movable, muscular organ on floor of oral cavity. **Extrinsic muscles** originate outside tongue (attach to bones in the area), insert into connective tissues, and move tongue from side to side and in and out to maneuver food for chewing and swallowing; **intrinsic muscles** originate in and insert into connective tissues within the tongue and alter shape and size of tongue for speech and swallowing.

k. *Lingual* (*lingua* = tongue) *frenulum.* Midline fold of mucous membrane on undersurface of tongue that aids in limiting the movement of the tongue posteriorly.

l. *Papillae* (pa-PIL-ē = nipple-shaped projections). Projections of lamina propria covered with keratinized epithelium; **filiform** (= threadlike) **papillae** are pointed, threadlike projections in parallel rows over anterior two-thirds of tongue; **fungiform** (= shaped like a mushroom) **papillae** are mushroomlike elevations distributed among filiform papillae and more numerous near tip of tongue (appear as red dots and most contain taste buds); **vallate (circumvallate)** (VAL-āt = wall-like) **papillae** are arranged in the form of an inverted V on the posterior surface of tongue (all contain taste buds); **foliate papillae** (FŌ-lē-āt = leaflike) are located in small trenches on the lateral margins of the tongue (most of their taste buds degenerate in early childhood).

Using a mirror, examine your mouth and locate as many of the structures (a through l) as you can. Label Figure 22.2.

2. Salivary Glands

Most saliva is secreted by the **salivary glands,** which lie outside the mouth and pour their contents into ducts that empty into the oral cavity. The carbohydrate-digesting enzyme in saliva is salivary amylase. There are three pairs of major salivary glands: **parotid** (*par* = near; *ot* = ear) **glands** (anterior and inferior to the ears), which secrete into the oral cavity vestibule through **parotid ducts; submandibular glands** (deep to

FIGURE 22.2 **Mouth (oral cavity).**

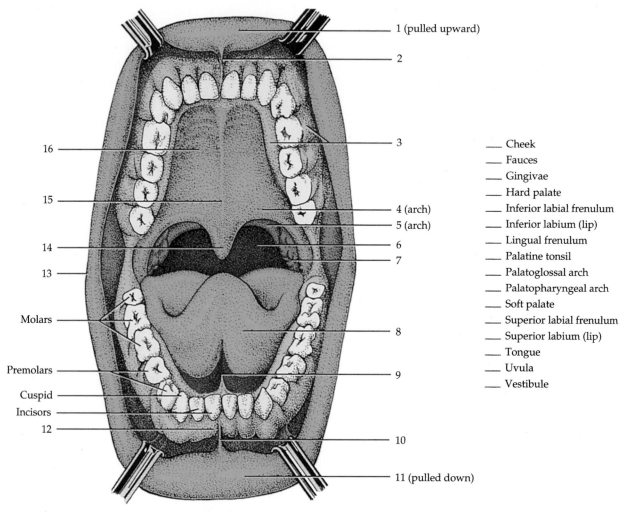

1 (pulled upward)
2
3
4 (arch)
5 (arch)
6
7
8
9
10
11 (pulled down)

16
15
14
13
Molars
Premolars
Cuspid
Incisors
12

___ Cheek
___ Fauces
___ Gingivae
___ Hard palate
___ Inferior labial frenulum
___ Inferior labium (lip)
___ Lingual frenulum
___ Palatine tonsil
___ Palatoglossal arch
___ Palatopharyngeal arch
___ Soft palate
___ Superior labial frenulum
___ Superior labium (lip)
___ Tongue
___ Uvula
___ Vestibule

Anterior view

the base of the tongue in the posterior part of the floor of the mouth), which secrete lateral to the lingual frenulum in the floor of the oral cavity through **submandibular ducts;** and **sublingual glands** (superior to the submandibular glands), which secrete into the floor of the oral cavity through **lesser sublingual ducts.**

Label Figure 22.3.

The parotid glands are compound tubuloacinar glands, whereas the submandibulars and sublinguals are compound acinar glands (see Figure 22.4).

Examine prepared slides of the three different types of salivary glands and compare your observations to Figure 22.4.

3. Teeth

Teeth (dentes) are located in the sockets of the alveolar processes of the mandible and maxillae. The alveolar processes are covered by **gingivae**

(JIN-ji-vē), or gums, which extend slightly into each socket to form the gingivae sulcus. The sockets are lined by a dense fibrous connective tissue called a **periodontal** (*peri* = around; *odont* = tooth) **ligament,** which anchors the teeth in position and acts as a shock absorber during chewing.

Following are the parts of a tooth:

a. *Crown.* Visible portion above level of gums.

b. *Root.* One to three projections embedded in socket.

c. *Neck.* Constricted junction of the crown and root near the gum line.

d. *Dentin.* Calcified connective tissue that gives teeth their basic shape and rigidity.

e. *Pulp cavity.* Enlarged part of the space in the crown within the dentin.

f. *Pulp.* Connective tissue containing blood vessels, lymphatic vessels, and nerves.

FIGURE 22.3 Location of salivary glands.

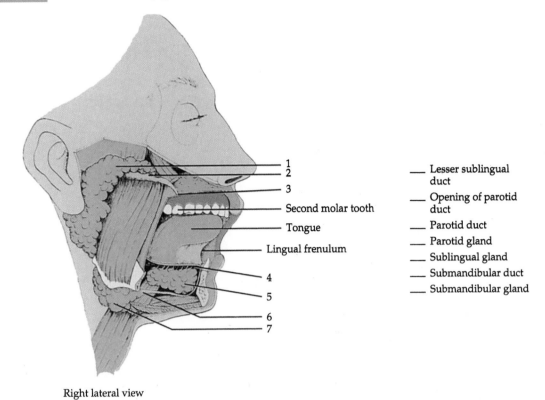

Right lateral view

____ Lesser sublingual duct

____ Opening of parotid duct

____ Parotid duct

____ Parotid gland

____ Sublingual gland

____ Submandibular duct

____ Submandibular gland

g. *Root canal.* Narrow extension of pulp cavity in root.

h. *Apical foramen.* Opening in base of root canal through which blood vessels, lymphatic vessels, and nerves extend.

i. *Enamel.* Covering of crown that consists primarily of calcium phosphate and calcium carbonate.

j. *Cementum.* Bonelike substance that covers and attaches root to periodontal ligament.

With the aid of your textbook, label the parts of a tooth shown in Figure 22.5.

4. Dentitions

Dentitions (sets of teeth) are of two types: **deciduous** (primary, milk, or baby) and **permanent** (secondary). Deciduous teeth begin to erupt at about 6 months of age, and one pair appears at about each month thereafter until all 20 are present. The deciduous teeth are as follows:

a. *Incisors.* Central incisors closest to midline, with lateral incisors on either side. Incisors

FIGURE 22.4   Histology of salivary glands.

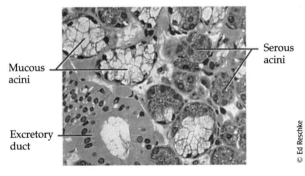

© Ed Reschke

are chisel-shaped, adapted for cutting into food, have only one root.

b. *Cuspids (canines).* Posterior to incisors. Cuspids have pointed surfaces (cusps) for tearing and shredding food, have only one root.

c. *Molars.* First and second molars posterior to canines. Molars crush and grind food. Upper (maxillary) molars have four cusps and three roots, lower (mandibular) molars have four cusps and two roots. ▪

FIGURE 22.5 **Parts of a tooth.**

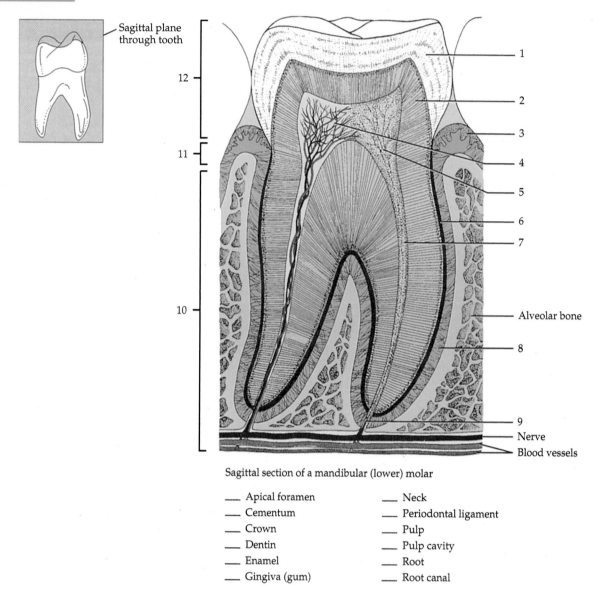

Sagittal plane through tooth

1
2
3
4
5
6
7
Alveolar bone
8
9
Nerve
Blood vessels

Sagittal section of a mandibular (lower) molar

___ Apical foramen	___ Neck
___ Cementum	___ Periodontal ligament
___ Crown	___ Pulp
___ Dentin	___ Pulp cavity
___ Enamel	___ Root
___ Gingiva (gum)	___ Root canal

All deciduous teeth are usually lost between 6 and 12 years of age and replaced by permanent dentition consisting of 32 teeth that appear between age 6 and adulthood. The permanent teeth are:

a. *Incisors.* Central incisors and lateral incisors replace those of deciduous dentition.

b. *Cuspids (canines).* These replace those of deciduous dentition.

c. *Premolars (bicuspids).* First and second premolars replace deciduous molars. Premolars crush and grind food, have two cusps and one root (upper first premolars have two roots).

d. *Molars.* These erupt behind premolars as jaw grows to accommodate them and do not re-

place any deciduous teeth. First molars erupt at age 6, second at age 12, and third (wisdom teeth) after age 17.

Using a mirror, examine your mouth and locate as many teeth of the permanent dentition as you can.

With the aid of your textbook, label the deciduous and permanent dentitions in Figure 22.6.

5. Pharynx

Through chewing, or **mastication** (mas'ti-KĀ-shun; *masticare* = to chew), the tongue manipulates food, the teeth grind it, and the food is

FIGURE 22.6   **Dentitions.**

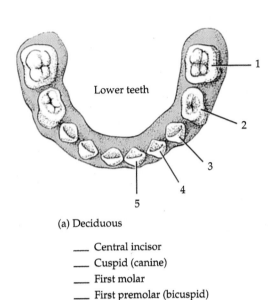

Lower teeth

(a) Deciduous

___ Central incisor
___ Cuspid (canine)
___ First molar
___ First premolar (bicuspid)

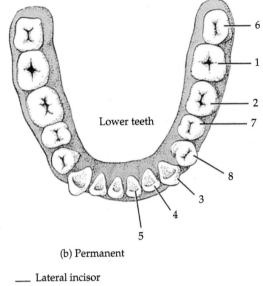

Lower teeth

(b) Permanent

___ Lateral incisor
___ Second molar
___ Second premolar (bicuspid)
___ Third molar (wisdom tooth)

mixed with saliva. As a result, the food is reduced to a soft, flexible mass called a **bolus** (= lump) that is easily swallowed. When food is first swallowed, it passes from the mouth into the pharynx.

The **pharynx** (= throat) is a funnel-shaped tube that extends from the internal nares to the esophagus posteriorly and the larynx anteriorly (see Figure 22.1). The pharynx is composed of skeletal muscle and lined by mucous membrane. Whereas the nasopharynx functions only in respiration, both the oropharynx and the laryngopharynx have digestive as well as respiratory functions. Food that is swallowed passes from the mouth into the oropharynx and laryngopharynx before passing into the esophagus and then into the stomach. Muscular contractions of the oropharynx and laryngopharynx help propel food into the esophagus.

6. Esophagus

The **esophagus** (e-SOF-a-gus = eating gullet) is a collapsible muscular tube that lies posterior to the trachea. The structure is about 25 cm (10 in.) long and begins at the inferior end of the laryngopharynx and passes through the mediastinum. Then, it pierces the diaphragm through an opening called the **esophageal hiatus** in the diaphragm and terminates in the superior portion of the stomach.

The esophagus conveys food from the pharynx to the stomach by peristalsis.

Histologically, the esophagus consists of a **mucosa** (nonkeratinized stratified squamous epithelium, lamina propria, muscularis mucosae), **submucosa** (areolar connective tissue, blood vessels, mucous glands), **muscularis** (superior third striated, middle third striated and smooth, inferior third smooth), and **adventitia** (ad-ven-TISH-a). The esophagus is not covered by a serosa.

Examine a prepared slide of a transverse section of the esophagus that shows its various coats. With the aid of your textbook, label Figure 22.7.

7. Stomach

The **stomach** is a J-shaped enlargement of the gastrointestinal tract inferior to the diaphragm (see Figure 22.1). It is in the epigastric, umbilical, and left hypochondriac regions of the abdomen. The superior part is connected to the esophagus; the inferior part empties into the duodenum, the first portion of the small intestine. The stomach is divided into four main regions: cardia, fundus, body, and pylorus. The **cardia** (CAR-dē-a) surrounds the superior opening of the stomach. A physiological sphincter, known as the **lower esophageal sphincter,** allows passage of the bolus from the esophagus into the cardia of the

FIGURE 22.7 Histology of esophagus (10×).

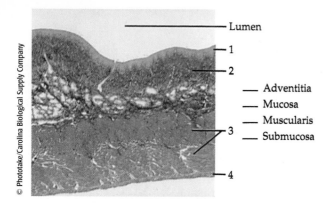

Lumen
1
2
Adventitia
Mucosa
Muscularis
3
Submucosa
4

secrete pepsinogen and gastric lipase. **Parietal cells** produce hydrochloric acid and intrinsic factor. The secretions of the mucous, chief, and parietal cells are collectively called **gastric juice.** In addition, gastric glands include one type of enteroendocrine cell. An **enteroendocrine** (*enteron* = intestine) **cell** is a hormone-producing cell in the gastrointestinal mucosa. One such cell is a **G cell,** located mainly in the pyloric antrum, that secretes the hormone gastrin into the blood stream.

8. Pancreas

The **pancreas** (*pan* = all; *creas* = flesh) is a retroperitoneal gland posterior to the greater curvature of the stomach (see Figure 22.1). The gland consists of a **head** (expanded portion near the curve of the duodenum), **body** (central portion), and **tail** (terminal tapering portion).

Histologically, the pancreas consists of **pancreatic islets (islets of Langerhans)** that contain (1) glucagon-producing **alpha cells,** (2) insulin-producing **beta cells,** (3) somatostatin-producing **delta cells,** and (4) pancreatic polypeptide-producing **F cells** (see Figure 15.5). The pancreas also consists of **acini** that produce pancreatic juice (see Figure 15.5). Pancreatic juice contains enzymes that assist in the chemical breakdown of carbohydrates, proteins, triglycerides, and nucleic acids.

Pancreatic juice is delivered from the pancreas to the duodenum by a large main duct, the **pancreatic duct (duct of Wirsung).** This duct unites with the common bile duct from the liver and pancreas and enters the duodenum in a common duct called the **hepatopancreatic ampulla (ampulla of Vater).** The ampulla opens on an elevation of the duodenal mucosa, the **major duodenal papilla.** An **accessory pancreatic duct (duct of Santorini)** may also lead from the pancreas and empty into the duodenum about 2.5 cm (1 in.) superior to the hepatopancreatic ampulla.

With the aid of your textbook, label the structures associated with the pancreas in Figure 22.10 on page 515.

9. Liver

The **liver** is located inferior to the diaphragm (see Figure 22.1). It occupies most of the right hypochondriac and part of the epigastric regions of the abdominopelvic cavity. The gland is divided into two principal lobes, a large **right lobe** and a smaller **left lobe,** separated by the **falciform ligament.** The falciform ligament attaches the liver to the anterior abdominal wall and diaphragm. Even

stomach. The rounded portion superior to and to the left of the cardia is the **fundus** (FUN-dus). Inferior to the fundus, the large central portion of the stomach is called the **body.** The narrow, inferior region that connects to the duodenum is the **pylorus** (pī-LOR-us; *pyl* = gate; *ouros* = guard). The pylorus consists of a **pyloric antrum** (AN-trum = cave), which is closer to the body of the stomach, and a **pyloric canal,** which is closer to the duodenum. The concave medial border of the stomach is called the **lesser curvature,** and the convex lateral border is the **greater curvature.** The pylorus communicates with the duodenum of the small intestine via a sphincter called the **pyloric sphincter.** The main chemical activity of the stomach is to begin the digestion of proteins.

Label Figure 22.8.

The stomach wall is composed of the same four layers as the rest of the gastrointestinal tract (mucosa, submucosa, muscularis, and serosa) with certain modifications. Examine a prepared slide of a section of the stomach that shows its various layers. With the aid of your textbook, label Figure 22.9a on page 514.

The surface of the **mucosa** is a layer of simple columnar epithelial cells called **surface mucous cells** (Figure 22.9). The mucosa contains a **lamina propria** (areolar connective tissue) and a **muscularis mucosae** (smooth muscle). Epithelial cells extend down into the lamina propria, forming many narrow channels called **gastric pits** and columns of secretory cells called **gastric glands.** Secretions from several gastric glands flow into each gastric pit and then into the lumen of the stomach. The gastric glands include three types of *exocrine gland* cells that secrete their products into the stomach lumen: mucous neck cells, chief cells, and parietal cells. Both mucous surface cells and **mucous neck cells** secrete mucus. The **chief cells**

FIGURE 22.8　Stomach. External and internal anatomy.

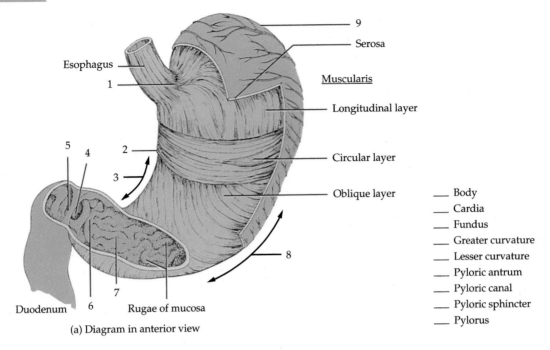

Esophagus

Serosa

Muscularis

Longitudinal layer

Circular layer

Oblique layer

1

2

3

5　4

7

6

Duodenum

Rugae of mucosa

8

9

___ Body
___ Cardia
___ Fundus
___ Greater curvature
___ Lesser curvature
___ Pyloric antrum
___ Pyloric canal
___ Pyloric sphincter
___ Pylorus

(a) Diagram in anterior view

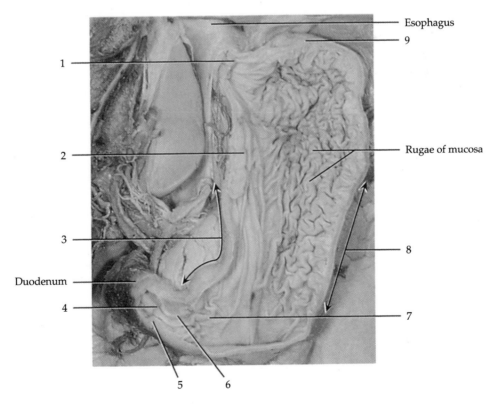

Esophagus

9

1

2

3

Duodenum

4

5　6

Rugae of mucosa

8

7

(b) Photograph in anterior view

FIGURE 22.9 Histology of stomach.

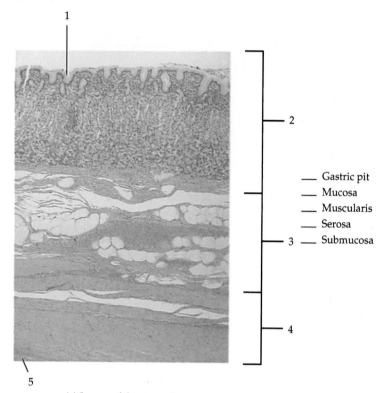

1

2

Gastric pit
Mucosa
Muscularis
Serosa
3 — Submucosa

4

5

(a) Layers of the stomach

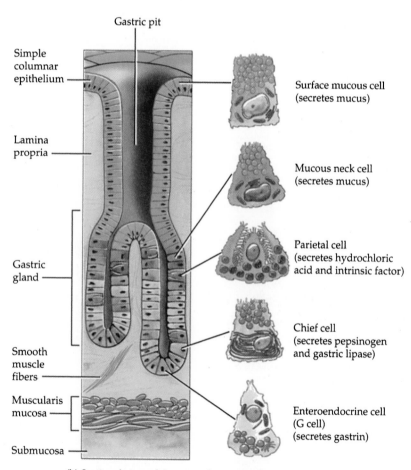

Gastric pit

Simple columnar epithelium

Surface mucous cell (secretes mucus)

Lamina propria

Mucous neck cell (secretes mucus)

Gastric gland

Parietal cell (secretes hydrochloric acid and intrinsic factor)

Chief cell (secretes pepsinogen and gastric lipase)

Smooth muscle fibers

Muscularis mucosa

Enteroendocrine cell (G cell) (secretes gastrin)

Submucosa

(b) Sectional view of the stomach mucosa showing gastric glands

Relations of the liver, gallbladder, duodenum, and pancreas.

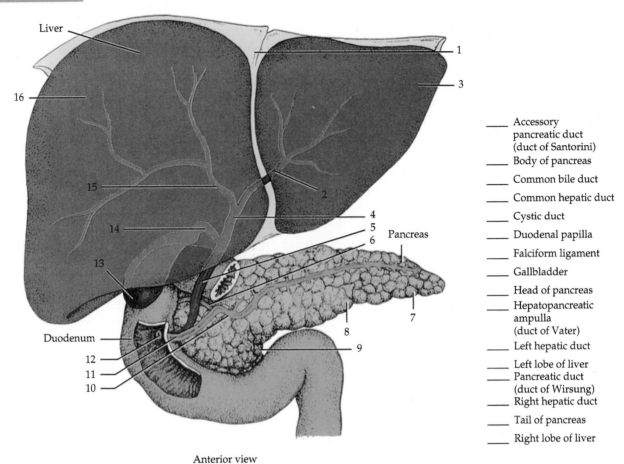

Liver

16

15

14

13

Duodenum

12

11

10

1

3

2

4
5
6

Pancreas

7

8

9

_____ Accessory
pancreatic duct
(duct of Santorini)
_____ Body of pancreas
_____ Common bile duct
_____ Common hepatic duct
_____ Cystic duct
_____ Duodenal papilla
_____ Falciform ligament
_____ Gallbladder
_____ Head of pancreas
_____ Hepatopancreatic
ampulla
(duct of Vater)
_____ Left hepatic duct
_____ Left lobe of liver
_____ Pancreatic duct
(duct of Wirsung)
_____ Right hepatic duct
_____ Tail of pancreas
_____ Right lobe of liver

Anterior view

though the right lobe is considered by many anatomists to include an inferior **quadrate lobe** and a posterior **caudate lobe,** based on internal morphology, these lobes more appropriately belong to the left lobe.

Each lobe is composed of functional units called **lobules.** Among the structures in a lobule are **hepatocytes (liver cells)** arranged in irregular, branching, interconnected plates around a **central vein; sinusoids,** endothelial lined spaces between hepatocytes through which blood flows; and **stellate reticuloendothelial (Kupffer) cells** that destroy worn-out leukocytes and blood cells, bacteria, and other foreign matter in the venous blood draining from the gastrointestinal tract.

Examine a prepared slide of several liver lobules. Compare your observations with Figure 22.11.

Bile is secreted by hepatocytes and functions in the emulsification of triglycerides in the small intestine. The liquid is passed to the small intestine as follows: Hepatocytes secrete bile into **bile canaliculi** (kan'-a-LIK-ū-lī = small canals) that

empty into small bile ductules. The ductules merge and eventually form the larger **right** and **left hepatic ducts,** one in each principal lobe of the liver. The right and left hepatic ducts unite outside the liver to form a single **common hepatic duct.** This duct joins the **cystic** (*cystis* = bladder) **duct** from the gallbladder to become the **common bile duct,** which empties into the duodenum at the hepatopancreatic ampulla (ampulla of Vater). When triglycerides are not being digested, a valve around the hepatopancreatic ampulla, the **sphincter of the hepatopancreatic ampulla (sphincter of Oddi),** closes, and bile backs up into the gallbladder via the cystic duct. In the gallbladder, bile is stored and concentrated.

With the aid of your textbook, label the structures associated with the liver in Figure 22.10.

10. Gallbladder

The **gallbladder** (*galla* = bile) is a pear-shaped sac in a depression of the posterior surface of the liver

FIGURE 22.11 Photomicrograph of liver lobule (100×).

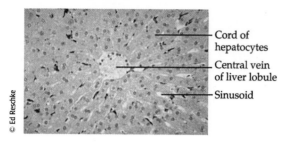

- Cord of hepatocytes
- Central vein of liver lobule
- Sinusoid

(see Figure 22.1). The gallbladder stores and concentrates bile. The cystic duct of the gallbladder and common hepatic duct of the liver merge to form the common bile duct. The **mucosa** of the gallbladder consists of simple columnar epithelium that contains rugae. The wall of the gallbladder lacks a submucosa. The **muscularis** consists of smooth muscle fibers and the outer coat consists of visceral peritoneum.

With the aid of your textbook, label the structures associated with the gallbladder in Figure 22.10.

11. Small Intestine

The major events of digestion and absorption occur in the **small intestine,** which begins at the pyloric sphincter of the stomach, coils through the central and inferior part of the abdomen, and joins the large intestine at the ileocecal sphincter (see Figure 22.1). The mesentery attaches the small intestine to the posterior abdominal wall. The small intestine is 2.5 cm (1 in) in diameter and is about 3 m (10 ft) in length in a living person. It is divided into three regions: **duodenum** (doo′-ō-DĒ-num), which begins at the pyloric sphincter of the stomach; **jejunum** (jē-JOO-num), the middle segment; and **ileum** (IL-ē-um), which terminates at the large intestine (Figure 22.12).

The wall of the small intestine is composed of the same four coats that make up most of the gastrointestinal tract. However, special features of both the mucosa and the submucosa facilitate the processes of digestion and absorption (Figure 22.13). The mucosa forms a series of fingerlike **villi** (= tuft of hair; **villus** is singular). These projections are 0.5 to 1 mm long and give the intestinal mucosa a velvety appearance. The large

FIGURE 22.12 Intestines. Note the parts of the small intestine on the right side of the illustration and the parts of the large intestine on the left side.

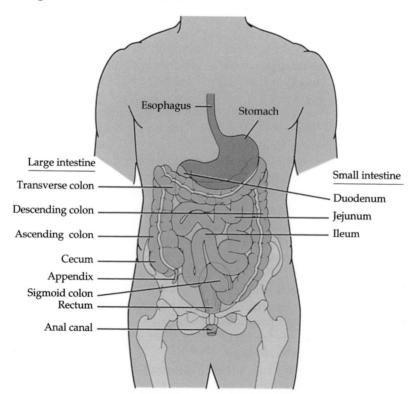

Esophagus
Stomach
Large intestine
Transverse colon
Descending colon
Ascending colon
Cecum
Appendix
Sigmoid colon
Rectum
Anal canal
Small intestine
Duodenum
Jejunum
Ileum

FIGURE 22.13 Histology of small intestine.

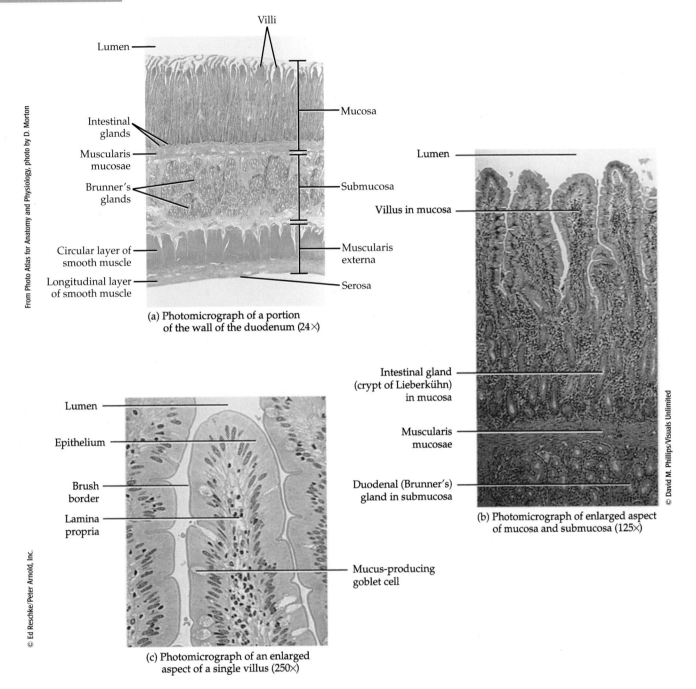

Villi

Lumen

Intestinal glands

Muscularis mucosae

Brunner's glands

Circular layer of smooth muscle

Longitudinal layer of smooth muscle

Mucosa

Submucosa

Muscularis externa

Serosa

(a) Photomicrograph of a portion of the wall of the duodenum (24×)

Lumen

Villus in mucosa

Intestinal gland (crypt of Lieberkühn) in mucosa

Muscularis mucosae

Duodenal (Brunner's) gland in submucosa

(b) Photomicrograph of enlarged aspect of mucosa and submucosa (125×)

Lumen

Epithelium

Brush border

Lamina propria

Mucus-producing goblet cell

(c) Photomicrograph of an enlarged aspect of a single villus (250×)

number of villi (20 to 40 per square millimeter) vastly increases the surface area of the epithelium available for absorption and digestion. Each villus has a core of lamina propria. Embedded in this connective tissue are an arteriole, a venule, a capillary network, and a **lacteal** (LAK-tē-al), which is a lymphatic capillary. Nutrients absorbed by the epithelial cells covering the villus pass through the wall of a capillary or a lacteal to enter blood or lymph, respectively.

The epithelium of the mucosa consists of simple columnar epithelium and contains absorptive cells, goblet cells, enteroendocrine cells, and Paneth cells. The apical (free) membrane of the absorptive cells features **microvilli** (mī′-krō-VIL-ī), which are microscopic, fingerlike projections of the plasma membrane that contain actin filaments. In a photomicrograph taken through a light microscope, the microvilli are too small to be seen individually. They form a fuzzy line, called

FIGURE 22.13 **Histology of small intestine. (Continued)**

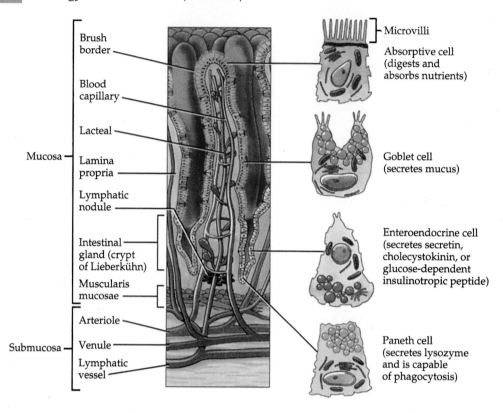

(d) Enlarged villus showing lacteal, capillaries, and intestinal gland

the **brush border,** at the apical surface of the absorptive cells, extending into the lumen of the small intestine. Larger amounts of digested nutrients can diffuse into the absorptive cells of the intestinal wall because the microvilli greatly increase the surface area of the plasma membrane. It is estimated that there are about 200 million microvilli per square millimeter of small intestine.

The mucosa contains many deep crevices lined with glandular epithelium. Cells lining the crevices form the **intestinal glands (crypts of Lieberkühn)** and secrete intestinal juice. **Paneth cells** are found in the deepest parts of the intestinal glands. They secrete lysozyme, a bactericidal enzyme, and are also capable of phagocytosis. They may have a role in regulating the microbial population in the intestines. The submucosa of the duodenum contains **duodenal (Brunner's) glands.** They secrete an alkaline mucus that helps neutralize gastric acid in the chyme. Some of the epithelial cells in the mucosa are goblet cells, which secrete additional mucus.

The lamina propria of the small intestine has an abundance of mucosa-associated lymphoid tissue (MALT) in the form of lymphatic nodules, masses of lymphatic tissue not surrounded by a capsule. **Solitary lymphatic nodules** are most numerous in the lower part of the ileum. Groups of lymphatic nodules, referred to as **aggregated lymphatic follicles (Peyer's patches),** are numerous in the ileum. The muscularis mucosae consist of smooth muscle.

The **muscularis** of the small intestine consists of two layers of smooth muscle. The outer, thinner layer contains longitudinally arranged fibers. The inner, thicker layer contains circularly arranged fibers.

Except for a major portion of the duodenum, the **serosa** (or visceral peritoneum) completely surrounds the small intestine.

Obtain prepared slides of the small intestine (section through its tunics and villi) and identify as many structures as you can, using Figure 22.13 and your textbook as references.

12. Large Intestine

The **large intestine** functions in the completion of absorption, the formation of feces, the expulsion of feces from the body, and the production of certain vitamins. The large intestine is about 1.5 m (5 ft) long and 6.5 cm (2.5 in.) in diameter, and extends from the ileum to the anus (see Figure 22.12). It is attached to the posterior abdominal wall by its **mesocolon,** which is a double layer of peritoneum. The large intestine is divided into four principal regions: cecum, colon, rectum, and anal canal.

The opening from the ileum into the large intestine is guarded by a fold of mucous membrane, the **ileocecal sphincter (valve).** Hanging inferior to the valve is a blind pouch, the **cecum,** to which is attached the **vermiform appendix** (*vermis* = worm-shaped; *appendix* = appendage) by an extension of visceral peritoneum called the **mesoappendix.** Inflammation of the vermiform appendix is called **appendicitis.** The open end of the cecum merges with the **colon** (= food passage). The first division of the colon is the **ascending colon,** which ascends on the right side of the abdomen

and turns abruptly to the left at the inferior surface of the liver **(right colic [hepatic] flexure).** The **transverse colon** continues across the abdomen, curves at the inferior surface of the spleen **(left colic [splenic] flexure),** and passes down the left side of the abdomen as the **descending colon.** The **sigmoid colon** begins near the iliac crest, projects medially toward the midline, and terminates at the rectum at the level of the third sacral vertebra. The **rectum** is the last 20 cm (8 in.) of the gastrointestinal tract and lies anterior to the sacrum and coccyx. Its terminal 2 to 3 cm (1 in.) is known as the **anal canal.** The opening of the anal canal to the exterior is the **anus.**

With the aid of your textbook, label the parts of the large intestine in Figure 22.14.

The wall of the large intestine differs from that of the small intestine in several respects. No villi or permanent circular folds are found in the mucosa. The **mucosa** consists of simple columnar ep-

FIGURE 22.14 **Large intestine.**

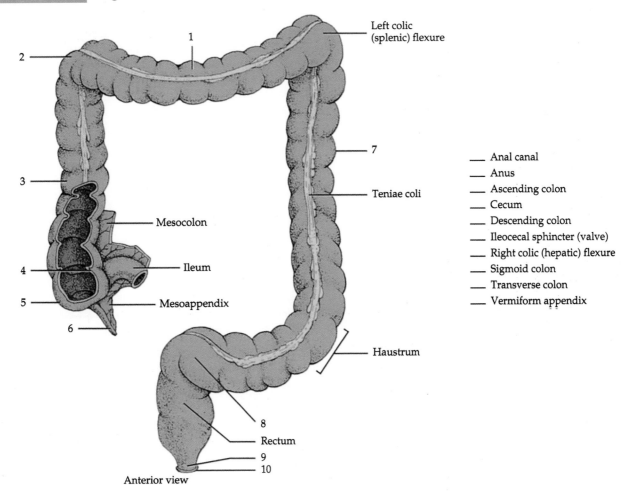

Left colic (splenic) flexure

Teniae coli

Mesocolon

Ileum

Mesoappendix

Haustrum

Rectum

Anterior view

___ Anal canal
___ Anus
___ Ascending colon
___ Cecum
___ Descending colon
___ Ileocecal sphincter (valve)
___ Right colic (hepatic) flexure
___ Sigmoid colon
___ Transverse colon
___ Vermiform appendix

ithelium, lamina propria (areolar connective tissue), and muscularis mucosae (smooth muscle). The epithelium contains mostly absorptive and goblet cells (Figure 22.15). The absorptive cells function primarily in water absorption. The goblet cells secrete mucus that lubricates the colonic contents as they pass through. Both absorptive and goblet cells are located in long, straight, tubular intestinal glands that extend the full thickness of the mucosa. Solitary lymphatic nodules are also found in the lamina propria of the mucosa.

The **submucosa** of the large intestine is similar to that found in the rest of the gastrointestinal tract.

The **muscularis** consists of an external layer of longitudinal muscles and an internal layer of circular muscles. Unlike other parts of the GI tract, portions of the longitudinal muscles are thickened, forming three conspicuous longitudinal bands called **teniae coli** (TĒ-nē-ē KŌ-lī; *tiniae* = flat band), separated by portions of the wall with less or no longitudinal muscle (see Figure 22.14). Each band runs the length of most of the large intestine. Tonic contractions of the bands gather the colon into a series of pouches called **haustra** (HAWS-tra; singular is **haustrum** = shaped like a pouch), which give the colon a puckered appearance. There is a single layer of circular muscle between teniae coli. The **serosa** of the large intestine is part of the visceral peritoneum. Small pouches of visceral peritoneum filled with fat are attached to teniae coli and are called **epiploic appendages.**

Examine a prepared slide of the large intestine showing its tunics. Compare your observations to Figure 22.15.

C. Deglutition

Swallowing, or **deglutition** (dē-gloo-TISH-un), is the mechanism that moves food from the mouth through the esophagus to the stomach. It is facilitated by saliva and mucus and involves a complex series of actions by the mouth, pharynx, and esophagus. Swallowing can be initiated voluntarily, but the remainder of the activity is almost completely under reflex control. Swallowing is a very orderly sequence of events involving the movement of food from the mouth to the stomach, while simultaneously involving the inhibition of respiratory activity and the prevention of entrance of foreign particles into the trachea.

Swallowing is divided into three stages: (1) the **voluntary stage,** when a **bolus** (soft, flexible mass of food) is passed into the oropharynx; (2) the **pharyngeal stage,** involving the movement of the bolus through the pharynx into the esophagus; and (3) the **esophageal stage,** which involves the movement of the bolus from the esophagus, through the lower esophageal sphincter, and into the stomach. Both the pharyngeal and esophageal phases are involuntary. The passage of solid or semisolid food from the mouth to the stomach takes 4 to 8 sec. Very soft foods and liquids pass through in about 1 sec.

FIGURE 22.15 Histology of large intestine.

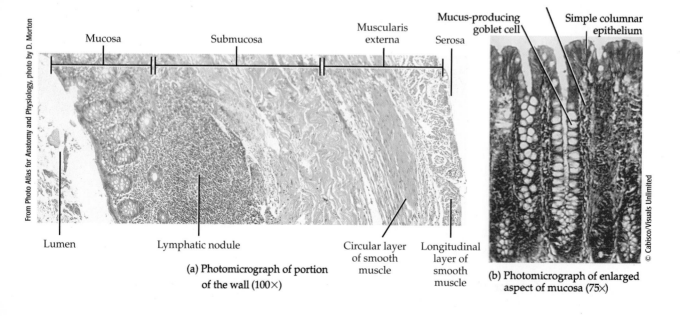

From Photo Atlas for Anatomy and Physiology, photo by D. Morton

© Cabisco/Visuals Unlimited

Mucosa   Submucosa   Muscularis externa   Serosa   Mucus-producing goblet cell   Simple columnar epithelium

Lumen   Lymphatic nodule   Circular layer of smooth muscle   Longitudinal layer of smooth muscle

(a) Photomicrograph of portion of the wall (100×)

(b) Photomicrograph of enlarged aspect of mucosa (75×)

FIGURE 22.15 ### Histology of large intestine. (Continued)

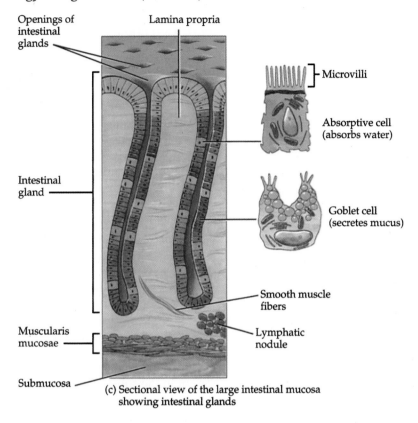

Openings of intestinal glands

Lamina propria

Microvilli

Absorptive cell (absorbs water)

Intestinal gland

Goblet cell (secretes mucus)

Smooth muscle fibers

Muscularis mucosae

Lymphatic nodule

Submucosa

(c) Sectional view of the large intestinal mucosa showing intestinal glands

PROCEDURE

1. Chew a cracker and note the movements of your tongue as you begin to swallow. Chew another cracker, lean over your chair, placing your mouth below the level of your pharynx, and attempt to swallow again.

2. Note the movements of your laboratory partner's larynx during the pharyngeal stage of deglutition.

3. *Clean the earpieces of a stethoscope with alcohol.* Using the stethoscope, listen to the sounds that occur when your partner swallows water. Place the stethoscope just to the left of the midline at the level of the sixth rib. You should hear the sound of water as it makes contact with the lower esophageal sphincter (valve), followed by the sound of water passing through the sphincter after the sphincter relaxes. The sphincter is at the junction of the esophagus and stomach.

4. Record your observations in Section C of the LABORATORY REPORT RESULTS at the end

of the exercise. At what point can deglutition be stopped voluntarily? _____

_____

D. Chemistry of Digestion

In order to understand how food is digested, let us examine the various processes that occur between absorption and final utilization of nutrients by the body's cells.

Nutrients are chemical substances in food that provide energy, act as building blocks to form new body components, or assist body processes. The six major classes of nutrients are carbohydrates, lipids, proteins, minerals, vitamins, and water. Cells break down carbohydrates, lipids, and proteins to release energy, or use them to build new structures and new regulatory substances, such as hormones and enzymes.

Enzymes are produced by living cells to catalyze or speed up many of the reactions in the body. Enzymes are proteins and act as **catalysts** to speed up reactions without being permanently altered by the reaction. Digestive enzymes function as catalysts to speed chemical reactions in the gastrointestinal tract.

Because digestive enzymes function outside the cells that produce them, they also are capable of reacting within a test tube and therefore provide an excellent means of studying enzyme activity.

▲ **CAUTION!** *Please reread Section A, "General Safety Precautions and Procedures," on page xi and Section C, "Precautions Related to Working with Reagents," on page xii at the beginning of the laboratory manual before you begin any of the following experiments. You should also read the experiments before you perform them to be sure that you understand all the procedures and safety precautions.*

1. Positive Tests for Sugar and Starch

Salivary amylase is an enzyme produced by the salivary glands. This enzyme starts starch digestion and **hydrolyzes** (splits using water) it into maltose (a disaccharide), maltotriose (a trisaccharide), and α-dextrins. We measure the amount of starch and sugar present before and after enzymatic activity. It is expected that the amount of starch should *decrease* and the sugar level should *increase* as a result of salivary amylase activity.

a. Test for Sugar

Benedict's test is commonly used to detect sugars. Glucose (monosaccharide), maltose (disaccharide), or any other reducing sugars react with **Benedict's solution,** forming insoluble red cuprous oxide. The precipitate of cuprous oxide can usually be seen in the bottom of the tube when standing. Benedict's solution turns green, yellow, orange, or red depending on the amount of reducing sugar present according to the following scale:

blue (−)
green (+)
yellow (++)
orange (+++)
red (++++)

Test for the presence of sugar (maltose) as follows:

PROCEDURE

1. Using separate medicine droppers, place 2 mL of maltose and 2 mL of Benedict's solution in a Pyrex test tube.

2. *Using a test tube holder,* place the test tube in a boiling water bath and heat for 5 min.

▲ **CAUTION!** *Make sure that the mouth of the test tube is pointed away from you and all other persons in the area.* Note the color change.

3. *Using a test tube holder,* remove the test tube from the water bath.

4. Repeat the same procedure using a starch solution instead of maltose. Notice that the color does not change, because the solution contains no sugar. Now you have a method for detecting sugar.

b. Test for Starch

Lugol's solution is a brown iodine solution used to test certain polysaccharides, especially starch. Starch, for example, gives a *deep blue* to *black* color with Lugol's solution (the black is really a concentrated blue color). Cellulose, monosaccharides, and disaccharides do not react. A negative test is indicated by a yellow to brown color of the solution itself, or possibly some other color (other than blue to black) resulting from pigments present in the substance being tested.

PROCEDURE

1. Using separate medicine droppers, place a drop of starch solution on a spot plate and then add a drop of Lugol's solution to it. Notice the black color that forms as the starch-iodine complex develops.

2. Repeat the test using a maltose solution in place of the starch solution. Note that there is no color change, because Lugol's solution and maltose do not combine. Now you have a method for detecting starch.

c. Digestion of Starch
PROCEDURE

1. Using separate medicine droppers, transfer 3 mL of a starch solution to a small beaker, add a fresh enzyme solution consisting of a pinch of amylase powder in 3 mL of water, and mix thoroughly with a glass rod.

2. Wait for 1 min, record the time, remove 1 drop of the mixture with a glass rod to the depression of a spot plate, and then test for starch with Lugol's solution.

3. At 1-min intervals, test 1-drop samples of the mixture until you no longer note a positive test for starch. Keep the glass rod in the mixture, stirring it from time to time.

4. After the starch test is seen to be negative, test the remaining mixture for the presence of glucose. Do this by adding 2 mL of the mixture with a medicine dropper to 2 mL of the Benedict's solution in a Pyrex test tube and heat in a boiling water bath as per the procedure outlined in a.2, "Test for Sugar." Answer questions a through c in Section D.1 of the LABORATORY REPORT RESULTS at the end of the exercise.

2. Effect of Temperature on Starch Digestion

In the following procedure, you will test starch digestion at five different temperatures to determine how temperature influences enzyme activity. Lugol's solution is again used for presence or absence of starch.

A fresh enzyme solution consisting of a pinch of amylase powder in 3 mL of water should be used as before. Five constant-temperature water baths should be available. Starting with the lowest temperature, these are: 0°C or cooler, 10°C, 40°C, 60°C, and boiling.

PROCEDURE

1. Prepare 10 Pyrex test tubes, 5 containing 1 mL each of enzyme solution and 5 containing 1 mL each of starch solution.

2. Using rubber bands, pair the tubes (i.e., a tube containing enzyme with one containing starch) and use a test tube holder to place one pair into each water bath.

▲ **CAUTION!** *Make sure that the mouth of the test tube is pointed away from you and all other persons in the area.*

3. Permit the tubes to adapt to the bath temperatures for about 5 min, then mix the enzyme and starch solutions of each pair together, and *using a test tube holder* place the single test tube in its respective water bath.

4. After 30 sec, *using a test tube holder,* remove the test tubes from the water baths and test all five tubes for starch on a spot plate using Lugol's solution.

5. Repeat every 30 sec until you have determined the time required for the starch to disappear (i.e., to be digested).

6. Record and graph your results in Section D.2 of the LABORATORY REPORT RESULTS at the end of the exercise.

3. Effect of pH on Starch Digestion

You can demonstrate the effectiveness of salivary amylase digestion at different pH readings.

PROCEDURE

1. Prepare three buffer solutions as follows:

   Solution A   pH 4.0
   Solution B   pH 7.0
   Solution C   pH 9.0

2. Once again a fresh enzyme solution consisting of a pinch of amylase powder in 3 mL of water should be used.

3. Using medicine droppers, mix 4 mL of a starch solution with 2 mL of buffer solution A in a Pyrex test tube.

4. Repeat this procedure with buffer solutions B and C.

5. You now have three test tubes of a starch-buffer solution, each at a different pH (4.0, 7.0, and 9.0).

6. Using separate medicine droppers, place one drop of starch-buffer solution A on a spot plate and immediately add one drop of the saliva.

7. Test for starch disappearance using Lugol's solution, and record the time when starch first disappears completely.

8. Repeat this test for the other two starch buffers (solutions B and C) and record the time when starch is no longer present at each pH.

9. Record and explain your results in Section D.3 of the LABORATORY REPORT RESULTS at the end of the exercise.

4. Action of Bile on Triglycerides

Bile is important in the process of lipid digestion because of its emulsifying effects (breaking down of large globules to smaller, uniformly distributed particles) on triglycerides (fats) and oils. *Bile does not contain any enzymes.* Emulsification of triglycerides by means of bile salts serves to increase the surface area of the triglyceride that will be exposed to the action of the lipase.

PROCEDURE

1. Place 5 mL of water into one Pyrex test tube and 5 mL of bile solution into a second.

2. Using a medicine dropper, add one to five drops of vegetable oil that has been colored with a fat-soluble dye, such as Sudan B, into each tube.

3. Place a stopper in both tubes. Shake them *vigorously,* and then let them stand in a test tube rack undisturbed for 10 min. Triglycerides or oils that are broken into sufficiently small droplets will remain suspended in water in the form of an **emulsion.** If emulsification has not occurred, the triglyceride or oil will lie on the surface of the water.

4. Answer questions in Section D.4 of the LABORATORY REPORT RESULTS at the end of the exercise.

5. Digestion of Triglycerides

You can demonstrate the effect of pancreatic juice on triglycerides by the use of pancreatin, which contains all the enzymes present in pancreatic juice. Because the optimum pH of the pancreatic enzymes ranges from 7.0 to 8.8, the pancreatin is prepared in sodium carbonate. The enzyme used in this test is **pancreatic lipase,** which digests triglycerides to fatty acids and glycerol. The fatty acid produced changes the color of *blue* litmus to *red.*

PROCEDURE

1. Using a medicine dropper, place 5 mL of litmus cream (heavy cream to which powdered litmus has been added to give it a blue color) in a Pyrex test tube and, *using a test tube holder,* place the test tube in a 40°C water bath.

2. Repeat the procedure with another 5-mL portion in a second tube, but put it in an ice bath.

3. When the tubes have adapted to their respective temperatures (in about 5 min), use a medicine dropper to add 5 mL of pancreatin to each tube and, *using a test tube holder,* replace them in their water baths until a color change occurs in one tube (approximately 20 min).

4. Summarize your results and your explanation in Section D.5 of the LABORATORY REPORT RESULTS at the end of the exercise.

6. Digestion of Protein

Here you demonstrate the effect of pepsin on protein and the factors affecting the rate of action of pepsin. **Pepsin,** a proteolytic enzyme, is secreted in inactive form (pepsinogen) by chief (zymogenic) cells in the lining of the stomach. Pepsin digests proteins (fibrin in this experiment) to peptides. The efficiency of pepsin activity depends on the pH of the solution, the optimum being 1.5 to 2.5. Pepsin is almost completely inactive in neutral or alkaline solutions.

PROCEDURE

1. Prepare and number the following five Pyrex test tubes. For this test, the quantity of each solution must be *measured carefully.*

   Tube 1—5 mL of 0.5% pepsin; 5 mL of 0.8% HCl
   Tube 2—5 mL of pepsin; 5 mL of water
   Tube 3—5 mL of pepsin, boiled for 10 min in a water bath; 5 mL of 0.8% HCl

▲ **CAUTION!** *Make sure that the mouth of the test tube is pointed away from you and all other persons in the area. Using a test tube holder, remove the test tube from the water bath.*

   Tube 4—5 mL of pepsin; 5 mL of 0.5% NaOH
   Tube 5—5 mL of water; 5 mL of 0.8% HCl

2. First determine the approximate pH of each test tube using Hydroin paper (range 1 to 11) and put the values in the table provided in Section D.6 of the LABORATORY REPORT RESULTS at the end of the exercise.

3. Using forceps, place a small amount of fibrin (the protein) in each test tube. An amount near the size of a pea will be sufficient. *Using a test tube holder,* put the tubes in a 40°C water bath and *carefully* shake occasionally. Maintain 40°C temperature closely. The tubes should remain in the water bath for *at least 1½ hr.*

4. Watch the changes that the fibrin undergoes. The swelling that occurs in some tubes should not be confused with digestion. Digested fibrin becomes transparent and disappears (dissolves) as the protein is digested to soluble peptides.

5. Finish the experiment when the fibrin is digested in one of the five tubes.

6. Record all your observations in the table provided in Section D.6 of the LABORATORY REPORT RESULTS at the end of the exercise.

ANSWER THE LABORATORY REPORT QUESTIONS AT THE END OF THE EXERCISE.

Digestive System

Name _____ Date _____

Laboratory Section _____ Score/Grade _____

SECTION C ■ Deglutition

1. Tongue movement at beginning of deglutition: _____
   _____

2. Description of swallowing with head below pharynx: _____
   _____

3. Laryngeal movements during pharyngeal deglutition: _____
   _____

4. Time lapse between two sounds of deglutition: _____
   _____

SECTION D ■ Chemistry of Digestion

1. Digestion of Starch

a. How long did it take for the starch to be digested? _____

b. Did your observation indicate the presence of maltose? _____

c. What is the meaning of this result? _____

2. Effect of Temperature on Starch Digestion

d. Record the time required for starch to disappear at the temperatures tested.

      _____ 0°C   _____ 40°C   _____ Boiling

      _____ 10°C   _____ 60°C

e. What is the optimum temperature for starch digestion? _____

f. What happens to amylase when it is boiled? _____

g. Is the effect of boiling reversible or irreversible? _____

h. Is the effect of freezing reversible or irreversible? _____

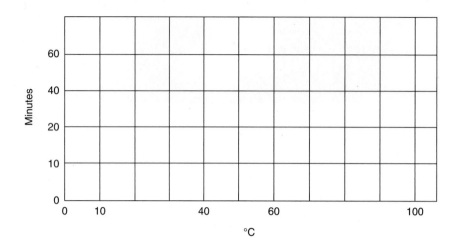

3. Effect of pH on Starch Digestion

i. Record the time required for starch to disappear at the pH readings tested.

    Solution A: pH 4.0 _____

    Solution B: pH 7.0 _____

    Solution C: pH 9.0 _____

j. Explain your results

    A: pH 4.0 _____

    B: pH 7.0 _____

    C: pH 9.0 _____

k. What is the optimum pH for the action of salivary amylase? _____

_____

4. Action of Bile on Triglycerides

l. What difference can you detect in the appearance of the mixtures in the two test tubes? _____

_____

m. How does emulsification of lipids by means of bile aid in the digestion of triglycerides? _____

_____

5. Digestion of Triglycerides

Record the results of the digestion of triglycerides by pancreatic lipase in pancreatic juice in the following table.

Tube no.	Temperature of water bath	Change in pH (color)	Explanation of results
1			
2			

6. Digestion of Protein

Record the results of the digestion of protein by pepsin in the following table.

Tube no.	Tube contents	pH	Digestion observed (yes or no)	Explanation of results
1				
2				
3				
4				
5				

EXERCISE 22	Digestive System

Name _____   Date _____

Laboratory Section _____   Score/Grade _____

PART 1 ■ Multiple Choice

_____ **1.** If an incision has to be made in the small intestine to remove an obstruction, the first layer of tissue to be cut is the (a) muscularis (b) mucosa (c) serosa (d) submucosa

_____ **2.** Mesentery, lesser omentum, and greater omentum are all directly associated with the (a) peritoneum (b) liver (c) esophagus (d) mucosa of the gastrointestinal tract

_____ **3.** Chemical digestion of carbohydrates is initiated in the (a) stomach (b) small intestine (c) mouth (d) large intestine

_____ **4.** A tumor of the villi and circular folds would interfere most directly with the body's ability to carry on (a) absorption (b) deglutition (c) mastication (d) peristalsis

_____ **5.** The main chemical activity of the stomach is to begin the digestion of (a) triglycerides (b) proteins (c) carbohydrates (d) all of the above

_____ **6.** Surgical cutting of the lingual frenulum would occur in which part of the body? (a) salivary glands (b) esophagus (c) nasal cavity (d) tongue

_____ **7.** To free the small intestine from the posterior abdominal wall, which of the following would have to be cut? (a) mesocolon (b) mesentery (c) lesser omentum (d) falciform ligament

_____ **8.** The cells of gastric glands that produce secretions directly involved in chemical digestion are the (a) mucous (b) parietal (c) chief (d) pancreatic islets (islets of Langerhans)

_____ **9.** An obstruction in the hepatopancreatic ampulla (ampulla of Vater) would affect the ability to transport (a) bile and pancreatic juice (b) gastric juice (c) salivary amylase (d) intestinal juice

_____ **10.** The terminal portion of the small intestine is known as the (a) duodenum (b) ileum (c) jejunum (d) pyloric sphincter (valve)

_____ **11.** The portion of the large intestine closest to the liver is the (a) right colic flexure (b) rectum (c) sigmoid colon (d) left colic flexure

_____ **12.** The lamina propria is found in which coat? (a) serosa (b) muscularis (c) submucosa (d) mucosa

_____ **13.** Which structure attaches the liver to the anterior abdominal wall and diaphragm? (a) lesser omentum (b) greater omentum (c) mesocolon (d) falciform ligament

_____ **14.** The opening between the oral cavity and pharynx is called the (a) vermilion border (b) fauces (c) vestibule (d) lingual frenulum

_____ **15.** All of the following are parts of a tooth *except* the (a) crown (b) root (c) cervix (d) neck

_____ **16.** Cells of the liver that destroy worn-out white and red blood cells and bacteria are termed (a) hepatocytes (b) stellate reticuloendothelial (Kupffer) cells (c) alpha cells (d) beta cells

_____ **17.** Bile is manufactured by which cells? (a) alpha (b) beta (c) hepatocytes (d) stellate reticuloendothelial (Kupffer)

_____ **18.** Which part of the small intestine secretes the intestinal digestive enzymes? (a) intestinal glands (b) duodenal (Brunner's) glands (c) lacteals (d) microvilli

_____ **19.** Structures that give the colon a puckered appearance are called (a) teniae coli (b) villi (c) rugae (d) haustra

PART 2 ■ Completion

20. An acute inflammation of the serous membrane lining the abdominal cavity and covering the abdominal viscera is referred to as _____.

21. The _____ is a sphincter (valve) between the ileum and large intestine.

22. The portion of the small intestine that is attached to the stomach is the _____.

23. The portion of the stomach closest to the esophagus is the _____.

24. The _____ forms the floor of the oral cavity and is composed of skeletal muscle covered with mucous membrane.

25. The convex lateral border of the stomach is called the _____.

26. The three special structures found in the wall of the small intestine that increase its efficiency in absorbing nutrients are the villi, circular folds, and _____.

27. The three pairs of salivary glands are the parotids, submandibulars, and _____.

28. The small intestine is divided into three segments: duodenum, ileum, and _____.

29. The large intestine is divided into four main regions: the cecum, colon, rectum, and

_____.

30. The enzyme that is present in saliva is called _____.

31. The transition of the lips where the outer skin and inner mucous membrane meet is called the

_____.

32. The _____ papillae are arranged in the form of an inverted V on the posterior surface of the tongue.

33. The portion of a tooth containing blood vessels, lymphatic vessels, and nerves is the

_____.

34. The teeth present in a permanent dentition, but not in a deciduous dentition, that replace the deciduous molars are the _____.

35. The portion of the gastrointestinal tract that conveys food from the pharynx to the stomach is the

    _____.

36. The inferior region of the stomach connected to the small intestine is the _____.

37. The clusters of cells in the pancreas that secrete digestive enzyme are called

    _____.

38. The caudate lobe, quadrate lobe, and central vein are all associated with the

    _____.

39. The common bile duct is formed by the union of the common hepatic duct and

    _____ duct.

40. The pear-shaped sac that stores bile is the _____.

41. _____ glands of the small intestine secrete an alkaline substance to protect the

    mucosa from excess acid.

42. The _____ attaches the large intestine to the posterior abdominal wall.

43. The _____ is the last 20 cm (8 in.) of the gastrointestinal tract.

44. A midline fold of mucous membrane that attaches the inner surface of each lip to its corresponding

    gum is the _____.

45. The bonelike substance that gives teeth their basic shape is called _____.

46. The _____ anchors teeth in position and helps to dissipate chewing forces.

47. The portion of the colon that terminates at the rectum is the _____ colon.

48. The palatine tonsils are between the palatoglossal and _____ arches.

49. The teeth closest to the midline are the _____.

50. The vermiform appendix is attached to the _____.

51. A soft, flexible mass of food that is easily swallowed is called a _____.

52. The cells in gastric glands that secrete pepsinogen are called _____.

PART 3 ■ **Matching**

_____ 53. Pancreatic lipase

_____ 54. Benedict's solution

_____ 55. Lugol's solution

_____ 56. Salivary amylase

_____ 57. Pepsin

A. Commonly used solution in the test for starch

B. Capable of digesting starch

C. Capable of digesting protein

D. Commonly used solution for detecting reducing sugars

E. Digests triglycerides into fatty acids and glycerol and changes the color of blue litmus to red

Urinary System

Objectives

At the completion of this exercise you should understand

A The anatomical characteristics and functions of the organs of the urinary system

B The anatomy of a sheep (or pig) kidney

C The physical and chemical characteristics of urine, and how these characteristics may vary under different physiological conditions

The **urinary system** functions to keep the body in homeostasis by controlling the composition and volume of the blood. The system accomplishes these functions by excreting wastes and foreign substances, regulating the blood ionic composition and pH, maintaining blood osmolarity, and regulating the blood pressure by secreting renin (which activates the renin-angiotensin-aldosterone pathway). The kidneys also assume a role in erythropoiesis by secreting *erythropoietin*, perform gluconeogenesis (synthesis of glucose molecules) during periods of fasting or starvation, and release *calcitriol*, the active form of vitamin D, to help regulate calcium homeostasis.

The urinary system consists of two kidneys, two ureters, one urinary bladder, and one urethra (Figure 23.1). Other systems that help in waste elimination are the respiratory, integumentary, and digestive systems.

The specialized branch of medicine that deals with the anatomy, physiology, and pathology of the kidney is known as **nephrology** (nef-ROL-ō-jē; *nephro* = kidney; *logy* = study of). The branch of medicine that deals with the male and female urinary systems and the male reproductive system is called **urology** (ū-ROL-ō-jē; *uro* = urine).

Using your textbook, charts, or models for reference, label Figure 23.1.

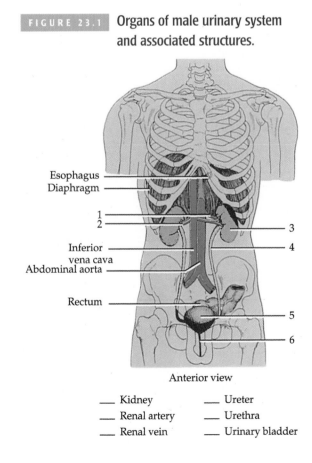

FIGURE 23.1 Organs of male urinary system and associated structures.

Esophagus
Diaphragm

1
2

Inferior vena cava
Abdominal aorta

Rectum

3
4

5
6

Anterior view

___ Kidney   ___ Ureter
___ Renal artery   ___ Urethra
___ Renal vein   ___ Urinary bladder

A. Organs of Urinary System

1. Kidneys

The paired **kidneys** are reddish-brown, kidney bean–shaped organs located just superior to the waist between the peritoneum and the posterior wall of the abdomen. Because their position is posterior to the peritoneum of the abdominal cavity, they are said to be **retroperitoneal** (re′-trō-per-i-tō-NĒ-al; *retro* = behind) organs. The kidneys are located between the levels of the last thoracic and third lumbar vertebrae, with the right kidney slightly lower than the left because the liver occupies considerable space on the right side superior to the kidney.

Three layers of tissue surround each kidney: the **renal** (*ren* = kidney) **capsule, adipose capsule,** and **renal fascia.** They function to protect the kidney and hold it firmly in place.

Near the center of the kidney's concave border is a deep vertical fissure called the **renal hilum,** through which the ureter leaves the kid-ney and through which blood vessels, lymphatics, and nerves enter and exit the kidney. The hilum expands into a cavity within the kidney called the **renal sinus.**

If a frontal section is made through a kidney, the following structures can be seen:

a. **Renal cortex** (*cortex* = rind or bark). Superficial, smooth-textured, reddish area.

b. **Renal medulla** (*medulla* = inner portion). Deep, reddish-brown area.

c. **Renal pyramids.** Cone-shaped structures, 8 to 18 in number, in the renal medulla. The bases (wider end) of the renal pyramids face the renal cortex, and the apices (narrower end), called **renal papillae,** are directed toward the renal hilum.

d. **Renal columns.** Portions of the renal cortex that extend between renal pyramids.

e. **Renal pelvis.** Single large cavity in the renal sinus, into which urine from the major calyces drains.

FIGURE 23.2 **Kidney.**

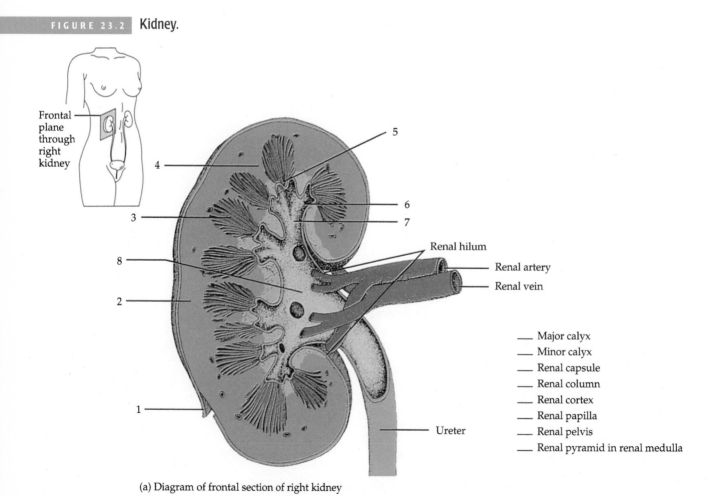

Frontal plane through right kidney

5

4

3

6

7

Renal hilum

Renal artery

Renal vein

8

2

1

Ureter

____ Major calyx
____ Minor calyx
____ Renal capsule
____ Renal column
____ Renal cortex
____ Renal papilla
____ Renal pelvis
____ Renal pyramid in renal medulla

(a) Diagram of frontal section of right kidney

FIGURE 23.2 **Kidney. (Continued)**

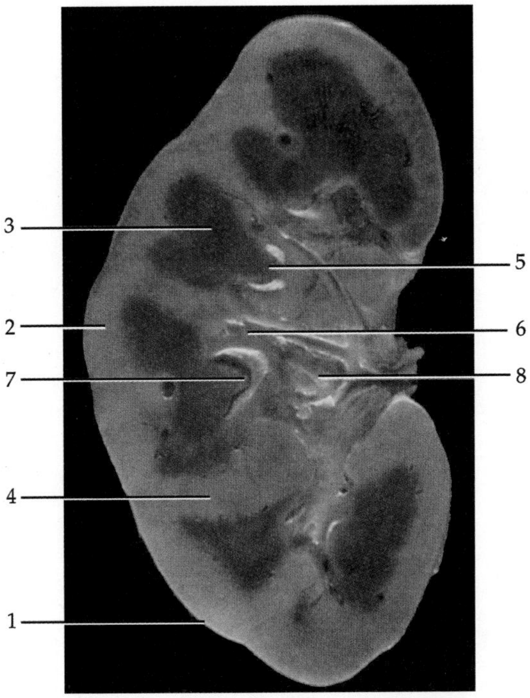

(b) Photograph of frontal section of right kidney

f. *Major calyces* (KĀ-li-sēz = cup). Consist of 2 or 3 cuplike extensions of the renal pelvis.

g. *Minor calyces.* Consist of 8 to 18 cuplike extensions of the major calyces.

Within the renal cortex and renal pyramids of each kidney are more than 1 million microscopic units called **nephrons,** the functional units of the kidneys (described shortly). As a result of their activity in regulating the volume and chemistry of the blood, they produce urine. Urine passes from the nephrons to the minor calyces, major calyces, renal pelvis, ureter, urinary bladder, and urethra.

Examine a specimen, model, or chart of the kidney and with the aid of your textbook, label Figure 23.2.

2. Nephrons

Basically, a **nephron** (NEF-ron) consists of two parts: (1) a renal corpuscle (KOR-pus-sul = tiny body) where plasma is filtered and (2) a renal tubule into which the filtered fluid (filtrate) passes. A **renal corpuscle** has two components: a capillary network called a **glomerulus** (glō-MER-ū-lus = small ball) surrounded by a double-walled epithelial cup, called a **glomerular (Bowman's) capsule,** lying in the renal cortex of

the kidney. The inner wall of the capsule, the **visceral layer,** consists of modified simple squamous epithelial cells called **podocytes** and surrounds the glomerulus. The **parietal layer,** which is composed of simple squamous epithelium, forms the outer wall of the capsule. A space called the **capsular (Bowman's) space** separates the two layers of the glomerular capsule.

The endothelial cells of the glomerular capillaries and podocytes form a leaky barrier known as the **filtration membrane.** This membrane permits filtration of water and small solutes but prevents filtration of most plasma proteins, blood cells, and platelets. Electron microscopy has determined that the membrane consists of the following components, given in the order in which substances are filtered (Figure 23.3).

a. *Glomerular endothelial cells.* The single layer of endothelial cells has large **fenestrations** (pores) that prevent filtration of blood cells and platelets but allow all components of blood plasma to pass through.

b. *Basal lamina.* This layer of acellular material lies between the endothelium and the podocytes. It consists of fibrils in a glycoprotein matrix and prevents filtration of larger plasma proteins.

FIGURE 23.3 Filtration membrane.

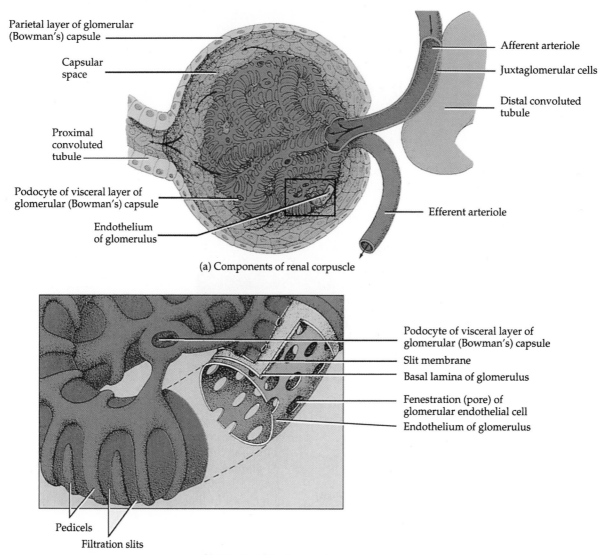

(a) Components of renal corpuscle

Parietal layer of glomerular (Bowman's) capsule

Capsular space

Proximal convoluted tubule

Podocyte of visceral layer of glomerular (Bowman's) capsule

Endothelium of glomerulus

Afferent arteriole

Juxtaglomerular cells

Distal convoluted tubule

Efferent arteriole

Podocyte of visceral layer of glomerular (Bowman's) capsule

Slit membrane

Basal lamina of glomerulus

Fenestration (pore) of glomerular endothelial cell

Endothelium of glomerulus

Pedicels

Filtration slits

(b) Details of filtration membrane

c. *Filtration slit membranes between pedicels.* The specialized epithelial cells that cover the glomerular capillaries are called **podocytes.** Extending from each podocyte are thousands of footlike structures called **pedicels** (PED-i-sels = little feet). The pedicels wrap around glomerular capillaries, except for spaces between them, which are called **filtration slits.** A thin membrane, the **slit membrane,** extends across filtration slits and permits the passage of molecules having a diameter smaller than 6–7 nm, including water, glucose, vitamins, amino acids, very small plasma proteins, ammonia, urea, and ions.

The filtration membrane filters blood passing through the kidney. Blood cells and large mole-cules, such as proteins, are retained by the filter and eventually are recycled into the blood. The filtered substances pass through the membrane and into the space between the parietal and visceral layers of the glomerular (Bowman's) capsule, and then enter the renal tubule (described shortly).

As the filtered fluid (filtrate) passes through the remaining parts of a nephron, substances are selectively added and removed. The end product of these activities is urine. Nephrons are frequently classified into two kinds. A **cortical nephron** usually has its glomerulus in the superficial renal cortex, and the remainder of the nephron penetrates only into the outer region of the renal medulla. A **juxtamedullary nephron** usually has its glomerulus deep in the cortex, close to the medulla, and other parts of the nephron

FIGURE 23.4 Nephrons. The colored arrows indicate the direction in which the filtrate flows.

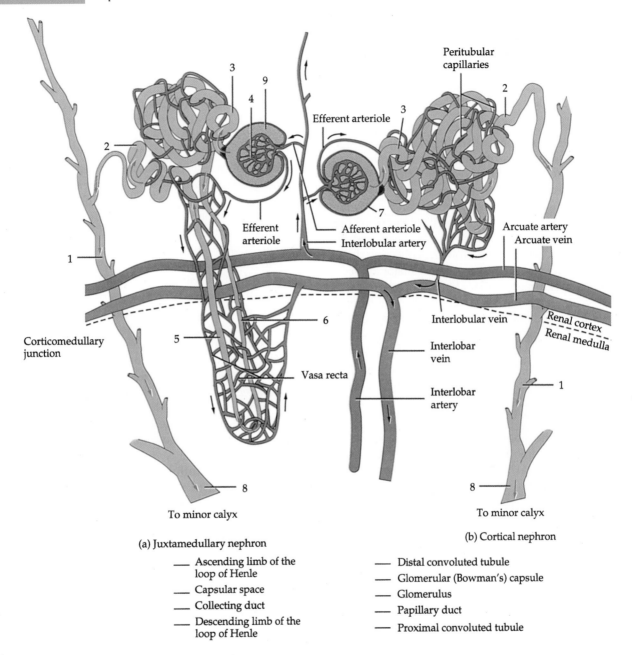

(a) Juxtamedullary nephron

___ Ascending limb of the
     loop of Henle
___ Capsular space
___ Collecting duct
___ Descending limb of the
     loop of Henle

(b) Cortical nephron

___ Distal convoluted tubule
___ Glomerular (Bowman's) capsule
___ Glomerulus
___ Papillary duct
___ Proximal convoluted tubule

extend into the deepest region of the renal medulla (see Figure 23.4). The following description of the remaining components of a nephron applies to juxtamedullary nephrons.

After the filtrate leaves the glomerular (Bowman's) capsule, it passes through the following parts of a renal tubule:

a. *Proximal convoluted tubule (PCT).* Coiled tubule in the renal cortex that originates at the glomerular (Bowman's) capsule; consists of simple cuboidal epithelium with a prominent brush border of microvilli.

b. *Loop of Henle (nephron loop).* U-shaped tubule that connects the proximal and distal convoluted tubules. It consists of a **descending limb of the loop of Henle** that dips down into the renal medulla and consists of simple squamous epithelium, and an **ascending limb of the loop of Henle** that ascends into the renal medulla and returns to the renal cortex and consists of simple cuboidal to low columnar epithelium; the ascending limb is wider in diameter than the descending limb.

c. *Distal convoluted tubule (DCT).* Coiled extension of the ascending limb in the renal cortex;

consists of simple cuboidal epithelium with fewer microvilli than the proximal convoluted tubule. Most of the simple cuboidal cells are **principal cells,** which have receptors for both antidiuretic hormone (ADH) and aldosterone, two hormones that regulate kidney functions. A smaller number are **intercalated cells,** which play a role in the homeostasis of blood pH.

Distal convoluted tubules of several nephrons terminate by merging into a single **collecting duct,** which also consists of principal and intercalated cells. In the renal medulla, collecting ducts receive distal convoluted tubules from several nephrons, pass through the renal pyramids, and converge into several hundred large **papillary ducts,** whose epithelium is simple columnar. The processed filtrate, called **urine,** passes from the collecting ducts to papillary ducts, minor calyces, major calyces, renal pelvis, ureter, urinary bladder, and urethra.

With the aid of your textbook, label the parts of a nephron and associated structures in Figure 23.4.

Examine prepared slides of various components of nephrons and compare your observations to Figure 23.5.

3. Blood and Nerve Supply

Nephrons are abundantly supplied with blood vessels, and although the kidneys constitute less than 0.5% of total body mass, they receive around 20% to 25% of the total cardiac output, or approximately 1200 mL, every minute. The blood supply originates in each kidney with the **renal artery,** which divides into many branches, eventually supplying the nephron and its complete tubule.

Before or immediately after entering the renal hilum, the renal artery divides into a larger anterior branch and a smaller posterior branch. From these branches, five **segmental arteries** originate. Each gives off several branches, the **interlobar arteries,** which pass between the renal pyramids in the renal columns. At the bases of the pyramids, the interlobar arteries arch between the renal medulla and renal cortex and here are known as **arcuate** (= shaped like a bow) **arteries.** Branches of the arcuate arteries, called **interlobular arteries,** enter the renal cortex. **Afferent** (*af* = toward; *ferrent* = to carry) **arterioles,** branches of the interlobular arteries, are distributed to tangled, ball-shaped capillary networks called the **glomeruli** (**glomerulus** = singular). Blood leaves the glomeruli via **efferent** (*ef* = out) **arterioles.**

The next sequence of blood vessels depends on the type of nephron. Around convoluted tubules, efferent arterioles of cortical nephrons divide to form capillary networks called **peritubular** (*peri* = around) **capillaries.** Efferent arterioles of juxtamedullary nephrons also form peritubular capillaries and, in addition, form long loops of blood vessels around medullary structures called **vasa recta** (VĀ-sa REK-ta; *vasa* = vessels; *recta* = straight). Peritubular capillaries eventually reunite to form **peritubular venules** and **interlobular veins.** Blood then drains into **arcuate veins,** and **interlobar veins.** Blood leaves the kidneys

---

FIGURE 23.5 Histology of a nephron and associated structures.

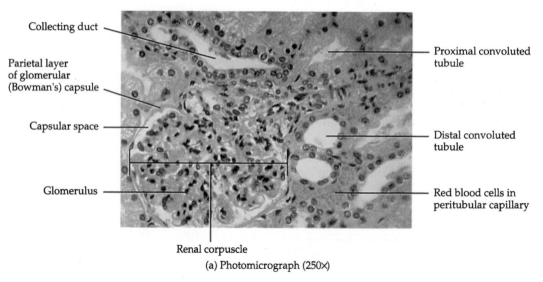

Collecting duct

Parietal layer of glomerular (Bowman's) capsule

Capsular space

Glomerulus

Proximal convoluted tubule

Distal convoluted tubule

Red blood cells in peritubular capillary

Renal corpuscle

(a) Photomicrograph (250×)

© Andrew Kuntzman, Ph.D

FIGURE 23.5   Histology of a nephron and associated structures. (Continued)

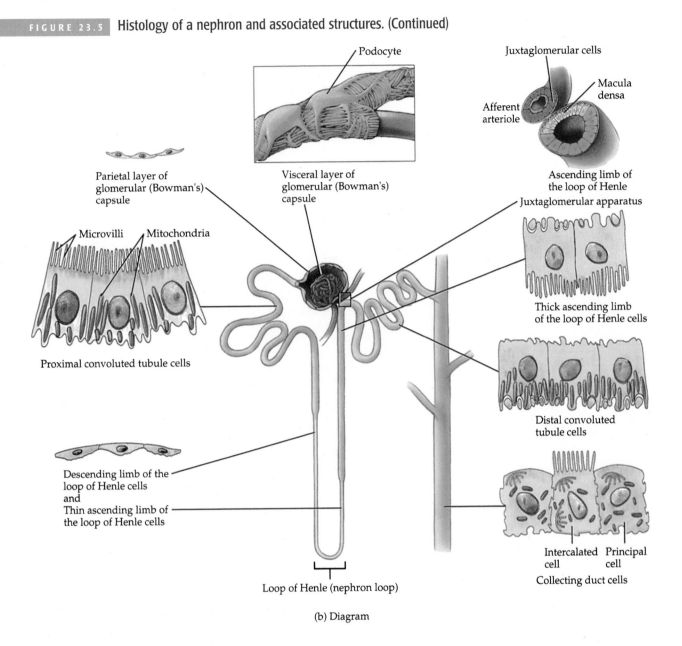

Podocyte

Juxtaglomerular cells

Macula densa

Afferent arteriole

Parietal layer of glomerular (Bowman's) capsule

Visceral layer of glomerular (Bowman's) capsule

Ascending limb of the loop of Henle

Microvilli   Mitochondria

Juxtaglomerular apparatus

Proximal convoluted tubule cells

Thick ascending limb of the loop of Henle cells

Distal convoluted tubule cells

Descending limb of the loop of Henle cells
and
Thin ascending limb of the loop of Henle cells

Intercalated cell   Principal cell

Collecting duct cells

Loop of Henle (nephron loop)

(b) Diagram

through the **renal vein** that exits at the hilum. (The vasa recta pass blood into the interlobular veins, arcuate veins, interlobar veins, and renal veins.)

In each nephron, the final part of the ascending limb of the loop of Henle makes contact with the afferent arteriole serving its own renal corpuscle. The cells of the renal tubule in this region are columnar and crowded together. Collectively, they are known as the **macula densa** (*macula* = spot; *densa* = dense). These cells monitor the Na^+ and Cl^- concentration of fluid in the tubule lumen. Alongside the macula densa, the wall of the afferent arteriole (and sometimes efferent arteriole) contains modified smooth muscle fibers called **juxtaglomerular (JG) cells.** Together with

the macula densa, they constitute the **juxtaglomerular apparatus,** or **JGA.** The JGA helps regulate blood pressure and the rate of blood filtration by the kidneys. The distal convoluted tubule begins a short distance past the macula densa.

Using Figures 23.4 and 23.6 as guides, trace a drop of blood from its entrance into the renal artery to its exit through the renal vein. As you do so, name in sequence each blood vessel through which blood passes for both cortical and juxtaglomerular nephrons.

Label Figure 23.7.

The nerve supply to the kidneys comes from the **renal plexus** of the autonomic system. The

FIGURE 23.6 Macroscopic blood vessels of the kidney. Macroscopic and microscopic blood vessels are shown in Figure 23.4.

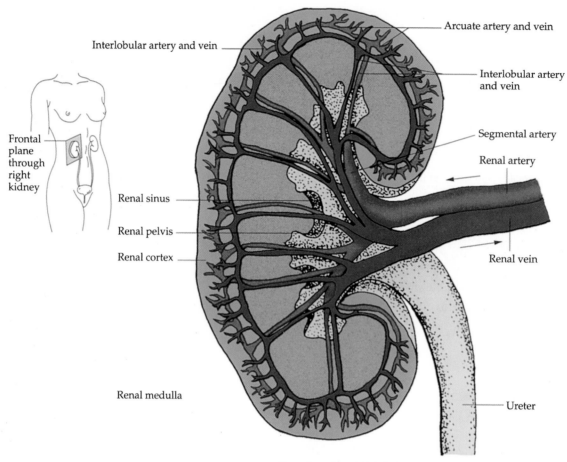

Frontal section of right kidney

nerves are vasomotor because they regulate the flow of blood in the kidney by causing vasodilation or vasoconstriction of arterioles.

4. Ureters

The body has two retroperitoneal **ureters** (Ū-re-ters), one for each kidney; each ureter is a continuation of the renal pelvis and runs to the urinary bladder (see Figure 23.1). Urine is transported through the ureters mostly by peristaltic contractions of the muscular layer of the ureters. Each ureter extends 25 to 30 cm (10 to 12 in.). At the base of the urinary bladder, the ureters curve medially and pass obliquely through the wall of the posterior aspect of the urinary bladder.

Histologically, the ureters consist of three layers: a deep coat, a **mucosa** of transitional epithelium and an underlying lamina propria (areolar

connective tissue); an intermediate coat, the **muscularis** (inner longitudinal and outer circular smooth muscle); and a superficial coat, the **adventitia** (areolar connective tissue).

Examine a prepared slide of the wall of the ureter showing its various layers. With the aid of your textbook, label Figure 23.8.

5. Urinary Bladder

The **urinary bladder** is a hollow, distensible muscular organ situated in the pelvic cavity posterior to the pubic symphysis (see Figure 23.1). In the male, the bladder is directly anterior to the rectum; in the female, it is anterior to the vagina and inferior to the uterus.

At the base of the interior of the urinary bladder is the **trigone** (TRĪ-gōn = triangle), a small triangular area bounded by the opening into the

FIGURE 23.7 Blood supply of the right kidney. This view is designed to show the *sequence* of blood flow, not the anatomical location of blood vessels, which is shown in Figure 23.6.

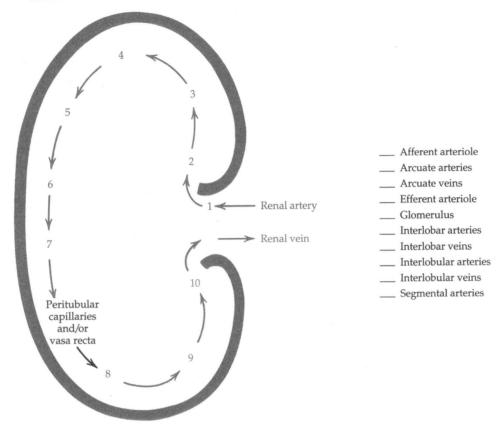

___ Afferent arteriole
___ Arcuate arteries
___ Arcuate veins
___ Efferent arteriole
___ Glomerulus
___ Interlobar arteries
___ Interlobar veins
___ Interlobular arteries
___ Interlobular veins
___ Segmental arteries

urethra (internal uretheral orifice) and ureteral openings into the bladder. Three coats make up the wall of the urinary bladder. The deepest is the **mucosa** that consists of transitional epithelium and an underlying lamina propria. Rugae (folds in the mucosa) are also present. Surrounding the mucosa is the intermediate **muscularis,** also called the **detrusor** (de-TROO-ser = to push down) **muscle,** which consists of three layers of smooth muscle: inner longitudinal, middle circular, and outer longitudinal layers. Around the opening to the urethra, the circular muscle fibers form an **internal urethral sphincter.** Inferior to this is the **external urethral sphincter,** which is composed of skeletal muscle and is a modification of the deep muscles of the perineum. The superficial coat of the urinary bladder is the **adventitia,** a layer of areolar connective tissue that is continuous with that of the ureters. Over the superior surface of the urinary bladder is the **serosa,** a layer of visceral peritoneum.

Urine is discharged from the bladder by an act called **micturition** (mik'-too-RISH-un; *mictur* = urinate), commonly known as urination, or voiding. The average capacity of the urinary bladder is 700 to 800 mL.

Using your textbook as a guide, label the external urethral sphincter, internal urethral sphincter, ureteral openings, and ureters in Figure 23.9.

Examine a prepared slide of the wall of the urinary bladder. With the aid of your textbook, label Figure 23.10.

6. Urethra

The **urethra** (ū-RĒ-thra) is a small tube leading from the internal urethral orifice in the floor of the urinary bladder to the exterior of the body. In females, this tube is posterior to the pubic symphysis and is directed obliquely, inferiorly, and anteriorly; its length is approximately 4 cm (1½ in.). The opening of the urethra to the exterior, the

FIGURE 23.8 **Histology of ureter (25×).**

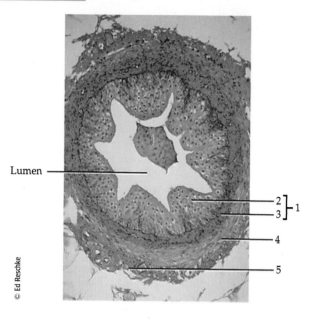

Lumen

2
3] 1
4
5

© Ed Reschke

___ Adventitia
___ Lamina propria of mucosa
___ Mucosa
___ Muscularis
___ Transitional epithelium of mucosa

external urethral orifice, is between the clitoris and vaginal orifice. In males, its length is around 20 cm (8 in.), and it follows a route different from that of the female. Immediately inferior to the urinary bladder, the urethra passes through the prostate (prostatic urethra), pierces the deep muscles of the perineum (membranous urethra), and finally through the penis (spongy urethra). The urethra is the terminal portion of the urinary system, and serves as the passageway for discharging urine from the body. In addition, in the male, the urethra serves as the duct through which semen is discharged from the body.

Label the urethra and urethral orifice in Figure 23.9.

B. Dissection of Sheep (or Pig) Kidney

The sheep kidney is very similar to the human kidney. You may use Figure 23.2 as a reference for this dissection.

▲ **CAUTION!** *Please reread Section D, "Precautions Related to Dissection," at the beginning of the labora-tory manual on page xiii before you begin your dissection.*

PROCEDURE

1. Examine the intact kidney and notice the renal hilum and the fatty tissue that normally surrounds the kidney. Strip away the fat.

2. As you peel the fat off, look carefully for the **adrenal (suprarenal) gland.** This gland is usually found attached to the superior surface of the kidney, as it is in the human. Most preserved kidneys do not have this gland. If it is present, remove it, cut it in half, and note its distinct outer **cortex** and inner **medulla.**

3. Look at the **renal hilum,** which is the concave area of the kidney. From here the **ureter, renal artery,** and **renal vein** enter and exit.

4. Differentiate these blood vessels by examining the thickness of their walls. Which vessel has the thicker wall?

5. With a sharp scalpel *carefully* make a frontal section through the kidney.

6. Identify the **renal capsule** as a thin, tough layer of connective tissue completely surrounding the kidney.

7. Immediately beneath this capsule is an outer light-colored area called the **renal cortex.** The inner dark-colored area is the **renal medulla.**

8. The **renal pelvis** is the large chamber formed by the expansion of the ureter inside the kidney. This renal pelvis divides into many smaller areas called **renal calyces,** each of which has a dark tuft of kidney tissue called a **renal pyramid.**

9. The bases of these pyramids face the cortical area. Their apices, called **renal papillae,** are directed toward the center of the kidney.

10. The calyces collect urine from collecting ducts and drain it into the renal pelvis and out through the ureter.

11. The renal artery divides into several branches that pass between the renal pyramids. These vessels are small and delicate and may be too difficult to dissect and trace through the renal medulla.

C. Urine

The kidneys function to maintain bodily homeostasis. This is accomplished by three processes: (1) filtration of the blood by the glomeruli, (2) tu-

FIGURE 23.9 Urinary bladder with female urethra.

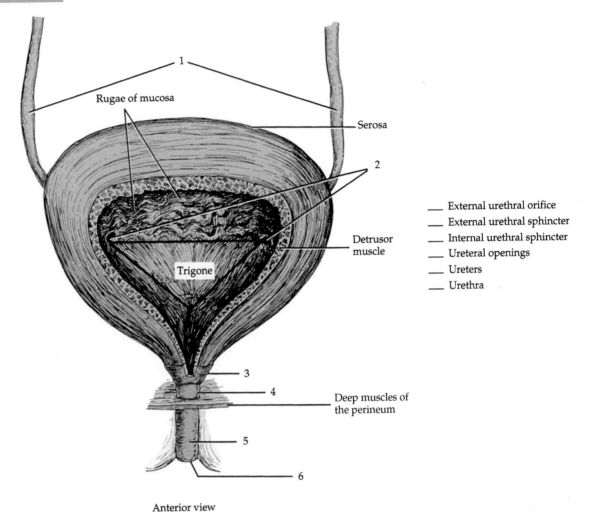

Rugae of mucosa

Serosa

Detrusor
muscle

Trigone

Deep muscles of
the perineum

Anterior view

___ External urethral orifice
___ External urethral sphincter
___ Internal urethral sphincter
___ Ureteral openings
___ Ureters
___ Urethra

bular reabsorption, and (3) tubular secretion. As a result of these three functions, **urine** is formed and eliminated from the body. Urine contains a high concentration of solutes, and in a healthy person, the volume, pH, and solute concentration of urine will vary with the needs of the internal environment. In certain pathological conditions, the characteristics of urine may change drastically. An analysis of the volume and the physical and chemical properties of urine tells us much about the state of the body.

1. Physical Characteristics

Normal urine usually varies between a straw yellow and an amber transparent color, and possesses a characteristic odor. Urine color varies considerably according to the ratio of solutes to water and according to an individual's diet.

Cloudy urine sometimes reflects the secretion of mucin from the urinary tract lining and is not necessarily an indicator of a pathological condition. The normal pH of urine ranges between 4.6 and 8.0 and averages 6.0. The pH of urine is also strongly affected by diet, with a high-protein diet lowering the pH (increase acidity), and a mostly vegetable diet increasing the pH (increase alkalinity) of the urine.

Specific gravity (density) is the ratio of the weight of a volume of a substance to the weight of an equal volume of distilled water. Water has a specific gravity of 1.000. The specific gravity of urine depends on the amount of solids in solution, and normally ranges from 1.001 to 1.035. The greater the concentration of solutes, the higher the specific gravity. In certain conditions, such as diabetes mellitus, specific gravity is high because of the high glucose content.

FIGURE 23.10 **Histology of urinary bladder (400×).**

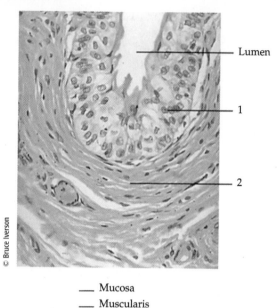

© Bruce Iverson

Lumen

1

2

—— Mucosa
—— Muscularis

2. Abnormal Constituents

When the body's metabolism becomes abnormal, many substances not normally found in urine may appear in varying amounts, while normal constituents may appear in abnormal amounts. **Urinalysis** is the analysis of the physical and chemical properties of urine, and is a vital tool in diagnosing pathological conditions.

a. *Albumin.* Usually appears in only very small amounts in urine because the molecules are too large to be filtered out of the blood through the capillary fenestrations. When excessive albumin is found in the urine, the condition is called **albuminuria.**

b. *Glucose.* Urine normally contains such small amounts of **glucose** that clinically glucose is considered to be absent from urine samples. Its presence in significant amounts is called **glucosuria,** and the most common cause is a high blood sugar (glucose) level seen in certain diseases such as diabetes mellitus. Occasionally it may be caused by stress.

c. *Erythrocytes hematuria.* The term utilized to describe the presence of red blood cells in a urine sample. Hematuria generally indicates the presence of a pathological condition within the kidneys.

d. *Leucocytes pyuria* (pī-Ū-rē-a). The condition that occurs when white blood cells and other components of pus are found in the urine, and this usually indicates an infection in the kidney or other urinary organs.

e. *Ketone bodies.* Normal urine contains a small amount of **ketone (acetone) bodies.** Their appearance in high levels in the urine produces the condition called **ketonuria** (kē-tō-NOO-rē-a) and may indicate diabetes mellitus, anorexia, starvation, or simply too little carbohydrate in the diet.

f. *Bilirubin.* When red blood cells are destroyed by macrophages, the globin portion of hemoglobin is split off and the heme is converted to biliverdin. Most of the biliverdin is converted to bilirubin, which gives bile its major pigmentation. An above-normal level of bilirubin in urine is called **bilirubinuria** (bil'-ē-rū-bi-NOO-rē-a).

g. *Urobilinogen.* The presence of urobilinogen (breakdown product of hemoglobin) in urine is called **urobilinogenuria** (ū-ro-bi-lin'-ō-je-NOO-rē-a). Traces are normal, but elevated urobilinogen may be due to hemolytic or pernicious anemia, infectious hepatitis, biliary obstruction, jaundice, cirrhosis, congestive heart failure, or infectious mononucleosis.

h. *Microbes.* The number and type of bacteria vary with specific infections in the urinary tract. One of the most common is *E. coli.* The most common fungus to appear in urine is *Candida albicans,* a cause of vaginitis. The most frequent protozoan seen is *Trichomonas vaginalis,* a cause of vaginitis in females and urethritis in males.

i. *Casts.* Tiny masses of material that have hardened and assumed the shape of the lumens of the nephron tubules. They are microscopic in size and are composed of many different substances.

j. *Calculi.* Insoluble **calculi** (stones) are various salts that have solidified in the urinary tract. They may be found anywhere from the kidney tubules to the external opening, and their presence causes considerable pain as they attempt to pass through the various lumens of the urinary system.

D. Urinalysis

In this exercise you will determine some of the characteristics of urine and perform tests for some abnormal constituents that may be present in urine. Some of these tests may be used in determining unknowns in urine specimens.

▲ **CAUTION!** *Please reread Section A, "General Safety Precautions and Procedures," on page xi and Section C, "Precautions Related to Working with Reagents," on page xii at the beginning of the laboratory manual before you begin any of the following experiments. Read the experiments before you perform them, to be sure that you understand all the procedures and safety precautions.*

When working with urine, avoid any kind of contact with an open sore, cut, or wound. Wear tight-fitting surgical gloves and safety goggles.

Work with your own urine only. After you have completed your experiments, place all glassware in a fresh household bleach solution or other comparable disinfectant, wash the laboratory tabletop with a fresh household bleach solution or comparable disinfectant, and dispose of the gloves and Multistix® in the appropriate biohazard container provided by your instructor.

1. Urine Collection

PROCEDURE

1. A specimen of urine may be collected at any time for routine tests; urine voided within 3 hr after meals, however, may contain abnormal constituents. For this reason the first voiding in the morning is preferred.

2. Both males and females should collect a midstream sample of urine in a sterile container. A midstream sample is essential to avoid contamination from the external genitalia, and to avoid the presence of pus cells and bacteria that are normally found in the urethra.

3. If not examined immediately, the specimen should be refrigerated to prevent unnecessary bacterial growth.

4. Before testing *always* mix urine by swirling, inverting the container, or stirring with a wooden swab stick.

5. *Keep all containers clean!*

6. Wrap all papers and sticks and put them in the garbage pail.

7. Rinse all test tubes and glass containers carefully with *cold water* after they have cooled.

8. Flush sinks well with cold water.

9. Obtain either your own freshly voided urine sample or a provided sample if one is available.

2. Multistix® Testing

Alternate methods can be employed for several of the tests you are about to perform. One alternate method is the use of plastic strips to which are attached paper squares impregnated with various reagents. These strips display a color reaction when dipped into urine with any abnormal constituents.

PROCEDURE

1. At this point, you should take a Multistix® and test your urine sample for the following: pH, protein, glucose, ketones, bilirubin, and blood (hemoglobin). Record your results below.

2. Determine which component(s) of urine are measured by the Multistix® you are using. If you are using a strip that tests for multiple substances, determine which squares measure which substances. Locate the color chart used to read the results. Note the appropriate time for reading each test.

3. Remove a test strip from the vial and *replace the cap.* Dip the test strip into your urine sample for no longer than 1 sec, being sure that all reagents on the strip are immersed.

4. Remove any excess urine from the strip by drawing the edge of the strip along the rim of the container holding your urine sample.

5. After the appropriate time, as indicated on the Multistix® vial, hold the strip close to the color blocks on the vial.

6. Make sure that the strip blocks are properly lined up with the color chart on the vial.

Test	Multistix® result
pH	_____
Protein	_____
Glucose	_____
Ketones	_____
Bilirubin	_____
Blood	_____
Leukocytes	_____
Specific gravity	_____

3. Physical Analysis

a. Color

Normal urine varies in color from straw yellow to amber because of the pigment **urochrome**, a by-product of hemoglobin destruction. Observe the color of your urine sample. Some abnormal colors are as follows:

Color	Possible cause
Silvery, milky	Pus, bacteria, epithelial cells
Smoky brown, rust	Blood
Orange, green, blue, red	Medications or liver disease

Record the color of your urine in Section D.3.a of the LABORATORY REPORT RESULTS at the end of the exercise.

How would sickle-cell anemia affect urine color? _____

_____

b. Transparency

A fresh urine sample should be clear. Cloudy urine may be due to substances such as mucin, phosphates, urates, fat, pus, mucus, microbes, crystals, and epithelial cells.

PROCEDURE

1. To determine transparency, cover the container and shake your urine sample and observe the degree of cloudiness.

2. Record your observations in Section D.3.b of the LABORATORY REPORT RESULTS at the end of the exercise.

c. pH

The pH of urine varies with several factors already indicated. You can test the pH of your urine by using either a Multistix® or pH paper. Because you have already determined the pH of your urine by using a Multistix®, you might want to verify the results using pH paper.

PROCEDURE

1. Place a strip of pH paper into your urine sample three consecutive times.

2. Shake off any excess urine.

3. Let the pH paper sit for 1 min and then compare it to the color chart provided.

4. Record your observations in Section D.3.c of the LABORATORY REPORT RESULTS at the end of the exercise.

How would the consumption of antacid (sodium bicarbonate) affect urine pH? _____

_____

_____

d. Specific Gravity

The specific gravity is easily determined using a urinometer (hydrometer). The urinometer is a float with a numbered scale near the top that indicates specific gravity directly.

PROCEDURE

1. Familiarize yourself with the scale on the urinometer neck. Determine the change in specific gravity represented by each calibration.

2. Allow your urine to reach room temperature. Urinometers are calibrated to read the specific gravity at 15°C (69°F). If the temperature differs, add or subtract 0.001 for each 3°C above or below 15°C.

3. Fill the cylinder ¾ full of urine and insert the urinometer. Make sure that it is free-floating; if not, spin the neck gently.

4. Read the scale at the bottom of the meniscus when the urinometer is at rest.

5. Record the specific gravity in Section D.3.d of the LABORATORY REPORT RESULTS at the end of the exercise.

6. Rinse the urinometer and cylinder. *Follow your instructor's directions for cleaning them.*

How would dehydration affect the specific gravity of urine? _____

_____

_____

4. Chemical Analysis

a. Chloride and Sodium Chloride

Most of the sodium chloride (NACl) present in the renal filtrate is reabsorbed; a small amount remains as a normal component of urine. The normal value for chloride (Cl^-) is 476 mg/100 mL; for sodium (Na^+) it is 294 mg/100 mL.

PROCEDURE

1. Using a medicine dropper, place 10 drops of urine into a Pyrex test tube.

2. Using a medicine dropper, add 1 drop of 20% potassium chromate and *gently* agitate the tube. It should be a yellow color.

3. Using a medicine dropper, add 2.9% silver nitrate solution *one drop at a time*, counting the drops and *gently* agitating the test tube during the time the solution is being added.

4. Count the number of drops needed to change the color of the solution from bright yellow to brown.

5. Determine the chloride and sodium chloride concentrations of the sample and record your results in Section D.4.a of the LABORATORY REPORT RESULTS at the end of the exercise. Each drop of silver nitrate added in step 4 is equivalent to (1) 61 mg of Cl^- per 100 mL of urine, and (2) 100 mg of NaCl per 100 mL of urine.

Thus, to determine the chloride concentration of the sample, multiply the number of drops times 61 to obtain the number of mg of Cl^- per 100 mL of urine.

To determine the sodium chloride concentration of the sample, multiply the number of drops times 100 to obtain the number of mg of NaCl per 100 mL of urine.

What would a high NaCl content in the urine indicate? _____

_____

_____

b. Glucose

Glucosuria (the presence of glucose in the urine) occurs in patients with diabetes mellitus or other disorders. Traces of glucose may occur in normal urine, but detection of these small amounts requires special tests.

(1) Benedict's Test

Benedict's solution is commonly used to detect reducing sugars in urine and is not specific for just glucose.

PROCEDURE

1. In a Pyrex test tube, combine 10 drops of urine with 5 mL of Benedict's solution. Mix the solution.

2. Using a test tube holder, place the test tube in a boiling water bath for 5 min.

▲ **CAUTION!** *Make sure that the mouth of the test tube is pointed away from you and all other persons in the area.*

3. *Using a test tube holder*, remove the test tube from the water bath and read the results according to the following chart.

4. Record your results in Section D.4.b of the LABORATORY REPORT RESULTS at the end of the exercise.

Color	Results
Blue	Negative
Greenish-yellow	1 + (0.5 g/100 mL)
Olive green	2 + (1 g/100 mL)
Orange-yellow	3 + (1.5 g/100 mL)
Brick red (with precipitate)	4 + (> 2 g/100 mL)

(2) Clinitest® Reagent Method
PROCEDURE

1. Using medicine droppers, place 10 drops of water and 5 drops of urine in a Pyrex test tube.

▲ **CAUTION!** *Place the test tube in a test-tube rack because it will become too hot to handle.*

2. With forceps, add one Clinitest® tablet.

▲ **CAUTION!** *The concentrated sodium hydroxide in the tablet generates enough heat to make the liquid in the test tube boil. Make sure that the mouth of the test tube is pointed away from you and all other persons in the area.*

3. The color of the solution is graded as in Benedict's test.

4. Fifteen seconds after boiling has stopped, shake the test tube *gently* and evaluate the color according to the preceding Benedict's test table. Disregard any color change that occurs after 15 sec.

▲ **CAUTION!** *Make sure the mouth of the test tube is pointed away from you and others.*

5. Record your results in Section D.4.b of the LABORATORY REPORT RESULTS at the end of the exercise.

How would a high-sugar diet affect the urine? Explain. _____

_____

_____

(3) Multistix® Method

Record your results in Section D.4.b of the LABORATORY REPORT RESULTS at the end of the exercise.

c. Protein

Normal urine contains traces of proteins that are hard to detect through regular laboratory procedures. Albumin is the most abundant serum protein and is the one usually detected. Because tests for albumin are determined by precipitating the protein either by heat (coagulation) or by adding a reagent, the urine sample should be either filtered or centrifuged. The test for protein will be done by the Albutest® reagent method and the Multistix® method. (Your lab instructor may use Albustix® in place of the Albutest.)

(1) Albutest® Reagent Method

PROCEDURE

1. Using forceps, place an Albutest® tablet on a clean, dry paper towel and add one drop of urine.

2. After the drop has been absorbed, add two drops of water and allow these to penetrate before reading.

3. Compare the color (in daylight or fluorescent light) on top of the tablet with the color chart provided in lab.

4. If albumin is present in the urine, a *blue-green* spot will remain on the surface of the tablet after the water is added. The amount of protein is indicated by the intensity of the blue-green color.

5. If the test is negative, the original color of the tablet will not be changed at the completion of the test.

6. Record your results in Section D.4.c of the LABORATORY REPORT RESULTS at the end of the exercise.

   Why is protein not normally found in urine?

   _____

   _____

(2) Multistix® Method

Record your results in Section D.4.c of the LABORATORY REPORT RESULTS at the end of the exercise.

d. Ketone (Acetone) Bodies

The presence of ketone (acetone) bodies in urine is a result of abnormal fat catabolism. Reagents such as sodium nitroprusside, ammonium sulfate, and ammonium hydroxide are available in the form of tablets. Ketones turn purple when added to these chemicals.

(1) Acetest® Tablet Method

PROCEDURE

1. Using forceps, place an Acetest® tablet on a clean, dry paper towel, and, using a medicine dropper, place one drop of urine on the tablet.

2. If acetone or ketone is present, a *lavender-purple* color develops within 30 sec. If the tablet becomes cream-colored from wetting, the results are negative. Compare results with the color chart that comes with the reagent.

3. Record your results in Section D.4.d of the LABORATORY REPORT RESULTS at the end of the exercise.

   Why would starvation cause ketones? _____

   _____

   _____

(2) Multistix® Method

Record your results in Section D.4.d of the LABORATORY REPORT RESULTS at the end of the exercise.

e. Bile Pigments

Bile pigments, biliverdin and bilirubin, are not normally present in urine. The presence of large quantities of bilirubin in the extracellular fluids produces jaundice, a yellowish tint to the body tissues, including yellowness of the skin and deep tissues.

(1) Shaken Tube Test for Bile Pigments

PROCEDURE

1. Fill a Pyrex test tube halfway with urine, stopper the tube, and shake it vigorously, being careful to keep the stopper in the tube.

2. A yellow color of foam indicates the presence of bile pigments.

3. Record your results in Section D.4.e of the LABORATORY REPORT RESULTS at the end of the exercise.

(2) Ictotest® for Bilirubin

PROCEDURE

1. Using a medicine dropper, place a drop of urine on one square of the special mat provided in the Ictotest® kit.

2. Using forceps, place one Ictotest® reagent tablet in the center of the moistened area.

▲ **CAUTION!** *Do not touch the tablets with your fingers. Recap the bottle.*

3. Add one drop of water directly to the tablet; after 5 sec, add another drop of water to the tablet so that the water runs off onto the mat. Observe the color of the mat around the tablet at 60 sec. The presence of bilirubin will turn the mat *blue* or *purple*. A slight *pink* or *red* color is negative for bilirubin.

4. Record your results in Section D.4.e of the LABORATORY REPORT RESULTS at the end of the exercise.

(3) Multistix® Method

Record your results in Section D.4.e of the LABORATORY REPORT RESULTS at the end of the exercise.

f. Hemoglobin

Hemoglobin is not normally found in urine.

(1) Multistix® Method

Record your results in Section D.4.f of the LABORATORY REPORT RESULTS at the end of the exercise.

5. Microscopic Analysis

Before you actually examine the sediment of urine microscopically, it will be helpful to read the following description of some of the major components of urinary sediment. For purposes of discussion, the constituents of sediment will be classified into three major groups: (1) cells, (2) casts, and (3) crystals.

Cells in urinary sediment are derived from the lining of the urinary tract as a result of normal wear and tear inflammations, and various disease processes. Using Figure 23.11a as a guide, see how many cells you can identify in your urine sample:

a. *Transitional cells.* Cells derived from the transitional epithelium that lines the renal pelvis, ureter, and urinary bladder.

b. *Squamous epithelial cells.* Cells derived from the distal portion of the urethra or inflamed areas of the urinary bladder.

c. *Red blood cells.* Usually not associated with normal urine. Their presence in urine (**hematuria**) may be associated with a variety of diseases.

d. *White blood cells.* More than a few indicates **pyuria** (the presence of white blood cells and other components of pus); indicates an infection.

Casts are hardened masses of material that assume the shape of the lumen of a renal tubule in which they form. They are formed by the precipitation of proteins and agglutination of cells. Casts form as a result of protein in urine passing through renal tubules, highly acidic urine, and highly concentrated urine. Casts are named after the cells or substances that compose them or on the basis of their appearance. Using Figure 23.11b as a guide, see how many casts you can identify in your urine sample:

a. *Hyaline casts.* Consist mostly of a microprotein derived from renal tubule epithelial cells.

b. *Epithelial casts.* Consist largely of tubular epithelial cells.

c. *Granular casts.* Considered one phase in the breakdown of cellular casts. Depending on the degree of breakdown, they may be classified as coarsely granular and finely granular.

d. *Waxy casts.* Considered the end stage of the breakdown of cellular casts.

e. *White blood cell casts.* Consist of aggregates of white blood cells.

f. *Red blood cell casts.* Consist of aggregates of red blood cells.

Crystals form from the end product of tissue metabolism and consumption of excessive amounts of certain foods and drugs. Identification of crystals is based on their shape. Using Figure 23.11c as a guide, see how many crystals you can identify in your urine sample:

a. *Calcium oxalate.* Range in shape from oval to dumbbell to octahedron (8-sided) to dodecahedron (12-sided); latter two shapes appear as "envelopes."

b. *Uric acid.* Yellow or red-brown rhombic prisms, hexagonal or square plates, or spheres.

c. *Triple phosphate.* Appear as six- to eight-sided prisms ("coffin lids") or feathery forms.

If you allow a urine specimen to stand undisturbed for a few hours, many suspended materials will settle to the bottom. A much faster method is to centrifuge a urine sample.

PROCEDURE

1. Place 5 mL of fresh urine in a centrifuge tube. Follow the directions of your instructor to

FIGURE 23.11   Photographs of some microscopic components of urinary sediment.

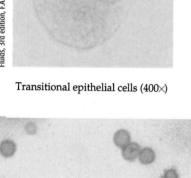

Courtesy of S.K. Strasinger, Urinalysis and Body Fluids, 3rd edition, F.A. Davis Co., Philadelphia

Transitional epithelial cells (400×)

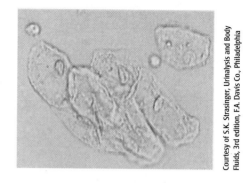

Courtesy of S.K. Strasinger, Urinalysis and Body Fluids, 3rd edition, F.A. Davis Co., Philadelphia

Squamous epithelial cells (400×)

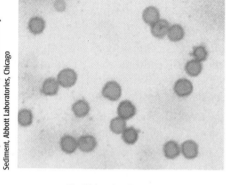

Courtesy of W. Jao M.D., et. al, An Atlas of Urinary Sediment, Abbott Laboratories, Chicago

Red blood cells (400×)

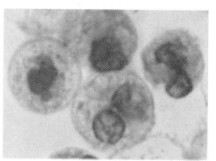

Courtesy of W. Jao M.D., et. al, An Atlas of Urinary Sediment, Abbott Laboratories, Chicago

Neutrophils (400×)

(a) Cells

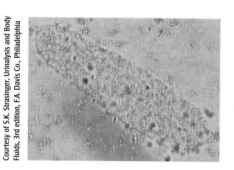

Courtesy of S.K. Strasinger, Urinalysis and Body Fluids, 3rd edition, F.A. Davis Co., Philadelphia

Hyaline cast (400×)

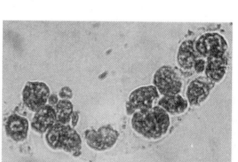

Courtesy of S.K. Strasinger, Urinalysis and Body Fluids, 3rd edition, F.A. Davis Co., Philadelphia

Epithelial cell cast (400×)

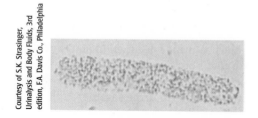

Courtesy of S.K. Strasinger, Urinalysis and Body Fluids, 3rd edition, F.A. Davis Co., Philadelphia

Finely granular cast (400×)

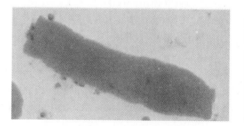

Courtesy of W. Jao M.D., et. al, An Atlas of Urinary Sediment, Abbott Laboratories, Chicago

Waxy cast (400×)

(b) Casts

Photographs of some microscopic components of urinary sediment. (Continued)

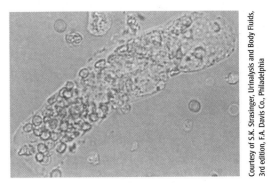

White blood cell cast (400×)

Red blood cell cast (400×)

(b) Casts (continued)

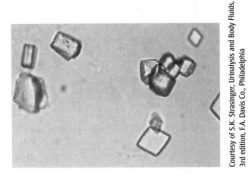

Calcium oxalate crystals (400×)

Uric acid crystals (400×)

Triple phosphate crystals (400×)

(c) Crystals

centrifuge the tube for 5 min at a slow speed (1500 revolutions per minute [rpm]).

2. Dispose of the supernatant (the clear urine) and mix the sediment by shaking the test tube.

3. Using a long medicine dropper or a Pasteur pipette with a bulb, place a small drop of sediment on a clean glass slide, add one drop of Sedi-stain® or methylene blue, and place a cover slip over the specimen.

4. Using low power and reduced light, examine the sediment for any of the microscopic elements pictured in Figure 23.11.

5. Increase the light and use high power or oil immersion to look for crystals.

Draw the results of your observation in Section D.5 of the LABORATORY REPORT RESULTS at the end of the exercise.

Why might some red blood cells in urinary sediment be crenated? _____

_____

Why might yeast be seen in the urine of a person with diabetes mellitus? _____

_____

6. Unknown Specimens

a. Unknowns Prepared by Instructor

When the composition of a substance has not been defined, it is called an **unknown.** In this ex-

ercise, the unknowns will be urine specimens to which the instructor has added glucose, albumin, or any other detectable substance. Each unknown contains only one added substance. Perform the previously outlined tests until you identify the substance in your unknown.

Record your results in Section D.6.a of the LABORATORY REPORT RESULTS at the end of the exercise.

b. Unknowns Prepared by Class (Optional)

The class should be divided into two groups. Each group adds certain substances, such as glucose, protein, starch, or fat, to normal, freshly voided urine, keeping accurate records as to what was added to each sample. The two groups then exchange samples, and each group does the basic chemical tests on urine to detect which substances were added. Each student should add one substance to one urine sample and see if another student can detect what was added.

Record your results in Section D.6.b of the LABORATORY REPORT RESULTS at the end of the exercise.

ANSWER THE LABORATORY REPORT QUESTIONS AT THE END OF THE EXERCISE.

Urinary System

Name _____   Date _____

Laboratory Section _____   Score/Grade _____

SECTION D ■ Urinalysis

3. Physical Analysis

Characteristic	Normal	Your sample
a. Color	Straw yellow to amber	_____
b. Sediment	None	_____
c. pH	4.6–8.0	_____
d. Specific gravity	1.001–1.035	_____

4. Chemical Analysis

Record your results in the spaces provided.

a. Chloride and Sodium Chloride

Chloride concentration _____ mg/100 mL

Sodium chloride concentration _____ mg/100 mL

b. Glucose

Benedict's test _____

Clinitest® tablet _____

Multistix® _____

c. Protein

Albutest® reagent tablets _____

Multistix® _____

d. Ketone (Acetone) Bodies

Acetest® tablet _____

Multistix® _____

e. Bile Pigments

Shaken tube (bile pigments) _____

Ictotest® (bilirubin) _____

Multistix® _____

f. Hemoglobin

Multistix® _____

5. Microscopic Analysis

Draw some of the substances (types of cells, types of crystals, or other elements) that you found in the microscopic examination of urinary sediment.

6. Unknown Specimens

a. What substance did you find in the unknown specimen that was prepared by your instructor?

_____

b. What substance did you find in the unknown specimen that was prepared by other students?

_____

Urinary System

Name _____   Date _____

Laboratory Section _____   Score/Grade _____

PART 1 ■ Multiple Choice

_____ 1. Beginning at the deepest layer and moving toward the superficial layer, identify the order of tissue layers surrounding the kidney. (a) renal capsule, renal fascia, adipose capsule (b) renal fascia, adipose capsule, renal capsule (c) adipose capsule, renal capsule, renal fascia (d) renal capsule, adipose capsule, renal fascia

_____ 2. The functional unit of the kidney is the (a) nephron (b) ureter (c) urethra (d) hilum

_____ 3. Substances filtered by the kidney must pass through the filtration membrane, which is composed of several parts. Which of the following choices lists the correct order of the parts as substances pass through the membrane? (a) epithelium of the visceral layer of glomerular (Bowman's) capsule, endothelium of the glomerulus, basal lamina of the glomerulus (b) endothelium of the glomerulus, basal lamina of the glomerulus, epithelium of the visceral layer of glomerular (Bowman's) capsule (c) basal lamina of the glomerulus, endothelium of the glomerulus, epithelium of the visceral layer of the glomerular (Bowman's) capsule (d) epithelium of the visceral layer of the glomerular (Bowman's) capsule, basal lamina of the glomerulus, endothelium of the glomerulus

_____ 4. In the glomerular (Bowman's) capsule, the afferent arteriole divides into a capillary network called a(n) (a) glomerulus (b) interlobular artery (c) peritubular capillary (d) efferent arteriole

_____ 5. Transport of urine from the renal pelvis into the urinary bladder is the function of the (a) urethra (b) calculi (c) casts (d) ureters

_____ 6. The terminal portion of the urinary system is the (a) urethra (b) urinary bladder (c) ureter (d) nephron

_____ 7. Damage to the renal medulla would interfere first with the functioning of which parts of a juxtamedullary nephron? (a) glomerular (Bowman's) capsule (b) distal convoluted tubule (c) collecting ducts (d) proximal convoluted tubules

_____ 8. An obstruction in the glomerulus would affect the flow of blood into the (a) renal artery (b) efferent arteriole (c) afferent arteriole (d) intralobular artery

_____ 9. Urine that leaves the distal convoluted tubule passes through the following structures in which sequence? (a) collecting duct, hilum, calyces, ureter (b) collecting duct, calyces, pelvis, ureter (c) calyces, collecting duct, pelvis, ureter (d) calyces, hilum, pelvis, ureter

_____ 10. The position of the kidneys posterior to the peritoneal lining of the abdominal cavity is described by the term (a) retroperitoneal (b) anteroperitoneal (c) ptosis (d) inferoperitoneal

_____ **11.** Of the following structures, the one to receive filtrate *last* as it passes through the nephron is the (a) proximal convoluted tubule (b) ascending limb of the loop of Henle (c) glomerulus (d) collecting duct

_____ **12.** Peristalsis of the ureter is a function of the (a) serosa (b) mucosa (c) submucosa (d) muscularis

_____ **13.** The trigone and the detrusor muscle are associated with the (a) kidney (b) urinary bladder (c) urethra (d) ureters

_____ **14.** The vertical fissure on the medial surface of the kidney through which blood vessels enter and exit is called the (a) renal medulla (b) major calyx (c) renal hilum (d) renal column

_____ **15.** Blood is drained from the kidneys by the (a) renal arteries (b) interlobar arteries (c) interlobular veins (d) renal veins

_____ **16.** The epithelium of the urinary bladder that permits distention is (a) stratified squamous (b) transitional (c) simple squamous (d) pseudostratified

_____ **17.** How many times a day is the entire volume of blood in the body filtered by the kidneys? (a) 100 times (b) 5 times (c) 30 times (d) 60 times

_____ **18.** The average urine capacity of the urinary bladder is (a) 1000 to 1200 mL (b) 50 to 100 mL (c) 700 to 800 mL (d) 200 to 300 mL

_____ **19.** The normal pH of urine is between (a) 4.6 and 8.0 (b) 2.0 and 4.8 (c) 10.0 and 12.0 (d) none of the above

_____ **20.** Normal urine has a specific gravity of approximately (a) 1.001 to 1.035 (b) 1.030 to 1.080 (c) 1.100 to 1.200 (d) none of the above

_____ **21.** The special chemical that may be used to detect glucose in the urine is (a) sulfosalicylic acid (b) Benedict's solution (c) Lugol's solution (d) nitric acid

PART 2 ■ Completion

22. In addition to the urinary system, other systems that help eliminate wastes are the respiratory, integumentary, and _____ systems.

23. The double-walled epithelial cup found in a nephron is called a(n) _____.

24. The special capillary network found inside of this double-walled cup is the

_____.

25. The major blood vessel that enters each kidney is the _____.

26. The nerve supply to the kidneys comes from the autonomic nervous system and is called the

_____.

27. Urine is expelled from the urinary bladder by an act called urination, voiding, or

_____.

28. The small tube in the urinary system that leads from the floor of the urinary bladder to the outside is the _____.

29. The apices of renal pyramids are referred to as renal _____.

30. Portions of the renal cortex that extend between renal pyramids are called renal

_____.

31. Cuplike extensions of the renal pelvis, usually two or three in number, are referred to as

_____.

32. Modified simple squamous epithelial cells of the visceral layer of the glomerular (Bowman's) capsule are called _____.

33. Distal convoluted tubules terminate by merging with _____.

34. Long loops of blood vessels around the medullary structures of juxtamedullary nephrons are called

_____.

35. Which blood vessel comes next in this sequence? Interlobar artery, arcuate artery, interlobular artery, _____.

36. The abnormal condition in which red blood cells are found in the urine in appreciable amounts is called _____.

37. Various salts that solidify in the urinary tract are called _____.

38. Various substances that have hardened and assumed the shape of the lumens of the nephron tubules are the _____.

39. The pH of urine in individuals on high-protein diets tends to be _____ than normal.

40. The greater the concentration of solutes in urine, the greater will be its _____.

pH and Acid-Base Balance

Objectives

At the completion of this exercise you should understand

A The concept of pH and how it is measured

B The concept of pH and physiological buffers as they apply to homeostasis

C The physiological roles, as well as the integration of the respiratory and urinary systems, in the regulation of pH

D The physiological processes involved in acid-base imbalances, and how the respiratory and urinary systems work together to correct these imbalances

A. The Concept of pH

When molecules of inorganic acids, bases, or salts dissolve in water, they undergo **ionization** (ī'-on-i-ZĀ-shun), or **dissociation** (dis'-sō-sē-Ā-shun); that is, they separate into ions and become surrounded by water molecules.

An **acid** can be defined as a substance that dissociates into one or more **hydrogen ions (H⁺)** and one or more **anions** (negative ions). Because H^+ is a single proton with a charge of +1, an acid can also be defined as a **proton donor.** A **base,** by contrast, dissociated into one or more **hydroxide ions (OH⁻)** and one or more **cations** (positive ions). A base can also be viewed as a **proton acceptor.** Hydroxide ions have a strong attraction for protons. A **salt,** when dissolved in water, dissociates into cations and anions, neither of which is H^+ or OH^-. Acids and bases react with one another to form salts. Body fluids must constantly contain balanced quantities of acids and bases. In solutions such as those found inside or outside body cells, acids dissociate into H^+ and anions. Bases, on the other hand, dissociate into OH^- and cations. The more hydrogen ions that exist in a solution, the more acidic the solution; conversely, the more hydroxide ions, the more basic (alkaline) the solution.

Biochemical reactions—those that occur in living systems—are very sensitive to even small changes in acidity or alkalinity. Any departure from the narrow limits of normal H^+ and OH^- concentrations may greatly modify cell functions and disrupt homeostasis. For this reason, the acids and bases that are constantly formed in the body must be kept in balance.

A solution's acidity or alkalinity is expressed on the **pH scale,** which runs from 0 to 14. This scale is based on the concentration of H^+ in moles per liter in a solution. The midpoint of the scale is 7, where the concentrations of H^+ and OH^- are equal. A substance with a pH of 7, such as pure water, is neutral. A solution that has more H^+ than OH^- is an **acidic solution** and has a pH below 7. A solution that has more OH^- than H^+ is a **basic (alkaline) solution** and has a pH above 7. A change of one whole number on the pH scale represents a 10-fold change from the previous concentration. A pH of 1 denotes 10 times more H^+ than a pH of 2. A pH of 3 indicates 10 times fewer H^+ than a pH of 2 and 100 times fewer H^+ than a pH of 1.

B. Measuring pH

1. Using Litmus Paper

PROCEDURE

⚠ **CAUTION!** *Please reread Section A, "General Safety Precautions and Procedures," on page xi and Section C, "Precautions Related to Working with Reagents," on page xii at the beginning of the laboratory manual before you begin any of the following experiments. Read the experiments before you perform them, to be sure that you understand all the procedures and safety precautions.*

1. Before you begin, it is important to know that *an acid solution will turn blue litmus paper red* and *a basic (alkaline) solution will turn red litmus paper blue.*

2. Using forceps, dip a strip of red litmus paper into each of the solutions to be tested. Use a *new strip* for each solution. Record your observations in the table in Section B.1 of the LABORATORY REPORT RESULTS at the end of the exercise.

3. Using forceps, dip a strip of blue litmus paper into each of the solutions to be tested. Use a *new strip* for each solution. Record your observations in the table in Section B.1 of the LABORATORY REPORT RESULTS at the end of the exercise.

4. Using your textbook as a guide, determine the pH of the following body fluids:

Body Fluid	pH
Bile	
Saliva	
Gastric juice	
Blood	
Pancreatic juice	
Semen	
Urine	

2. Using pH Paper

PROCEDURE

▲ **CAUTION!** *Please reread Section A, "General Safety Precautions and Procedures," on page xi and Section C, "Precautions Related to Working with Reagents," on page xii at the beginning of the laboratory manual before you begin any of the following experiments. Read the experiments before you perform them, to be sure that you understand all the procedures and safety precautions.*

1. Using forceps, dip a strip of pH paper into each of the solutions to be tested. Use a *new strip* for each solution. Use wide-range pH paper to determine the approximate pH and narrow-range pH paper to determine a more accurate pH.

2. Compare the color of the strip of pH paper to the color chart on the pH paper container.

3. Record your observations in the table in Section B.2 of the LABORATORY REPORT RESULTS at the end of the exercise.

3. Using a pH Meter

Because there are different types of pH meters, your instructor will demonstrate how to use the pH meter in your laboratory.

PROCEDURE

▲ **CAUTION!** *Please reread Section A, "General Safety Precautions and Procedures," on page xi and Section C, "Precautions Related to Working with Reagents," on page xii at the beginning of the laboratory manual before you begin any of the following experiments. Read the experiments before you perform them, to be sure that you understand all the procedures and safety precautions.*

1. Examine the pH meter that has been made available to you and identify the following parts: (1) electrodes, (2) pH dial or digital display, (3) temperature control, and (4) calibration control.

2. Plug in the pH meter and turn it on.

NOTE *Some models take up to one-half hour to warm up.*

3. Using the temperature control knob, adjust the pH meter for the temperature of the solutions to be tested.

4. To calibrate the pH meter, place the electrode(s) in a beaker that contains a pH 7 buffer solution. The electrode(s) should be immersed at least 1 in. into the solution. Adjust the pH meter with the appropriate controls so that the meter will show a pH value of 7. Now the instrument is calibrated.

5. Depress the standby button and remove the electrode(s) from the buffer solution. *The electrode(s) should not touch anything.* Rinse the electrode(s) with distilled water, using a wash bottle. The rinse water can be collected in an empty beaker.

6. Immerse the electrode(s) into the first solution to be tested. Release the standby button and note the pH. Record the pH in the table in Section B.3 of the LABORATORY REPORT RESULTS at the end of the exercise.

7. Depress the standby button, remove the electrode(s) from the solution, and rinse the distilled water. Test the pH of the remaining solutions and record each pH in the table in Section B.3 of the LABORATORY REPORT RESULTS at the end of the exercise.

C. Acid-Base Balance

A very important electrolyte in terms of the body's acid-base balance is H^+. Although some H^+ enter the body in ingested foods, most are produced as a result of the cellular metabolism of substances such as glucose, fatty acids, and amino acids. One of the major challenges to homeostasis is keeping the H^+ concentration at an appropriate level to maintain proper acid-base balance.

The balance of acids and bases is maintained by controlling the H^+ concentration of body fluids, particularly extracellular fluid. In a healthy person, the pH of the systemic arterial blood remains between 7.35 and 7.45. Metabolism typically produces a huge excess of H^+. If there were no mechanisms for disposal of H^+, the rising concentration of H^+ in body fluids would quickly lead to death. Homeostasis of H^+ concentration within a narrow pH range is essential to survival and depends on three major mechanisms.

1. *Buffer systems.* Buffers act quickly to bind H^+ temporarily, which removes excess H^+ from solution but not from the body.

2. *Exhalation of carbon dioxide.* By increasing the rate and depth of breathing, more carbon dioxide can be exhaled. Within minutes this reduces the level of carbonic acid in blood, which raises blood pH (reduces blood H^+ level).

3. *Kidney excretion.* The slowest mechanism, taking hours or days, but the only way to eliminate acids other than carbonic acid is through their excretion in urine.

In the following experiments, you will note the relationship between buffers and the exhalation of carbon dioxide to pH.

1. Buffers and pH

Most **buffer systems** of the body consist of a weak acid and the salt of that acid, which functions as a weak base. Buffers prevent rapid, drastic changes in the pH of a body fluid by converting strong acids and bases into weak acids and bases. Buffers work within fractions of a second. Strong acids lower pH more than weak ones because strong acids release H^+ more readily and thus contribute more free hydrogen ions. Similarly, strong bases raise pH more than weak ones. The principal buffer systems of the body fluids are the protein buffer system, the carbonic acid–bicarbonate system, and the phosphate system.

a. Protein Buffer System

The **protein buffer system** is the most abundant buffer in intracellular fluid and blood plasma. Inside red blood cells the protein hemoglobin is an especially good buffer. Proteins are composed of amino acids, an organic compound that contains at least one carboxyl group (COOH) and at least one amino group (NH_2). The free carboxyl group at one end of a protein acts like an acid by releasing H^+ when pH rises and can dissociate in this way:

$$NH_2-\overset{\overset{\textstyle R}{|}}{\underset{\underset{\textstyle H}{|}}{C}}-COOH \rightarrow NH_2-\overset{\overset{\textstyle R}{|}}{\underset{\underset{\textstyle H}{|}}{C}}-COO + H^+$$

The H^+ is then able to react with any excess OH^- in the solution to form water.

The free amine group at the other end of a protein can act as a base by combining with hydrogen ions when pH falls as follows:

$$COOH-\overset{\overset{\textstyle R}{|}}{\underset{\underset{\textstyle H}{|}}{C}}-NH_2 + H^+ \rightarrow COOH-\overset{\overset{\textstyle R}{|}}{\underset{\underset{\textstyle H}{|}}{C}}-NH_3^+$$

Thus, proteins act as both acidic and basic buffers.

The following exercise will demonstrate how buffers resist changes in pH.

PROCEDURE

▲ **CAUTION!** *Please reread Section A, "General Safety Precautions and Procedures," on page xi and Section C, "Precautions Related to Working with Reagents," on page xii at the beginning of the laboratory manual before you begin any of the following experiments. Read the experiments before you perform them, to be sure that you understand all the procedures and safety precautions.*

1. Using a pH meter, immerse the electrode(s) in a beaker containing distilled water and determine the pH. _____

2. Drop by drop, slowly add 0.05 M hydrochloric acid (HCl) to the distilled water. Gently swirl the beaker after each drop is added. Note how many drops it takes for the pH of the solution to change one whole number.

   _____

   What is the pH of the solution? _____

3. Remove the electrode(s) from the solution and rinse with distilled water.

4. Now immerse the electrode(s) in a pH 7 buffer solution. Drop by drop, slowly add 0.05 M HCl to the solution. Gently swirl the beaker after each drop is added. Note how many drops it takes for the pH of the solution to change one whole number. _____

   What conclusion can you draw from this observation? _____

   _____

   _____

5. Remove the electrode(s) from the solution and rinse with distilled water.

6. Immerse the electrode(s) in a beaker of fresh distilled water and determine the pH. _____

   _____

7. Drop by drop, slowly add 0.05 M sodium hydroxide (NaOH) to the distilled water. Gently swirl the beaker after each drop is added. Note how many drops it takes for the pH of the solution to change one whole number.

   _____

   What is the pH of the solution? _____

8. Remove the electrode(s) from the solution and rinse with distilled water.

9. Now immerse the electrode(s) in a pH 7 buffer solution. Drop by drop, slowly add 0.05 M NaOH to the solution. Gently swirl the beaker after each drop is added. Note how many drops it takes for the pH of the solution to change one whole number. _____

   What conclusion can you draw from this observation? _____

b. Carbonic Acid–Bicarbonate Buffer System

The **carbonic acid–bicarbonate buffer system** is based on the *bicarbonate ion* (HCO_3^-), which can act as a weak base, and *carbonic acid* (H_2CO_3), which can act as a weak acid. Thus, the buffer system can compensate for either an excess or a shortage of H^+. For example, if there is an excess of H^+ (an acid condition), HCO_3^- can function as a weak base and remove the excess H^+ as follows:

$$H^+ \ + \ HCO_3^- \rightarrow H_2CO_3 \rightarrow H_2O + CO_2$$

Hydrogen ion   Bicarbonate ion (weak base)   Carbonic acid   Water   Carbon dioxide

On the other hand, if there is a shortage of H^+ (an alkaline condition), H_2CO_3 can function as a weak acid and provide H^+ as follows:

$$H_2CO_3 \rightarrow \ H^+ \ + \ HCO_3^-$$

Carbonic acid (weak acid)   Hydrogen ion   Bicarbonate ion

A typical bicarbonate buffer system consists of a mixture of carbonic acid (H_2CO_3) and its salt, sodium bicarbonate ($NaHCO_3$). The carbonic acid–bicarbonate buffer system is an important regulator of blood pH. When a strong acid, such as hydrochloric acid (HCl), is added to a buffer solution containing sodium bicarbonate, which behaves like a weak base, the following reaction occurs:

$$HCl \ + \ NaHCO_3 \rightarrow NaCl + \ H_2CO_3$$

Hydrochloric acid (strong acid)   Sodium bicarbonate (weak base)   Sodium chloride   Carbonic acid (weak acid)

If a strong base, such as sodium hydroxide (NaOH), is added to a buffer solution containing a weak acid, such as carbonic acid, the following reaction occurs:

$$NaOH \ + \ H_2CO_3 \ \rightarrow H_2O + NaHCO_3$$

Sodium hydroxide (strong base)   Carbonic acid (weak acid)   Water   Sodium bicarbonate (weak base)

Normal metabolism produces more acids than bases and thus tends to acidify the blood rather than make it more alkaline. Accordingly, the body needs more bicarbonate salt than it needs carbonic acid. Bicarbonate molecules outnumber carbonic acid molecules 20:1.

c. Phosphate Buffer System

The **phosphate buffer system** acts via a similar mechanism as the carbonic acid–bicarbonate buffer system. The components of the phosphate buffer system are the sodium salts of dihydrogen phosphate ($H_2PO_4^-$) and sodium monohydrogen phosphate ions ($H_2PO_4^{2-}$). The dihydrogen phosphate ion acts as the weak acid and is capable of buffering strong bases, such as OH^-.

$$NaOH + NaH_2PO_4 \rightarrow H_2O + Na_2HPO_4$$

Sodium hydroxide (strong base) Sodium dihydrogen phosphate (weak acid) Water Sodium monohydrogen phosphate (weak base)

The monohydrogen phosphate ion acts as the weak base and is capable of buffering the H^+ released by strong acids such as hydrochloric acid (HCl).

$$HCl + Na_2HPO_4 \rightarrow NaCl + NaH_2PO_4$$

Hydrochloric acid (strong acid) Sodium monohydrogen phosphate (weak base) Sodium chloride (salt) Sodium dihydrogen phosphate (weak acid)

Because the concentration of phosphates is highest in intracellular fluid, the phosphate buffer system is an important regulator of pH in the cytosol. It also is present at a lower level in extracellular fluids and acts to buffer acids in urine. $Na_2H_2PO_4$ is formed when excess H^+ in the kidney tubules combines with Na_2HPO_4. In this reaction, Na^+ released from Na_2HPO_4 forms sodium bicarbonate ($NaHCO_3$) and passes into the blood. The H^+ that replaces Na^+ becomes part of the NaH_2PO_4 that passes into the urine. This reaction is one of the mechanisms by which the kidneys help maintain blood pH by excreting H^+ in the urine.

2. Respirations and pH

Breathing also plays a role in maintaining the pH of body fluids. An increase in the carbon dioxide (CO_2) concentration in body fluids increases H^+ concentration and thus lowers the pH (makes it more acidic). This is illustrated by the following reactions:

$$CO_2 + H_2O \rightleftharpoons H_2CO_3 \rightleftharpoons H^+ + HCO_3^-$$

Conversely, a decrease in the CO_2 concentration of body fluids raises the pH (makes it more basic).

The pH of body fluids can be adjusted, usually within a couple of minutes, by a change in the rate and depth of breathing. With increased ventilation, more CO_2 is exhaled, the preceding reaction is driven to the left, H^+ concentration falls, and the blood pH rises. Because carbonic acid (H_2CO_3) can be eliminated by exhaling CO_2, it is called a **volatile acid.** If ventilation is slower than normal, less carbon dioxide is exhaled, and the blood pH falls. Doubling the ventilation increases the pH by about 0.23, from 7.4 to 7.63. Reducing ventilation to one-quarter its normal rate lowers the pH by 0.4, from 7.0. These examples show the powerful effect of alterations in breathing on pH of body fluids.

The pH of body fluids and the rate and depth of breathing interact via a negative feedback loop. If, for example, the blood becomes more acidic, the increase in H^+ is detected by chemoreceptors which stimulate the inspiratory area in the medulla oblongata. As a result, the diaphragm and other muscles of respiration contract more forcefully and frequently—the rate and depth of breathing increase so more CO_2 is exhaled.

The same effect is achieved if the blood level of CO_2 increases. Ventilations increase, which removes more CO_2 from blood to reduce the H^+ concentration, and blood pH increases. On the other hand, if the pH of the blood increases, the respiratory center is inhibited and the rate and depth of breathing decrease. A decrease in the CO_2 concentration of blood has the same effect. When ventilation decreases CO_2 accumulates in blood and the H^+ concentration increases. The respiratory mechanism normally can eliminate more acid or base than can all the buffers combined, but it is limited to eliminating only the sole volatile acid, carbonic acid.

The following exercise will demonstrate the relationship of exhalation of carbon dioxide to pH.

PROCEDURE

▲ **CAUTION!** *Please reread Section A, "General Safety Precautions and Procedures," on page xi and Section C, "Precautions Related to Working with Reagents," on page xii at the beginning of the laboratory manual before you begin any of the following experiments. Read the experiments before you perform them, to be sure that you understand all the procedures and safety precautions.*

1. Fill a large beaker with 100 mL of distilled water.

2. Add 5 mL of 0.10 normal sodium hydroxide (NaOH) solution and 5 drops of phenol red.

3. Phenol red is a pH indicator. It remains red in a basic solution, changes to orange in a neutral solution, and changes to yellow in an acidic solution.

4. While at rest, exhale through a straw into the solution. Your partner should determine how long it takes for the solution to change from

   orange to yellow. _____

▲ **CAUTION!** *Perform step 5 only if you have no known or apparent cardiac or other health problems and are capable of such an activity.*

5. Run in place for about 100 steps.

6. Exhale through a straw into a fresh solution as prepared in steps 1 and 2. Your partner should determine how long it takes for the solution to change from orange to yellow.

7. Change places with your partner and repeat the experiment. Explain the difference in time it took for the solution to change from orange to yellow at rest and following exercise. ____

_____

_____

_____

D. Acid-Base Imbalances

The normal pH range of systemic arterial blood is between 7.35 to 7.45. **Acidosis** (or **acidemia**) is a condition in which blood pH is below 7.35. **Alkalosis** (or **alkalemia**) is a condition in which blood pH is higher than 7.45.

A change in blood pH that leads to acidosis or alkalosis can be countered by **compensation,** the physiological response to an acid-base imbalance. If a person has an altered pH due to metabolic causes, respiratory mechanisms (hyperventilation or hypoventilation) can help compensate for the alteration. **Respiratory compensation** occurs within minutes and is maximized within hours. On the other hand, if a person has an altered pH due to respiratory causes, metabolic mechanisms (kidney excretion) can compensate for the alteration. **Renal compensation** may begin in minutes but takes days to reach a maximum.

1. Physiological Effects

The major physiological effect of acidosis is depression of the CNS through depression of synaptic transmission. If the systemic arterial blood pH falls below 7, depression of the nervous system is so severe that the individual becomes disoriented, then comatose and may die. Patients with severe acidosis usually die while in a state of coma. On the other hand, the major physiological effect of alkalosis is overexcitability in both CNS and peripheral nerves. Neurons conduct impulses repetitively, even when not stimulated by normal stimuli, resulting in nervousness, muscle spasms, and even convulsions and death.

In the discussion that follows, note that both respiratory acidosis and alkalosis are primary disorders of blood pCO_2 (normal range 35–45 mm Hg). On the other hand, both metabolic acidosis and alkalosis are primary disorders of bicarbonate (HCO_3^-) concentration (normal range 22–26 mEq/liter in systemic arterial blood).

2. Respiratory Acidosis

The hallmark of **respiratory acidosis** is an elevated pCO_2 of systemic arterial blood above 45 mm Hg. Inadequate exhalation of CO_2 causes the blood pH to drop; any condition that decreases the movement of CO_2 from the blood to the alveoli of the lungs to the atmosphere causes a buildup of carbon dioxide, carbonic acid, and hydrogen ions. Such conditions include emphysema, pulmonary edema, injury to the respiratory center of the medulla oblongata, airway obstruction, or disorders of the muscles involved in breathing. Renal compensation involves increased excretion of H^+ and increased reabsorption of HCO_3^- by the kidneys. Treatment of respiratory acidosis aims to increase the exhalation of CO_2. Excessive secretions can be suctioned out of the respiratory tract, and artificial respiration can be given. In addition, intravenous administration of bicarbonate and ventilation therapy may be helpful.

3. Respiratory Alkalosis

In **respiratory alkalosis** arterial blood pCO_2 is decreased below 35 mm Hg. Hyperventilation causes the pH to increase. It occurs in conditions that stimulate the respiratory center. Such conditions include oxygen deficiency due to high altitude or pulmonary disease, cerebrovascular accident (stroke), and severe anxiety. Renal compensation may bring blood pH into normal range if the kidneys decrease excretion of H^+ and reabsorption of HCO_3^-. Treatment of respiratory alkalosis is aimed at increasing the level of CO_2 in the body. One simple measure is to have the person inhale and exhale into a paper bag.

4. Metabolic Acidosis

In **metabolic acidosis,** the systemic arterial plasma HCO_3^- level drops below 22 mEq/liter. The decrease in pH is caused by three situations that lower the plasma level of bicarbonate, such as may occur with severe diarrhea or renal dysfunction; accumulation of an acid, other than carbonic acid, as may occur in ketosis; or failure of the kidneys to excrete H^+ derived from metabolism of dietary proteins. Respiratory compensation by hyperventilation can help bring blood pH into the normal range. Treatment of metabolic acidosis consists of administering intravenous solutions of sodium bicarbonate and correcting the cause of acidosis.

5. Metabolic Alkalosis

In **metabolic alkalosis** the blood HCO_3^- concentration is elevated above 26 mEq/liter. A nonrespiratory loss of acid by the body or excessive intake of alkaline drugs causes the pH to increase above 7.45. Excessive vomiting of gastric contents results in a substantial loss of hydrochloric acid and is probably the most frequent cause of metabolic alkalosis. Other causes of metabolic alkalosis include gastric suctioning, use of certain diuretics, endocrine disorders, excessive administration of alkaline drugs (antacids), and dehydration. Respiratory compensation through hypoventilation may bring blood pH into the normal range. Treatment of metabolic alkalosis consists of fluid therapy to replace chloride, potassium, and other electrolyte deficiencies and correcting the cause of alkalosis.

A summary of acidosis and alkalosis is presented in Table 24.1.

Based on an analysis of respiratory gases, you can determine if a person has acidosis or alkalosis and whether the acidosis or alkalosis is respiratory or metabolic, as reflected by the change in pH.

In the following table (bottom of page), note the normal ranges for pH, pCO_2, and HCO_3^-. Also note how values above and below normal relate to acidosis and alkalosis.

If a change in pH is a result of an abnormal pCO_2 value, then the condition is respiratory in nature. If, instead, a change in pH is a result of an abnormal HCO_3^- value, the condition is metabolic in nature.

Problems related to acid-base imbalance are reflected in each of the following conditions. Determine whether each is (1) acidosis or alkalosis and (2) metabolic or respiratory:

a. pH = 7.32
HCO_3^- = 10 mEq/liter

b. pH = 7.48
pCO_2 = 32 mm Hg

c. pH = 7.52
HCO_3^- = 28 mEq/liter

d. pH = 7.30
pCO_2 = 48 mm Hg

e. pH = 7.32
HCO_3 = 10 mEq/liter
pCO_2 = 33 mm Hg

f. pH = 7.48
HCO_3 = 27 mEq/liter
pCO_2 = 32 mm Hg

g. pH = 7.52
HCO_3 = 28 mEq/liter

h. pH = 7.30
HCO_3 = 24 mEq/liter
pCO_2 = 48 mm Hg

i. pH = 7.49
HCO_3 = 28 mEq/liter
pCO_2 = 47 mm Hg

j. pH = 7.53
HCO_3 = 18 mEq/liter
pCO_2 = 30 mm Hg

k. pH = 7.35
HCO_3 = 28 mEq/liter
pCO_2 = 48 mm Hg

ANSWER THE LABORATORY REPORT QUESTIONS AT THE END OF THE EXERCISE.

	pH	pCO_2	HCO_3^-
Normal range	7.35–7.45	35–45 mm Hg	22–26 mEq/liter
Acidosis	Below 7.35	Above 45 mm Hg	Below 22 mEq/liter
Alkalosis	Above 7.45	Below 35 mm Hg	Above 26 Eq/liter

TABLE 24.1 Summary of acidosis and alkalosis

Condition	Definition	Common cause	Compensatory mechanism
Respiratory acidosis	Increased pCO_2 (above 45 mm Hg) and decreased pH (below 7.35) if there is no compensation.	Hypoventilation due to emphysema, pulmonary edema, trauma to respiratory center, airway obstructions, dysfunction of muscles of respiration.	Renal: increased excretion of H^+; increased reabsorption of HCO_3^-. If compensation is complete, pH will be within normal range, but pCO_2 will be high.
Respiratory alkalosis	Decreased pCO_2 (below 35 mm Hg) and increased pH (above 7.45) if there is no compensation.	Hyperventilation due to oxygen deficiency, pulmonary disease, cerebrovascular accident (CVA), or severe anxiety.	Renal: decreased excretion of H^+; decreased reabsorption of HCO_3^-. If compensation is complete, pH will be within normal range, but pCO_2 will be low.
Metabolic acidosis	Decreased HCO_3^- (below 22 mEq/liter) and decreased pH (below 7.35) if there is no compensation.	Loss of bicarbonate due to diarrhea, accumulation of acid (ketosis) renal dysfunction.	Respiratory: hyperventilation, which increases loss of CO_2. If compensation is complete, pH will be within normal range, but HCO_3^- will be low.
Metabolic alkalosis	Increased HCO_3^- (above 26 mEq/liter) and increased pH (above 7.45) if there is no compensation.	Loss of acid or excessive intake of alkaline drugs; due to vomiting, gastric suctioning, use of certain diuretics, and intake of alkaline drugs.	Respiratory: hypoventilation, which slows loss of CO_2. If compensation is complete, pH will be within normal range, but HCO_3^- will be high.

pH and Acid-Base Balance

Name _____   Date _____

Laboratory Section _____   Score/Grade _____

SECTION B ■ Measuring pH

1. Using Litmus Paper

Solution	Color of red litmus paper	Color of blue litmus paper	Is the solution acidic or basic?
Milk of magnesia			
Vinegar			
Coffee			
Carbonated soft drink			
Orange juice			
Distilled water			
Baking soda			
Lemon juice			

2. Using pH Paper

Solution	pH
Milk of magnesia	
Vinegar	
Coffee	
Carbonated soft drink	
Orange juice	
Distilled water	
Baking soda	
Lemon juice	

3. Using a pH Meter

Solution	pH
Milk of magnesia	
Vinegar	
Coffee	
Carbonated soft drink	
Orange juice	
Distilled water	
Baking soda	
Lemon juice	

EXERCISE 24

pH and Acid-Base Balance

Name _____ Date _____

Laboratory Section _____ Score/Grade _____

PART 1 ▪ Multiple Choice

_____ 1. An acid is a substance that dissociates into (a) OH^- (b) H^+ (c) HCO_3^- (d) Na^+

_____ 2. Which of the following pHs is more acidic? (a) 6.89 (b) 6.91 (c) 7.00 (d) 6.83

_____ 3. The pH of distilled (pure) water is (a) 4.2 (b) 6.35 to 6.85 (c) 7.0 (d) 3.0 to 3.5

_____ 4. Which mechanism is quickest to restore pH? (a) exhalation of CO_2 (b) buffers (c) kidney excretion (d) inhalation of oxygen

_____ 5. In the carbonic acid–bicarbonate buffer system, which substance functions to buffer a strong base? (a) NaOH (b) $NaHCO_3$ (c) H_2CO_3 (d) NaCl

_____ 6. The most abundant buffer in body cells and plasma is the (a) protein buffer (b) phosphate buffer (c) hemoglobin buffer (d) carbonic acid–bicarbonate buffer

_____ 7. Doubling the ventilation rate increases pH by about (a) 0.75 (b) 2.21 (c) 1.86 (d) 0.23

PART 2 ▪ Completion

8. Bases dissociate into _____ ions and cations.

9. A change of one whole number on the pH scale represents a _____ -fold change from the previous concentration.

10. A(n) _____ solution will turn red litmus paper blue.

11. The pH of systemic arterial blood is _____.

12. Most H^+ in the body is produced as a result of _____.

13. In the protein buffer system, the carboxyl group acts as a(n) _____.

14. If the rate and depth of ventilation increase, pH will _____.

15. If blood becomes more basic, the rate and depth of ventilation will _____.

16. A pH higher than 7.45 is referred to as _____.

17. Metabolic acidosis and alkalosis are disorders of _____ concentration.

18. The principal physiological effect of _____ is depression of the CNS.

19. _____ acidosis is characterized by a pH below 7.35 and an elevated pCO_2.

20. Compensation for metabolic acidosis is _____.

Reproductive Systems

Objectives

At the completion of this exercise you should understand

A The anatomical characteristics and functions of the organs of the male and female reproductive systems

B The phases of the female reproductive cycle

C The parts of a fetus-containing pig uterus

Sexual reproduction is the process by which new individuals of a species are produced and the genetic material is passed from generation to generation. This maintains continuation of the species. Cell division in a multicellular organism is necessary for growth as well as repair, and it involves passing of genetic material from parent cells to daughter cells. In somatic cell division a parent cell produces two identical daughter cells. This process is involved in replacing cells and growth. In reproductive cell division, sperm and egg cells are produced for continuity of the species.

The male and female reproductive organs may be grouped by function. (1) The testes and ovaries, also called **gonads,** function in the production of gametes—sperm cells and secondary oocytes, respectively. The gonads also secrete sex hormones. (2) The **ducts** of the reproductive systems store and transport gametes. (3) **Accessory sex glands** produce materials that support gametes and facilitate their movements. (4) Finally, **supporting structures,** such as the penis and the uterus, assist the delivery and joining of gametes and, in females, the growth of the embryo and fetus during pregnancy. In this exercise, you will study the structure of the male and female reproductive organs and associated structures.

The specialized branch of medicine concerned with the diagnosis and treatment of diseases of the female reproductive system is **gynecology** (gī-ne-KOL-ō-jē; *gyneco* = woman). As mentioned in a previous exercise, **urology** (u-ROL-ō-jē) is the study of the urinary system. Urologists also diagnose and treat diseases and disorders of the male reproductive system. The branch of medicine that deals with male disorders, especially infertility and sexual dysfunction, is called **andrology** (an-DROL-ō-jē; *andro* = masculine).

A. Organs of Male Reproductive System

The **male reproductive system** includes (1) the testes, or male gonads, which produce sperm and secrete hormones; (2) a system of ducts that transports and stores sperm, assists in their maturation, and conveys them to the exterior; (3) accessory sex glands, whose secretions contribute to semen; and (4) several supporting structures, including the scrotum and penis.

1. Testes

The **scrotum** (SKRŌ-tum = bag) is the supporting structure for the testes. It is a sac consisting of loose skin and superficial fascia that hangs from the root (attached portion) of the penis. Externally, the scrotum looks like a single pouch of skin separated into lateral portions by a median ridge called the **raphe** (RĀ-fē = seam). Internally, the **scrotal septum** divides the scrotum into two sacs, each containing a single testis. The septum consists of superficial fascia and muscle tissue called the **dartos muscle** (DAR-tōs = skinned), which consists of bundles of smooth muscle fibers. Dartos muscle is also found in the subcutaneous tissue of the scrotum. When it contracts, the dartos muscle causes wrinkling of the skin of the scrotum and elevates the testes.

The location of the scrotum and contraction of its muscle fibers regulate the temperature of the testes. Because the scrotum is outside the pelvic cavity, it maintains the temperature of the testes about 2–3°C below body temperature. This cooler temperature is required for reproduction and survival of sperm. The **cremaster** (krē-MAS-ter = suspender) **muscle** is a small band of skeletal muscle in the spermatic cord that is a continuation of the internal oblique muscle. It elevates the testes during sexual arousal and upon exposure to cold. This action moves the testes closer to the pelvic cavity where they can absorb body heat. Exposure to warmth reverses the process. The dartos muscle also contracts in response to cold and relaxes in response to warmth.

The **testes,** or **testicles,** are paired oval glands that lie in the pelvic cavity for most of fetal life. They usually begin their descent into the scrotum during the latter half of the seventh month of fetal development; full descent is not complete until just before birth. If the testes do not descend, the condition is called **cryptorchidism** (krip-TOR-ki-dizm; *crypt* = hidden; *orchid* = testis). Bilateral cryptorchidism results in sterility, because the cells that stimulate the initial development of sperm cells are destroyed by the higher temperature of the pelvic cavity. The chance of testicular cancer is 30% to 50% greater in cryptorchid testes.

Each testis is partially covered by a serous membrane called the **tunica** (*tunica* = sheath) **vaginalis,** which is derived from the peritoneum. Internal to the tunica vaginalis is a capsule of dense irregular connective tissue, the **tunica albuginea** (al'-bū-JIN-ē-a; *albu* = white), which extends inward forming septa that divide the testis into a series of 200 to 300 internal compartments called **lobules.** Each lobule contains one to three tightly coiled **seminiferous** (*semin* = seed; *fer* = to carry) **tubules** where sperm are produced. The process by which the seminiferous tubules of the testes produce sperm is called **spermatogenesis.**

Label the structures associated with the testes in Figure 25.1.

The seminiferous tubules contain two types of cells: spermatogenic cells and Sertoli cells. **Spermatogenic cells** are sperm-forming cells in various stages that undergo mitosis and differentiation to eventually produce sperm. Together with supporting cells, they line the seminiferous tubules. Sperm

production begins at the periphery of the seminiferous tubules in stem cells called **spermatogonia** (sper'-ma-tō-GŌ-nē-a; *gonia* = generation or offspring; singular is **spermatogonium**). Toward the lumen of the tubule are layers of progressively more mature cells. In order of advancing maturity, these are **primary spermatocytes** (SPER-ma-tō-sītz), **secondary spermatocytes, spermatids,** and **sperm.** By the time a **sperm cell,** or **spermatozoon** (sper'-ma-tō-ZŌ-on; *zoon* = life; plural is **sperm,** or **spermatozoa**), has nearly reached maturity, it is released into the lumen of the tubule. **Spermatogenesis** takes 65 to 75 days.

Embedded among the spermatogenic cells in the tubules are large **Sertoli** or **sustentacular** (sus'-ten-TAK-ū-lar) **cells** which extend from the basement membrane to the lumen of the tubule. Sertoli cells support and protect developing spermatogenic cells; nourish spermatocytes, spermatids, and sperm; phagocytize excess spermatid cytoplasm as development proceeds; and mediate the effects of testosterone and follicle-stimulating hormone (FSH). Sustentacular cells also control movements of spermatogenic cells and the release of sperm into the lumen of the seminiferous tubules. They produce fluid for sperm transport, secrete androgen-binding protein, and secrete the hormone inhibin, which helps regulate sperm production by inhibiting the secretion of FSH. In the spaces between adjacent seminiferous tubules are clusters of cells called **Leydig (interstitial) cells.** These cells secrete testosterone, the most prevalent androgen (male sex hormone). Because they produce both sperm and hormones, the testes are both exocrine and endocrine glands.

Using your textbook, charts, or models as reference, label Figure 25.2.

Sperm are produced at the rate of about 300 million per day. Once ejaculated, most do not survive more than 48 hr within the female reproductive tract. The parts of a sperm are as follows:

a. *Head.* Contains the **nucleus** (containing the haploid number of chromosomes) and **acrosome** (containing enzymes, which help penetration of the sperm cell into a secondary oocyte).

b. *Tail.* Propels the sperm along its way. Subdivided into four parts: neck, middle piece, principal piece, and end piece. The **neck** is the constricted region just behind the head. The **middle piece** contains numerous mitochondria which provide ATP for locomotion. The **principal piece** is the longest portion of the tail, and the **end piece** is the terminal, tapering portion of the tail.

FIGURE 25.1 Testis showing its system of ducts.

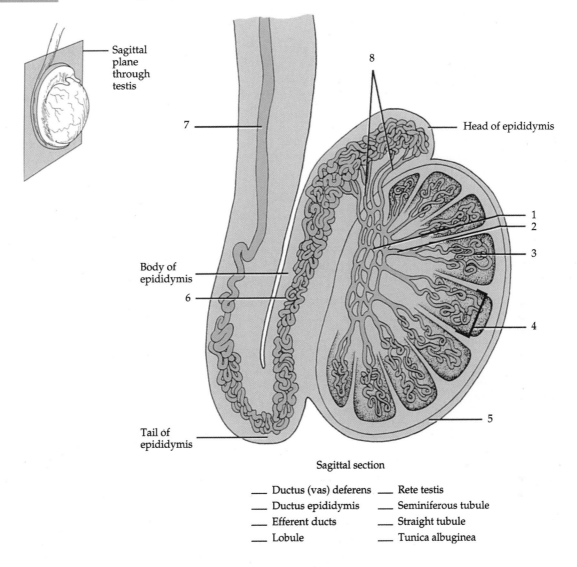

Sagittal plane through testis

8

7

Head of epididymis

1
2
3
4
5

Body of epididymis

6

Tail of epididymis

Sagittal section

___ Ductus (vas) deferens   ___ Rete testis
___ Ductus epididymis   ___ Seminiferous tubule
___ Efferent ducts   ___ Straight tubule
___ Lobule   ___ Tunica albuginea

With the aid of your textbook, label Figure 25.3.

2. Ducts

Pressure generated by fluid secreted by Sertoli cells pushes sperm and fluid along the lumen of the seminiferous tubules into tubes called **straight tubules,** from which they are transported into a network of ducts, the **rete** (RĒ-tē = network) **testis.** The sperm cells are moved into a series of coiled **efferent ducts** that empty into a single tube called the **ductus epididymis** (ep'-i-DID-i-mis; *epi* = above; *didymis* = testis). From there, they are passed into the **ductus (vas) deferens,** which ascends along the posterior border of the epididymis, passes through the inguinal canal, enters the pelvic cavity, and loops over the ureters and passes over the side and down the posterior sur-

face of the urinary bladder. The ductus (vas) deferens and duct from the seminal vesicle together form the **ejaculatory** (e-JAK-ū-la-tō'-rē; *ejacul* = to expel) **duct,** which ejects the sperm and seminal vesicle secretions into the **urethra,** the terminal duct of the system. The male urethra is divisible into (1) a *prostatic urethra,* which passes through the prostate; (2) a *membranous urethra,* which passes through the deep muscles of the perineum; and (3) a *spongy (penile) urethra,* which passes through the corpus spongiosum of the penis (see Figure 25.6). The **epididymis** (*epi* = above; *didymis* = testis) is a comma-shaped organ that is divisible into a head, body, and tail. The head is the large superior portion where the efferent ducts from the testis join the ductus epididymis; the body is the narrower middle portion of the epididymis; and the tail is the smaller inferior portion in which the epididymis continues as the ductus (vas) deferens.

FIGURE 25.2   Seminiferous tubules showing various stages of spermatogenesis.

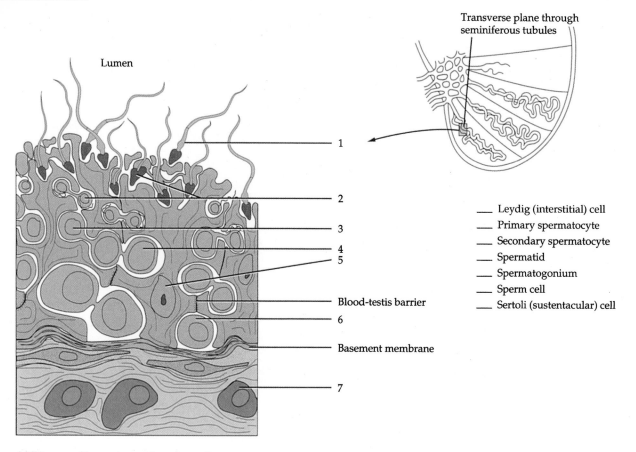

(a) Diagram of transverse section of a portion
    of a seminiferous tubule

___ Leydig (interstitial) cell
___ Primary spermatocyte
___ Secondary spermatocyte
___ Spermatid
___ Spermatogonium
___ Sperm cell
___ Sertoli (sustentacular) cell

___ Leydig (interstitial) cell
___ Primary spermatocyte
___ Secondary spermatocyte
___ Spermatid
___ Spermatogonium
___ Sperm cell

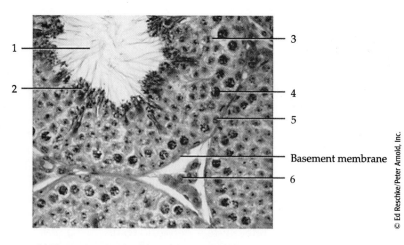

(b) Photomicrograph of transverse section of a
    portion of a seminiferous tubule (400×)

© Ed Reschke/Peter Arnold, Inc.

FIGURE 25.3 **Parts of a sperm cell.**

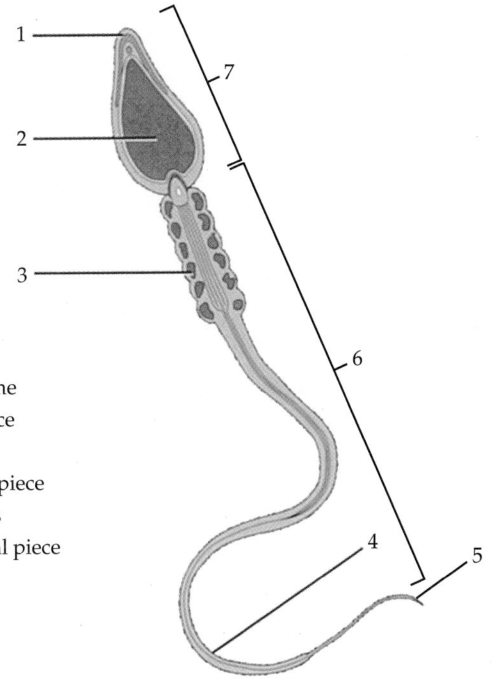

1

7

2

3

6

4

5

___ Acrosome
___ End piece
___ Head
___ Middle piece
___ Nucleus
___ Principal piece
___ Tail

Label the various ducts of the male reproductive system in Figure 25.1.

The ductus epididymis is lined with **pseudostratified columnar epithelium.** The free surfaces of the cells contain long, branching microvilli called **stereocilia.** The muscularis deep to the epithelium consists of smooth muscle. Functionally, the ductus epididymis is the site where sperm matures (acquires motility and ability to fertilize an ovum). This occurs over a 10- to 14-day period. The ductus epididymis also stores sperm and propels them by peristaltic contraction of its smooth muscle into the ductus (vas) deferens. Sperm cells may remain in storage in the ductus epididymis up to a month or more. After that, they are expelled from the epididymis or reabsorbed in the epididymis.

Obtain a prepared slide of the ductus epididymis showing its mucosa and muscularis. Compare your observations to Figure 25.4.

Histologically, the **ductus (vas) deferens** is also lined with **pseudostratified columnar epithelium,** and its muscularis consists of three layers of smooth muscle. Peristaltic contractions of the muscularis propel sperm cells toward the urethra during ejaculation. One method of sterilization in males, **vasectomy,** involves removal of a portion of each ductus (vas) deferens.

Obtain a prepared slide of the ductus (vas) deferens showing its mucosa and muscularis. Compare your observations to Figure 25.5.

3. Accessory Sex Glands

Whereas the ducts of the male reproductive system store and transport sperm, the **accessory sex glands** secrete most of the liquid portion of **semen.** Semen is a mixture of sperm cells and the secretions of the seminal vesicles, the prostate, and the bulbourethral glands.

The **seminal vesicles** (VES-i-kuls) are paired, convoluted, pouchlike structures lying posterior to and at the base of the urinary bladder anterior to the rectum. The glands secrete the alkaline viscous fluid that contains fructose, prostaglandins, and clotting proteins. Fluid secreted by the seminal vesicles normally constitutes about 60% of the volume of semen.

FIGURE 25.4 **Histology of ductus epididymis (250×).**

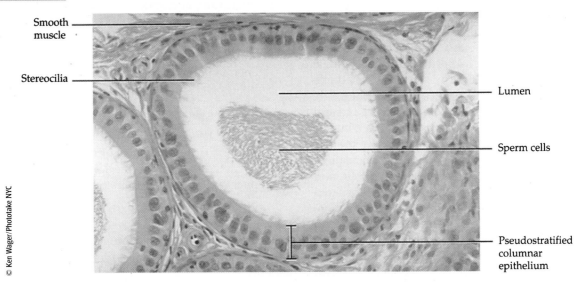

Smooth muscle

Stereocilia

Lumen

Sperm cells

Pseudostratified columnar epithelium

FIGURE 25.5 **Histology of ductus (vas) deferens (75×).**

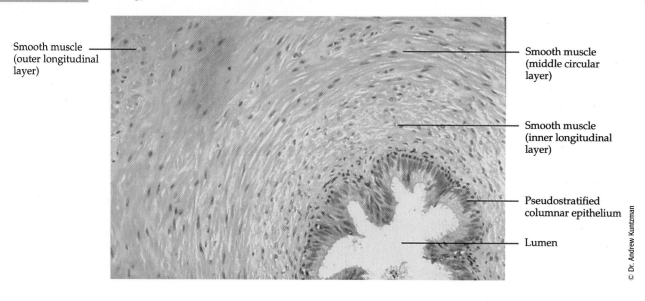

Smooth muscle (outer longitudinal layer)

Smooth muscle (middle circular layer)

Smooth muscle (inner longitudinal layer)

Pseudostratified columnar epithelium

Lumen

The **prostate** (PROS-tāt) is a single doughnut-shaped gland about the size of a golf ball. It is inferior to the urinary bladder and surrounds the prostatic urethra. The prostate secretes a milky white, slightly acidic fluid into the prostatic urethra. The prostatic secretion makes up about 25% of the volume of semen.

The paired, pea-sized **bulbourethral** (bul′-bō-ū-RĒ-thrall), or **Cowper's glands,** are located inferior to the prostate on either side of the membranous urethra. They secrete an alkaline fluid into the urethra that protects the passing sperm by neutralizing acids from urine in the urethra.

With the aid of your textbook, label the accessory glands and associated structures in Figure 25.6.

4. Penis

a. *Body.* Main portion of penis composed of three cylindrical masses of tissue, each bound by fibrous tissue called the **tunica albuginea.** The two dorsolateral masses are called the **corpora cavernosa penis** (*corpora* = body; *cavernosa* = hollow). The smaller midventral mass, the **corpus spongiosum penis,** contains the spongy urethra and functions in keeping the spongy urethra open during ejaculation. All three masses are enclosed by fascia and skin and consist of erectile tissue permeated by blood sinuses. Upon sexual stimulation, the arteries supplying the penis dilate, and large quantities of blood enter the blood sinuses. Expansion of these spaces compresses the veins draining the penis, so more blood

that enters is trapped. These vascular changes, due to local release of nitric oxide and a parasympathetic reflex result in an **erection.** The penis returns to its flaccid state when the arteries constrict and pressure on the veins is relieved.

b. *Root.* The attached portion of the penis. It consists of the **bulb of the penis,** the expanded portion of the base of the corpus spongiosum penis, and the **crura** (*crus* = resembling a leg; singular is *crus*) **of the penis,** the two separated and tapered portions of the corpora cavernosa penis. The bulb of the penis is attached to the inferior surface of the deep muscles of the perineum and enclosed by the bulbospongiosus muscle. Each crus of the penis is attached to the ischial and inferior pubic rami and surrounded by the ischiocavernosus muscle. Contraction of these skeletal muscles aids ejaculation.

c. *Glans* (*glandes* = acorn) *penis.* The slightly enlarged acorn-shaped distal end of the corpus spongiosum penis. The margin of the glans penis is termed the **corona.** The distal urethra enlarges within the glans penis and forms a terminal slitlike opening, the **external urethral orifice.** Covering the glans in an uncircumcised penis is the loosely fitting **prepuce** (PRĒ-poos), or **foreskin.**

With the aid of your textbook, label the parts of the penis in Figure 25.7.

Now that you have completed your study of the organs of the male reproductive system, label Figure 25.8.

FIGURE 25.6 Relationships of some male reproductive organs.

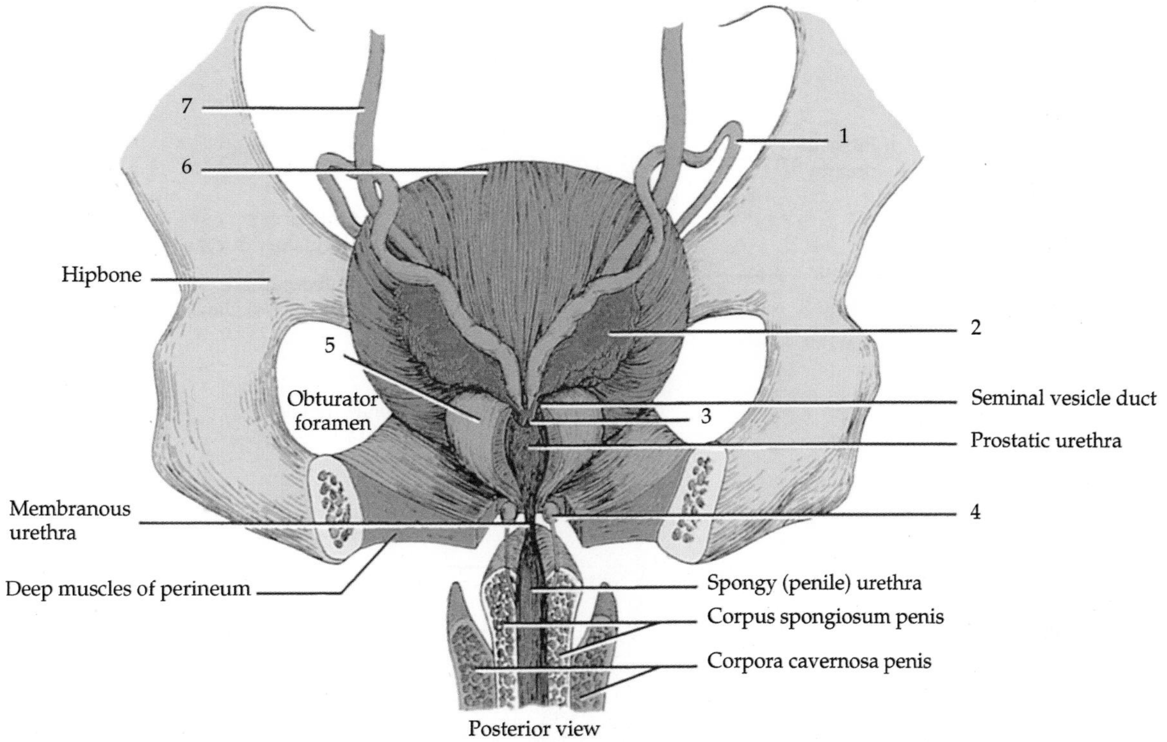

___ Bulbourethral (Cowper's) gland
___ Ductus (vas) deferens
___ Ejaculatory duct
___ Prostate
___ Seminal vesicle
___ Ureter
___ Urinary bladder

B. Organs of Female Reproductive System

The **female reproductive system** includes the ovaries (female gonads), which produce secondary oocytes; uterine (fallopian) tubes, or oviducts, which transport secondary oocytes and fertilized ova to the uterus; vagina; external organs collectively called the vulva or pudendum; and the mammary glands.

1. Ovaries

The **ovaries** (= egg receptacle) or female gonads are paired glands that resemble unshelled almonds in size and shape. Functionally, the ovaries produce gametes, secondary oocytes, and discharge them about once a month by a process called ovulation, and secrete estrogens, progesterone (female sex hormones), relaxin, and inhibin. The ovaries are positioned in the superior pelvic cavity, one on each side of the uterus, and are maintained by a series of ligaments:

a. ***Broad ligament.*** Part of the parietal peritoneum, attaches to the ovaries by a double-layered fold of peritoneum called the **mesovarium.**
b. ***Ovarian ligament.*** Anchors ovary to uterus.
c. ***Suspensory ligament.*** Attaches ovary to pelvic wall.

The point of entrance for blood vessels and nerves is the **hilum.** With the aid of your textbook, label the ovarian ligaments in Figure 25.9.

Histologically, the ovaries consist of the following parts:

1. ***Germinal epithelium.*** A layer of simple epithelium (low cuboidal or squamous) that covers the surface of the ovary and is continuous with the mesothelium that covers the mesovarium. The term *germinal epithelium* is a misnomer since it does not give rise to oocytes, although at one time it was believed that it did.
2. ***Tunica albuginea.*** A whitish capsule of dense, irregular connective tissue immediately deep to the germinal epithelium.

FIGURE 25.7 Internal structure of frontal section of penis.

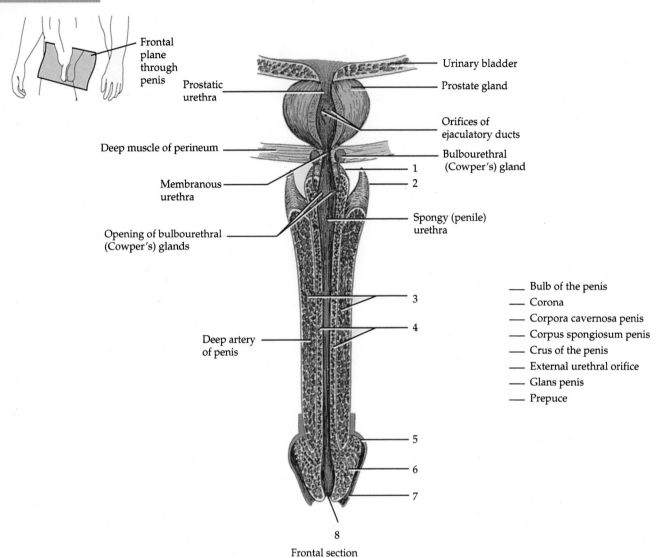

Frontal plane through penis

Prostatic urethra

Deep muscle of perineum

Membranous urethra

Opening of bulbourethral (Cowper's) glands

Deep artery of penis

Urinary bladder

Prostate gland

Orifices of ejaculatory ducts

Bulbourethral (Cowper's) gland

1

2

Spongy (penile) urethra

3

4

5

6

7

8

Frontal section

___ Bulb of the penis
___ Corona
___ Corpora cavernosa penis
___ Corpus spongiosum penis
___ Crus of the penis
___ External urethral orifice
___ Glans penis
___ Prepuce

3. *Ovarian cortex.* A region just deep to the tunica albuginea that consists of dense connective tissue and contains ovarian follicles (described shortly).

4. *Ovarian medulla.* A region deep to the ovarian cortex that consists of loose connective tissue and contains blood vessels, lymphatic vessels, and nerves.

5. *Ovarian follicles* (*folliculus* = little bag). Lie in the cortex and consist of **oocytes** in various stages of development and their surrounding cells. When the surrounding cells form a single layer, they are called **follicular cells.** Later in development, when they form several layers, they are referred to as **granulosa cells.** The surrounding cells nourish the developing oocyte and begin to secrete estrogens as the

follicle grows larger. Ovarian follicles undergo a series of changes prior to ovulation, progressing through several distinct stages. The most numerous and peripherally arranged follicles are termed **primordial follicles.** If a primordial follicle progresses to ovulation (release of a mature ovum), it will sequentially transform into a **primary follicle,** then a **secondary follicle,** and finally a **mature (graafian) follicle.**

6. *Mature (graafian) follicle.* A large, fluid-filled follicle that is preparing to rupture and expel a secondary oocyte, a process called **ovulation.**

7. *Corpus luteum* (= yellow body). Contains the remnants of a mature follicle after ovulation. The corpus luteum produces proges-

FIGURE 25.8 **Male organs of reproduction and surrounding structures.**

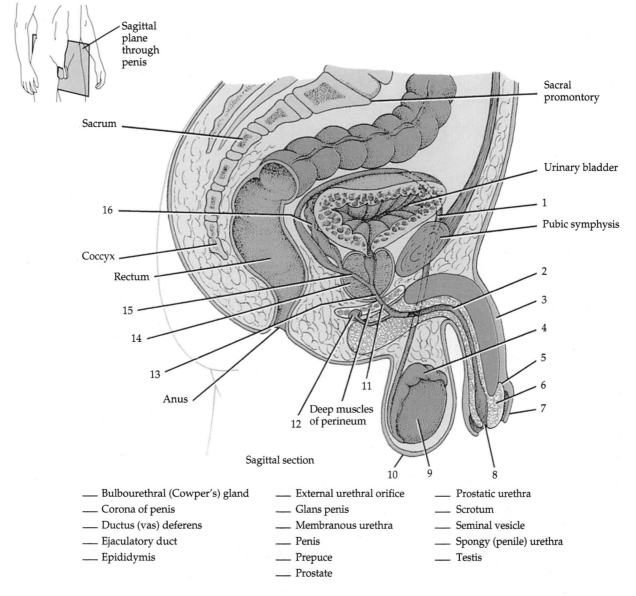

Sagittal plane through penis

Sacrum

16

Coccyx

Rectum

15

14

13

Anus

12

Deep muscles of perineum

Sagittal section

Sacral promontory

Urinary bladder

1

Pubic symphysis

2

3

4

5

6

7

10 9 8

11

—— Bulbourethral (Cowper's) gland
—— Corona of penis
—— Ductus (vas) deferens
—— Ejaculatory duct
—— Epididymis

—— External urethral orifice
—— Glans penis
—— Membranous urethra
—— Penis
—— Prepuce
—— Prostate

—— Prostatic urethra
—— Scrotum
—— Seminal vesicle
—— Spongy (penile) urethra
—— Testis

terone, estrogens, relaxin, and inhibin until it degenerates and turns into fibrous tissue called a **corpus albicans** (= white body).

With the aid of your textbook, label the parts of an ovary in Figure 25.10.

Obtain prepared slides of the ovary, examine them, and compare your observations to Figure 25.11.

2. Uterine (Fallopian) Tubes

The **uterine (fallopian) tubes,** or **oviducts,** extend laterally from the uterus and transport sec-

ondary oocytes and fertilized ova from the ovaries to the uterus. Fertilization normally occurs in the uterine tubes. The tubes lie between folds of the broad ligaments of the uterus. The funnel-shaped, open distal end of each uterine tube, called the **infundibulum,** ends in a fringe of fingerlike projections called **fimbriae** (FIM-bre-ē = fringe). The **ampulla** of the uterine tube is the widest, longest portion, making up about the lateral two-thirds of its length. The **isthmus** is the more medial short, narrow, thick-walled portion that joins the uterus.

With the aid of your textbook, label the parts of the uterine tubes in Figure 25.9.

FIGURE 25.9   Uterus and associated female reproductive structures. The left side of figure (a) has been sectioned to show internal structures.

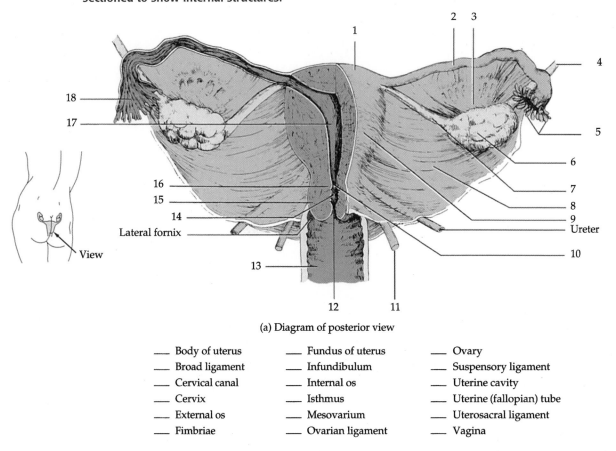

(a) Diagram of posterior view

___ Body of uterus	___ Fundus of uterus	___ Ovary
___ Broad ligament	___ Infundibulum	___ Suspensory ligament
___ Cervical canal	___ Internal os	___ Uterine cavity
___ Cervix	___ Isthmus	___ Uterine (fallopian) tube
___ External os	___ Mesovarium	___ Uterosacral ligament
___ Fimbriae	___ Ovarian ligament	___ Vagina

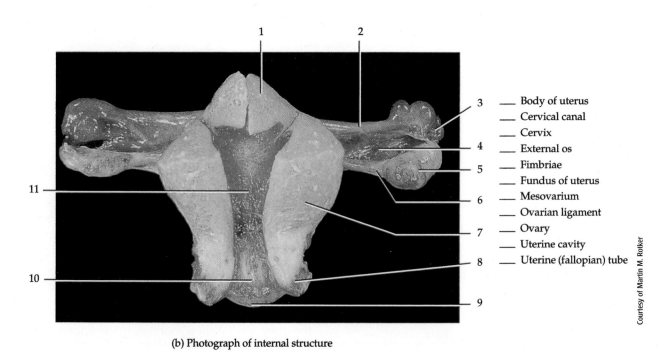

___ Body of uterus
___ Cervical canal
___ Cervix
___ External os
___ Fimbriae
___ Fundus of uterus
___ Mesovarium
___ Ovarian ligament
___ Ovary
___ Uterine cavity
___ Uterine (fallopian) tube

(b) Photograph of internal structure

Courtesy of Martin M. Rotker

FIGURE 25.10 Histology of ovary. Arrows indicate sequence of developmental stages that occur as part of ovarian cycle.

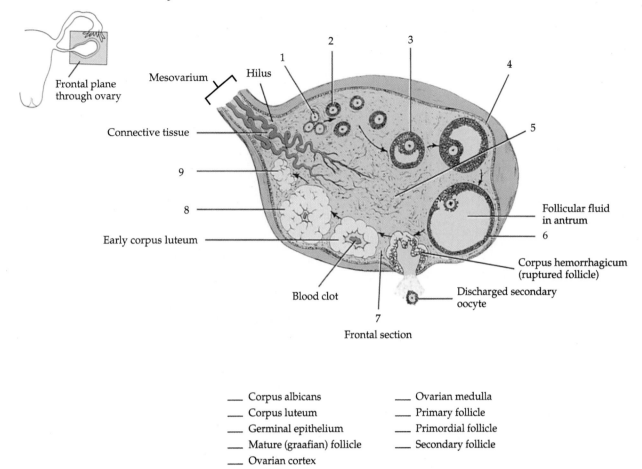

Frontal plane through ovary

Mesovarium

Hilus

Connective tissue

9

8

Early corpus luteum

Blood clot

7

Frontal section

Follicular fluid in antrum

6

Corpus hemorrhagicum (ruptured follicle)

Discharged secondary oocyte

___ Corpus albicans
___ Corpus luteum
___ Germinal epithelium
___ Mature (graafian) follicle
___ Ovarian cortex

___ Ovarian medulla
___ Primary follicle
___ Primordial follicle
___ Secondary follicle

Histologically, the mucosa of the uterine tubes contains ciliated columnar cells and secretory cells. The muscularis is composed of an inner circular and an outer longitudinal layers of smooth muscle. Peristaltic contractions of the muscularis and the ciliary action of the mucosa cells help move the oocyte or fertilized ovum toward the uterus. The serosa is the outer covering.

Examine a prepared slide of the wall of the uterine tube, and compare your observations to Figure 25.12.

3. Uterus

The **uterus (womb)** serves as part of the pathway for sperm deposited in the vagina to reach the uterine tubes. It is also the site of implantation of a fertilized ovum, development of the fetus during pregnancy, and labor. The uterus is also the source of menstrual flow when implantation of a fertilized ovum fails to occur. Situated between the urinary bladder and the rectum, the uterus is the size and shape of an inverted pear. The uterus is subdivided into the following regions:

a. *Fundus.* Dome-shaped portion superior to the uterine tubes.

b. *Body.* Tapering central portion.

c. *Cervix.* Inferior narrow opening into vagina.

d. *Isthmus* (IS-mus). Constricted region between body of the uterus and cervix.

e. *Uterine cavity.* Interior of the body of the uterus.

f. *Cervical canal.* Interior of the cervix.

g. *Internal os.* Site where cervical canal opens into uterine cavity.

h. *External os.* Site where cervical canal opens into vagina.

Label these structures in Figure 25.9.

FIGURE 25.11 **Histology of ovary.**

Tunica
albuginea

Cortex

Primary (preantral)
follicle granulosa
cells

Primary oocyte

Primordial
follicle

Theca folliculi

Zona pellucida

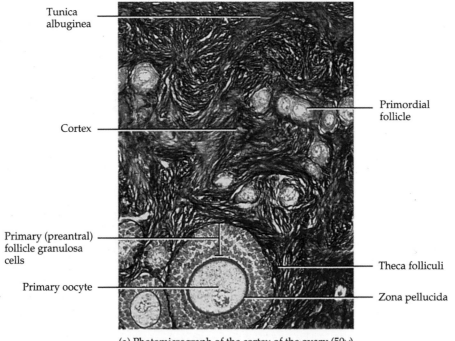

(a) Photomicrograph of the cortex of the ovary (50×)

Corona
radiata

Antrum

Layer of
granulosa
cells

Theca interna

Zona
pellucida

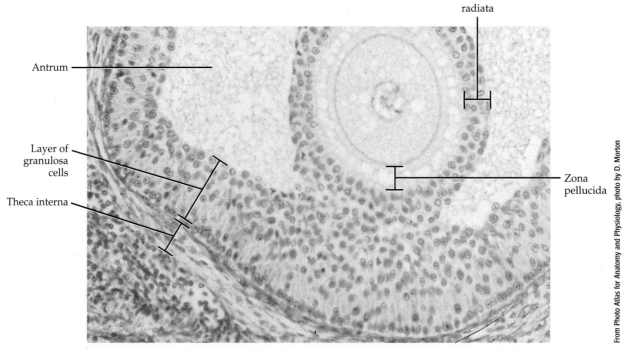

(b) Photomicrograph of a secondary follicle (400×)

FIGURE 25.12 Histology of uterine (fallopian) tube (30×). (From *Photo Atlas for Anatomy and Physiology*, photo by D. Morton.)

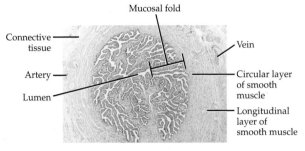

Connective tissue — Mucosal fold — Vein — Artery — Lumen — Circular layer of smooth muscle — Longitudinal layer of smooth muscle

The uterus is maintained in position by the following ligaments:

a. *Broad ligaments.* Double folds of peritoneum attaching the uterus to either side of the pelvic cavity.

b. *Uterosacral ligaments.* Peritoneal extensions that connect the uterus to the sacrum.

c. *Cardinal (lateral cervical) ligaments.* These ligaments are located inferior to the bases of the broad ligaments and extend from the pelvic wall to the cervix and vagina.

d. *Round ligaments.* Bands of fibrous connective tissue between folds of broad ligaments; they extend from uterus to external genitals (labia majora).

Label the uterine ligaments in Figure 25.9.

Histologically, the uterus consists of three layers of tissue: endometrium, myometrium, and perimetrium (serosa). The inner **endometrium** (*endo* = within) is a mucous membrane that consists of simple columnar epithelium, an underlying endometrial stroma (thick lamina propria) composed of connective tissue and endometrial (uterine) glands. The endometrium is divided into two layers: (1) **stratum functionalis** (*functional layer),* the layer that lines the uterine cavity and sloughs off during menstruation; and (2) **stratum basalis** (*basal layer),* the permanent layer that gives rise to a new stratum functionalis after each menstruation. The middle **myometrium** (*myo* = muscle) consists of three layers of smooth muscle and is thickest in the fundus and thinnest in the cervix. During labor and childbirth coordinated contractions help to expel the fetus. The outer layer is the **perimetrium** (*peri* = around; *metrium* = uterus), or **serosa,** part of the visceral peritoneum. It is composed of simple squamous epithelium and areolar connective tissue. Laterally, it becomes the broad ligament. Anteriorly, it covers the urinary bladder and forms a shallow pouch, the **vesicouterine** (ves´-i-kō-Ū-ter-in; *vesico* = bladder) **pouch.** Posteriorly, it covers the rectum and forms a deep pouch, the **rectouterine** (rek-tō-Ū-ter-in; *recto* = rectum) **pouch (pouch of Douglas)**—the most inferior point in the pelvic cavity.

4. Vagina

The **vagina** (= sheath) is a tubular, long fibromuscular canal lined with mucous membrane that extends from the exterior of the body to the uterine cervix. It is the outlet for menstrual flow, the receptacle for the penis during sexual intercourse, and the passageway for childbirth. The vagina is situated between the urinary bladder and rectum; it is directed superiorly and posteriorly, where it attaches to the uterus. A recess called the **fornix** (arch or vault) surrounds the vaginal attachment to the cervix (see Figure 25.9). When properly inserted a contraceptive diaphragm rests on the fornix covering the cervix. The opening of the vagina to the exterior, the **vaginal orifice,** may be bordered by a thin fold of vascularized membrane, the **hymen** (membrane).

Label the vagina in Figure 25.9.

Histologically, the mucosa of the vagina consists of nonkeratinized stratified squamous epithelium and areolar connective tissue that lies in a series of transverse folds, the **rugae.** The muscularis is composed of an outer circular layer and an inner longitudinal layer of smooth muscle that can stretch considerably to accommodate the penis during sexual intercourse and a child during birth.

The adventitia is the superficial layer of the vagina. It consists of areolar connective tissue and anchors the vagina to adjacent organs such as the urethra and urinary bladder anteriorly and the rectum and anal canal posteriorly.

5. Vulva

The **vulva** (VUL-va = to wrap around), or **pudendum** (pū-DEN-dum), refers to the external genitals of the female. The following components comprise the vulva:

a. *Mons pubis* (MONZ PŪ-bis). Elevation of adipose tissue covered by skin and coarse pubic hair that cushions the pubic symphysis.

b. *Labia majora* (LĀ-bē-a ma-JŌ-ra; *labia* = lip). Two longitudinal folds of skin that extend inferiorly and posteriorly from the mons pubis. Singular is **labium majus.** The folds, covered by pubic hair, contain an abundance of adipose tissue and sebaceous (oil) and sudoriferous (sweat) glands. They are homologous to the scrotum.

c. **Labia minora** (MĪ-nō-ra). Two smaller folds of mucous membrane medial to labia majora. Singular is **labium minus.** The folds have numerous sebaceous glands but few sudoriferous glands and are devoid of fat and pubic hair. The labia minora are homologous to the spongy urethra.

d. **Clitoris** (KLI-to-ris). Small, cylindrical mass of erectile tissue and nerves located at anterior junction of labia minora. The exposed portion is called the **glans;** the covering is called the **prepuce** (foreskin). The clitoris is homologous to the glans penis in males.

e. **Vestibule.** Region between labia minora containing vaginal orifice, hymen (if present), ex-

ternal urethral orifice, and openings of the ducts of several glands:

1. **Vaginal orifice.** Opening of vagina to exterior.
2. **Hymen.** Thin fold of vascularized membrane that borders vaginal orifice.
3. **External urethral orifice.** Opening of urethra to exterior.
4. **Orifices of paraurethral (Skene's) glands.** Located on either side of external urethral orifice. The glands secrete mucus. The paraurethral glands are homologous to the prostate.
5. **Orifices of ducts of greater vestibular** (ves-TIB-ū-lar), or **Bartholin's, glands.** Located in a groove between hymen and labia mi-

FIGURE 25.13   **Vulva.**

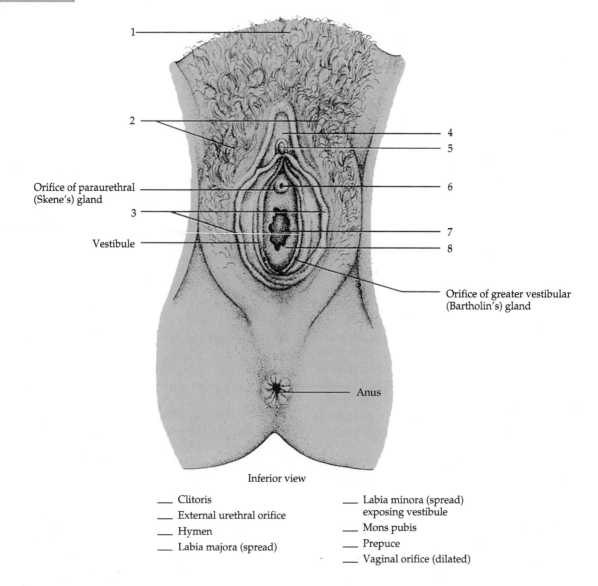

Inferior view

___ Clitoris
___ External urethral orifice
___ Hymen
___ Labia majora (spread)

___ Labia minora (spread) exposing vestibule
___ Mons pubis
___ Prepuce
___ Vaginal orifice (dilated)

nora. These glands produce a small quantity of mucus during sexual arousal and intercourse that adds to cervical mucus and provides lubrication. The greater vestibular glands are homologous to the bulbourethral glands in males.

6. ***Orifices of ducts of lesser vestibular glands.*** Microscopic orifices opening into vestibule.

f. ***Bulb of the vestibule.*** Two elongated masses of erectile tissue just deep to the labia on either side of the vaginal orifice. It engorges with blood during sexual arousal, narrowing the vaginal orifice and placing pressure on the penis during intercourse. The bulb of the vestibule is homologous to the corpus spongiosum penis and bulb of the penis in males.

Using your textbook as an aid, label the parts of the vulva in Figure 25.13.

6. Mammary Glands

The **mammary** (*mamma* = breast) **glands** are modified sudoriferous (sweat) glands that produce milk. They lie over the pectoralis major muscles and are attached to them by a layer of deep fascia. They consist of the following structures:

a. ***Lobes.*** Around 15 to 20 compartments separated by adipose tissue.

b. ***Lobules.*** Smaller compartments in lobes that contain grape-like clusters of milk-secreting glands called **alveoli** (= small cavity).

c. ***Secondary tubules.*** Receive milk from alveoli.

d. ***Mammary ducts.*** Receive milk from secondary tubules.

e. ***Lactiferous*** (*lact* = milk) ***sinuses.*** Expanded distal portions of mammary ducts that store milk.

f. ***Lactiferous ducts.*** Receive milk from lactiferous sinuses.

g. ***Nipple.*** Projection on anterior surface of mammary gland that contains lactiferous ducts.

h. ***Areola*** (a-RĒ-ō-la = small space). Circular pigmented skin surrounding the nipple.

With the aid of your textbook, label the parts of the mammary gland in Figure 25.14.

FIGURE 25.14 **Mammary glands.**

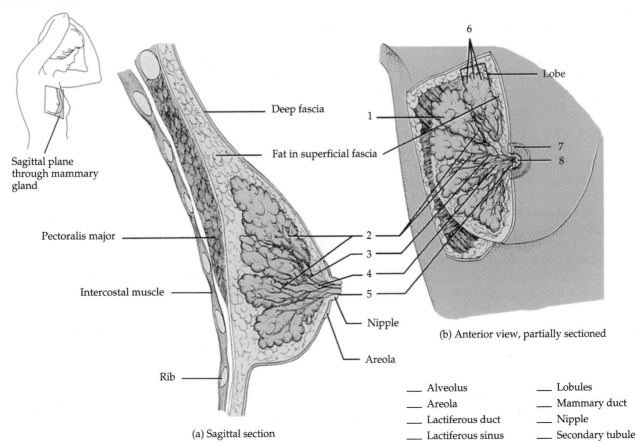

Sagittal plane through mammary gland

Deep fascia

Fat in superficial fascia

Pectoralis major

Intercostal muscle

Rib

Nipple

Areola

6

Lobe

1

7
8

2
3
4
5

(a) Sagittal section

(b) Anterior view, partially sectioned

___ Alveolus   ___ Lobules
___ Areola   ___ Mammary duct
___ Lactiferous duct ___ Nipple
___ Lactiferous sinus ___ Secondary tubule

FIGURE 25.15 Histology of mammary gland showing alveoli (100×). (From *Photo Atlas for Anatomy and Physiology*, photo by D. Morton.)

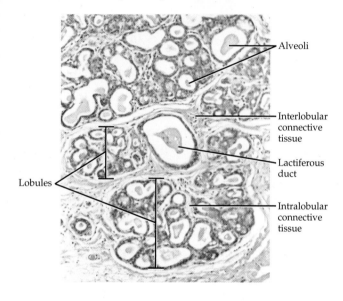

Examine a prepared slide of alveoli of the mammary gland and compare your observations to Figure 25.15.

Now that you have completed your study of the organs of the female reproductive system, label Figure 25.16.

7. Female Reproductive Cycle

During their reproductive years, nonpregnant females normally experience a cyclical sequence of changes in the ovaries and uterus. Each cycle takes about a month and involves both oogenesis and preparation of the uterus to receive a fertilized ovum. Hormones secreted by the hypothalamus, anterior pituitary, and ovaries control the main events. The **ovarian cycle** is a series of events in the ovaries that occur during and after the maturation of an oocyte. The **uterine (menstrual) cycle** is a concurrent series of changes in the endometrium of the uterus to prepare it for the arrival of a fertilized ovum that will develop in the uterus until birth. If fertilization does not occur, the stratum functionalis portion of the endometrium sloughs off. The general term **female reproductive cycle** encompasses the ovarian and uterine cycles, the hormonal changes that regulate them, and cyclical changes in the breasts and cervix.

a. Hormonal Regulation

The uterine cycle and ovarian cycle are controlled by gonadotropin-releasing hormone (GnRH) from the hypothalamus. See Figure 25.17. GnRH stimulates the release of follicle-stimulating hormone (FSH) and luteinizing hormone (LH) from the anterior pituitary. FSH, in turn, initiates follicular growth and stimulates the secretion of estrogens by the growing follicles. LH stimulates the further development of ovarian follicles and their full secretion of estrogens, brings about ovulation, promotes formation of the corpus luteum, and stimulates the production of estrogens, progesterone, relaxin, and inhibin by the corpus luteum.

b. Phases of the Female Reproductive Cycle

The duration of the female reproduction cycle typically is 24 to 35 days. For this discussion, we shall assume a duration of 28 days, divided into four phases: the menstrual phase, preovulatory phase, ovulation, and postovulatory phase (Figure 25.17).

(1) Menstrual Phase (Menstruation)

The **menstrual** (MEN-stroo-al) **phase,** also called **menstruation** (men'-stroo-Ā-shun) or **menses** (= month), lasts for roughly the first five days of the cycle. (By convention, the first day of menstruation marks the first day of a new cycle.)

Events in the ovaries. Under the influence of FSH, several primordial follicles develop into primary follicles and then into secondary follicles. This developmental process may take several months to occur. Therefore a follicle that begins to develop at the beginning of a particular menstrual cycle may not reach maturity and ovulate until several menstrual cycles later (see Figure 25.10).

Events in the uterus. Menstrual flow from the uterus consists of 50 to 150 mL of blood, tissue, fluid, mucus, and epithelial cells shed from the endometrium. This discharge occurs because the declining levels of progesterone and estrogens stimulate the release of prostaglandins that cause the uterine spiral arterioles to constrict. As a result, the cells they supply become oxygen-deprived and start to die. Eventually, the entire stratum functionalis sloughs off. At this time the endometrium is very thin, about 2 to 5 mm, because only the stratum basalis remains. The menstrual flow passes from the uterine cavity through the cervix and vagina to the exterior.

(2) Preovulatory Phase

The **preovulatory phase,** the second phase of the female reproductive cycle, is the time between the end of menstruation and ovulation. The preovula-

FIGURE 25.16 Female organs of reproduction and surrounding structures.

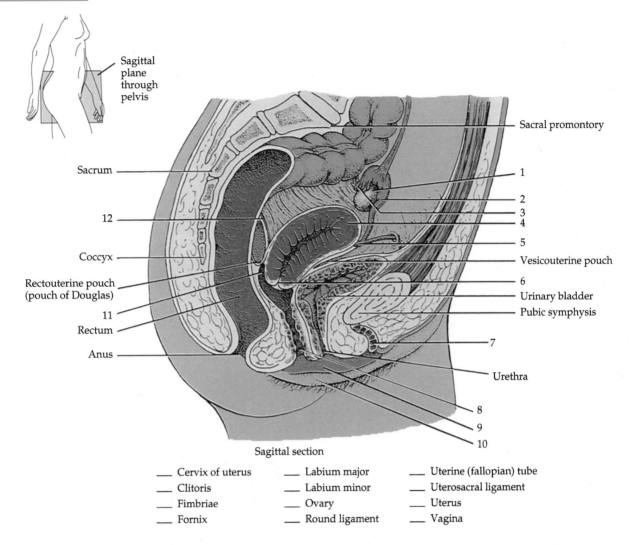

Sagittal
plane
through
pelvis

Sacrum

12

Coccyx

Rectouterine pouch
(pouch of Douglas)

11

Rectum

Anus

Sacral promontory

1
2
3
4
5

Vesicouterine pouch

6

Urinary bladder
Pubic symphysis

7

Urethra

8
9
10

Sagittal section

___ Cervix of uterus   ___ Labium major   ___ Uterine (fallopian) tube
___ Clitoris   ___ Labium minor   ___ Uterosacral ligament
___ Fimbriae   ___ Ovary   ___ Uterus
___ Fornix   ___ Round ligament   ___ Vagina

tory phase of the cycle is more variable in length than the other phases and accounts for most of the difference when cycles are shorter or longer than 28 days. It lasts from days 6 to 13 in a 28-day cycle.

Events in the ovaries. Under the influence of FSH, some secondary follicles continue to grow and begin to secrete estrogens and inhibin. By about day 6, a single follicle in one of the two ovaries has outgrown all the others and is called the **dominant follicle.** Estrogens and inhibin secreted by the dominant follicle decrease the secretion of FSH, which causes the other less well-developed follicles to stop growing and undergo atresia. Fraternal (nonidentical) twins may result if two secondary follicles achieve co-dominance and later are ovulated and fertilized at the same time.

Normally, the one dominant follicle becomes the **mature (graafian) follicle** that continues to enlarge until it is more than 20 mm in diameter and ready for ovulation (see Figure 25.10). This follicle forms a blisterlike bulge due to the swelling antrum on the surface of the ovary. During the final maturation process, the dominant follicle continues to increase its production of estrogens under the influence of an increasing level of LH. Although estrogens are the main ovarian hormones before ovulation, small amounts of progesterone are produced by the mature follicle a day or two before ovulation.

With reference to the ovaries, the menstrual phase and preovulatory phase together are termed the **follicular** (fō-LIK-ū-lar) **phase** because ovarian follicles are growing and developing.

FIGURE 25.17 Menstrual cycle.

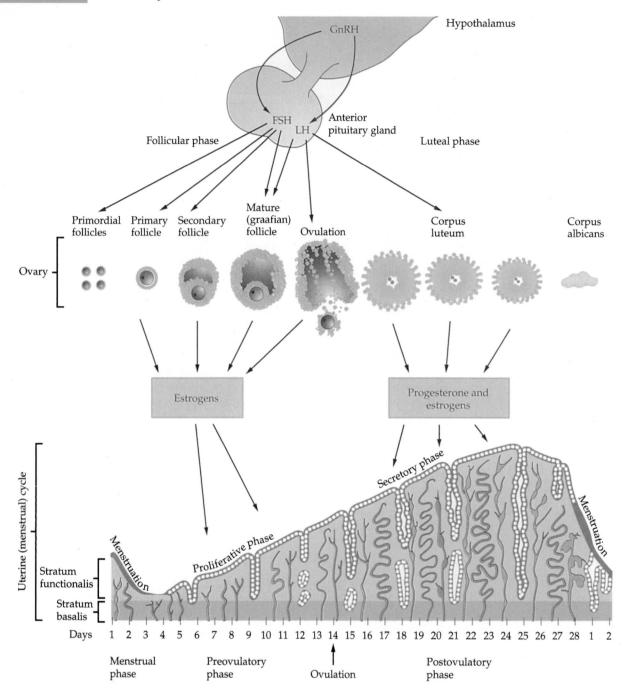

Events in the uterus. Estrogens liberated into the blood by growing ovarian follicles stimulate the repair of the endometrium. Cells of the stratum basalis undergo mitosis and produce a new stratum functionalis. As the endometrium thickens, the short, straight endometrial glands develop and the arterioles coil and lengthen as they penetrate the stratum functionalis. The thickness of the endometrium approximately doubles to about 4 to 10 mm. With reference to the uterus, the preovulatory phase is also termed the **proliferative phase** because the endometrium is proliferating.

(3) Ovulation

The rupture of the mature (graafian) follicle and release of the secondary oocyte into the pelvic cavity, called **ovulation,** usually occurs on day 14 in a 28-day cycle. During ovulation, the secondary oocyte remains surrounded by its zona pellucida and corona radiata. It generally takes a total of about 20 days (spanning the last 6 days of the previous cycle and the first 14 days of the current cycle) for a secondary follicle to develop into a fully mature follicle. During this time the developing ovum completes meiosis I to become a secondary oocye, which begins meiosis II but halts in metaphase. The *high* levels of estrogens during the last part of the preovulatory phase exert a *positive feedback* effect on both LH and GnRH and cause ovulation.

An over-the-counter home test that detects the rising level of LH can be used to predict ovulation a day in advance. FSH also increases at this time, but not as dramatically as LH because FSH is stimulated only by the increase in GnRH. The positive feedback effect of estrogens on the hypothalamus and anterior pituitary gland does not occur if progesterone is present at the same time.

(4) Postovulatory Phase

The **postovulatory phase** of the female reproductive cycle is the time between ovulation and the onset of the next menses. In duration, it is the most constant part of the female reproductive cycle. It lasts for 14 days, from days 15 to 28 in a 28-day cycle.

Events in one ovary. After ovulation, the mature follicle collapses, and blood within it forms a clot due to minor bleeding during rupture and collapse of the follicle to become the **corpus hemorrhagicum** (*hemo* = blood; *rrhagic* = bursting forth) (see Figure 25.10). The clot is eventually absorbed by the remaining follicular cells. In time, the follicular cells enlarge, change character, and form the **corpus luteum** under the influence of LH. Stimulated by LH, the corpus luteum secretes progesterone, estogens, relaxin, and inhibin. If the secondary oocyte is *fertilized* and begins to divide, the corpus luteum persists past its normal 2-week lifespan. It is "rescued" from degeneration by **human chorionic** (kō-rē-ON-ik) **gonadotropin (hCG),** a hormone produced by the chorion of the embryo beginning about 8 days after fertilization. The chorion eventually develops into the placenta, and the presence of the hCG in maternal blood or urine is an indication of pregnancy. As the pregnancy progresses, the placenta itself begins to secrete estrogens to support pregnancy and progesterone to support pregnancy and breast development for lactation. Once the placenta begins its secretion, the role of the corpus luteum becomes minor. With reference to the ovaries, this phase of the cycle is also called the **luteal phase.**

If hCG does not rescue the corpus luteum, after 2 weeks its secretions decline and it degenerates into a corpus albicans (see Figure 25.10). The lack of progesterone and estrogens due to degeneration of the corpus luteum then causes menstruation. In addition, the decreased levels of progesterone, estrogens, and inhibin promote the release of GnRH, FSH, and LH, which stimulate follicular growth, and a new ovarian cycle begins.

Events in the uterus. Progesterone and estrogens produced by the corpus luteum are responsible for preparing the endometrium to receive a fertilized ovum. Preparatory activities include growth and coiling of the endometrial glands, which begin to secrete glycogen, vascularization of the superficial endometrium, and thickening of the endometrium. These preparatory changes peak about 1 week after ovulation, corresponding to the time of possible arrival of a fertilized ovum. With reference to the uterus, this phase of the cycle is called the **secretory phase** because of the secretory activity of the endometrial glands.

Obtain microscope slides of the endometrium showing the menstrual, preovulatory, and postovulatory phases of the menstrual cycle. See if you can note the differences in thickness of the endometrium, distribution of blood vessels, and distribution and size of endometrial glands.

C. Dissection of Fetus-Containing Pig Uterus

Examination of the uterus of a pregnant pig reveals that the fetuses are equally spaced in the two uterine horns. Each fetus produces a local enlargement of the horn. The litter size normally ranges from 6 to 12. Your instructor may have you dissect the fetus-containing uterus of a pregnant pig or have one available as a demonstration (Figure 25.18). If you do a dissection, use the following directions. Also examine a chart or model of a human fetus and pregnant uterus if they are available.

▲ **CAUTION!** *Please reread Section D, "Precautions Related to Dissection," at the beginning of the laboratory manual on page xiii before you begin your dissection.*

FIGURE 25.18 **Fetal pig in opened chorionic vesicle.**

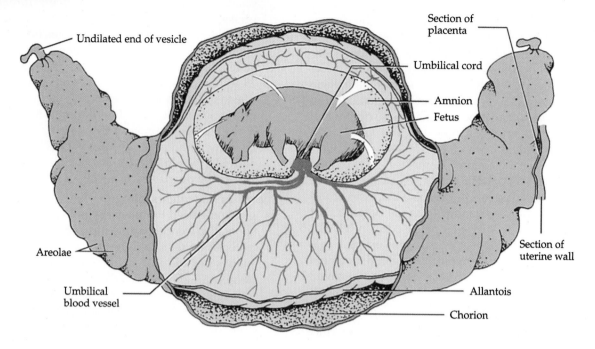

PROCEDURE

1. Using a sharp scissors, cut open one of the enlargements of the horn and you will see that each fetus is enclosed together with an elongated, sausage-shaped **chorionic vesicle.**

2. You will also notice many round bumps called **areolae** located over the chorionic surface.

3. The lining of the uterus together with the wall of the chorionic vesicle forms the **placenta.**

4. Carefully cut open the chorionic vesicle, avoiding cutting or breaking the second sac lying within that surrounds the fetus itself.

5. The vesicle wall is the fusion of two extraembryonic membranes, the outer **chorion** (KOR-ē-on) and the inner **allantois** (a-LAN-tō-is). The allantois is the large sac growing out from the fetus, and the umbilical cord contains its stalk.

6. The **umbilical blood vessels** are seen in the allantoic wall spreading out in all directions and are also seen entering the **umbilical cord.**

7. A thin-walled nonvascular **amnion** surrounds the fetus. This membrane is filled with **amniotic fluid,** which acts as a protective water cushion and prevents adherence of the fetus and membranes.

ANSWER THE LABORATORY REPORT QUESTIONS AT THE END OF THE EXERCISE.

EXERCISE 25

Reproductive Systems

Name _____   Date _____

Laboratory Section _____   Score/Grade _____

PART 1 ■ Multiple Choice

_____ 1. Structures of the male reproductive system where sperm are produced are the (a) efferent ducts (b) seminiferous tubules (c) seminal vesicles (d) rete testis

_____ 2. The superior portion of the male urethra is encircled by the (a) epididymis (b) testes (c) prostate (d) seminal vesicles

_____ 3. Cryptorchidism is a condition associated with the (a) prostate gland (b) testes (c) seminal vesicles (d) bulbourethral (Cowper's) glands

_____ 4. Weakening of the suspensory ligament would directly affect the position of the (a) mammary glands (b) uterus (c) uterine (fallopian) tubes (d) ovaries

_____ 5. Organs in the female reproductive system responsible for transporting secondary oocytes and ova from the ovaries to the uterus are the (a) uterine (fallopian) tubes (b) seminal vesicles (c) inguinal canals (d) none of the above

_____ 6. The name of the process that is responsible for the actual production of sperm cells is called (a) cryptorchidism (b) oogenesis (c) spermatogenesis (d) spermatogonia

_____ 7. Fertilization normally occurs in the (a) uterine (fallopian) tubes (b) vagina (c) uterus (d) ovaries

_____ 8. The portion of the uterus that assumes an active role during labor and childbirth is the (a) serosa (b) endometrium (c) peritoneum (d) myometrium

_____ 9. Stereocilia are associated with the (a) ductus (vas) deferens (b) oviduct (c) epididymis (d) rete testis

_____ 10. The major portion of the volume of semen is contributed by the (a) bulbourethral (Cowper's) glands (b) testes (c) prostate gland (d) seminal vesicles

_____ 11. These ligaments are located inferior to the bases of the broad ligaments and extend from the pelvic wall to the cervix and vagina (a) cardinal ligament (b) round ligament (c) broad ligament (d) ovarian ligament

_____ 12. Which sequence, from inside to outside, best represents the histology of the uterus? (a) stratum basalis, stratum functionalis, myometrium, perimetrium (b) myometrium, perimetrium, stratum functionalis, stratum basalis (c) stratum functionalis, stratum basalis, myometrium, perimetrium (d) stratum basalis, stratum functionalis, perimetrium, myometrium

_____ 13. The capsule of dense irregular connective tissue that divides the testis into lobules is called the (a) dartos (b) raphe (c) tunica albuginea (d) germinal epithelium

_____ **14.** Which sequence best represents the course taken by sperm cells from their site of origin to the exterior? (a) seminiferous tubules, efferent ducts, epididymis, ductus (vas) deferens, ejaculatory duct, urethra (b) seminiferous tubules, efferent ducts, epididymis, ductus (vas) deferens, urethra, ejaculatory duct (c) seminiferous tubules, efferent ducts, ductus (vas) deferens, epididymis, ejaculatory duct, urethra (d) seminiferous tubules, epididymis, efferent ducts, ductus (vas) deferens, ejaculatory duct, urethra

_____ **15.** The ovaries are anchored to the uterus by the (a) ovarian ligament (b) broad ligament (c) suspensory ligament (d) mesovarium

_____ **16.** The terminal duct for the male reproductive system is the (a) urethra (b) ductus (vas) deferens (c) inguinal canal (d) ejaculatory duct

_____ **17.** The site of sperm cell maturation is the (a) ductus (vas) deferens (b) spermatic cord (c) epididymis (d) testes

_____ **18.** Which of the following is the site of menstruation, implantation of a fertilized ovum, development of the fetus during pregnancy, and labor? (a) uterus (b) uterine (fallopian) tubes (c) vagina (d) cervix

_____ **19.** Glands lying over the pectoralis major muscles are the (a) lesser vestibular glands (b) adrenal (suprarenal) glands (c) mammary glands (d) greater vestibular (Bartholin's) glands

PART 2 ▪ Completion

20. Discharge of a secondary oocyte from the ovary about once each month is a process referred to as

_____.

21. The inferior, narrow portion of the uterus that opens into the vagina is the

_____.

22. The clusters of milk-secreting cells of the mammary glands are referred to as

_____.

23. The slightly enlarged acorn-shaped distal end of the penis is called the _____.

24. Covering the slightly enlarged region of the penis is a loosely fitting skin called the

_____.

25. The circular pigmented area surrounding each nipple of the mammary glands is the

_____.

26. After a secondary oocyte leaves the ovary, it enters the open, funnel-shaped distal end of the uterine (fallopian) tube called the _____.

27. The portion of a sperm cell that contains the nucleus and acrosome is the

_____.

28. Vasectomy refers to removal of a portion of the _____.

29. The mass of erectile tissue in the penis that contains the spongy urethra is the

_____.

30. Both the mature (graafian) follicle and _____ of the ovary secrete hormones.

31. The superior dome-shaped portion of the uterus is called the _____.

32. The _____ anchor the uterus to either side of the pelvic cavity.

33. The outlet for menstrual flow and passageway for childbirth is the _____.

34. Two longitudinal folds of skin that extend inferiorly and posteriorly from the mons pubis and are covered with pubic hair are the _____.

35. The _____ is a small cylindrical mass of erectile tissue at the anterior junction of the labia minora.

36. The thin fold of vascularized membrane that borders the vaginal orifice is the _____.

37. Complete the following sequence for the passage of milk: alveoli, secondary tubules, _____ , lactiferous sinuses, lactiferous ducts, nipple.

38. The layer of simple epithelium covering the free surface of the ovary is the _____.

39. The phase of the menstrual cycle between days 6 and 13 during which endometrial repair occurs is the _____ phase.

40. During menstruation, the stratum _____ of the endometrium is sloughed off.

41. The most immature spermatogenic cells are called _____.

42. The _____ contains the remnants of an ovulated mature follicle.

43. The hypothalamic hormone that controls the uterine and ovarian cycles is _____.

44. High levels of estrogens exert a positive feedback on LH and GnRH that cause _____.

Development

Objectives

At the completion of this exercise you should understand

A The histological characteristics of the structures involved in the processes of spermatogenesis and spermiogenesis

B The histological characteristics of the structures involved in the process of oogenesis

C The events involved in embryonic development

D The major changes associated with embryonic and fetal development

Developmental biology is the study of the sequence of events from the fertilization of a secondary oocyte to the formation of an adult organism. As we look at the sequence from fertilization to birth, we will consider fertilization, implantation, placental development, embryonic development, and fetal growth. From fertilization through the eighth week of development, the developing human is called an **embryo** (-*bryo* = grow) and this is the period of **embryological development.** The study of development from the fertilized egg through the eighth week *in utero* is termed **embryology** (em-brē-OL-ō-jē). **Fetal development** begins at week nine and continues until birth. During this time the developing human is called a **fetus** (*FĒ-tus* = offspring).

A. Spermatogenesis

The process by which the testes produce haploid (*n*) sperm cells involves several phases, including meiosis, and is called **spermatogenesis** (sper'-ma-tō-JEN-e-sis; *spermato* = sperm; *genesis* = to produce). In order to understand spermatogenesis, review the following concepts.

1. In sexual reproduction, each new organism is the result of the joining of two different **gametes,** one produced by each parent. Male gametes, produced in the testes, are called sperm, and female gametes, produced in the ovaries, are called oocytes.

2. The cell resulting from joining of the gametes, called a **zygote** (*zygon* = yolk), contains two full sets of chromosomes (DNA), one set from each parent. Through repeated mitotic cell divisions, a zygote develops into a new organism.

3. Gametes differ from all other body cells (somatic cells) in that they contain the **haploid** (one-half) **chromosome number,** symbolized as *n*. In humans, this number is 23, which composes a single set of chromosomes. The nucleus of a somatic cell contains the **diploid chromosome number,** symbolized as 2*n*. In humans, this number is 46, which composes two sets of paired chromosomes. One set of 23 chromosomes comes from the mother, and the other set comes from the father.

4. In a diploid cell, two chromosomes that make up each pair are called **homologous** (*homo* = same) **chromosomes (homologues).** In human diploid cells, the members of 22 of the 23 pairs of chromosomes are morphologically similar and are called **autosomes.** The other pair, termed X and Y chromosomes, are called the **sex chromosomes** because they determine one's gender. In the female, the homologous pair of sex chromosomes consists of X chromosomes; in the male, the pair consists of an X and a Y chromosome.

5. If each gamete had the same number of chromosomes as somatic cells, the number of

chromosomes would double with each fertilization, and with every succeeding generation the chromosome number would continue to double and normal development could not occur.

6. This continual doubling of the chromosome number does not occur at fertilization because of **meiosis** (*meio* = less), a process by which gametes receive the haploid chromosome number. Thus, when haploid (*n*) gametes fuse, the zygote contains the diploid chromosome number (2*n*) and can undergo normal development.

In humans, spermatogenesis takes 65 to 75 days. The seminiferous tubules are lined with immature cells called **spermatogonia** (sper'-ma-tō-GŌ-nē-a; *sperm* = seed; *gonium* = generation or offspring), or stem cells (see Figure 26.2). Singular is **spermatogonium**. These cells develop from **primordial** (*primordialis* = primitive or early form) **germ cells** that arise from yolk sac endoderm and enter the testes early in development. In the embryonic testes, the primordial germ cells differentiate into spermatogonia but remain dormant until they begin to undergo mitotic proliferation at puberty. Spermatogonia contain the diploid (2*n*) chromosome number. Some spermatogonia remain relatively undifferentiated and capable of extensive mitotic division. Following division, some of the daughter cells remain undifferentiated and serve as a reservoir of precursor cells to prevent depletion of the stem cell population. Such cells remain near the basement membrane of the seminiferous tubule. The remainder of the daughter cells differentiate into spermatogonia that lose contact with the basement membrane of the seminiferous tubule, squeeze through the tight junctions of the blood-testis barrier, undergo certain developmental changes, and differentiate into **primary spermatocytes** (SPER-ma-tō-sītz'). Primary spermatocytes, like spermatogonia, are diploid (2*n*); that is, they have 46 chromosomes.

1. Meiosis I

Each primary spermatocyte enlarges and then begins meiosis. Two nuclear divisions take place as part of meiosis. In the first, called **meiosis I**, DNA is replicated and 46 chromosomes (each made up of two identical chromatids from the replicated DNA) form and move toward the equatorial plane of the cell. There they line up in homologous pairs so that there are 23 pairs of duplicated chromosomes in the center of the cell. This pairing of homologous chromosomes is called **synapsis.** The resulting four chromatids form a structure called a **tetrad.** In a tetrad, portions of one chromatid may be exchanged with portions of another. This exchange between parts of genetically different (non-sister) chromatids is called **crossing over** (Figure 26.1), which results in **genetic recombination.** Thus, the sperm eventually produced are genetically unlike each other and unlike the cell that produced them—one reason for the great variation among humans. Next, the meiotic spindle forms and the kinetochore microtubules organized by the centromeres extend toward the poles of the cell. As the pairs separate, one member of each pair migrates to opposite poles of the dividing cell. The random arrangement of chromosome pairs on the spindle is another reason for variation among humans. The cells formed by the first nuclear division (meiosis I) are called **secondary spermatocytes.** Each cell has 23 chromosomes—the haploid number. Each chromosome of the secondary spermatocytes, however, is made up of two chromatids (two copies of the DNA) still attached by a centromere (since the centromere did not split). Moreover, the genes of the chromosomes of secondary spermatocytes may be rearranged as a result of crossing over.

2. Meiosis II

The second stage of meiosis is **meiosis II.** There is no replication of DNA. The chromosomes (each composed of two chromatids) line up in single file along the metaphase plate, and the two chromatids of each chromosome separate. The haploid cells formed from meiosis II are called **spermatids.** Each contains half the original chromosome number, or 23 chromosomes, and is haploid. Each primary

---

FIGURE 26.1 Crossing over within a tetrad, resulting in genetic recombination.

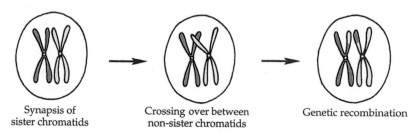

Synapsis of sister chromatids

Crossing over between non-sister chromatids

Genetic recombination

spermatocyte therefore produces four spermatids by meiosis I and meiosis II. Spermatids lie close to the lumen of the seminiferous tubule.

3. Spermiogenesis

The final stage of spermatogenesis, called **spermiogenesis** (sper'-mē-ō-JEN-e-sis), involves the maturation of haploid spermatids into sperm. Each spermatid embeds in a Sertoli cell and develops a head with an acrosome (enzyme-containing granule) and a flagellum (tail). Sertoli cells extend from the basement membrane to the lumen of the semi-niferous tubule, where they nourish the developing spermatids. Since there is no cell division in spermiogenesis, each spermatid develops into a single **sperm cell (spermatozoon).** The release of a sperm cell from a Sertoli cell is known as **spermiation.**

Sperm enter the lumen of the seminiferous tubule and migrate to the ductus epididymis, where they complete their maturation and become capable of fertilizing a secondary oocyte. Sperm are also stored in the ductus (vas) deferens. Here, they can retain their fertility for up to several months.

With the aid of your textbook, label Figure 26.2.

FIGURE 26.2   **Spermatogenesis.**

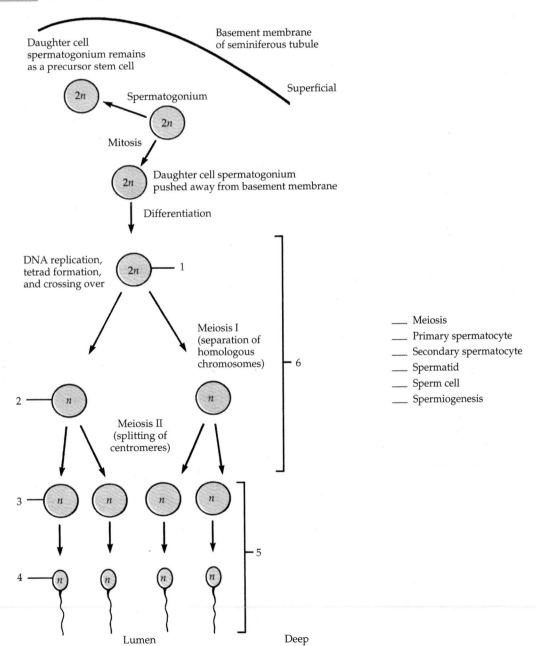

_____ Meiosis

_____ Primary spermatocyte

_____ Secondary spermatocyte

_____ Spermatid

_____ Sperm cell

_____ Spermiogenesis

B. Oogenesis

The formation of haploid (*n*) secondary oocytes in the ovary involves several phases, including meiosis, and is referred to as **oogenesis** (ō'-ō-JEN-e-sis; *oo* = egg; *genesis* = to produce). With some important exceptions, oogenesis occurs in essentially the same manner as spermatogenesis.

1. Meiosis I

During early fetal development, primordial (primitive) germ cells migrate from the endoderm of the yolk sac to the ovaries. There, germ cells differentiate within the ovaries into **oogonia** (ō'-ō-GŌ-nē-a); singular is **oogonium** (ō'-ō-GŌ-nē-um). Oogonia are diploid (*2n*) cells that divide mitotically to produce millions of germ cells. Even before birth, many of these germ cells degenerate, a process known as **atresia** (a-TRĒ-zē-a). A few develop into larger cells called **primary oocytes** (Ō'-ō-sītz) that enter prophase of meiosis I during fetal development but do not complete it until after puberty. At birth 200,000 to 2,000,000 oogonia and primary oocytes remain in each ovary. Of these, about 40,000 remain at puberty, and only 400 will mature and ovulate during a woman's reproductive lifetime; the remaining 99.98% undergo atresia.

Each primary oocyte is surrounded by a single layer of follicular cells, and the entire structure is called **primordial follicle** (see Figure 25.10). Although the stimulating mechanism is unclear, a few primordial follicles start to grow, even during childhood. They become **primary follicles,** which are surrounded first by one layer of cuboidal follicular cells and then by six to seven layers of cuboidal and low-columnar cells called **granulosa cells.** As a follicle grows, it forms a clear glycoprotein layer, called the **zona pellucida** (pe-LOO-si-da), between the primary oocyte and the granulosa cells. The innermost layer of granulosa cells becomes firmly attached to the zona pellucida and is called the **corona radiata** (*corona* = crown; *radiata* = radiation). The outermost granulosa cells rest on a basement membrane. Encircling the basement membrane is a region called the **theca folliculi.** As the primary follicle continues to grow, the theca differentiates into two layers; (1) the **theca interna,** a vascularized internal layer of cuboidal secretory cells, and (2) the **theca externa,** an outer layer of connective tissue cells and collagen fibers. The granulosa cells begin to secrete follicular fluid, which builds up in a cavity called the **antrum** in the center of the follicle. The follicle is now termed a **secondary follicle.** During early childhood, primordial and developing follicles continue to undergo atresia.

Each month after puberty, the gonadotropins secreted by the anterior pituitary stimulate the resumption of oogenesis. The diploid primary oocyte completes meiosis I, and two haploid cells of unequal size, both with 23 chromosomes (*n*) of two chromatids each, are produced. The follicle in which these events are taking place, termed the **mature (graafian) follicle,** will soon rupture and release its oocyte, a process known as **ovulation.**

The smaller cell produced by meiosis I, called the **first polar body,** is essentially a packet of discarded nuclear material. The larger cell, known as the **secondary oocyte,** receives most of the cytoplasm. Once a secondary oocyte is formed, it proceeds to the metaphase of meiosis II and then stops at this stage.

2. Meiosis II

At ovulation, the secondary oocyte (with the first polar body and corona radiate) is expelled into the pelvic cavity. Normally, the cells are swept into the uterine (fallopian) tube. If fertilization does not occur, the cells degenerate. If sperm cells are present in the uterine tube and one penetrates the secondary oocyte (fertilization), however, meiosis II resumes. The secondary oocyte splits into two haploid (*n*) cells of unequal size. The larger cell is the **ovum,** or mature egg; the smaller one is the **second polar body.** The nuclei of the sperm cell and the ovum then unite, forming a diploid (*2n*) **zygote.** The first polar body may also undergo another division to produce two polar bodies. If it does, the primary oocyte ultimately gives rise to a single haploid (*n*) ovum and three haploid (*n*) polar bodies, which all degenerate. Thus, one primary oocyte gives rise to a single gamete (ovum), whereas a primary spermatocyte produces four gametes (sperm).

With the aid of your textbook, label Figure 26.3.

C. Embryonic Period

The **embryonic period** is the first two months of development.

1. Fertilization

During **fertilization** (fer-til-i-ZĀ-shun; *fertil* = fruitful), the genetic material from a haploid sperm cell (spermatozoon) and a haploid second-

FIGURE 26.3 **Oogenesis.**

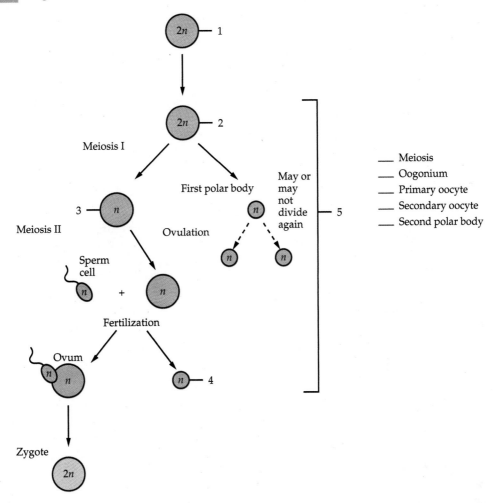

ary oocyte merges into a single diploid nucleus (Figure 26.4a). Of the about 200 million sperm introduced into the vagina, less than 2 million (1%) reach the cervix of the uterus and only about 200 reach the secondary oocyte. Fertilization normally occurs in the uterine (fallopian) tube within 12 to 24 hours after ovulation. Sperm can remain viable in the vagina for about 48 hours though a secondary oocyte is viable for only about 24 hours after ovulation. Thus, pregnancy is *most likely* to occur during a 3-day "window" from 2 days before ovulation to 1 day after ovulation. Peristaltic contractions and the action of cilia transport the oocyte through the uterine tube. Sperm swim from the vagina into the cervical canal propelled by whiplike movements of their tails (flagella). The passage of sperm through the rest of the uterus and into the uterine tube results mainly from muscular contractions of the walls of these organs. Prostaglandins in semen are believed to stimulate uterine mobility at the time of inter-

course and to aid in the movement of sperm through the uterus and into the uterine tube.

Besides contributing to sperm cell movement, the female reproductive tract also confers on sperm cells the capacity to fertilize a secondary oocyte. Although sperm cells undergo maturation in the epididymis, they are still not able to fertilize an oocyte until they have been in the female reproductive tract for several hours.

Capacitation (ka-pas'-i-TĀ-shun) refers to a series of functional changes that causes the sperm's tail to beat even more vigorously and prepare its plasma membrane to fuse with the oocyte's plasma membrane. During this process, sperm are acted upon by secretions in the female reproductive tract that result in removal of cholesterol, glycoproteins, and proteins from the plasma membrane around the head of the sperm. It requires the collective action of many sperm cells to have just one penetrate the secondary oocyte. For fertilization to occur, a sperm cell first must penetrate the corona radiata

FIGURE 26.4 **Fertilization.**

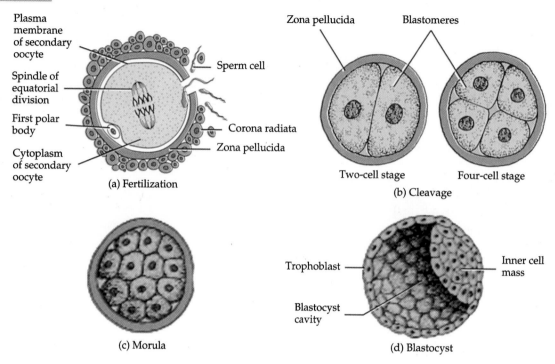

(a) Fertilization

Plasma membrane of secondary oocyte
Spindle of equatorial division
First polar body
Cytoplasm of secondary oocyte
Sperm cell
Corona radiata
Zona pellucida

(b) Cleavage

Zona pellucida
Blastomeres
Two-cell stage
Four-cell stage

(c) Morula

(d) Blastocyst

Trophoblast
Blastocyst cavity
Inner cell mass

and zona pellucida around the oocyte. Sperm cells bind to receptors in the zona pellucida. This binding triggers the **acrosomal reaction,** the release of the contents of the acrosome. The acrosomal enzymes digest a path through zone pellucida. Normally only one sperm cell fuses with a secondary oocyte. This event is called **syngamy** (*syn* = coming together; *gamy* = marriage). Syngamy causes depolarization, of the oocyte and triggers the intracellular release of calcium ions. The release of calcium ions in turn stimulates exocytosis of secretory vesicles from the oocyte. The molecules released by exocytosis promote changes in the zona pellucida to block entry of other sperm cells. This prevents **polyspermy,** fertilization by more than one sperm cell.

Once a sperm cell has entered a secondary oocyte, the oocyte completes meiosis II. It divides into a larger ovum (mature egg) and a smaller second polar body that fragments and disintegrates (see Figure 26.3). When a sperm cell has entered a secondary oocyte, the tail is shed and the nucleus in the head of the sperm develops into a structure called the **male pronucleus.** The nucleus of the fertilized ovum develops into a **female pronucleus.** After the pronuclei are formed, they fuse to produce a single diploid nucleus that contains 23 chromosomes (*n*) from each pronu-

cleus. Thus the fusion of the haploid (*n*) pronuclei restores the diploid number (2*n*) of 46 chromosomes. The fertilized ovum, now is called a **zygote** (ZĪ-gōt = yolk).

2. Formation of the Morula

After fertilization, rapid mitotic cell divisions of the zygote take place. These early divisions of the zygote are called **cleavage** (Figure 26.4b). Although cleavage increases the number of cells, it does not increase the size of the embryo, which is still contained within the zona pellucida.

The first division of the zygote begins about 24 hours after fertilization and is completed about 30 hours after fertilization, and each succeeding division takes slightly less time. By the second day after fertilization, the second cleavage is completed. By the end of the third day, there are 16 cells. The progressively smaller cells produced by cleavage are called **blastomeres** (BLAS-tō-mērz; *blast* = germ or sprout; *meres* = part). Successive cleavages eventually produce a solid sphere of cells, still surrounded by the zona pellucida, called the **morula** (MOR-ū-la; *morula* = mulberry) (Figure 26.4c). A few days after fertilization, the morula is about the same size as the original zygote.

3. Development of the Blastocyst

By the end of the fourth day, the number of cells in the morula increases, as it continues to move through the uterine (fallopian) tube toward the uterine cavity. The morula receives nourishment from glycogen-rich secretions of endometrial (uterine) glands, sometimes called **uterine milk.** When the morula enters the uterine cavity on day 4 or 5, the dense cluster of cells has reorganized into a large fluid-filled cavity called the **blastocyst cavity,** and the collection of blastomeres which is now called a **blastocyst** (*cyst* = bag) (Figure 26.4d).

Further rearrangement of the blastomeres results in the formation of the distinct structures: an outer superficial layer of cells called the **trophoblast** (TROF-ō-blast; *tropho* = nourish) and the **inner cell mass (embryoblast).** The trophoblast ultimately forms part of the membranes composing the fetal portion of the placenta while the inner cell mass develops into the embryo.

PROCEDURE

1. Obtain prepared slides of the embryonic development of the sea urchin. First try to find a zygote. This will appear as a single cell surrounded by an inner fertilization membrane and an outer, jellylike membrane. Draw a zygote in the spaces provided.

2. Now find several cleavage stages. See if you can isolate 2-cell, 4-cell, 8-cell, and 16-cell stages. Draw the various stages in the spaces provided.

3. Try to find a blastula (called a blastocyst in humans), a hollow ball of cells with a lighter center due to the presence of the blastocele. Draw a blastula in the space provided.

4. Implantation

The blastocyst remains free within the cavity of the uterus for about 2 days before it attaches to the uterine wall. The endometrium is in its secretory phase. About 6 days after fertilization the blastocyst loosely attaches to the endometrium, a process called **implantation** (Figure 26.5).

As the blastocyst implants, usually on the posterior portion of the fundus or body of the uterus, it orients so that the inner cell mass is toward the endometrium. About 8 days after fertilization, the trophoblast develops two layers in the region of contact between the blastocyst and endometrium. These layers are a **syncytiotrophoblast** (sin-sīt′-ē-ō-TRŌF-ō-blast) that contains no cell boundaries

Zygote

2-cell stage

4-cell stage

8-cell stage

16-cell stage

Blastocyst

and a **cytotrophoblast** (sī-tō-TRŌF-ō-blast) between the inner cell mass and syncytiotrophoblast that is composed of distinct cells (Figure 26.5c). These two layers of trophoblast become part of the chorion (one of the fetal membranes) as they undergo further growth. During implantation, the syncytiotrophoblast secretes enzymes that enable the blastocyst to penetrate the uterine lining by digesting and liquefying the endometrial cells. The endometrial secretions further nourish the burrowing blastocyst for about a week after implantation. Eventually, the blastocyst becomes buried in the endometrium and inner one-third of the myometrium. The trophoblast also secretes human chorionic gonadotropin (hCG) that rescues the corpus luteum from degeneration and sustains its secretion of progesterone and estrogens.

FIGURE 26.5 Implantation.

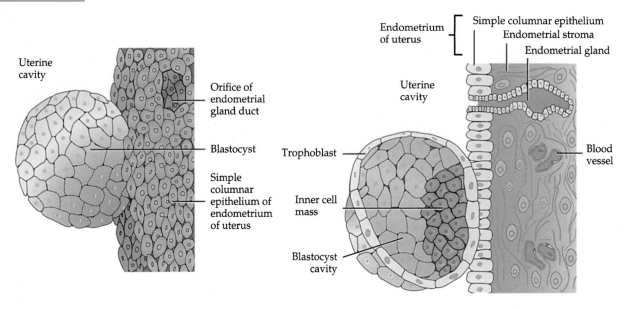

(a) External view of blastocyst, about 6 days after fertilization

(b) Frontal section of blastocyst, about 6 days after fertilization

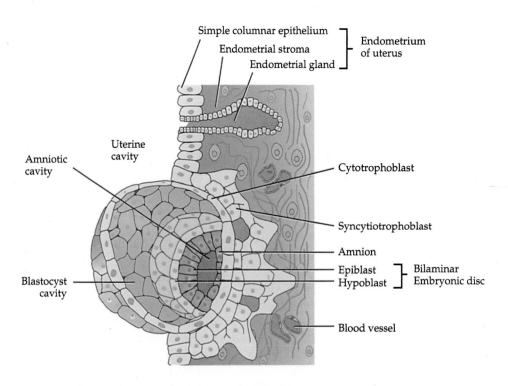

(c) Frontal section of blastocyst, about 7 days after fertilization

5. Primary Germ Layers

Within 8 days after fertilization, the cells of the inner cytotrophoblast proliferate and form the amnion (a fetal membrane) and a space, the **amniotic** (am-nē-OT-ik; *amnion* = lamb) **cavity,** adjacent to the inner cell mass.

About the 12th day after fertilization, striking changes appear (Figure 26.6a). The cells of the endodermal layer have been dividing rapidly, so that groups of them now extend around the yolk sac, another fetal membrane (described shortly). Cells of the cytotrophoblast give rise to a loose connective tissue, the **extraembryonic mesoderm**

FIGURE 26.6 Formation of the primary germ layers and associated structures.

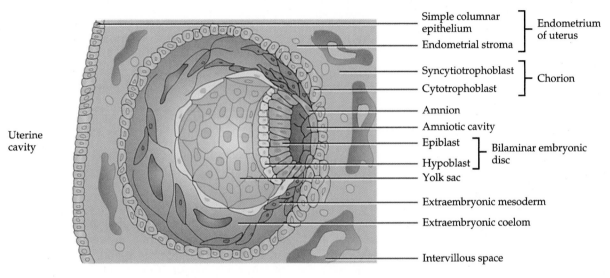

(a) Transverse section of blastocyst, about 12 days after fertilization

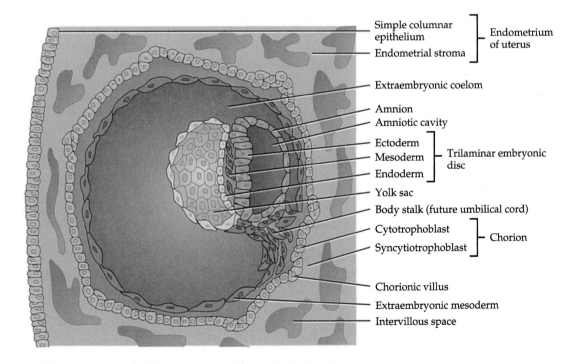

(b) Transverse section of blastocyst, about 14 days after fertilization

FIGURE 26.6 Formation of the primary germ layers and associated structures. (Continued)

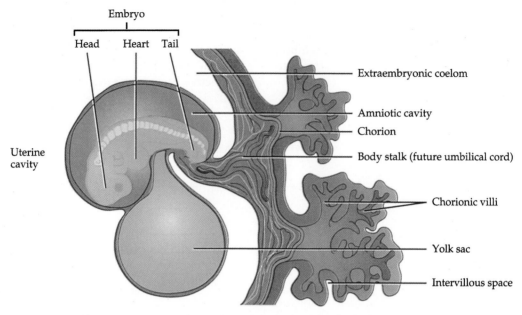

(c) Transverse section of an embryo, about 25 days after fertilization

(*meso* = middle). This completely fills the space between the cytotrophoblast and yolk sac. Soon a number of large cavities develop in the extraembryonic mesoderm and come together to form a single, larger cavity called the **extraembryonic coelom** (SĒ-lōm; *koiloma* = cavity).

The first major event of the third week of development occurs about 15 days after fertilization. The inner cell mass of the blastocyst, now called the bilaminar embryonic disc, begins to differentiate into the three **primary germ layers:** ectoderm, endoderm, and mesoderm. These are the major embryonic tissues from which all tissues and organs of the body will develop. The process by which the two-dimensional bilaminar (two-layered) embryonic disc, consisting of epiblast and hypoblast, transforms into a two-dimension trilaminar (three-layered) embryomic disc consisting of the three germ layers is called **gastrulation** (gas'-troo-LĀ-shun; *gastrula* = little belly). The layer of cells of the inner cell mass that is closer to the amniotic cavity develops into the **ectoderm** (*ecto* = outside). The layer of the inner cell mass that borders the yolk sac develops into the **endoderm** (*endo* = inside). As the amniotic cavity forms, the inner cell mass at this stage is called the **trilaminar embryonic disc.** It will form the embryo (Figure 26.6b). As the embryo develops (Figure 26.6c), the endoderm becomes the epithelial lining of most of the gastrointestinal tract, urinary

bladder, gallbladder, liver, pharynx, larynx, trachea, bronchi, lungs, vagina, urethra, and thyroid, parathyroid, and thymus glands, among other structures. The mesoderm develops into muscle; cartilage, bone and other connective tissues; red bone marrow, lymphoid tissue, endothelium of blood and lymphatic vessels, gonads, dermis of the skin, and other structures. The ectoderm develops into the entire nervous system, epidermis of skin, epidermal derivatives of the skin, and portions of the eye and other sense organs.

6. Embryonic Membranes

A second major event that occurs during the embryonic period is the formation of the **embryonic (extraembryonic) membranes** (Figure 26.7). These membranes lie outside the embryo and protect and nourish the embryo and, later, the fetus. The membranes are the yolk sac, amnion, chorion, and allantois.

In many species (such as birds), the **yolk sac** is a membrane that is the primary source of nourishment for the developing embryo. A human embryo receives its nutrients from the endometrium; the yolk sac is empty, small, and decreases in size as development progresses. Nevertheless, the yolk sac has several important functions. It nourishes the embryo during the second and third weeks and is the source of blood

FIGURE 26.7 **Embryonic membranes.**

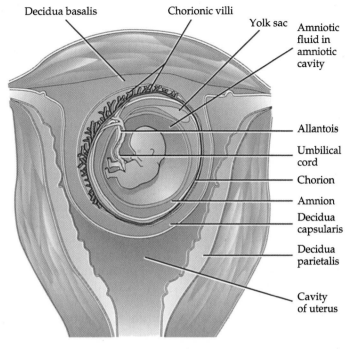

Frontal section

cells from the third to the sixth weeks. The yolk sac also contains cells that migrate into the developing gonads and differentiate into the primitive germ cells (spermatogonia and oogonia); forms part of the gut; functions as a shock absorber; and helps prevent drying out of the embryo.

The **amnion** is a thin, protective membrane that forms by the eighth day after fertilization and initially overlies the embryonic disc. As the embryo grows, the amnion eventually surrounds the entire embryo, creating the amniotic cavity that becomes filled with **amniotic fluid.** Most amniotic fluid is initially derived from a filtrate of maternal blood. Later, the fetus makes daily contributions to the fluid by excreting urine into the amniotic cavity. Amniotic fluid serves as a shock absorber for the fetus, helps regulate fetal body temperature, prevents dessication, and prevents adhesions between the skin of the fetus and surrounding tissues. Embryonic cells are sloughed off into amniotic fluid; they can be examined in the procedure called **amniocentesis** (am′-nē-ō-sen-TĒ-sis). The amnion usually ruptures just before birth and with its fluid constitutes the "bag of waters."

The **chorion** (KŌ-rē-on) is derived from the trophoblast of the blastocyst and the mesoderm that lines the trophoblast. It surrounds the embryo and, later, the fetus. Eventually, the chorion

becomes the principal embryonic part of the placenta, the structure for exchange of materials between the mother and fetus. It also produces human chorionic gonadotropin (hCG) and protects the embryo and fetus from the immune response of the mother. The inner layer of the chorion eventually fuses with the amnion.

The **allantois** (a-LAN-tō-is; *allant* = sausage) is a small, vascularized outpouching of the hindgut. It functions in early formation of blood and blood vessels and is associated with the development of the urinary bladder. It is not a prominent structure in humans.

7. Placenta and Umbilical Cord

Development of the **placenta** (pla-SEN-ta = flat cake), the third major event of the embryonic period, is accomplished by the third month of pregnancy. The placenta is formed by the chorionic villi of the chorion of the embryo and a portion of the endometrium (decidua basalis) of the mother (see Figure 26.7). Functionally, the placenta is the site of exchange of nutrients and wastes between the mother and fetus. It allows oxygen and nutrients to diffuse from maternal blood while carbon dioxide and wastes diffuse from fetal blood into maternal blood.

The placenta also is a protective barrier since most microorganisms cannot cross it. However, certain viruses, such as those that cause AIDS, German measles, chickenpox, measles, encephalitis, and poliomyelitis, may pass through the placenta. The placenta also stores nutrients such as carbohydrates, proteins, calcium, and iron, which are released into fetal circulation as required. Finally, the placenta produces several hormones that are necessary to maintain pregnancy. Almost all drugs, including alcohol, and many substances that can cause birth defects pass freely through the placenta.

If implantation occurs, a portion of the endometrium becomes modified and is called the **decidua** (dē-SID-ū-a; *deciduus* = falling off). The decidua includes all but the stratum basalis layer of the endometrium and separates from the endometrium after the fetus is delivered. Different regions of the decidua are named based on their positions relative to the site of the implanted blastocyst (see Figure 26.7). The **decidua basalis** is the portion of the endometrium between the embryo and the stratum basalis of the uterus. It becomes the maternal part of the placenta. The **decidua capsularis** is the portion of the endometrium located between the embryo and the uterine cavity. The **decidua parietalis** (par-rī-e-TAL-is) is the remaining modified endometrium that lines the noninvolved areas of the rest of the uterus. As the embryo and later the fetus enlarge, the decidua capsularis bulges into the uterine cavity and fuses with the decidua parietalis, thereby obliterating the uterine cavity. By about 27 weeks, the decidua capsularis degenerates and disappears.

During embryonic life, fingerlike projections of the chorion, called **chorionic villi** (kō'-rē-ON-ik VIL-ī), grow into the decidua basalis of the endometrium (see Figure 26.7). These will contain fetal blood vessels of the allantois. They continue growing until they are bathed in maternal blood sinuses called **intervillous** (in-ter-VIL-us) **spaces.** Thus maternal and fetal blood vessels are brought into proximity. It should be noted, however, that maternal and fetal blood do not normally mix. Instead, oxygen and nutrients in the blood of the mother's intervillous spaces, the spaces between chorionic villi, diffuse across the cell membranes into the capillaries of the villi while waste products diffuse in the opposite direction. From the capillaries of the villi, nutrients and oxygen enter the fetus through the umbilical vein. Wastes leave the fetus through the umbilical arteries, pass into the capillaries of the villi, and diffuse into the maternal blood. A few materials, such as IgG antibodies, pass from the blood of the mother into the capillaries of the villi.

The **umbilical** (um-BIL-i-kul) **cord** is the actual connection between the placenta and embryo. It consists of two umbilical arteries that carry deoxygenated fetal blood to the placenta, one umbilical vein that carries oxygenated blood into the fetus, and a mass of supporting mucous connective tissue called **Wharton's jelly** which derived from the allantois. The entire umbilical cord is surrounded by a layer of amnion giving it a shiny appearance (see Figure 26.7).

After the birth of the baby the placenta detaches from the uterus and is termed the **afterbirth.** At this time, the umbilical cord is tied off and then severed, leaving the baby on its own. The small portion (about an inch) of the cord that remains attached to the infant begins to wither and falls off, usually within 12 to 15 days after birth. The area where the cord was attached becomes covered by a thin layer of skin and scar tissue forms. The scar is the **umbilicus (navel).**

Pharmaceutical companies use human placentas to harvest hormones, drugs, and blood and portions of placentas are also used for burn coverage. The placental and umbilical cord veins can also be used in blood vessel grafts.

D. Fetal Period

During the **fetal period,** the months of development after the second month, all the organs of the body grow rapidly from the original primary germ layers, and the organism takes on a human appearance. Some of the principal changes associated with embryonic and fetal growth are summarized in Table 26.1.

ANSWER THE LABORATORY REPORT QUESTIONS AT THE END OF THE EXERCISE.

TABLE 26.1	Summary of changes during embryonic and fetal development

Time	Approximate size and weight	Representative changes
Embryonic period		
1–4 weeks	0.6 cm (3/16 in.)	Blood vessel formation begins and blood forms in yolk sac, allantois, and chorion. Heart forms and begins to beat. Chorionic villi develop and placental formation begins. The embryo folds. The primitive gut, pharyngeal arches, and limb buds develop. Eyes and ears begin to develop, tail forms, and body systems begin to form.
5–8 weeks	3 cm (1.25 in.) 1 g (1/30 oz)	Limbs become distinct and digits appear. Heart becomes four-chambered. Eyes are far apart and eyelids are fused. Nose develops and is flat. Face is more human-like. Ossification begins. Blood cells start to form in liver. External genitals begin to differentiate. Tail disappears. Major blood vessels form. Many internal organs continue to develop.
Fetal period		
9–12 weeks	7.5 cm (3 in.) 30 g (1 oz)	Head constitutes about half the length of the fetal body, and fetal length nearly doubles. Brain continues to enlarge. Face is broad, with eyes fully developed, closed, and widely separated. Nose develops a bridge. External ears develop and are low set. Ossification continues. Upper limbs almost reach final relative length but lower limbs are not quite as well developed. Heartbeat can be detected. Gender is distinguishable from external genitals. Urine secreted by fetus is added to amniotic fluid. Red bone marrow, thymus, and spleen participate in blood cell formation. Fetus begins to move, but its movements cannot be felt yet by the mother. Body systems continue to develop.
13–16 weeks	18 cm (6.5–7 in.) 100 g (4 oz)	Head is relatively smaller than rest of body. Eyes move medially to their final positions, and ears move to their final positions on the sides of the head. Lower limbs lengthen. Fetus appears even more human-like. Rapid development of body systems occurs.
17–20 weeks	25–30 cm (10–12 in.) 200–450 g (0.5–1 lb)	Head is more proportionate to rest of body. Eyebrows and head hair are visible. Growth shows but lower limbs continue to lengthen. Vernix caseosa (fatty secretions of sebaceous glands and dead epithelial cells) and lanugo (delicate fetal hair) cover fetus. Brown fat forms and is the site of heat production. Fetal movements are commonly felt by mother (quickening).
21–25 weeks	27–35 cm (11–14 in.) 550–800 g (1.25–1.5 lb)	Head becomes even more proportionate to rest of body. Weight gain is substantial, and skin is pink and wrinkled. By 24 weeks, Type II alveolar cells begin to produce surfactant.
26–29 weeks	32–42 cm (13–17 in.) 1110–1350 g (2.5–3 lb)	Head and body are more proportionate and eyes are open. Toenails are visible. Body fat is 3.5% of total body mass and additional subcutaneous fat smoothes out some wrinkles. Testes begin to descend toward scrotum at 28 to 32 weeks. Red bone marrow is major site of blood cell production. Many fetuses born prematurely during this period survive if given intensive care because lungs can provide adequate ventilation and central nervous system is developed enough to control breathing and body temperature.
30–34 weeks	41–45 cm (16.5–18 in.) 2000–2300 g (4.5–5 lb)	Skin is pink and smooth. Fetus assumes upside-down position. Pupillary reflex is present by 30 weeks. Body fat is 8% of total body mass. Fetuses 33 weeks and older usually survive if born prematurely.
35–38 weeks	50 cm (20 in.) 3200–3400 g (7–7.5 lb)	By 38 weeks, circumference of fetal abdomen is greater than that of head. Skin is usually bluish-pink, and growth slows as birth approaches. Body fat is 16% of total body mass. Testes are usually in scrotum in full-term infants. Even after birth, infant is not completely developed; an additional year is required, especially for complete development of the nervous system.

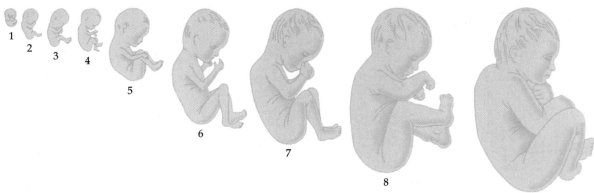

1 2 3 4 5 6 7 8 9 (Months)

Development

Name _____ Date _____

Laboratory Section _____ Score/Grade _____

PART 1 ■ Multiple Choice

_____ 1. The basic difference between spermatogenesis and oogenesis is that (a) two more polar bodies are produced in spermatogenesis (b) the secondary oocyte contains the haploid chromosome number, whereas the mature sperm cell contains the diploid number (c) in oogenesis, one secondary oocyte is produced, and in spermatogenesis four mature sperm cells are produced (d) both mitosis and meiosis occur in spermatogenesis, but only meiosis occurs in oogenesis

_____ 2. The union of a sperm cell nucleus and a secondary oocyte nucleus resulting in formation of a zygote is referred to as (a) implantation (b) fertilization (c) gestation (d) parturition

_____ 3. The most advanced stage of development for these stages is the (a) morula (b) zygote (c) ovum (d) blastocyst

_____ 4. Damage to the mesoderm during embryological development would directly affect the formation of (a) muscle tissue (b) the nervous system (c) the epidermis of the skin (d) hair, nails, and skin glands

_____ 5. The placenta, the organ of exchange between mother and fetus, is formed by union of the endometrium with the (a) yolk sac (b) amnion (c) chorion (d) umbilicus

_____ 6. One oogonium produces (a) one ovum and three polar bodies (b) two ova and two polar bodies (c) three ova and one polar body (d) four ova

_____ 7. Implantation is defined as (a) attachment of the blastocyst to the uterine (fallopian) tube (b) attachment of the blastocyst to the endometrium (c) attachment of the embryo to the endometrium (d) attachment of the morula to the endometrium

_____ 8. Epithelium lining most of the gastrointestinal tract and a number of other organs is derived from (a) ectoderm (b) mesoderm (c) endoderm (d) mesophyll

_____ 9. The nervous system is derived from the (a) ectoderm (b) mesoderm (c) endoderm (d) mesophyll

_____ 10. Which of the following is *not* an embryonic membrane? (a) amnion (b) placenta (c) chorion (d) allantois

PART 2 ■ Completion

11. A normal human sperm cell, as a result of meiosis, contains _____

    chromosomes.

12. The process that permits an exchange of genes resulting in their recombination and a part of the variation among humans is called _____.

13. The result of meiosis in spermatogenesis is that each primary spermatocyte produces four

    _____.

14. The stage of spermatogenesis that results in maturation of spermatids into sperm cells is called

    _____.

15. The afterbirth expelled in the final stage of delivery is the _____.

16. After the second month, the developing human is referred to as a(n) _____.

17. Embryonic tissues from which all tissues and organs of the body develop are called the

    _____.

18. The cells of the inner cell mass divide to form two cavities: amniotic cavity and

    _____.

19. Somatic cells that contain two sets of chromosomes are referred to as _____.

20. Rapid mitotic cell divisions of the zygote after fertilization are called _____.

21. In oogenesis, primordial follicles develop into _____ follicles.

22. The clear, glycoprotein layer between the oocyte and granulosa cells is called the

    _____.

23. _____ refers to the functional changes that sperm cells undergo in the female reproductive tract that allow them to fertilize a secondary oocyte.

24. The _____ of a blastocyst develops into an embryo.

25. The decidua _____ is the portion of the endometrium between the embryo and stratum basalis of the uterus.

26. The process by which the bilaminar embryonic disc transforms into a trilaminar embryonic disc is called _____.

Genetics

Objectives

At the completion of this exercise you should understand

A The difference between genotype and phenotype

B How to utilize a Punnett square

C Sex-linked inheritance and some associated sex-linked traits

D The Mendelian Law of Segregation

Genetics (je-NET-iks) is the branch of biology that studies inheritance. **Inheritance** is the passage of hereditary traits from one generation to the next. It is through the passage of hereditary traits that you acquired your characteristics from your parents and will transmit your characteristics to your children. If all individuals were brown-eyed, we could learn nothing of the hereditary basis of eye color. However, because some people are blue-eyed and marry brown-eyed people, we can gain some knowledge of how hereditary traits are transmitted. We constantly analyze the genetic bases of the *differences* between individuals. Some of these differences occur normally, such as differences in eye color, blood groups, or ability to taste PTC (phenylthiocarbamide). Other differences are abnormal, such as physical abnormalities and abnormalities in the processes of metabolism.

A. Genotype and Phenotype

The vast majority of human cells, except gametes, contain 23 pairs of chromosomes (diploid number) in their nuclei. One chromosome from each pair comes from the mother, and the other comes from the father. The two chromosomes that belong to a pair are called **homologous** (hō-MOL-ō-gus) **chromosomes,** and these homologues contain genes that control the same traits. The homologue of a chromosome that contains a gene for height also contains a gene for height.

The relationship of genes to heredity is illustrated by the disorder called **phenylketonuria,** or **PKU** (see Figure 27.1). People with PKU are unable to manufacture the enzyme phenylalanine hydroxylase. PKU results from the presence of an abnormal gene symbolized as *p*. The normal gene is symbolized as *P. P* and *p* are said to be alleles. An **allele** is one of the many alternative forms of a gene, occupying the same location (position of a gene on a chromosome) on homologous chromosomes. The chromosome that has the gene that directs phenylalanine hydroxylase production will have either *p* or *P* on it. Its homologue will also have either *p* or *P*. Thus every individual will have one of the following genetic makeups, or **genotypes** (JĒ-nō-tīps): *PP, Pp,* or *pp*. Although people with genotypes of *Pp* have the abnormal gene, only those with genotype *pp* suffer from the disorder because the normal gene masks the abnormal one. An allele that masks the expression of another allele is called the **dominant allele,** and the trait expressed is said to be a dominant trait. The allele that is masked is called the **recessive allele.** The trait expressed when two recessive genes are present is called the recessive trait. Several dominant and recessive traits inherited in human beings are listed in Table 27.1.

Traditionally, the dominant gene is symbolized with a capital letter and the recessive one with a lowercase letter. When the same genes appear on homologous chromosomes, as in *PP* or

FIGURE 27.1   Inheritance of phenylketonuria (PKU).

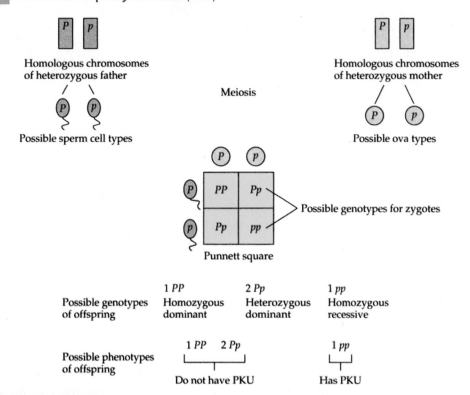

TABLE 27.1	Selected hereditary traits in humans

Dominant	Recessive
Coarse body hair	Fine body hair
Male pattern baldness	Baldness
Normal skin pigmentation	Albinism
Freckles	Absence of freckles
Astigmatism	Normal vision
Near- or farsightedness	Normal vision
Normal hearing	Deafness
Broad lips	Thin lips
Tongue roller	Inability to roll tongue into a U shape
PTC taster	PTC nontaster
Large eyes	Small eyes
Polydactylism (extra digits)	Normal digits
Brachydactylism (short digits)	Normal digits
Syndactylism (webbed digits)	Normal digits
Feet with normal arches	Flat feet
Hypertension	Normal blood pressure
Diabetes insipidus	Normal excretion
Huntington's chorea	Normal nervous system
Normal mentality	Schizophrenia
Migraine headaches	Normal
Widow's peak	Straight hairline
Curved (hyperextended) thumb	Straight thumb
Normal Cl^- transport	Cystic fibrosis
Hypercholesterolemia (familial)	Normal cholesterol level

pp, the person is said to be **homozygous** for a trait. When the genes on homologous chromosomes are different, however, as in *Pp*, the person is said to be **heterozygous** for the trait. **Phenotype** (FĒ-nō-tīp; *pheno* = showing) refers to how the genetic composition is expressed in the body. An individual with *Pp* has a different genotype from one with *PP*, but both have the same phenotype—which in this case is normal production of phenylalanine hydroxylase.

B. Punnett Squares

To determine how gametes containing haploid chromosomes unite to form diploid fertilized eggs, special charts called **Punnett squares** are used. The Punnett square is merely a device that helps one visualize all the possible combinations of male and female gametes, and is invaluable as a learning exercise in genetics. Usually, the possible paternal alleles in sperm cells are placed at the side of the chart and the possible maternal alleles in secondary oocytes are placed at the top (Figure 27.1). The spaces in the chart represent the possible genotypes for that trait in fertilized ova formed by the union of the male and female gametes. Possible combinations are determined simply by dropping the female gamete on the left into the two boxes below it and dropping the female gamete on the right into the two spaces under it. The upper male gamete is then moved across to the two spaces in line with it, and the lower male gamete is moved across to the two spaces in line with it.

C. Sex Inheritance

Lining up human chromosomes in pairs reveals that the last pair (the twenty-third pair) differs in males and in females (Figure 27.2a). In females, the pair consists of two rod-shaped chromosomes designated as X chromosomes. One X chromosome is also present in males, but its mate is hook-shaped and called a Y chromosome. The XX pair in the female and the XY pair in the male are called the **sex chromosomes,** and all other pairs of chromosomes are called **autosomes.**

The sex of an individual is determined by the sex chromosomes (Figure 27.2b). When a spermatocyte undergoes meiosis to reduce its chromosome number from diploid to haploid, one

FIGURE 27.2 Inheritance of sex. In (a), note the sex chromosomes, X and Y.

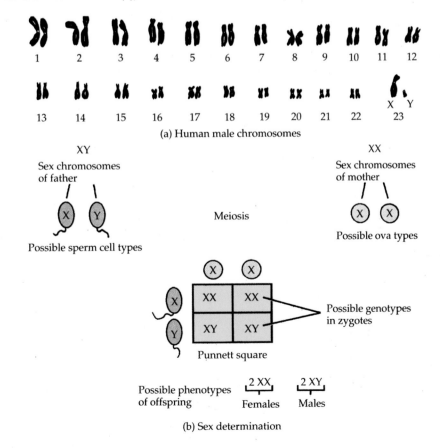

(a) Human male chromosomes

(b) Sex determination

daughter cell will contain the X chromosome and the other will contain the Y chromosome. When the secondary oocyte is fertilized by an X-bearing sperm, the offspring normally will be a female (XX). Fertilization by a Y sperm cell normally produces a male (XY).

Occasionally chromosomes fail to move toward opposite poles of a cell in meiotic anaphase. This is called **nondisjunction** and results in one sex cell having two members of a chromosome pair while the other receives none. Thus, eggs can contain two X's or no X (symbolized as 0), and sperm cells may contain both an X and a Y chromosome, two X's or two Y's, or no sex chromosomes at all.

Because the X chromosome contains so many genes unrelated to sex that are necessary for development, a zygote must contain at least one X chromosome to survive. Thus, Y0 and YY zygotes do not develop. However, other zygotes with sex chromosome anomalies do develop. Examples are Turner's syndrome (the presence of only one X chromosome, X0) and Kleinfelter's syndrome (an extra Y chromosome, XYY).

"Extra" X chromosomes (more than two in the female and more than one in the male) have a surprisingly minor effect on the individual, compared with the significant effect of other additional chromosomes. Studies show that only one X chromosome is active in any cell. Any additional X chromosomes are randomly inactivated early in development via a mechanism called **x-chromosome inactivation** (lyonization), and do not express the genes contained on them.

These inactivated X chromosomes remain tightly coiled against the cell membrane and can be seen as what are called **Barr bodies** (Figure 27.3). Since XY males do not have inactivated X chromosomes, no Barr bodies will be seen in normal male cells.

PROCEDURE

1. Make a buccal smear by *gently* scraping the inside of your cheek with the flat end of a toothpick. Discard the toothpick and *gently* scrape the same area again. This will produce more live cells from a deeper layer of the epithelium of the mucous membrane.

2. Spread the material over a clean glass slide.

3. Promptly place the slide in a Coplin jar filled with fixative for 1 min.

4. Remove the slide and wash gently under running tap water.

5. Place the slide in a Coplin jar filled with Giemsa stain for 10 to 20 min. Fresh solutions of stain prepared within 2 hr require 10 min; older stains require a longer time.

6. Wash the slide *gently* under running water and air-dry.

7. Examine the slide under high power and look for interphase nuclei. Identify Barr bodies, small disc-shaped chromatin bodies lying against the nuclear membrane (see Figure 27.3). Depending on the position of the nuclei and the staining technique, Barr bodies should be seen in 30% to 70% of the cells of a normal female.

8. Examine a slide prepared from the buccal epithelium of a class member not of your sex.

9. Draw a cell containing a Barr body in the space provided.

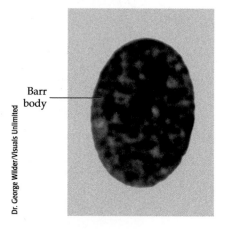

Barr body

D. Sex-Linked Inheritance

As do the other 22 pairs of chromosomes, the sex chromosomes contain genes that are responsible for the transmission of a number of nonsexual

| FIGURE 27.3 | Barr body in cell from buccal mucosa of human female. Feulgen stain; magnification 1000×.

Barr body

Dr. George Wilder/Visuals Unlimited

traits. Genes for these traits appear on X chromosomes, but many of these genes are absent from Y chromosomes. Traits transmitted by genes on the X chromosome are called **sex-linked traits.** This pattern of heredity is different from the pattern described earlier. About 150 sex-linked traits are known in humans. Examples of sex-linked traits are red–green color blindness and hemophilia.

1. Red–Green Color Blindness

Let us consider the most common type of color blindness, called red–green color blindness. In this condition, there is a deficiency in either red or green cones and red and green are seen as the same color, either red or green, depending on which cone is present. The gene for **red–green color blindness** is a recessive one designated *c*. Normal color vision, designated *C*, dominates. The *C/c* genes are located on the X chromosome. The Y chromosome does not contain these genes. Thus the ability to see colors depends entirely on the X chromosomes. The possible combinations are:

Genotype	Phenotype
$X^C X^C$	Normal female
$X^C X^c$	Normal female (carrying the recessive gene)
$X^c X^c$	Red–green color-blind female

$X^C Y$	Normal male
$X^c Y$	Red–green color-blind male

Only females who have two X^c genes are red–green color-blind. This rare situation can result only from the mating of a color-blind male and a color-blind or carrier female. In $X^C X^c$ females, the trait is masked by the normal, dominant gene. Males, on the other hand, do not have a second X chromosome that would mask the trait. Therefore all males with an X^c gene will be red–green color blind. The inheritance of red–green color-blindness is illustrated in Figure 27.4.

2. Hemophilia

Hemophilia is a hereditary blood disorder in which there is a deficient production of certain factors involved in blood clotting. Hemophilia is a much more serious defect than color blindness because individuals with severe hemophilia can bleed to death from even a small cut. Hemophilia is caused by a recessive gene, as is color blindness. If *H* represents normal clotting and *h* represents abnormal clotting, $X^h X^h$ females will be hemophiliacs. Males with $X^H Y$ will be normal and males with $X^h Y$ will be hemophiliacs. Other sex-linked traits in humans are fragile X syndrome, nonfunctional sweat glands, certain forms of diabetes, some types of deafness, uncontrollable rolling of the eyeballs, absence of central incisors, night

FIGURE 27.4 Inheritance of red–green color blindness.

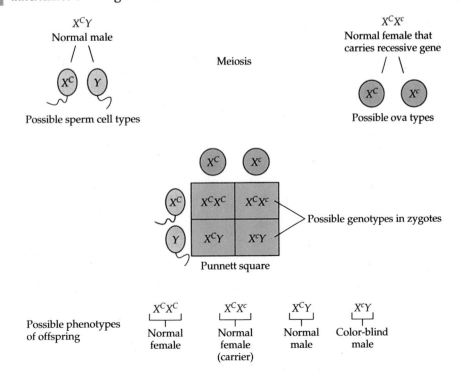

blindness, one form of cataract, juvenile glaucoma, and juvenile muscular dystrophy.

E. Mendelian Laws

In any genetic cross, all the offspring in the first (that is, the parental, or P_1) generation are symbolized as F_1. The F is from the Latin word *filial*, which means progeny. The second generation is symbolized as F_2, the third as F_3, and so on. The recognized "father" of genetics is Gregor Mendel, whose basic experiments were performed on garden peas. As a result of his tests, Gregor Mendel postulated what are now called **Mendelian Laws,** or **Mendelian Principles.** The **First Mendelian Law,** or the **Law of Segregation,** asserts that, in cells of individuals, genes occur in pairs, and that when those individuals produce germ cells, each germ cell receives only one member of the pair.

This law applies equally to pollen grains (or sperm) and to ova. The genetic cross is represented as follows:

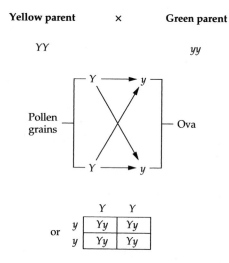

All possible combinations of pollen grains and ova are indicated by the arrows. Notice that all combinations yield the genotype Yy. All these F_1 seeds were yellow, yellow being dominant to green, or, in genetic terms, Y being dominant to y. These F_1 individuals resembled the yellow parent in phenotype (being yellow) but not in genotype (Yy as opposed to YY). Both parents were homozygous. Both members of that pair of alleles were the same. The yellow parent was homozygous for Y and the green parent for y. The F_1 individuals were heterozygous, having one Y and one y.

When the F_1 plants were self-fertilized, the F_2 seeds appeared in the ratio of 3 yellow/1 green.

Mendel found similar 3:1 ratios for the other traits he studied, and this type of result has been reported in many species of animals and plants for a variety of traits. Not only does the recessive trait reappear in the F_2, but also in a definite proportion of the individuals, one-fourth of the total. If the sample is small, the ratio may deviate considerably from 3:1, but as the progeny or sampling numbers get larger, the ratio usually comes closer and closer to an exact 3:1 ratio. The reason is that the ratio depends on the random union of gametes. The result is a 3:1 phenotypic ratio, or a 1:2:1 genotypic ratio.

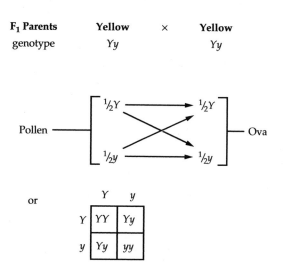

Thus, the four combinations of pollen and ova are expected to occur as follows:

$$
\begin{array}{lll}
\frac{1}{4}\,YY & = & \text{yellow} \\
\frac{1}{4}\,Yy & = & \text{yellow} \\
\frac{1}{4}\,Yy & = & \text{yellow}
\end{array} \bigg\} \; \frac{3}{4}
$$

$$
\frac{1}{4}\,yy \;\; = \;\; \text{green} \; \Big\} \; \frac{1}{4}
$$

It is important to realize that these fractions depend on the operation of the laws of probability. A model using coins will emphasize the point. This model consists of two coins, a nickel and a penny, tossed at the same time. The penny may represent the pollen (male parent). At any given toss, the chances are equal that the penny will come up "heads" or that it will come up "tails." Similarly, at any given fertilization the chances are equal that a Y-bearing pollen grain or that a y-bearing one will be transmitted. The nickel represents the ovum. Again, the chances are equal for "heads" or "tails," just as the chances are equal that in any fertilization a Y-bearing or a y-bearing ovum will take part. If we toss the two coins together and do it many times, we will obtain approximately the following:

¼ nickel heads; penny heads	(= YY)
¼ nickel heads; penny tails	(= Yy)
¼ nickel tails; penny heads	(= yY)
¼ nickel tails; penny tails	(= yy)

If we assume "heads" as dominant, we find that three-fourths of the time there is at least one "head" and one-fourth of the time no "heads" (both coins "tails"). Hence this gives us a model of the 3:1 ratio dependent on the laws of probability.

The genetic cross just demonstrated considers only one pair of alleles (yellow versus green or "heads" versus "tails") and is therefore called a **monohybrid cross.** Mendel's second principle applies to genetic crosses in which two traits or two pairs of alleles are considered. These **dihybrid crosses** enabled him to postulate his second principle, the **Principle of Independent Assortment.** This principle states that the segregation of one pair of traits occurs independently of the segregation of a second pair of traits. This is the case only if the traits are caused by genes located on nonhomologous chromosomes.

When Mendel crossed garden peas with round yellow seeds with garden peas with wrinkled green seeds, his F_1 generation showed that yellow and round were dominant. If self-fertilization then occurred, the F_2 generation resulted as follows:

Round yellow	¾ × ¾ = 9/16
Round green	¾ × ¼ = 3/16
Wrinkled yellow	¼ × ¾ = 3/16
Wrinkled green	¼ × ¼ = 1/16

Therefore, in a dihybrid cross, the expected phenotypic ratio was 9:3:3:1, with 9/16 of the F_2 being doubly dominant, and only 1/16 being doubly recessive.

F. Multiple Alleles

In the genetics examples we have considered to this point, we have discussed only two alleles of each gene. However, many, and possibly all, genes have **multiple alleles;** that is, they exist in more than two allelic forms even though a diploid cell cannot carry more than two alleles.

One example of multiple alleles in humans involves ABO blood groups (Exercise 16). The four basic blood types (phenotypes) of the ABO system are determined by three alleles: I^A, I^B, and i. Alleles

I^A and I^B are not dominant over each other. Rather, they are **codominant;** that is, both genes are expressed equally. Both I^A and I^B alleles, however, are dominant over allele i. These three alleles can give rise to six genotypes, as follows:

Genotype	Phenotype (blood type)
$I^A I^A$ or $I^A i$	A
$I^B I^B$ or $I^B i$	B
$I^A I^B$	AB
ii	O

Given this information, is it possible for a child with type O blood to have a mother with type O blood and a father with type AB blood?

Explain _____

_____

_____

If two children in a family have type O blood, the mother has type B blood and the father has type A blood, what is the genotype of the father?

_____

What is the genotype of the mother? _____

G. Genetics Exercises

1. Karyotyping

A group of cytogeneticists meeting in Denver, Colorado, in 1960 adopted a system for classifying and identifying human chromosomes. Chromosome **length** and **centromere position** were the bases for classification. The Denver classification has become a standard for human chromosome studies. By the early 1970s, most human chromosomes could be identified microscopically.

Every chromosome pair could not be identified consistently until chromosome **banding techniques** finally distinguished all 46 human chromosomes. Bands are defined as parts of chromosomes that appear lighter or darker than adjacent regions with particular staining methods.

A **karyotype** is a chart made from a photograph of the chromosomes in metaphase. The chromosomes are cut out and arranged in matched pairs according to length (see Figure 27.2a). Their comparative size, shape, and morphology are then examined to determine if they are normal.

Karyotyping helps scientists to visualize chromosomal abnormalities. For example, individuals with Down syndrome typically have 47 chromosomes, instead of the usual 46, with chromosome 21 being represented three times rather than only twice. The syndrome is characterized by mental retardation, retarded physical development, and distinctive facial features (round head, broad skull, slanting eyes, and large tongue). With chronic myelogenous leukemia, part of the long arm of a chromosome 22 is missing, resulting in the blood disease. The chromosome is referred to as the Philadelphia chromosome, named for the city where it was first detected.

2. PKU Screening

Phenylketonuria (PKU), an inherited metabolic disorder that occurs in approximately 1 in 16,000 births, is transmitted by an autosomal recessive gene (see Figure 27.1). Individuals with this condition do not have the enzyme phenylalanine hydroxylase, which converts the amino acid phenylalanine to tyrosine. As a result, phenylalanine and phenylpyruvic acid accumulate in the blood and urine. These substances are toxic to the central nervous system and can produce irreversible brain damage. Most states in the United States require routine screening for this disorder at birth.

The procedure for testing for PKU is as follows.

PROCEDURE

1. A Phenistix® test strip is made specifically for testing urine for phenylpyruvic acid. Dip this test strip in freshly voided urine.

2. Compare the color change with the color chart on the Phenistix® bottle. The test is based on the reaction of ferric ions with phenylpyruvic acid to produce a gray-green color.

3. Record your results in Section G.2 of the LABORATORY REPORT RESULTS at the end of the exercise.

3. PTC Inheritance

The ability to taste the chemical compound known as phenylthiocarbamide, commonly called PTC, is inherited. On the average, 7 out of 10 people, on chewing a small piece of paper treated with PTC, detect a definite bitter or sweet taste. Others do not taste anything.

Individuals who can taste something (bitter or sweet) are called "tasters" and have the dominant allele *T*, either as *TT* or *Tt*. A nontaster is a homozygous recessive and is designated as *tt*.

Determine your phenotype for tasting PTC and record your results in Section G.3 of the

LABORATORY REPORT RESULTS at the end of the exercise.

NOTE *If PTC paper is not available, a 0.5% solution of phenylthiourea (PTU) can be substituted because the capacity to taste PTU is also inherited as a dominant.*

4. Corn Genetics

Genetic corn may be purchased and used in this exercise. Each ear of corn represents a family of offspring. Mark a starting row with a pin to avoid repetition. Count the kernels (individuals) for each trait (color, wrinkled, or smooth). Record your results in Section G.4 of the LABORATORY REPORT RESULTS at the end of the exercise.

Develop a ratio by using your lowest number as "1" and dividing it into the others to determine what multiples of it they are. See how close you come to Mendel's ratios. Figure out the probable genotype and phenotype of the parent plants if you can. Monohybrid crosses, test crosses, dihybrid crosses, and trihybrid crosses are available.

5. Color Blindness

Using either Stilling or Ishihara test charts, test the entire class for red–green color blindness. Tests for color blindness depend on the person's ability to distinguish various colors from one another and also on his or her ability to judge correctly the degree of contrast between colors.

Of all men, 2% are color-blind to red and 6% to green, so 8% of all men are red–green color-blind. Red–green color blindness is rare in the female, occurring in only 1 of every 250 women. Record your results in Section G.5 of the LABORATORY REPORT RESULTS at the end of the exercise.

6. Mendelian Laws of Inheritance

Follow the procedure outlined in the explanation of the Mendelian Law of Segregation, tossing a nickel and a penny simultaneously to prove the law and determine ratios.

PROCEDURE

1. Toss the nickel and the penny together 10 times to get the genotypes of a family of 10. Repeat this procedure for a total of five times to obtain five families of 10 offspring. Record all of the results on the chart in Section G.6.1 of the LABORATORY REPORT RESULTS at the end of the exercise.

NOTE *Use the following symbols for the following exercises.*

  G = gene for yellow
  g = gene for green
  GG = the genotype of an individual pure (homozygous) for yellow
  gg = the genotype of an individual pure (homozygous) for green
  Gg = the genotype of the hybrid (heterozygous) individual, phenotypically yellow
  ♀ = symbol for female
  ♂ = symbol for male

2. Obeying the Mendelian Law of Segregation and using the Punnett square shown in Section G.6.2 of the LABORATORY REPORT RESULTS at the end of the exercise, cross yellow garden peas with green garden peas (a monohybrid cross). Show the P_1, F_1, and F_2 generations and all the different phenotypes and genotypes.

3. Obeying the Mendelian Law of Independent Assortment and using the Punnett square shown in Section G.6.4 of the LABORATORY REPORT RESULTS at the end of the exercise, cross the round yellow seeds with the wrinkled green seeds (a dihybrid cross). Show the P_1, F_1, and F_2 generations and all the different phenotypes and genotypes. The F_2 generation can be generated from a Punnett square comparable to that for the monohybrid cross, but with 16 rather than 4 squares.

7. Observing Phenotypes

The pattern of inheritance of many human traits is complex and involves many genes; the inheritance of other traits is controlled by single genes. You are asked to record your phenotype and your genotype, if it can be determined, for several traits controlled by single genes. For example, if you have the phenotypically dominant trait A, your genotype will be $A-$, indicating that the allele symbolized as "$-$" is not known, since you could be homozygous dominant *(AA)* or heterozygous *(Aa)* for the trait. If you have the recessive trait $a-$, your genotype will be aa. Record

your phenotype and genotype for the following traits in Section G.6.6 of the LABORATORY REPORT RESULTS at the end of the exercise.

1. **Attached earlobes.** The dominant gene E causes earlobes that develop free from the neck; ee results in adherent earlobes connected to the cervical skin.

2. **Tongue rolling.** The dominant gene R causes the development of muscles that allow the tongue to be rolled into a U shape. The rr genotype prohibits such rolling.

3. **Hair whorl direction.** The dominant gene W causes the hair whorl on the cranial surface of the scalp to turn in a clockwise direction; the genotype ww determines a counterclockwise whorl.

4. **Little-finger bending.** The dominant gene B causes the distal segment of the little finger to bend laterally. The genotype bb results in a straight distal segment.

5. **Double-jointed thumbs.** The dominant gene J results in loose ligaments that allow the thumb to be bent out of the constricted orientation caused by the recessive genotype jj.

6. **Widow's peak.** The dominant gene W causes the hairline to extend caudally in the midline of the forehead. The recessive genotype ww results in a straight hairline.

7. **Rh factor.** A dominant gene Rh results in the presence of the Rh antigen on red blood cells. This antigen is not present with the recessive genotype $rhrh$. Use the results obtained in Exercise 16.I.4 to determine your phenotype.

   What additional information would you need to determine your complete genotype if you have the dominant phenotype for these traits?

   _____

   _____

   _____

ANSWER THE LABORATORY REPORT QUESTIONS AT THE END OF THE EXERCISE.

EXERCISE 27

Genetics

Name _____ Date _____

Laboratory Section _____ Score/Grade _____

SECTION G ■ Genetics Exercises

2. PKU Screening

_____ negative—cream color

_____ 15 mg%—light green

_____ 40 mg%—medium green

_____ 100 mg%—dark green

3. PTC Inheritance

_____ bitter taste

_____ sweet taste

_____ negative (no taste)

4. Corn Genetics

Ratio _____ monohybrid cross

Ratio _____ test cross

Ratio _____ dihybrid cross

Ratio _____ trihybrid cross

5. Color Blindness

	Male students	Female students
Red color-blind	_____	_____
Green color-blind	_____	_____

6. Mendelian Laws of Inheritance

1. Record the results of tossing a nickel and a penny together 10 times.

	Female (nickel)	Male (penny)	1	2	3	4	5	Total	Class total
Dominant offspring	A Heads	A Heads							
	A Heads	a Heads	Tails						
	a Tails	A Heads							
Recessive offspring	a Tails	a Tails							
Ratio dominant to recessive									

2. Complete the following monohybrid cross. Fill in genotypes (within circles) and phenotypes (under circles).

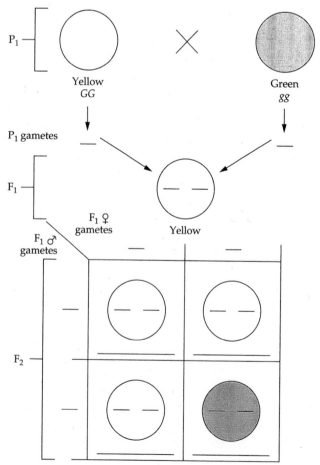

Monohybrid cross in the garden pea (*Pisum sativum*). G = allele for yellow, g = allele for green, P_1 = parental generation, F_1 = first filial generation, F_2 = second filial generation.

3. What is the phenotype ratio of the F$_2$ generation? _____ yellow / _____ green.

4. Complete the following dihybrid cross. Fill in genotypes (within circles) and phenotypes (under circles).

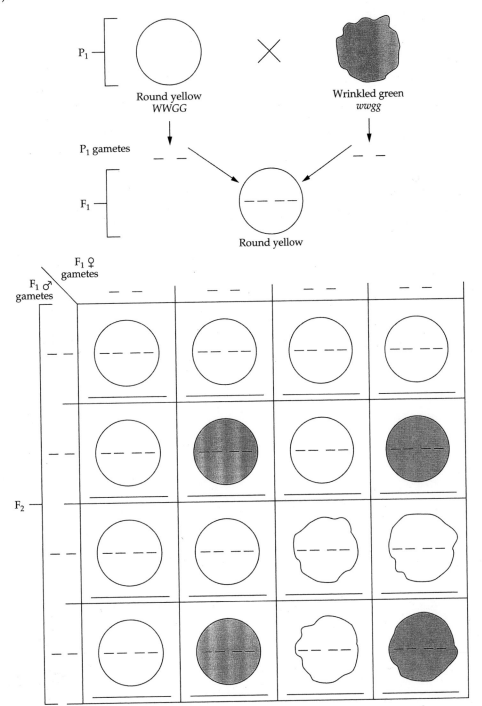

Dihybrid cross in the garden pea *(Pisum sativum)*. W = allele for round, *w* = allele for wrinkled, G = allele for yellow, *g* = allele for green.

5. What is the phenotype ratio of the F$_2$ generation?

_____ round yellow / _____ wrinkled yellow / _____ round green / _____ wrinkled green.

6.   Observe phenotypes.

Trait	Phenotype	Genotype
Earlobes		
Tongue rolling		
Hair whorl		
Little-finger bending		
Double-jointed thumbs		
Widow's peak		
Rh factor		

EXERCISE 27

Genetics

Name _____   Date _____

Laboratory Section _____   Score/Grade _____

PART 1 ■ Multiple Choice

_____ 1. Using the symbols *Aa* to represent genes, which of the following is true? (a) the trait is homozygous for the dominant characteristic (b) the trait is homozygous for the recessive characteristic (c) the trait is heterozygous (d) sex-linked inheritance is in operation

_____ 2. Which statement concerning the normal inheritance of sex is correct? (a) all zygotes contain a Y chromosome (b) some ova contain a Y chromosome (c) all ova and all sperm cells contain an X chromosome (d) all ova have an X chromosome, some sperm cells have an X chromosome, and some sperm cells have a Y chromosome

_____ 3. The genotype that will express characteristics associated with hemophilia (assume that *H* represents the gene for normal blood) is (a) $X^H X^h$ (b) $X^h Y$ (c) $X^H X^H$ (d) $X^H Y$

_____ 4. The exact position of a gene on a chromosome is called the (a) homologue (b) locus (c) triad (d) allele

PART 2 ■ Completion

5. When the same genes appear on homologous chromosomes, as in *PP* or *pp*, the individual is said to be _____ for the trait.

6. If different genes appear on homologous chromosomes, as in *Pp*, the individual is _____ for the trait.

7. Genetic composition expressed in the body or morphologically is called the body's _____.

8. The device that helps one visualize all the possible combinations of male and female gametes is called the _____.

9. The twenty-third pair of human chromosomes are the sex chromosomes. All of the other pairs of chromosomes are called _____.

10. Red–green color blindness is an inherited trait that is specifically called a(n) _____ trait.

11. The recognized "father" of genetics is _____.

12. A genetic cross that involves only one pair of alleles (or traits) is called a(n)

_____ cross.

13. Passage of hereditary traits from one generation to another is called _____.

14. The genetic makeup of an individual is called the person's _____.

15. One of the many alternative forms of a gene is called its _____.

16. Two chromosomes that belong to a pair are called _____ chromosomes.

17. A gene that masks the expression of its allele is called a(n) _____ gene.

18. Failure of chromosomes to move to opposite poles of a cell during meiotic prophase is called

_____.

19. A normal female who carries the recessive gene for red–green color blindness would have the following genotype: _____.

20. Genes that are expressed equally are said to be _____.

Some Important Units of Measurement

English Units of Measurement

Fundamental or derived unit	Units and equivalents
Length	12 inches (in.) = 1 foot (ft) = 0.333 yard (yd)
	3 ft = 1 yd
	1760 yd = 1 mile (mi)
	5280 ft = 1 mi
Mass	1 ounce (oz) = 28.35 grams (g); 1 g = 0.0353 oz
	1 pound (lb) = 453 g = 16 oz; 1 kilogram (kg) = 2.205 lb
	1 ton = 2000 lb = 907 kg
Time	1 second (sec) = 1/86,400 of a mean solar day
	1 minute (min) = 60 sec
	1 hour (hr) = 60 min = 3600 sec
	1 day = 24 hr = 1440 min = 86 400 sec
Volume	1 fluid dram (fl dr) = 0.125 fluid ounce (fl oz)
	1 fl oz = 8 fl dr = 0.0625 quart (qt) = 0.008 gallon (gal)
	1 qt = 256 fl dr = 32 fl oz = 2 pints (pt) = 0.25 gal
	1 gal = 4 qt = 128 fl oz = 1024 fl dr

Metric Units of Length and Some English Equivalents

Metric unit	Meaning of prefix	Metric equivalent	English equivalent
1 kilometer (km)	kilo = 1000	1000 m	3280.84 ft or 0.62 mi; 1 mi = 1.61 km
1 hectometer (hm)	hector = 100	100 m	328 ft
1 dekameter (dam)	deka = 10	10 m	32.8 ft
1 meter (m)	Standard unit of length		39.37 in. or 3.28 ft or 1.09 yd
1 decimeter (dm)	deci = $\frac{1}{10}$	0.1 m	3.94 in.
1 centimeter (cm)	centi = $\frac{1}{100}$	0.01 m	0.394 in.; 1 in. = 2.54 cm
1 millimeter (mm)	milli = $\frac{1}{1000}$	0.001 m = $\frac{1}{100}$ cm	0.0394 in.
1 micrometer (μm) [formerly micron (μ)]	micro = $\frac{1}{1,000,000}$	0.0000001 m = $\frac{1}{10,000}$ cm	3.94×10^{-5} in.
1 nanometer (nm) [formerly millimicron (mμ)]	nano = $\frac{1}{1,000,000,000}$	0.0000000001 m = $\frac{1}{10,000,000}$ cm	3.94×10^{-8} in.

Temperature

Unit	K	°F	°C
1 degree Kelvin (K)	1	$\frac{9}{5}$(K) − 459.7	K + 273.16*
1 degree Fahrenheit (°F)	$\frac{5}{9}$(°F) + 255.4	1	$\frac{5}{9}$(°F − 32)
1 degree Celsius (°C)	°C − 273	$\frac{9}{5}$(°C) + 32	1

* Absolute zero (K) = −273.16°C

Volume

Unit	mL	cm³	qt	oz
1 milliliter (mL)	1	1	1.06×10^{-3}	3.392×10^{-2}
1 cubic centimeter (cm³)	1	1	1.06×10^{-3}	3.392×10^{-2}
1 quart (qt)	943	943	1	32
1 fluid ounce (fl oz)	29.5	29.5	3.125×10^{-2}	1

Periodic Table of the Elements

KEY

6	Atomic Number
C	Symbol
12.01	Atomic Weight
Carbon	Name

1																	2
H 1.0080 Hydrogen																	**He** 4.003 Helium
3 **Li** 6.940 Lithium	4 **Be** 9.013 Berilium											5 **B** 10.82 Boron	6 **C** 12.011 Carbon	7 **N** 14.008 Nitrogen	8 **O** 16.000 Oxygen	9 **F** 19.00 Fluorine	10 **Ne** 20.183 Neon
11 **Na** 22.991 Sodium	12 **Mg** 24.32 Magnesium											13 **Al** 26.98 Aluminum	14 **Si** 28.09 Silicon	15 **P** 30.975 Phosphorus	16 **S** 32.066 Sulfur	17 **Cl** 35.457 Chlorine	18 **Ar** 39.944 Argon
19 **K** 39.100 Potassium	20 **Ca** 40.08 Calcium	21 **Sc** 44.96 Scandium	22 **Ti** 47.90 Titanium	23 **V** 50.95 Vanadium	24 **Cr** 52.01 Chromium	25 **Mn** 54.94 Manganese	26 **Fe** 55.85 Iron	27 **Co** 58.94 Cobalt	28 **Ni** 58.71 Nickel	29 **Cu** 63.54 Copper	30 **Zn** 65.38 Zinc	31 **Ga** 69.72 Gallium	32 **Ge** 72.60 Germanium	33 **As** 74.91 Arsenic	34 **Se** 78.96 Selenium	35 **Br** 79.916 Bromine	36 **Kr** 83.80 Krypton
37 **Rb** 85.48 Rubidium	38 **Sr** 87.63 Strontium	39 **Y** 88.92 Yttrium	40 **Zr** 91.22 Zirconium	41 **Nb** 92.91 Niobium	42 **Mo** 95.95 Molybdenum	43 **Tc** (99) Technetium	44 **Ru** 101.1 Ruthenium	45 **Rh** 102.91 Rhodium	46 **Pd** 106.4 Palladium	47 **Ag** 107.880 Silver	48 **Cd** 112.41 Cadmium	49 **In** 114.82 Indium	50 **Sn** 118.70 Tin	51 **Sb** 121.76 Antimony	52 **Te** 127.61 Tellurium	53 **I** 126.91 Iodine	54 **Xe** 131.30 Xenon
55 **Cs** 132.91 Cesium	56 **Ba** 137.36 Barium	57 **La** 138.92 Lanthanum	72 **Hf** 178.50 Hafnium	73 **Ta** 180.95 Tantalum	74 **W** 183.86 Wolfram	75 **Re** 186.22 Rhenium	76 **Os** 190.2 Osmium	77 **Ir** 192.2 Iridium	78 **Pt** 195.09 Platinum	79 **Au** 197.0 Gold	80 **Hg** 200.61 Mercury	81 **Tl** 204.39 Thallium	82 **Pb** 207.21 Lead	83 **Bi** 209.00 Bismuth	84 **Po** (210) Polonium	85 **At** (210) Astatine	86 **Rn** (222) Radon
87 **Fr** (223) Francium	88 **Ra** (226) Radium	89 **Ac** (227) Actinium	104 **Rf** (261) Rutherfordium	105 **Db** (262) Dubnium	106 **Sg** (263) Seaborgium	107 **Bh** (262) Bohrium	108 **Hs** (265) Hassium	109 **Mt** (266) Meitnerium									

58 **Ce** 140.13 Cerium	59 **Pr** 140.92 Praseodymium	60 **Nd** 144.27 Neodymium	61 **Pm** (147) Promethium	62 **Sm** 150.35 Samarium	63 **Eu** 152.0 Europium	64 **Gd** 157.26 Gadolinium	65 **Tb** 158.93 Terbium	66 **Dy** 162.51 Dysprosium	67 **Ho** 164.94 Holmium	68 **Er** 167.27 Erbium	69 **Tm** 168.94 Thulium	70 **Yb** 173.04 Ytterbium	71 **Lu** 174.99 Lutetium
90 **Th** (232) Thorium	91 **Pa** (231) Protactinium	92 **U** 238.07 Uranium	93 **Np** (237) Neptunium	94 **Pu** (242) Plutonium	95 **Am** (243) Americium	96 **Cm** (247) Curium	97 **Bk** (249) Berkelium	98 **Cf** (251) Californium	99 **Es** (254) Einsteinium	100 **Fm** (253) Fermium	101 **Md** (256) Mendelevium	102 **No** (253) Nobelium	103 **Lr** 257 Lawrencium

Eponyms Used in This Laboratory Manual

An **eponym** is a term that includes reference to a person's name; for example, you may be more familiar with *Achilles tendon* than you are with the technical, but correct term *calcaneal tendon*. Because eponyms remain in frequent use, this glossary indicates which current terms replace eponyms in this manual. In the body of the text eponyms are cited in parentheses, immediately following the current terms where they are used for the first time in a chapter or later in the laboratory manual. In addition, although eponyms are included in the index, they have been cross-referenced to their current terminology.

Eponym	Current terminology
Achilles tendon	calcaneal tendon
Adam's apple	thyroid cartilage
ampulla of Vater (VA-ter)	hepatopancreatic ampulla
Bartholin's (BAR-tō-linz) gland	greater vestibular gland
Billroth's (BIL-rōtz) cord	splenic cord
Bowman's (BŌ-manz) capsule	glomerular capsule
Bowman's (BŌ-manz) gland	olfactory gland
Broca's (BRŌkaz) area	motor speech area
Brunner's (BRUN-erz) gland	duodenal gland
bundle of His (HISS)	atrioventricular (AV) bundle
canal of Schlemm (SHLEM)	scleral venous sinus
circle of Willis (WIL-is)	cerebral arterial circle
Cooper's (KOO-perz) ligament	suspensory ligament of the breast
Cowper's (KOW-perz) gland	bulbourethral gland
crypt of Lieberkühn (LĒ-ber-kyūn)	intestinal gland
duct of Rivinus (ri-VĒ-nus)	lesser sublingual duct
duct of Santorini (san'-tō-RĒ-nē)	accessory duct
duct of Wirsung (VĒR-sung)	pancreatic duct
end organ of Ruffini (roo-FĒnē)	type-II cutaneous mechanoreceptor
eustachian (ū-STĀ-kē-an) tube	auditory tube
fallopian (fal-LŌ-pē-an) tube	uterine tube
gland of Zeis (ZĪS)	sebaceous ciliary gland
Golgi (GOL-jē) tendon organ	tendon organ
graafian (GRAF-ē-an) follicle	mature follicle
Hassall's (HAS-alz) corpuscle	thymic corpuscle
haversian (ha-VĒR-shun) canal	central canal
haversian (ha-VĒR-shun) system	osteon
interstitial cell of Leydig (LĪ-dig)	interstitial endocrinocyte
islet of Langerhans (LANG-er-hanz)	pancreatic islet
Kupffer (KOOP-fer) cells	stellate reticuloendothelial cell
loop of Henle (HEN-lē)	loop of the nephron
Malpighian (mal-PIG-ē-an) corpuscle	splenic nodule
Meibomian (mī-BŌ-mē-an) gland	tarsal gland
Meissner's (MĪS-nerz) corpuscle	corpuscle of touch
Merkel (MER-kel) disc	tactile disc
Müller's (MIL-erz) duct	paramesonephric duct
Nissl (NISS-l) bodies	chromatophilic substance
node of Ranvier (ron-VĒ-ā)	neurofibral node
organ of Corti (KOR-tē)	spiral organ
pacinian (pa-SIN-ē-an) corpuscle	lamellated corpuscle
Peyer's (PĪ-erz) patches	aggregated lymphatic follicles
plexus of Auerbach (OW-er-bak)	myenteric plexus
plexus of Meissner (MĪS-ner)	submucous plexus
pouch of Douglas	rectouterine pouch
Purkinje (pur-KIN-jē) fiber	conduction myofiber
Rathke's (RATH-kēz) pouch	hypophyseal pouch
Schwann (SCHVON) cell	neurolemmocyte
Sertoli (ser-TŌ-lē) cell	sustentacular cell
Skene's (SKĒNZ) gland	paraurethral gland
sphincter of Oddi (OD-dē)	sphincter of the hepatopancreatic ampulla
Stensen's (STEN-senz) duct	parotid duct
Volkmann's (FŌLK-manz) canal	perforating canal
Wharton's (HWAR-tunz) duct	submandibular duct
Wharton's (HWAR-tunz) jelly	mucous connective tissue
Wormian (WER-mē-an) bone	sutural bone

Credits

This page constitutes an extension of the copyright page. We have made every effort to trace the ownership of all copyrighted material and to secure permission from copyright holders. In the event of any question arising as to the use of any material, we will be pleased to make the necessary corrections in future printings. Thanks are due to the following authors, publishers, and agents for permission to use the material indicated.

Exercise 1. 2: Courtesy of Olympus America, Inc.

Exercise 3. 42: All Photos © Photo Researchers, Inc.

Exercise 4. 53: © Biophoto Associates/Photo Researchers, Inc. **53:** © M.I. Walker/Photo Researchers, Inc. **53:** © Biophoto Associates/Photo Researchers, Inc. **54:** © Biophoto Associates/Photo Researchers, Inc. **54:** © Fred Hossler/Visuals Unlimited **54:** © Ed Reschke **55:** © Ed Reschke **60:** © Ed Reschke **60:** © Biophoto Associates/Photo Researchers, Inc. **60:** © Robert Brons/Biological Photo Service **61:** © Biophoto Associates/Photo Researchers, Inc. **61:** © Ed Reschke **61:** © Bruce Iverson/Visuals Unlimited. **62:** © Fred Hossler/Visuals Unlimited **62:** © Frederick C. Skvara **62:** © Chuck Brown/Photo Researchers, Inc.

Exercise 5. 74: Reproduced by permission from R.G. Kessel and R.H. Kardon, *Tissues and Organs: A Text Atlas of Scanning Electron Microscopy*, W.H. Freeman, 1979

Exercise 6. 86: From Photo Atlas for Anatomy and Physiology, photo by D. Morton **88:** © Biophoto Associates/Photo Researchers, Inc.

Exercise 8. 136: © John Eads

Exercise 9. 142: © Visuals Unlimited **148:** © Visuals Unlimited **149:** © Phototake/Carolina Biological Supply Company

Exercise 11. All photos © Joel Gordon

Exercise 12. 241: From Photo Atlas for Anatomy and Physiology, photo by D. Morton

Exercise 13. 250: © John D. Cunningham/Visuals Unlimited **265:** © John D. Cunningham/Visuals Unlimited **272:** © Fred Hossler/Visuals Unlimited **277:** © Martin M. Rotker

Exercise 14. 305: © James R. Smail & Russell A. Whitehead, Macalester College **306:** © Ed Reschke **312:** © Andrew Kuntzman, Wright State University **321:** Both photos Science VU/Visuals Unlimited

Exercise 15. 339: © Bruce Iverson **340:** © Project Masters, Inc./The Bergman Collection **340:** © G.W. Wills, MD/Biological Photo Service **342:** © R. Calentine/Visuals Unlimited **343:** Photographs courtesy of Fisher Scientific

Exercise 16. 351: All photos from Photo Atlas for Anatomy and Physiology, photo by D. Morton **355:** © John Eads **358:** Courtesy of Fisher Scientific **358:** Courtesy of Fisher Scientific **359:** © Lukes Medical Corporation **361:** © Leica Microsystems, Inc. **361:** © Leica Microsystems, Inc.

Exercise 17. 383: © Lukens Medical Corporation **384:** © Martin M. Rotker/Photo Researchers, Inc.

Exercise 18. 394: © Martin Rotker/Photo Researchers, Inc. **394:** © Carolina Biological Supply/Phototake NYC **394:** From Photo Atlas for Anatomy and Physiology, photo by D. Morton

Index